For Reference

Not to be taken from this room

A Dictionary of Genetics

A DICTIONARY
OF GENETICS
FOURTH EDITION

Robert C. King
Northwestern University

William D. Stansfield
California Polytechnic State University

New York ∎ Oxford
OXFORD UNIVERSITY PRESS
1990

Oxford University Press

Oxford   New York   Toronto
Delhi   Bombay   Calcutta   Madras   Karachi
Petaling Jaya   Singapore   Hong Kong   Tokyo
Nairobi   Dar es Salaam   Cape Town
Melbourne   Auckland

and associated companies in
Berlin   Ibadan

Published by Oxford University Press, Inc.,
200 Madison Avenue, New York, New York 10016

Library of Congress Cataloging-in-Publication Data
King, Robert C.
A dictionary of genetics
Robert C. King, William D. Stansfield.
— 4th ed.   p.   cm.   Includes bibliographical references.
ISBN 0-19-506370-8. — ISBN 0-19-506371-6 (pbk.)
1. Genetics—Dictionaries.   I. Stansfield, William D. 1930–
II. Title.   QH427.K55   1990 575.1′03—dc20
89-27279

2 4 6 8 9 7 5 3
Printed in the United States of America
on acid-free paper

PREFACE

Genetics is the most rapidly advancing of the life sciences, and no other field has stimulated so many diverse disciplines in both the natural and social sciences. The fact that genetics has attracted mathematicians, physicists, chemists, physicians, anthropologists, and other scientists of diverse backgrounds to contribute to its development is one of the chief reasons for its prodigious growth. Such growth is, of course, accompanied by a proliferation in terminology, and this terminology constitutes a problem both to beginning students and to scientists from other disciplines who read papers by geneticists.

Geneticists use many words and abbreviations that are not found in collegiate dictionaries or dictionaries of biology. This is so because various terms, especially from molecular genetics, are newly coined; others, like those used in quantitative genetics are from other sciences, such as statistics, geology, medicine, and physics. Thus, to be truly useful, a dictionary for students of genetics needs to define not only words like **balanced lethal system, crossing over, mutation,** and **operon,** but terms and abbreviations such as **chi-square test, continental drift, retinoblastoma,** and **rep.** Therefore, our dictionary is broader than its name implies, since we attempt to define both strictly genetic words and also a variety of nongenetic terms that are often encountered in the genetics literature.

Scientific papers are peppered with the taxonomic names of species and genera that are studied by geneticists, but the student often may have no idea whether the author is referring to a bacterium, fungus, grass, or insect. Even when informed that "*Oenothera* is a genus of angiosperms," a reader unsophisticated in taxonomy may not know where these species fit in the Plant Kingdom. Therefore, we have placed, in alphabetical order in the body of the dictionary, the scientific names of the species that have been investigated by geneticists. Each species is identified by a common name, and its economic importance is described, if appropriate. Often, an organism with little or no economic importance has been investigated because it has certain useful advantages for studying specific genetic problems. In such cases, a few sentences are given to elucidate these advantages. Then, in Appendix A, a classification of living organisms is presented in which are placed all of the species cited in the dictionary, and these species entries in turn are cross referenced to Appendix A. For example, the entry on **_Chlamydomonas reinhardi_** is cross referenced so that the student can find it within the phylum Chlorophyta of the Kingdom Protoctista in Appendix A. Appendix B gives a list of the scientific names of roughly 240 domesticated species, but groups them alphabetically by common name.

In earlier editions of our dictionary, we have provided in Appendix C a chronological listing of discoveries and inventions that led to advances in genetics, cytology, or the study of evolution. While we have added new entries to the Chronology, we have also tried to make it more useful by citing Appendix C entries within the definitions themselves. Thus, the definition of **electrophoresis** contains a reference to a 1933 entry describing the invention of the technique by Tiselius, and the definition of **heat shock puffs** cites the first report of the phenomenon by Ritossa in 1962.

The Fourth Edition has grown considerably. The number of definitions now stands at 7,100, with 20 percent of them new or updated. The number of illustrations has risen from 225 to 250. The number of entries in Appendix C has grown from 560 to 635, and the index now lists 1,010 of the scientists cited. The bibliography of major sources for the chronology now contains 91 references (40 percent of them new). The periodical list presented as Appendix D has grown from 393 to 452 entries.

Certain rules have been followed regarding the arrangement of definition entries. Each is placed in alphabetical order using the boldface title letter by letter, while ignoring spaces between words. Thus, **S phase** is placed between *Sphaerocarpus* and **spheroplast.** Identical alphabetical listings are entered so that lowercase letters precede uppercase letters. Thus, the **r** entry comes before the **R** entry. In titles beginning with a Greek letter, the letter is spelled out. Thus, λ **phage** is found under **lambda phage.** In titles starting with a number or containing numbers, the numbers are ignored in the alphabetical placement. Thus, **M5 technique** is treated as **M technique** and **T24 oncogene** as **T oncogene.** However, numbers are used to determine the order in the series, e.g., **4S, 5S,** and **5.8S RNA.** When looking for a two-word term, if you don't find the definition listed under the first word, try the second. For example, definitions for **genetic recombination** and **beta thalassemia** occur under **recombination** and **thalassemias,** respectively.

The authors will welcome suggestions for any improvements, but are particularly anxious to receive advice concerning additional entries for Appendix C, so that these can be included subsequent editions.

Acknowledgments: Drs. Pamela Mulligan and Lynn Margulis were especially generous with advice and provided many useful definitions. They also made valuable suggestions as to scientific advances suitable for inclusion in the Chronology, and Drs. Susanne Gollin, Adam Wilkins, Eliot Spiess, Roger Melvold, Hans Noll, Robert Holmgren, Susan Pierce, and James Douglas Engel were also helpful in this regard.

Evanston, Ill.   R.C.K.
San Luis Obispo, Calif.   W.D.S.
October 1989

CONTENTS

A Dictionary of Genetics

A

A 1. mass number of an atom; 2. haploid set of autosomes; 3. ampere; 4. adenine or adenosine.

Å Angstrom unit (*q.v.*).

A_2 *See* hemoglobin.

A 23187 *See* ionophore.

AA-AMP amino acid adenylate.

A, B antigens mucopolysaccharides responsible for the ABO blood group system. The A and B antigens reside on the surface of erythrocytes, and differ only in the sugar attached to the penultimate monosaccharide unit of the carbohydrate chain. This minor chemical difference makes the macromolecule differentially active antigenically. The I^A and I^B genes presumably control the formation or functioning of the enzymes that add the specific sugar units to the carbohydrate chains in a preformed mucopolysaccharide molecule. The I^O allele is inactive in this regard and when homozygous results in the O phenotype. Glycoproteins with properties antigenically identical to the A, B antigens are ubiquitous, having been isolated from bacteria and plants. Every human being more than 6 months old possesses those antibodies of the A, B system that are not directed against its own blood-group antigens. These "preexisting natural" antibodies probably result from immunization by the ubiquitous antigens mentioned above. The *I* alleles reside on autosome 9. *See* **Appendix C**, 1901, Landsteiner; 1925, Bernstein; **blood group, H substance, secretor gene.**

aberrations *See* **chromosomal aberration, radiation-induced chromosomal aberration.**

ABM paper aminobenzyloxy methyl cellulose paper, which when chemically activated, reacts covalently with single-stranded nucleic acids.

ABO blood group system a system of alleles residing on human chromosome 9 that specifies certain red cell antigens. *See* **AB antigens, blood groups, Bombay blood group.**

abortive transduction failure of a transducing exogenote to become integrated into the host chromosome, but rather existing as a nonreplicating particle in only one cell of a clone. *See* **transduction.**

abortus a dead fetus born prematurely, whether the abortion was artificially induced or spontaneous.

absolute plating efficiency the percentage of individual cells that give rise to colonies when inoculated into culture vessels. *See* **relative plating efficiency.**

absorbance (*also* absorbency) a measure of the loss of intensity of radiation passing through an absorbing medium. It is defined in spectrophotometry by the relation: $\log (I_O/I)$, where I_O = the intensity of the radiation entering the medium and the I = the intensity after traversing the medium. *See* **Beer-Lambert law, OD$_{260}$ unit.**

abundance in molecular biology, the average number of molecules of a specific mRNA in a given cell, also termed *representation.* The abundance, A = NRf/M, where N = Avogadro's number, R = the RNA content of the cell in grams, f = the fraction the specific RNA represents of the total RNA, and M = the molecular weight of the specific RNA in daltons.

abzymes catalytic antibodies. A class of monoclonal antibodies that bind to and stabilize molecules in the transition state through which they must pass to form products. *See* **enzyme.**

acatalasemia the hereditary absence of catalase (*q.v.*) in man; inherited as an autosomal recessive.

acatalasia synonym for acatalasemia (*q.v.*).

acceleration *See* **heterochrony.**

accelerator an apparatus that imparts kinetic energy to charged subatomic particles to produce a high-energy particle stream for analyzing the atomic nucleus.

acceptor stem the double-stranded branch of a tRNA molecule to which an amino acid is attached (at the 3′, CCA terminus) by a specific aminoacyl-tRNA synthetase. *See* **transfer RNA.**

accessory chromosomes *See* **B chromosomes.**

accessory nuclei bodies resembling small nuclei that occur in the oocytes of most Hymenoptera and those of some Hemiptera, Coleoptera, Lepidoptera, and Diptera. Accessory nuclei are covered by a double membrane possessing annulate pores. They are originally derived from the oocyte nucleus, but they subsequently form by the amitotic division of other accessory nuclei.

***Ac, Ds* system** Activator–Dissociation system (*q.v.*).

ace *See* **symbols used in human cytogenetics.**

acentric designating a chromatid or a chromosome that lacks a centromere. *See* **chromosome bridge.**

Acer the genus of maple trees. *A. rubrum,* the red maple, and *A. saccharum,* the sugar maple, are studied genetically because of their commercial importance.

Acetabularia a genus of large, unicellular green algae. Grafting experiments between species of this genus have provided information on the nuclear control of cytoplasmic differentiation.

Acetobacter a genus of aerobic bacilli which secure energy by oxidizing alcohol to acetic acid.

aceto-carmine a stain used in the preparation of chromosome squashes consisting of a 5% solution of carmine in 45% acetic acid. Largely supplanted by aceto-orcein.

aceto-orcein a fluid consisting of 1% orcein (*q.v.*) dissolved in 45% acetic acid, used in making squash preparations of chromosomes. *See* **Appendix C,** 1925, Bernstein; **salivary gland squash preparation.**

acetylcholine a biogenic amine that plays an important role in the transmission of nerve impulses across synapses and from nerve endings to the muscles innervated. Here it changes the permeability of the sarcolemma and causes contraction. Acetylcholine is evidently a very ancient hormone, since it is present even in protists.

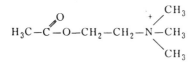

acetylcholinesterase the enzyme that catalyses the hydrolysis of acetylcholine (*q.v.*) into choline and acetate. Also called *cholinesterase.*

acetyl-coenzyme A *See* **coenzyme A.**

acetyl serine *See* **N-acetyl serine.**

achiasmate referring to meiosis without chiasmata. In those species in which crossing over is limited to one sex, the achiasmate meiosis generally occurs in the heterogametic sex.

achondroplasia a hereditary dwarfism due to retarded growth of the long bones. In humans it is inherited as an autosomal dominant trait. Homozygotes die at an early age. *See* **bovine achondroplasia, fowl achondroplasia.**

achromatic figure the mitotic apparatus (*q.v.*).

A chromosomes *See* **B chromosomes.**

acid fuchsin an acidic dye used in cytochemistry.

acidic amino acid an amino acid (*q.v.*) having a net negative charge at neutral *p*H. Those universally found in proteins are aspartic acid and glutamic acid, which bear negatively charged side chains in the *p*H range generally found in living systems.

acidic dye an organic anion that binds to and stains positively charged macromolecules.

Acinonyx jubatus the cheetah, a carnivore that has the distinction of being the world's fastest land animal. Cheetahs are of genetic interest because, while most other species of cats show heterozygosity levels of 10–20%, cheetahs have levels close to zero. This high degree of homozygosity is correlated with low fecundity, high mortality of cubs, and low disease resistance.

Acoelomata a subdivision of the Protostomia containing species in which the space between the epidermis and the digestive tube is occupied by a cellular parenchyma. *See* **classification.**

acquired characteristics, inheritance of inheritance by offspring of characteristics that arose in their parents as responses to environmental influences and are not the result of gene action. *See* **Lamarckism.**

acquired immunodeficiency syndrome *See* **AIDS, HIV.**

Acraniata a subphylum of Chordata containing animals without a true skull. *See* **Appendix A.**

acrasin a chemotactic agent produced by *Dictyostelium discoideum* that is responsible for the aggregation of the cells. Acrasin has been shown to be cyclic AMP (*q.v.*).

Acrasiomycota the phylum containing the cellular slime molds. These are protoctists that pass through a unicellular stage of amoebae that feed on bacteria. Subsequently these amoebae aggregate to form a fruiting structure that produces spores. The two most extensively studied species from this phylum are *Dictyostelium discoideum* and *Polysphondylium pallidum.*

acridine dye any of a class of organic molecules that bind to DNA and in bacteriophages act as mutagenic agents by causing additions or deletions in the base sequences.

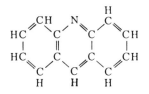

acridine orange an acridine dye that functions both as a fluorochrome and a mutagen.

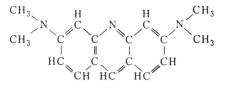

acriflavin an acridine dye that produces reading frame shifts *(q.v.).*

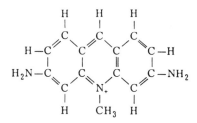

acritarchs spherical bodies thought to represent the earliest eukaryotic cells, estimated to begin in the fossil record about 1.6 billion years ago. Most acritarchs were probably thick-walled, cyst-forming protists. *See* **Proterozoic.**

acrocentric designating a chromosome or chromatid with a nearly terminal centromere. *See* **telocentric chromosome.**

acromycin *See* **tetracycline.**

acron the anterior nonsegmented portion of the embryonic arthropod that produces eyes and antennae. *See* **maternal polarity mutants.**

acrosome an apical organelle in the sperm head that is secreted by the Golgi material and that digests the egg coatings to permit fertilization.

acrostical hairs one or more rows of small bristles along the dorsal surface of the thorax of *Drosophila.*

acrosyndesis telomeric pairing by homologs during meiosis.

acrotrophic *See* **meroistic.**

acrylamide *See* **polyacrylamide gel.**

ACTH adrenocorticotropic hormone *(q.v.).*

actidione cycloheximide.

actin a protein that is the major constituent of the 7-nm–wide microfilaments of cells. Actin microfilaments (F actin) are polymers of a globular subunit (G actin) of Mr 42,000. Each G actin molecule has a defined polarity, and during polymerization the subunits align "head to tail," so that all G actins point in the same direction. F actin grows by the addition of G actin to its ends, and cytochalasin B *(q.v.)* inhibits this process. All the actins that have been studied, from sources as diverse as slime molds, fruit flies, and vertebrate muscle cells, are similar in size and amino acid sequence, suggesting that they evolved from a single ancestral gene. In mammals and birds, there are four different muscle actins. α_1 is unique to skeletal muscle; α_2, to cardiac muscle; α_3, to smooth vascular muscle; and α_4, to smooth enteric muscle. Two other actins (β and γ) are found in the cytoplasm of both muscle and nonmuscle cells. *See* **alternative splicing, contractile ring, fibronectin, isoforms, myosin, spectrin, stress fibers, tropomyosin, vinculin.**

actin genes genes encoding the various isoforms of actin. In *Drosophila,* for example, actin genes have been localized at six different chromosomal sites. Two genes encode cytoplasmic actins, while the other four encode muscle actins. The amino acid–encoding segments of the different actin genes have very similar compositions, but the segments specifying the trailers *(q.v.)* differ considerably in nucleotide sequences.

actinomycete any prokaryote placed in the phylum actinobacteria *(see* **Appendix A***).* Actinomycetes belonging to the genus *Streptomyces* produce a large number of the antibiotics, of which actinomycin D *(q.v.)* is an example.

actinomycin D an antibiotic produced by *Streptomyces chrysomallus* that prevents the transcription of messenger RNA. *See* **RNA polymerase.**

activated macrophage a macrophage that has been stimulated (usually by a lymphokine) to enlarge, to increase its enzymatic content, and to increase its nonspecific phagocytic activity.

activating enzyme an enzyme that catalyzes a reaction involving ATP and a specific amino acid. The product is an activated complex that subsequently reacts with a specific transfer RNA.

activation analysis a method of extremely sensitive analysis based on the detection of characteristic radionuclides produced by neutron activation.

activation energy the energy required for a chemical reaction to proceed. Enzymes (*q.v.*) combine transiently with a reactant to produce a new complex that has a lower activation energy. Under these circumstances the reaction can take place at the prevailing temperature of the biological system. Once the product is formed, the enzyme is released unchanged.

activator a molecule that converts a repressor into a stimulator of operon transcription; e.g., the repressor of a bacterial arabinose operon becomes an activator when combined with the substrate.

Activator-Dissociation system controlling elements in maize. *Ac* is an *autonomous* element that is inherently unstable. It has the ability to excise itself from one chromosomal site and to transpose to another. *Ac* is detected by its activation of *Ds*. *Ds* is *nonautonomous* and is not capable of excision or transposition by itself. *Ac* need not be adjacent to *Ds* or even on the same chromosome in order to activate *Ds*. When *Ds* is so activated, it can alter the level of expression of neighboring genes, the structure of the gene product, or the time of development when the gene expresses itself, as a consequence of nucleotide changes inside or outside of a given cistron. An activated *Ds* can also cause chromosome breakage, which may yield deletions or generate a bridge-breakage-fusion-bridge cycle (*q.v.*). *See* **Appendix C**, 1950, McClintock; **transposable elements.**

active center in the case of enzymes, a flexible portion of the protein that binds to the substrate and converts it into the reaction product. In the case of carrier and receptor proteins, the active center is the portion of the molecule that interacts with the specific target compounds.

active immunity immunity conferred on an organism by its own exposure and response to antigen. In the case of immunity to disease-causing agents, the antigenic pathogens may be adminis-

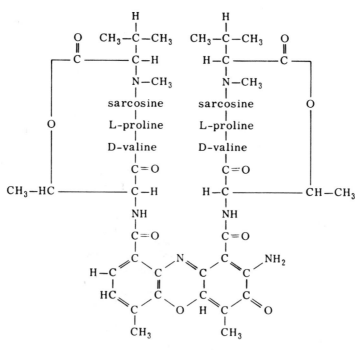

Actinomycin D.

6

tered in a dead or attenuated form. *See also* **passive immunity.**

active site   that portion(s) of a protein that must be maintained in a specific shape and amino acid content to be functional. Examples: **1.** in an enzyme, the substrate-binding region; **2.** in histones or repressors, the parts that bind to DNA; **3.** in an antibody, the part that binds antigen; **4.** in a hormone, the portion that recognizes the cell receptor.

actomyosin   *See* **myosin.**

acute transfection   infection of cells with DNA for a short period of time.

acylated tRNA   a transfer RNA molecule to which an amino acid is covalently attached. Also referred to as an activated tRNA, a charged tRNA, or a loaded tRNA.

adaptation   **1.** the process by which organisms undergo modification so as to function more perfectly in a given environment. **2.** any developmental, behavioral, anatomical, or physiological characteristic of an organism that, in its environment, improves its chances for survival and of leaving descendants.

adaptive enzyme   an enzyme that is formed by an organism in response to an outside stimulus. The term has been replaced by the term *inducible enzyme.* The discovery of adaptive enzymes led eventually to the elucidation of the mechanisms that switch gene transcription on and off. *See* **Appendix C,** 1937, Karström; **regulator gene.**

adaptive landscape   a three-dimensional graph that shows the frequencies of two genes, each present in two allelic forms (aA and bB in the illustration) plotted against average fitness for a given set of environmental conditions, or a comparable conceptual plot in multidimensional space to accommodate more than two loci.

adaptive norm   the array of genotypes (compatible with the demands of the environment) possessed by a given population of a species.

adaptive peak   a high point (perhaps one of several) on an adaptive landscape (*q.v.*), from which movement in any planar direction (changed gene frequencies) results in lower average fitness.

adaptive radiation   evolution from a generalized, primitive species of diverse, specialized species, each adapted to a distinct mode of life.

adaptive surface, adaptive topography   synonyms for adaptive landscape (*q.v.*).

adaptive value   the property of a given genotype when compared with other genotypes that confers fitness (*q.v.*) to an organism in a given environment.

adaptor   a short, synthetic DNA segment containing a restriction site that is coupled to both ends of a blunt-ended restriction fragment. The adaptor is used to join one molecule with blunt ends to a second molecule with cohesive ends. The restriction site of the adaptor is made identical to that of the other molecule so that when cleaved by the same restriction enzyme both DNAs will contain mutually complementary cohesive ends.

adaptor hypothesis   the proposal that polynucleotide adaptor molecules exist that can recognize specific amino acids and also the regions of the RNA templates that specify the placement of amino acids in a newly forming polypeptide. *See* **Appendix C,** 1958, Crick; **transfer RNA.**

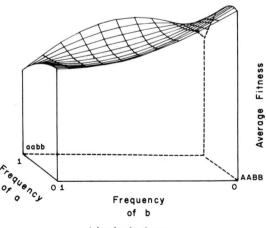

Adaptive landscape.

7

ADCC antibody-dependent cellular cytotoxicity; also known as antibody-dependent cell-mediated cytotoxicity. Cell-mediated cytotoxicity requires prior binding of antibody to target cells for killing to occur. It does not involve the complement cascade. *See* **K cells.**

additive factor one of a group of non-allelic genes affecting the same phenotypic characteristics and each enhancing the effect of the other in the phenotype. *See* **quantitative inheritance.**

additive gene action 1. a form of allelic interaction in which dominance is absent; the heterozygote is intermediate in phenotype between homozygotes for the alternative alleles. 2. the cumulative contribution made by all loci (of the kind described above) to a polygenic trait.

additive genetic variance genetic variance attributed to the average effects of substituting one allele for another at a given locus, or at the multiple loci governing a polygenic trait. It is this component of variance that allows prediction of the rate of response for selection of quantitative traits. *See* **quantitative inheritance.**

adduct the product of a chemical reaction that results in the addition of a small chemical group to a relatively large recipient molecule. Thus the alkylating agent ethyl methane sulfonate (*q.v.*) can add ethyl groups to the guanine molecules of DNA. These ethylated guanines would be examples of DNA adducts.

adenine *See* **bases of nucleic acids.**

adenine deoxyriboside *See* **nucleoside.**

adenohypophysis the anterior, intermediate, and tuberal portions of the hypophysis, which originate from the buccal lining in the embryo.

adenohypophysis hormone *See* **growth hormone.**

adenosine *See* **nucleoside.**

adenosine phosphate any of three compounds in which the nucleoside adenosine is attached through its ribose group to one, two, or three phosphoric acid molecules. They are illustrated below.
AMP, ADP, and ATP are interconvertible. ATP upon hydrolysis yields the energy used to drive a multitude of biological processes (muscle contraction, photosynthesis, bioluminescence, and the biosynthesis of proteins, nucleic acids, polysaccharides and lipids).

adenovirus any of a group of spherical DNA viruses characterized by a shell containing 252 capsomeres. Adenoviruses infect a number of mammalian species including man; some are oncogenic. Alternative splicing (*q.v.*) was discovered in adenovirus-2.

adenylcyclase the enzyme that catalyzes the conversion of adenosine triphosphate (ATP) into cyclic adenosine monophosphate (AMP). Also called adenylate cyclase.

adenylic acid *See* **nucleotide.**

adhesion plaques *See* **vincullin.**

adjacent disjunction, adjacent segregation *See* **translocation heterozygote.**

adjuvant a mixture injected together with an antigen that serves to intensify unspecifically the immune response. *See* **Freund's adjuvant.**

adoptive immunity, adoptive transfer the transfer of an immune function from one organism to another through the transfer of immunologically active or competent cells.

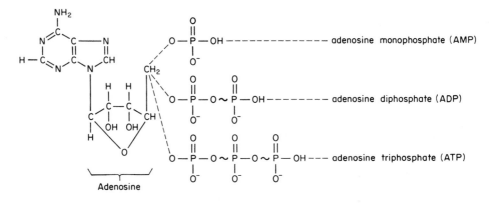

Adenosine phosphate.

ADP adenosine diphosphate. *See* **adenosine phosphate.**

adrenal corticosteroid a family of steroid hormones formed in the adrenal cortex. There are more than 30 of these hormones, and all are synthesized from cholesterol by cortical cells that have been stimulated by the adrenocorticotropic hormone (*q.v.*).

adrenocorticotropic hormone a single-chain peptide hormone (39 amino acids long) stimulating secretion by the adrenal cortex. It is produced by the adenohypophysis of vertebrates. Abbreviated ACTH. Also called *corticotropin.*

Adriamycin an antibiotic produced by *Streptomyces peucetius* that interacts with topoisomerase. DNA isolated from Adriamycin-poisoned cells contains single- and double-strand breaks. *See* **gyrase.**

advanced in systematics, the later or derived stages or conditions within a lineage that exhibits an evolutionary advance; the opposite of primitive.

adventitious embryony the production by mitotic divisions of an embryonic sporophyte from the tissues of another sporophyte without a gametophytic generation intervening.

Aedes a genus of mosquitoes containing over 700 species, several of which transmit important human diseases. *A. aegypti,* the vector of yellow fever, has a diploid chromosome number 6, and about 60 mutations have been mapped among its three linkage groups. Among these are genes conferring resistance to insecticides such as DDT and pyrethrins (*both of which see*).

Aegilops a genus of grasses including several species of genetic interest, especially *A. umbellulata,* a wild Mediterranean species resistant to leaf rust. A gene for rust resistance has been transferred from *A. umbellulata* to *Triticum vulgare.*

aerobe a cell that lives in air and utilizes oxygen. A strictly aerobic cell cannot live in the absence of oxygen.

aestivate to pass through a hot, dry season in a torpid condition. *See also* **hibernate.**

afferent leading toward the organ or cell involved. In immunology, the events or stages involved in activating the immune system.

affinity in immunology, the inate binding power of an antibody combining site with a single antigen binding site. *Compare with* **avidity.**

affinity chromatography a technique for separating molecules by their affinity to bind to ligands (e.g., antibodies) attached to an insoluble matrix (e.g., Sepharose). The bound molecules can subsequently be eluted in a relatively pure state.

afibrinogenemia an inherited disorder of the human blood-clotting system characterized by the inability to synthesize fibrinogen; inherited as an autosomal recessive.

aflatoxins a family of toxic compounds synthesized by *Aspergillus flavus* and other fungi belonging to the same genus. Aflatoxins bind to purines, making base pairing impossible, and they inhibit both DNA replication and RNA transcription. These mycotoxins are highly toxic and carcinogenic, and they often are contaminants of grains and oilseed products that are stored under damp conditions. The structure of aflatoxin G_1 is shown.

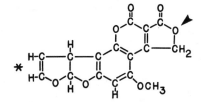

Aflatoxin B_1 has a CH_2 substituted for the O at the position marked by the arrow. Aflatoxin B_2 and G_2 are identical to B_1 and G_1, except that the ring labeled with an asterisk lacks a double bond.

African bees *Apis mellifera scutellata,* a race of bees, originally from South Africa, that was accidentally introduced into Brazil in 1957 and has spread as far as Mexico. It threatens to invade the southern United States. African bees are poor honey producers and tend to sting much more often than European bees. Because of daily differences in flight times of African queens and European drones, hybridization is rare. *See* **Apis mellifera.**

African Eve *See* **mtDNA lineages**

African green monkey *See* **Cercopithecus aethiops.**

agamete a haploid, asexual reproductive cell resulting from meiosis in an agamont. Agametes disperse and grow into gamonts (*q.v.*).

agammaglobulinemia the inability in humans to synthesize certain immunoglobulins. The most common form is inherited as an X-linked recessive trait, and plasma cells are absent. *See* **antibody.**

agamogony the series of cellular or nuclear divisions that generates agamonts.

agamont the diploid adult form of a protoctist that also has a haploid adult phase in its life cycle. An agamont undergoes meiosis and produces agametes. *See* **gamont.**

agamospermy the formation of seeds without fertilization. The male gametes, if present, serve only to stimulate division of the zygote. *See* **apomixis.**

Agapornis a genus of small parrots. The nest building of various species and their hybrids has provided information on the genetic control of behavior patterns.

agar a polysaccharide extract of certain seaweeds used as a solidifying agent in culture media.

agarose a linear polymer of alternating D-galactose and 3,6-anhydrogalactose molecules. The polymer, fractionated from agar, is often used in gel electrophoresis because few molecules bind to it, and therefore it does not interfere with electrophoretic movement of molecules through it.

agar plate count the number of bacterial colonies that develop on an agar-containing medium in a petri dish seeded with a known amount of inoculum. From the count the concentration of bacteria per unit volume of inoculum can be determined.

age-dependent selection selection in which the values for relative fitness of different genotypes vary with the age of the individual.

agglutination the clumping of viruses or cellular components in the presence of a specific immune serum.

agglutinin any antibody capable of causing clumping of erythrocytes, or more rarely other types of cells.

agglutinogen an antigen that stimulates the production of agglutinins.

aggregation chimera a mammalian chimera made through the mingling of cells of two embryos. The resulting composite embryo is then transferred into the uterus of a surrogate mother where it comes to term.

agonistic behavior any social interaction between members of the same species that involves aggression or threat and conciliation or retreat.

agouti the grizzled color of the fur of mammals resulting from alternating bands of light (phaeomelanin) and dark (eumelanin) pigments in the individual hairs. It is also the name given to the genes (usually dominant to the dark color genes) responsible for the insertion of the light pigment bands into the hairs. *See* **melanin.**

agranular reticulum endoplasmic reticulum devoid of attached ribosomes.

agranulocytes white blood cells whose cytoplasm contains few or no granules and that possess an unlobed nucleus; mononuclear leucocytes including lymphocytes and monocytes.

agriculturally important species *See* **Appendix B.**

Agrobacterium tumefaciens the bacterium responsible for crown gall disease in a wide range of dicotyledonous plants. The bacterium enters only dead, broken plant cells and then may transmit a tumor-inducing plasmid into adjacent living plant cells. This infective process is a natural form of genetic engineering. Strains of *A. tumefaciens* carrying the plasmid may be artificially genetically engineered to introduce foreign genes of choice into plant cells, and then by growing the cells in tissue culture, whole plants can be regenerated, every cell of which contains the foreign gene. *See* **Appendix C,** 1907, Smith; **Ti plasmid.**

Agropyron elongatum a weed related to crabgrass noted for its resistance to stem rust. Genes conferring rust resistance have been transferred from this species to *Triticum aestivum.*

AHF antihemophilic factor. *See* **blood clotting.**

AI, AID, AIH *See* **artificial insemination.**

AIA anti-immunoglobulin antibodies, produced in response to foreign antibodies introduced into an experimental animal.

AIDS the *a*cquired *i*mmuno*d*eficiency *s*yndrome, a disease caused by the human immunodeficiency virus (HIV). This virus attacks lymphocytes of helper T subclass and macrophages. The depletion of these cells makes the patient susceptible to pathogens that would easily be controlled by a healthy immune system. The infection is transmitted by sexual intercourse, by direct contamination of the blood (as when virus-contaminated drug paraphernalia is shared), or by passage of the virus from an infected mother to her fetus or to a suckling baby. *See* **HIV, lymphocyte, retroviruses.**

akinetic acentric (*q.v.*).

ala alanine. *See* **amino acid.**

albinism 1. deficiency of chromoplasts in plants. 2. the inability to form melanin (*q.v.*) in the eyes, skin and hair, due to a tyrosinase deficiency. In humans, the condition is inherited as an autosomal recessive. Amelanic melanocytes are present in the skin of albinos. In a wide variety of animals, the albino gene has pleiotropic effects on the visual pathways, resulting in nystagmus and crossed eyes. *See* **ocular albinism.**

albino   1. a plant lacking chromoplasts. **2.** an animal lacking pigmentation.

albomaculatus   referring to a variegation consisting of irregularly distributed white and green regions on plants resulting from the mitotic segregation of genes or plastids.

albumin   a water-soluble 70 kd protein that represents 40–50% of the plasma protein in adult mammals. It is important both as an osmotic and as a *p*H buffer and also functions in the transport of metal ions and various small organic molecules. Albumin is synthesized and secreted by the liver. In the mouse the albumin gene resides on chromosome 5, separated from the alpha fetoprotein gene by a DNA segment about 13.5 kb long. In humans, these two genes are in the long arm of chromosome 4. *See* **Appendix C,** 1967, Sarich and Wilson; **alpha fetoprotein.**

alcaptonuria   alkaptonuria (*q.v.*).

alcohol   any hydrocarbon that carries one or more hydroxyl groups. The term is often used to refer specifically to ethyl alcohol, the product of yeast-based fermentations. Hereditary differences in alcohol preference are known to exist in mice. *See* **Appendix C,** 1962, Rodgers and McClearn.

alcohol dehydrogenase   (ADH) a zinc-containing enzyme found in bacteria, yeasts, plants, and animals that reversibly oxidizes primary and secondary alcohols to the corresponding aldehydes and ketones. In the case of yeast, ADH functions as the last enzyme in alcoholic fermentation. In *Drosophila melanogaster,* ADH is a dimeric protein. By suitable crosses between null activity mutants it is possible to generate heteroallelic individuals that exhibit partial restoration of enzyme activity. This is often due to the production of a heterodimer with improved functional activity. *See* **allelic complementation.**

aldehyde   any of a class of organic compounds having the general formula $C_nH_{2n}O$ and containing a terminal $-C\diagdown{}^{H}_{O}$ group.

aldosterone   an adrenal corticosteroid hormone that controls the sodium and potassium balance in the vertebrates.

aleurone   the outer layer of the endosperm of a seed.

aleurone grain   a granule of protein occurring in the aleurone.

Aleutian mink   an autosomal recessive mutation in *Mustela vison* producing diluted pigmentation of the fur and eyes. The homozygotes show a lysosomal defect similar in humans to the Chédiak-Steinbrinck-Higashi syndrome (*q.v.*).

alga (*plural* **algae**)   any of a large group of aquatic, chlorophyll-bearing organisms ranging from single cells to giant seaweeds. *See* **Appendix A:** Cyanobacteria, Dinoflagellata, Euglenophyta, Xanthophyta, Bacillariophyta, Phaecophyta, Rhodophyta, Gamophyta, Chlorophyta.

alien addition monosomic   a genome that contains a single chromosome from another species in addition to the normal complement of chromosomes.

alien substitution   replacement of one or more chromosomes of a species by those from a different species.

aliphatic   designating molecules made up of linear chains of carbon atoms.

aliquot   a part, such as a representative sample, that divides the whole without a remainder. Two is an aliquot of six because it is contained exactly three times. Loosely used for any fraction or portion.

alkali metal   any of five elements in Group IA of the periodic table: lithium (Li), sodium (Na), potassium (K), rubidium (Rb), and cesium (Cs).

alkaline earth   any element of Group IIA of the periodic table: beryllium (Be), magnesium (Mg), calcium (Ca), strontium (Sr), barium (Ba), and radium (Ra).

alkaline phosphatase   an enzyme that removes 5′-P termini of DNA and leaves 5′-OH groups.

alkaloids   a group of nitrogen-containing organic substances found in plants; many are pharmacologically active (e.g., caffeine, cocaine, nicotine).

alkapton   2,5-dihydroxyphenylacetic acid. *See* **homogentisic acid.**

alkaptonuria (*also* **alcaptonuria**)   a relatively benign hereditary disease in man due to an autosomal recessive gene. Alkaptonurics cannot make the liver enzyme homogentisic acid oxidase. Therefore, homogentisic acid (*q.v.*) is not broken down to simpler compounds but is excreted in the urine. Since the colorless homogentisic acid is readily oxidized to a black pigment, the urine of alkaptonurics darkens when exposed to air. This disease enjoys the historic distinction of being the first metabolic disease studied *See* **Appendix C,** 1909, Garrod.

alkylating agent   a compound causing the substitution of an alkyl group (usually methyl or ethyl) for an active hydrogen atom in an organic compound. According to the number of reactive groups they contain, alkylating agents are classified as mono-, bi-, or polyfunctional. Many chemical mutagens are alkylating agents. *See* **busulfan, chlorambucil, cyclophosphamide, epoxide, ethylmethane sulfonate, melphalan, Myleran, nitrogen mustard, sulfur mustard, TEM, Thio-tepa, triethylenethiophosphoramide.**

alkyl group   a univalent radical having the general formula C_nH_{2n+1} derived from a saturated aliphatic hydrocarbon by removal of one atom of hydrogen. Named by replacing the ending *-ane* of the hydrocarbon with *-yl* (e.g., meth*ane* becomes meth*yl*).

allantois   a saclike outgrowth of the ventral side of the hindgut present in the embryos of reptiles, birds, and mammals. The allantois represents a large and precocious development of the urinary bladder.

allatum hormones   hormones synthesized by the insect corpus allatum. The titer of allatum hormones influences the qualitative properties of each molt in holometabolous insects. At high concentrations, larval development ensues; at lower levels, the insect undergoes pupal metamorphosis, and in the absence of the allatum hormones adult differentiation takes place. The allatum hormones thus have a juvenilizing action and for this reason have also been called juvenile hormones (JHs). The structural formulas for three of the juvenile hormones are shown below. In adult females, the allatum hormone is required for vitellogenesis. The JH analogue, ZR515 (*q.v.*), is often used as a substitute for natural JHs in *Drosophila* experiments. *See* **Appendix C,** 1966, Röller *et al.;* **ring gland.**

allele   a shorthand form of **allelomorph;** one of a series of possible alternative forms of a given gene (cistron, *q.v.*), differing in DNA sequence, and affecting the functioning of a single product (RNA and/or protein). If more than two alleles have been identified in a population, the locus is said to show *multiple allelism. See* **heteroallele, homoallele.**

allelic complementation   the production of nearly normal phenotype in an organism carrying two different mutant alleles in trans configuration. Such complementation is sometimes caused by the reconstruction in the cytoplasm of a functional protein from the inactive products of the two alleles. When such a phenomenon can be demonstrated by mixing extracts from individuals homozygous for each allele, the term *in vitro complementation* is used. Synonymous with **intrallelic complementation.** *See* **alcohol dehydrogenase.**

allelic exclusion   in a clone of plasma cells heterozygous for an immunoglobulin gene, the detection of only one allelic product. The mechanism that excludes activation of the other allele is under study.

allelic frequency   the percentage of all alleles at a given locus in a population gene pool represented by a particular allele. For example, in a population containing 20 *AA*, 10 *Aa*, and 5 *aa*, the frequency

JH 1 $CH_3-CH_2-\underset{\underset{O}{\diagdown\diagup}}{\overset{\overset{CH_3}{|}}{C}}-CH-CH_2-CH_2-\underset{\underset{CH_3}{\underset{|}{CH_2}}}{\overset{}{C}}=CH-CH_2-CH_2-\underset{}{\overset{\overset{CH_3}{|}}{C}}=CH-\overset{\overset{O}{||}}{C}-O-CH_3$

JH 2 $CH_3-CH_2-\underset{\underset{O}{\diagdown\diagup}}{\overset{\overset{CH_3}{|}}{C}}-CH-CH_2-CH_2-\underset{\underset{CH_3}{|}}{\overset{}{C}}=CH-CH_2-CH_2-\underset{}{\overset{\overset{CH_3}{|}}{C}}=CH-\overset{\overset{O}{||}}{C}-O-CH_3$

JH 3 $CH_3-\underset{\underset{O}{\diagdown\diagup}}{\overset{\overset{CH_3}{|}}{C}}-CH-CH_2-CH_2-\underset{\underset{CH_3}{|}}{\overset{}{C}}=CH-CH_2-CH_2-\underset{}{\overset{\overset{CH_3}{|}}{C}}=CH-\overset{\overset{O}{||}}{C}-O-CH_3$

ZR 515 $CH_3-\underset{\underset{O-CH_3}{|}}{\overset{\overset{CH_3}{|}}{C}}-CH_2-CH_2-CH_2-\underset{\underset{CH_3}{|}}{CH}-CH_2-CH=CH-\overset{}{\underset{}{C}}=CH-\overset{\overset{O}{||}}{C}-O-\underset{\diagdown CH_3}{\overset{\diagup CH_3}{CH_3}}$

Allatum hormones.

of the *A* allele is $[2(20) + 1(10)]/2(35) = 5/7 = 0.714$. *See* **gene frequency.**

allelism test   complementation test (*q.v.*).

allelomorph   commonly shortened to allele (*q.v.*) *See* **Appendix C,** 1902–09, Bateson.

allelopathy   an interaction involving two different species in which chemicals introduced into the environment by one suppress the growth or reproduction of the other.

allelotype   the frequency of alleles in a breeding population.

Allen's rule   the generalization that the extended body parts of a warm-blooded species (tail, ears, and limbs) are relatively shorter in the colder regions of the species range than in the warmer.

allergen   a substance inducing hypersensitivity.

allergy   an immune hypersensitivity response to an agent that is nonantigenic to most of the individuals in a population.

allesthetic trait   any individual characteristic that has an adaptive function only via the nervous systems of other organisms. For example, odors, display of color patterns, mating calls, etc., that are important components of courtship in various species.

Allium   the genus that includes *A. cepa,* the onion; *A. porrum,* the leek; *A. sativum,* the garlic; and *A. schoenoprasum,* the chive—all classic subjects for cytological studies of mitotic chromosomes.

alloantigen   an antigen (*q.v.*) that elicits an immune response (*q.v.*) when introduced into a genetically different individual of the same species. Antibodies produced in response to alloantigens are called *alloantibodies. See* **histocompatibility molecules.**

allochromacy   the formation of other coloring agents from a given dye which is unstable in solution. Nile blue (*q.v.*) exhibits allochromacy.

allocycly   a term referring to differences in the coiling behavior shown by chromosomal segments or whole chromosomes. Allocyclic behavior characterizes the pericentric heterochromatin, the nucleolus organizer, and in some species entire sex chromosomes. If a chromosome or chromosomal segment is tightly condensed in comparison with the rest of the chromosomal complement, the chromosome or chromosomal segment is said to show *positive heteropycnosis* (*q.v.*). Allocycly is also used to describe asynchronous separation of bivalents during the first anaphase in meiosis. In man, for example, the X and Y chromosomes segregate ahead of the autosomes and are said to show *positive allocycly.*

allogeneic   referring to genetically different members of the same species, especially with regard to alloantigens (*q.v.*).

allogeneic disease   *See* **graft-versus-host reaction.**

allograft   a graft of tissue from a donor of one genotype to a host of a different genotype but of the same species.

allolactose   *See* **lactose.**

allometry   the relation between the rate of growth of a part of an individual and the growth rate of the whole or of another part. In the case of isometry, the relative proportions of the body parts remain constant as the individual grows; in all other cases, the relative proportions change as total body size increases. *See* **heterauxesis.**

allomone   any chemical secreted by an organism that influences behavior in a member of another species, benefiting only the producer. If both species benefit, it is a *synamone.* If only the receiver benefits, it is a *kairomone.*

alloparapatric speciation   a mode of gradual speciation in which new species originate through populations that are initially allopatric, but later become parapatric before completely effective reproductive isolation has evolved. Natural selection may enhance incipient reproductive isolating mechanisms in the zone of contact by character displacement (*q.v.*), and other mechanisms. *Compare with* **parapatric speciation.**

allopatric speciation   the development of distinct species through differentiation of populations in geographic isolation.

allopatry   referring to species living in different geographic locations and separated by distance alone or by some barrier to migration such as a mountain range, river, or desert. *Compare with* **sympatry.**

allophene   a phenotype not due to the mutant genetic constitution of the cells of the tissue in question. Such a tissue will develop a normal phenotype if transplanted to a wild-type host. *See* **autophene.**

allophenic mice   chimeric mice produced by removing cleaving eggs from mice of different genotypes, fusing the blastomeres *in vitro,* and reimplanting the fused embryos into the uterus of another mouse to permit embryogenesis to continue. Viable mice containing cells derived from two or more embryos have been obtained and used in cell lineage studies. *See* **Appendix C,** 1967, Mintz.

allopolyploid (*also* **alloploid**) a polyploid organism arising from the combination of genetically distinct chromosome sets.

alloprocoptic selection a mode of selection in which association of opposites increases the fitness of the associates. An example involves the loci governing alcohol dehydrogenase in *Drosophila melanogaster*. The fertility is greater than expected when two mating individuals are homozygous for different alleles and smaller than expected when they are homozygous for the same allele.

allostery the reversible interaction of a small molecule with a protein molecule, which leads to changes in the shape of the protein and a consequent alteration of the interaction of that protein with a third molecule.

allosteric effectors small molecules that reversibly bind to allosteric proteins at a site different from the active site, causing an allosteric effect.

allosteric enzyme a regulatory enzyme whose catalytic activity is modified by the noncovalent attachment of a specific metabolite to a site on the enzyme other than the catalytic site.

allosteric protein a protein showing allosteric effects.

allosteric site a region on a protein other than its active site (*q.v.*), to which a specific effector molecule may bind and influence (either positively or negatively) the functional activity of the protein. For example, in the lactose system of *Escherichia coli*, the *lac* repressor becomes inactive (cannot bind to the *lac* operator) when allolactose is bound to the allosteric site of the repressor molecule. *See lac* **operon.**

allosyndesis the pairing of homoeologous chromosomes (*q.v.*) in an allopolyploid (*q.v.*). Thus if the genetic composition of an alloploid is given by AABB, where AA represent the chromosomes derived from one parent species and BB the chromosomes derived from the other parent species, then during meiotic prophase, A undergoes allosyndetic pairing with B. Such pairing indicates that the A and B chromosomes have some segments that are homologous, presumably because the two parent species have a common ancestry. In the case of *autosyndesis*, A pairs only with A, and B with B. *Segmental alloploids* form both bivalents and multivalents during meiosis because of allosyndesis.

allotetraploid an organism that is diploid for two genomes, each from a different species; synonymous with amphidiploid.

allotypes antigenic markers on immunoglobulin chains (*q.v.*) or other serum proteins that are not common to all normal members of a species. For example, in humans 3 genetic variants of the kappa light chains are known that result from amino acid substitution at positions 153 and/or 191. *Compare with* **idiotypes, isotypes.**

allotype suppression the systematic and long-term suppression of the expression of an immunoglobulin allotype in an animal induced by treatment with antibodies against the allotype.

allotypic differentiation *See in vivo* culturing of **imaginal discs.**

allozygote an individual homozygous at a given locus, whose two homologous genes are of independent origin, as far as can be determined from pedigree information. *See* **autozygote.**

allozymes allelic forms of an enzyme that can be distinguished by electrophoresis, as opposed to the more general term *isozyme* (*q.v.*). *See* **Appendix C, 1966, Lewontin and Hubby.**

alpha amanitin *See* **amatoxins.**

alpha chain one of the two polypeptides found in adult and fetal hemoglobin (*q.v.*).

alpha fetoprotein the major plasma protein of fetal mammals. AFP is a 70 kd glycoprotein that is synthesized and secreted by the liver and the yolk sac. The genes encoding AFP and serum albumen arose in evolution as the result of a duplication of an ancestral gene $(3–5) \times 10^8$ years ago. *See* **albumen.**

alpha galactosidase an enzyme that catalyzes the hydrolysis of substrates that contain α-galactosidic residues, including glycosphingolipids and glycoproteins. In humans, α-galactosidase exists in two forms, A and B. The A form is encoded by a gene on the X chromosome. Fabry disease (*q.v.*) is caused by mutations at this locus. The B form is encoded by a gene on chromosome 22.

alpha helix a characteristic helical secondary structure seen in many proteins. The alpha helix allows maximum intramolecular hydrogen bond formation between $C=O$ and $H-N$ groups. One turn of the helix occurs for each 3.6 amino acid residues. *See* **protein structure.**

alpha particle a helium nucleus, consisting of two protons and two neutrons, and having a double positive charge.

alpha tocopherol vitamin E. (*q.v.*)

alphoid sequences a complex family of repetitive

DNA sequences found in the centromeric heterochromatin of human chromosomes. The alphoid family is composed of tandem arrays of 170-bp segments. The segments isolated from different chromosomes show a consensus sequence, but also differences with respect to individual bases, so that the 170-bp units may vary in sequence by as much as 40%. The repeats are organized in turn into groups containing several units in tandem, and these groups are further organized into larger sequences 1 to 6 kb in length. These large segments are then repeated to generate segments 0.5 to 10 mbps in size. Such larger, or "macro," DNA repeats are chromosome-specific. Since alphoid sequences are not transcribed, they play an as yet undefined structural role in the chromosome cycle. The variation in the sequences within the alphoid DNA results in a high frequency of RFLPs. These are inherited and can be used to characterize the DNAs of specific individuals and their relatives. *See* **DNA fingerprint technique, restriction fragment length polymorphisms.**

alteration enzyme   a protein of phage T4 that is injected into a host bacterium along with the phage DNA; this protein modifies host RNA polymerase by linking it to ADP-ribose. RNA polymerase modified in this way renders it incapable of binding to sigma factor and thus unable to initiate transcription at host promoters. *See* **RNA polymerase.**

alternate disjunction, alternate segregation   *See* **translocation heterozygote.**

alternation of generations   reproductive cycles in which a haploid phase alternates with a diploid phase. In mosses and vascular plants, the haploid phase is the gametophyte, the diploid the sporophyte.

alternative splicing   a mechanism for generating multiple protein isoforms from a single gene that involves the splicing together of nonconsecutive exons during the processing of some, but not all, transcripts of the gene. This is illustrated in the diagram, where a gene is made up of five exons joined by introns i^1–i^4. The exons may be spliced by the upper pathway shown by the dotted lines to generate a mature transcript containing all five exons. This type of splicing is termed *constitutive*. The alternative mode of splicing shown generates a mature transcript that lacks exon 4. If each exon encodes 20 amino acids, the constitutive splicing path would result in a polypeptide made up of 100 amino acids. The alternative path would produce a polypeptide only 80 amino acids long. If the amino acid sequences of the two proteins were determined, the first 60 and the last 20 would be identical. More than 50 genes are known to generate protein diversity through alternative splicing in organisms including *Drosophila,* chickens, rats, mice, and humans. *See* **Appendix C,** 1977, Weber *et al.;* **fibronectin, isoforms, posttranscriptional modification, myosin genes, tropomyosin.**

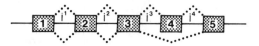

altricial   referring to the type of ontogeny seen in vertebrate species characterized by large litters, short gestations, and the birth of relatively undeveloped, helpless young. *Compare with* **precocial.**

altruism   behavior of an individual that benefits others. To the extent that the "others" are related to the altruist (the one exhibiting altruistic behavior), such actions may actually be an expression of fitness. *See* **inclusive fitness.**

Alu family   the most common dispersed repeated DNA sequence in the human genome, consisting of about 300 base pairs in perhaps half a million copies, accounting for about 5% of human DNA. Each segment apparently is made up of two 140 base-pair sequences joined head to tail with a 31-base-pair insert in the right-hand monomer. These sequences appear to be readily transposable. The family name is derived from the fact that these sequences are cleaved by the restriction endonuclease *Alu I.*

Amanita phalloides   a poisonous mushroom which is the source of amatoxins and phallotoxins *(both of which see).*

amatoxins   a group of bicyclic octapeptides which are among the poisons produced by *Amanita phalloides (q.v.).* These poisons inhibit transcription in eukaryotic cells because of their interaction with RNA polymerase II. However, they do not affect the RNA polymerases of mitochondria or chloroplasts. Alpha amanitin (formula, p. 16) is an amatoxin most commonly used experimentally to inhibit transcription. *See* **phallotoxins, RNA polymerase.**

amaurosis   blindness occurring without an obvious lesion in the eye, as from a disease of the optic nerve or brain.

amber codon   the mRNA triplet UAG that causes termination of protein translation; one of three "stop" codons. The terms *amber* and *ochre (q.v.)* originated from a private laboratory joke and have nothing to do with colors.

Amberlite   trade name for a family of ion-exchange resins.

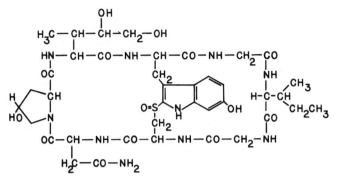

Alpha amanitin.

amber mutation a mutation in which a polypeptide chain is terminated prematurely. Amber mutations are the result of a base substitution that converts a codon specifying an amino acid into UAG, which signals chain termination. In certain strains of *E. coli* amber mutations are suppressed. These strains contain a tRNA with a AUC anticodon, which inserts an amino acid at the UAG site and hence permits translation to continue. *See* **ochre mutation, nonsense mutation.**

amber suppressor any mutant gene coding for a tRNA whose anticodon can respond to the UAG stop codon by the insertion of an amino acid that renders the gene product at least partially functional. For example, a mutant tyrosine-tRNA anticodon 3'AUC would recognize 5'UAG, tyrosine would be inserted, and chain growth would continue.

Ambystoma mexicanum the Mexican axolotl, a widely used laboratory species. The urodele for which the most genetic information is available. *See* **neoteny.**

amensalism a species interaction in which one is adversely affected and the other is unaffected.

Ames test a bioassay for detecting mutagenic and possibly carcinogenic compounds, developed by Bruce N. Ames in 1974. Reverse mutants to histidine independence are scored by growing *his⁻ Salmonella typhimurium* on plates deficient in histidine in the presence of the chemical (test) and in its absence (control).

AMINO ACID	ONE-LETTER SYMBOL	THREE-LETTER SYMBOL	mRNA CODE DESIGNATION
alanine	A	ala	GCU, GCC, GCA, GCG
arginine	R	arg	CGU, CGC, CGA, CGG, AGA, AGG
asparagine	N	asn	AAU, AAC
aspartic acid	D	asp	GAU, GAC
cysteine	C	cys	UGU, UGC
glutamic acid	E	glu	GAA, GAG
glutamine	Q	gln	CAA, CAG
glycine	G	gly	GGU, GGC, GGA, GGG
histidine	H	his	CAU, CAC
isoleucine	I	ile	AUU, AUC, AUA
leucine	L	leu	UUA, UUG, CUU,CUC, CUA, CUG
lysine	K	lys	AAA, AAG
methionine	M	met	AUG
phenylalanine	F	phe	UUU, UUC
proline	P	pro	CCU, CCC, CCA, CCG
serine	S	ser	UCU, UCC, UCA, UCG, AGU, AGC
threonine	T	thr	ACU, ACC, ACA, ACG
tryptophan	W	trp	UGG
tyrosine	Y	tyr	UAU, UAC
valine	V	val	GUU, GUC, GUA, GUG

Amino acids.

amethopterin methotrexate (*q.v.*).

amino acid any of the subunits that are polymerized to form proteins. The twenty amino acids universally found in proteins are listed below together with their abbreviations and their messenger RNA code designations. *Also see* **genetic code.**

amino acid activation a coupled reaction catalyzed by a specific aminoacyl synthetase that attaches a specific amino acid (AA) to a specific transfer RNA (tRNA) in preparation for translation (*q.v.*).

$$AA + ATP \rightarrow AA\text{-}AMP + 2P$$
$$AA\text{-}AMP + tRNA \rightarrow AA\text{-}tRNA + AMP$$

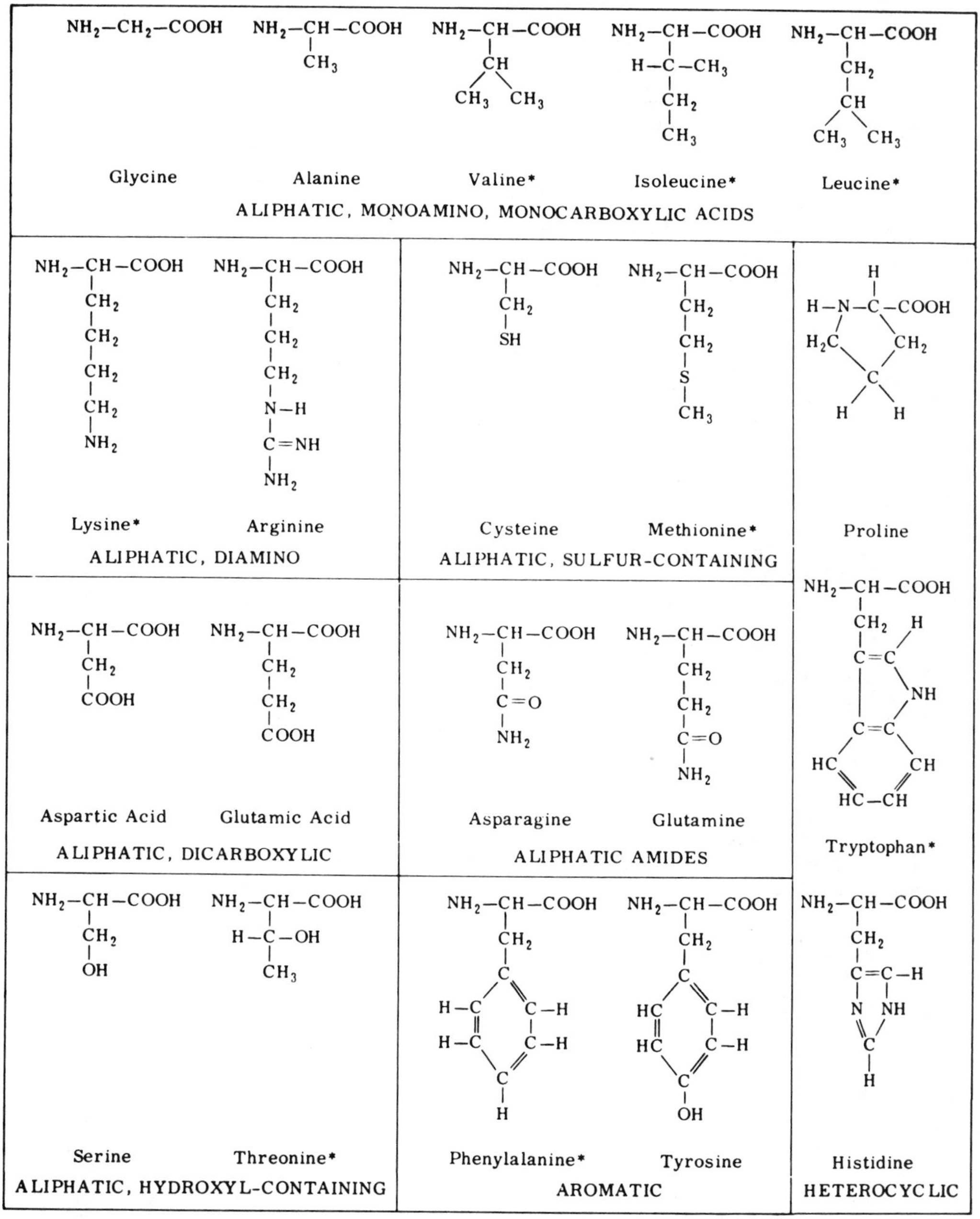

NH_2-CH_2-COOH	$NH_2-CH-COOH$ $\quad\quad\quad CH_3$	$NH_2-CH-COOH$ $\quad\quad CH$ $\quad CH_3\ CH_3$	$NH_2-CH-COOH$ $\quad H-C-CH_3$ $\quad\quad CH_2$ $\quad\quad CH_3$	$NH_2-CH-COOH$ $\quad\quad CH_2$ $\quad\quad CH$ $\quad CH_3\ CH_3$
Glycine	Alanine	Valine*	Isoleucine*	Leucine*
ALIPHATIC, MONOAMINO, MONOCARBOXYLIC ACIDS				

ALIPHATIC, DIAMINO — Lysine*, Arginine

ALIPHATIC, SULFUR-CONTAINING — Cysteine, Methionine*

Proline

ALIPHATIC, DICARBOXYLIC — Aspartic Acid, Glutamic Acid

ALIPHATIC AMIDES — Asparagine, Glutamine

Tryptophan*

ALIPHATIC, HYDROXYL-CONTAINING — Serine, Threonine*

AROMATIC — Phenylalanine*, Tyrosine

HETEROCYCLIC — Histidine

*required in the diet of mammals.

Structural formulas of the universal amino acids.

amino acid attachment site   the 3′ end of a tRNA molecule to which an amino acid is covalently attached by an aminoacyl bond. *See* **amino acid activation, aminoacyl-tRNA synthetases, transfer RNA.**

amino acid sequence   the linear order of the amino acids in a peptide or protein.

amino acid side chain   a group attached to an amino acid, represented by **R** in the general formula for an amino acid:

$$NH_2-CH-COOH$$
$$|$$
$$R$$

aminoaciduria   the presence of one or more amino acids in the urine in abnormal quantities because of a metabolic defect.

aminoacyl adenylate   the activated compound that is an intermediate in the formation of a covalent bond between an amino acid and its specific transfer RNA; abbreviated AA-AMP. *See* **AMP, transfer RNA.**

aminoacyl site   one of two binding sites for tRNA molecules on a ribosome; commonly called the *A site. See* **translation.**

aminoacyl-tRNA   an aminoacyl ester of a transfer RNA molecule.

aminoacyl-tRNA binding site   *See* **translation.**

aminoacyl-tRNA synthetases   enzymes that activate amino acids and attach each activated amino acid to its own species of tRNA. These enzymes catalyze: (1) the reaction of a specific amino acid (AA) with adenosine triphosphate (ATP) to form AA-AMP, and (2) the transfer of the AA-AMP complex to a specific transfer RNA, forming AA-tRNA and free AMP (adenosine monophosphate). *See* **adenosine phosphate.**

amino group   a chemical group ($-NH_2$) which with the addition of a proton can form $-NH_3^+$.

p amino benzoic acid   a component of folic acid (*q.v.*).

aminopeptidase   an enzyme (in both prokaryotes and eukaryotes) that removes the formylated methionine (fMet) or methionine from the NH_2 terminus of growing or completed polypeptide chains.

aminopterin   *See* **folic acid.**

aminopurine   a base analogue producing transitions (*q.v.*).

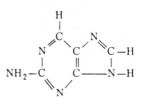

amino terminal end   the end of a polypeptide chain that has a free amino group.

amitosis   the division of a nucleus into two parts by constriction without the participation of a mitotic apparatus. Accessory nuclei (*q.v.*) grow by amitosis.

amixis   a reproductive cycle lacking meiosis and fertilization. Asexual reproduction. *Contrast with* **amphimixis apomixis.**

amniocentesis   sampling of amniotic fluid for the prenatal diagnosis of fetal disorders. During the procedure, a hollow needle is inserted through the skin and muscle of the mother's abdomen, through the uterus, and into the amniotic sac that surrounds the fetus. Cells that have sloughed from the fetus are suspended in the fluid. Cells in the sample are cultured for about three weeks to raise their numbers to the point where chromosomal and biochemical analyses can be made. Amniocentesis cannot be done until about the 16th week of pregnancy, since the sac containing the embryo is not large enough to permit safe withdrawal of the fluid until this time. *See* **Appendix C**, 1967, Jacobson and Barter; **chorionic villi sampling.**

amniocytes   cells obtained by amniocentesis (*q.v.*).

amnion   A fluid-filled sac within which the embryos of reptiles, birds, and mammals develop. The wall of this sac has a two-layered epithelium. The inner epithelium of the wall is the amnion, although the term is sometimes applied to the whole sac. The outer epithelium is usually called the chorion. Amniotic fluid within the sac provides a liquid environment for the embryo.

amniote   a land-living vertebrate (reptile, bird, or mammal) whose embryos have an amnion and allantois.

Amoeba proteus   a common species of rhizopod; a giant protozoan used for microsurgical nuclear transplantations. *See* **Appendix C**, 1967, Goldstein and Prescott.

amoeboid movement cellular motility involving cytoplasmic streaming into cellular extensions called *pseudopodia*.

amorph a mutant gene having no effect compared with that of the wild-type allele.

AMP adenosine monophosphate.

amphidiploid an organism that is diploid for two genomes, each from a different species; synonymous with allotetraploid. *See* **Appendix C,** 1925, Goodspeed and Clausen.

amphimixis sexual reproduction resulting in an individual having two parents: synonymous with mixis. *Contrast with* **amixis, apomixis.**

Amphioxus See Branchiostoma.

amphipathic descriptive of a molecule that has distinct polar and nonpolar segments (e.g., membrane phospholipids).

amphoteric compound (*also* **ampholyte**) a substance that can act both as an acid and a base. Thus a protein is amphoteric because it tends to lose protons on the more alkaline side of its isoelectric point and to gain protons on the acid side of its isoelectric point.

ampicillin *See* **penicillin.**

amplification *See* **gene amplification.**

amylase an enzyme that hydrolyzes glucosidic bonds in polyglucosans such as glycogen.

amyloplast a starch-rich plastid.

anabolism the metabolic synthesis of complex molecules from simpler precursors, usually requiring the expenditure of energy and specific anabolic enzymes. *Contrast with* **catabolism.**

anaerobe a cell that can live without oxygen. A strict anaerobe cannot live in the presence of oxygen.

anagenesis phyletic evolution within a single lineage without subdivision or splitting; the opposite of cladogenesis.

analogous referring to structures or processes that have evolved convergently, as opposed to the term *homologous* (*q.v.*). Analogous structures have similar functions but are different in evolutionary origin: e.g., the wing of a butterfly and of a bat.

analogue a compound related to, but slightly different structurally from a biologically significant molecule, such as an amino acid (*see* **azaserine**), a pyrimidine or purine (*see* **base analogues**), or a hormone (*see* **ZR515**).

analysis of variance a statistical technique that allows the partitioning of the total variation observed in an experiment among several statistically independent possible causes of the variation. Among such causes are treatment effects, grouping effects, and experimental errors. Checking the absence of an effect due to the treatment is often the purpose of the inquiry. The statistical test of the hypothesis that the treatment had no effect is the F test or variance-ratio test. If the ratio of the mean square for treatments to the mean square for error exceeds a certain constant that depends on the respective degrees of freedom of the two mean squares at a chosen significance level, then the treatments are inferred to have been effective. Analysis of variance is particularly useful in judging which sources of uncontrolled variation in an experiment need to be allowed for in testing treatment effects.

anamnestic response *See* **immune response, immunological memory.**

anaphase *See* **mitosis.**

anaphase lag delay in the movement of one or more chromosomes from the metaphase plate during anaphase, often resulting in chromosome loss (*q.v.*).

anaphylaxis a systemic allergic or hypersensitivity response leading to immediate respiratory and/or vascular difficulties.

Anas platyrhyncha the mallard duck, ancestor to the domestic or Pekin duck, *A.p. domestica.*

anastomosis the joining of two or more cell processes or tubular vessels to form a branching system.

anastral mitosis the type of mitosis characteristically found in plants. A spindle forms, but no centrioles or asters are observed.

anautogenous insect an adult female insect that must feed for egg maturation. *See* **autogenous insect.**

anchorage-dependent cells cells (or *in vitro* cell cultures) that will grow, survive, or maintain function only when attached to an inert surface such as glass or plastic; also known as *substrate-dependent cells. See* **microcarriers, suspension culture.**

Anderson disease a hereditary glycogen storage disease of humans arising from a deficiency of the enzyme amylo-(1,4:1,6)-transglucosidase. Inherited as an autosomal recessive. Prevalence 1/500,000.

androdioecy a sexual dimorphism in plants having bisexual and separate male individuals.

androecious referring to plants having only male flowers.

androecium the aggregate of the stamens in a flower.

androgen any compound with male sex hormone activity. In mammals, the most active androgens are synthesized by the interstitial cells of the testis. *See* **testosterone.**

androgenesis development from a fertilized egg followed by disintegration of the maternal nucleus prior to syngamy so that the resulting individual possesses only paternal chromosomes and is haploid.

androgenic gland a gland found in most crustaceans belonging to the subclass Malacostraca. When implanted into maturing females, the gland brings about masculinization of primary and secondary sex characters.

androgenital syndrome masculinization of genitals in humans with XX chromosomal constitution, caused by overactivity of the adrenal cortex; inherited as an autosomal recessive trait.

anemia a disorder characterized by a decrease in hemoglobin per unit volume of blood. In the case of *hemolytic* anemia there is a destruction of red blood cells. In the case of *hypochromic* anemia, there is a reduction in the hemoglobin content of the erythrocyte.

anemophily pollination by the wind.

anergy the lack of an expected immune response.

aneucentric referring to an aberration generating a chromosome with more than one centromere.

aneuploidy a condition in which the chromosome number of an individual is not an exact multiple of the typical haploid set for the species. The suffix -somic is used in the nomenclature (e.g., monosomic, double trisomic, tetrasomic, etc.).

aneurin vitamin B$_1$; more commonly known as *thiamine.*

aneusomatic referring to organisms whose cells contain variable numbers of individual chromosomes.

angiosperm a flowering plant. Any species in the Superclass Angiospermae (*see* **Appendix A,** Kingdom Plantae) characterized by having seeds enclosed in an ovary.

Angstrom unit a unit of length equal to one ten-thousandth of a micron (10^{-4} micron; a micron being 10^{-6} meter); convenient for describing atomic dimensions; also equivalent to 10^{-1} milli-microns (mμ) or 10^{-10} meter. Abbreviated A, A°, Å, Å.U., or A.U. Named in honor of the Swedish physicist Anders Jonas Ångstrom.

Animalia the kingdom containing animals (heterotrophic organisms developing from a blastula). *See* **Appendix A,** Eukaryotes.

animal pole that pole of an egg which contains the most cytoplasm and the least yolk.

anion a negatively charged ion. *Contrast with* **cation.**

anisogamy that mode of sexual reproduction in which one of the sex cells, the egg, is large and nonmobile, whereas the other (the sperm) is small and motile. *See* **isogamy.**

anisotropy a directional property of crystals and fibers having a high degree of molecular orientation. Anisotropic substances have different physical properties when tested in different directions. When a ray of plane polarized light passes through anisotropic material, it is split into two rays polarized in mutually perpendicular planes. This property of anisotropic material is called *birefringence.* Muscle fibers and the metaphase spindle are examples of living materials exhibiting birefringence. Materials showing no birefringence are said to be *isotropic. See* **polarization microscope.**

ankylosing spondylitis an arthritic disease resulting in a stiffening and bending of the spine; inherited as an autosomal dominant with reduced penetrance. Over 90% of patients with this disease carry the B27 HLA antigen. *See* **histocompatiblity.**

anlage the embryonic primordium from which a specific part of the organism develops.

anneal to subject first to heating then to cooling. In molecular genetics experiments, annealing is used to produce hybrid nucleic acid molecules containing paired strands, each from a different source. Heating results in the separation of the individual strands of any double-stranded, nucleic-acid helix, and cooling leads to the pairing of any molecules that have segments with complementary base pairs.

annidation the phenomenon where a mutant is maintained in a population because it can flourish in an available ecological niche that the parent organisms cannot utilize. A wingless mutant insect, for example, might be poorly adapted in its ancestral habitat but able to live in tunnels and crevices that a winged form could not occupy.

annulate lamellae paired membranes arranged in stacks and possessing annuli resembling those of the nuclear membrane. Annulate lamellae may serve to transfer nuclear material to the cytoplasm by the replication of the nuclear envelop and may be a mechanism for storing gene-derived information to be used for cytoplasmic differentiation during early embryogenesis. During insect oogenesis, annulate lamellae occur alongside nurse cell nuclei and germinal vesicles, and they are abundant in the ooplasm.

annulus a ring. Applied to any of a number of ring-shaped parts of animals and plants. Used in cytology to refer to the ring-shaped nuclear pores.

anode the positive electrode; the electrode to which negative ions are attracted. *Contrast with* **cathode.**

anodontia the congenital absence of teeth. *Hypodontia* is currently the preferred term.

Anopheles a genus containing about 150 species of mosquitoes, many of which are of medical importance. Africa's principal malaria vector is *A. gambiae.* Other vector species are *A. funestus, A. quadrimaculatus, A. atroparvus, A. nili, A. moucheti,* and *A. pharoensis.* Polytene chromosomes occur in both larval salivary gland cells and adult ovarian nurse cells. Sibling species can often be separated by differences in the banding patterns of their polytene chromosomes.

Anser anser the Gray Lag goose, a favorite experimental organism for students of animal behavior and its hereditary components.

antagonist a molecule that bears sufficient structural similarity to a second molecule to compete with that molecule for binding sites on a third molecule. *See* **competition.**

antenatal before birth; during pregnancy.

antennae the first paired appendages on the head of arthropods.

Antennapedia a gene residing at 47.9 on the genetic map and within segment 84B of the salivary map of *Drosophila melanogaster.* The *Antp* gene is one of a cluster of three genes that specify the type of differentiation that cells in the segments from the head to the anterior portion of the second thoracic segment will undergo. *See* **Appendix C**, 1983, Scott *et al.;* **bithorax,** homeotic mutations, *Polycomb,* segment identity genes.

anther the terminal portion of a stamen bearing pollen sacs.

anther culture a technique that utilizes anthers or pollen cells to generate haploid tissue cultures or even plants. *See* **Appendix C**, 1973, Debergh and Nitsch.

antheridium the male gametangium of algae, fungi, bryophytes, and pteridophytes. *Contrast with* **oogonium.**

anthesis the time of flowering.

anthocyanin a red, violet, or blue glycosidic pigment occurring in solution in the cell-sap of flowers, fruits, stems, and leaves. These pigments are responsible for some autumn coloring of leaves and the tinting of young shoots and buds in the spring. *See* **pelargonin.**

anthropocentrism also called *anthropomorphism.* **1.** explanation of natural phenomena or processes in terms of human values. **2.** assuming humans to be of central importance in the universe or ultimate end of creation. **3.** ascribing human characteristics to a non-human organism.

anthropoid designating the great apes of the family Pongidae, including the gibbons, orangutans, gorillas, and chimpanzees.

anthropometry the science that deals with the measurements of the human body and its parts.

antiauxin a molecule that competes with an auxin (*q.v.*) for auxin receptor sites. A well-known antiauxin is 2,6-dichlorophenoxyacetic acid.

antibiotic a bacteriocidal or bacteriostatic substance produced by certain microorganisms, especially species of the genera *Penicillium, Cephalosporium,* and *Streptomyces. See* **actinomycin D, ampillicin, chloramphenicol, cyclohexamide, erythromycin, kanamycin, neomycin, novobiocin, penicillin, puromycin, semisynthetic antibiotic, streptomycin,** and **tetracycline.**

antibiotic resistance the acquisition of tolerance to a specific antibiotic by a microorganism that was previously adversely affected by the drug. Such resistance generally results from a mutation or the acquisition of R plasmids (*q.v.*) by the microorganism.

antibody a protein produced by lymphoid cells (plasma cells) in response to foreign substances (antigens) and capable of coupling specifically with its homologous antigen (the one that stimulated the immune response) or with substances that are chemically very similar to that same antigen. *See* **Appendix C**, 1890, von Behring; 1939, Tiselius and Kabat; **immunoglobulin.**

antibody-antigen reaction *See* **antigen-antibody reaction.**

anticlinal referring to a layer of cells cutting the circumference of a cylindrical plant organ at right angles. *Compare* **periclinal.**

anticoding strand *See* **coding strand.**

anticodon   the triplet of nucleotides in a transfer RNA molecule which associates by complementary base pairing with a specific triplet (codon) in the messenger RNA molecule during its translation in the ribosome.

antidiuretic hormone   *See* **vasopressin.**

antigen   **1.** a foreign substance that, upon introduction into a vertebrate animal, stimulates the production of homologous antibodies; a complete antigen or immunogen. A complex antigenic molecule may carry several antigenically distinct sites (determinants). **2.** a substance that is chemically similar to certain parts of an immunogen and can react specifically with its homologous antibody, but is too small to stimulate antibody synthesis by itself; an incomplete antigen or hapten (*q.v.*).

antigen-antibody reaction   the formation of an insoluble complex between an antigen and its specific antibody. In the case of soluble antigens, the complex precipitates, while cells that carry surface antigens are agglutinated. *See* **Appendix C,** 1900, Ehrlich; 1986, Amit *et al.*

antigenic conversion   **1.** appearance of a specific antigen(s) on cells as a consequence of virus infection. **2.** antibody-induced shift in certain protozoans or parasites to a new cell-surface antigen and cessation of expression of another antigen; resulting from switches in gene activity; synonymous in this sense with *serotype transformation; compare with* **antigenic modulation.**

antigenic determinant   a small chemical complex (relative to the size of the macromolecule or cell of which it is a part) that determines the specificity of an antigen-antibody interaction. That portion of an antigen that actually makes contact with a particular antibody or T cell receptor. An epitope. *Contrast with* **paratope.**

antigenic mimicry   acquisition or production of host antigens by a parasite, enabling it to escape detection by the host's immune system (as occurs in *Schistosoma*). *See* **schistosomiasis.**

antigenic modulation   suppression of cell-surface antigens in the presence of homologous antibodies.

antigen variation   the sequential expression of a series of variable surface glycoproteins (VSGs) by trypanosomes while in the bloodstream of a mammalian host. The production of VSGs allows the parasite to evade the immune defenses of the host by keeping one step ahead of the antibodies the host raises against them. Trypanosomes contain hundreds of different genes for the individual VSGs, but in a single trypanosome only one of these genes is expressed at a given time. The switch from one

VSG to another is accompanied by rearrangements of the DNA that generates additional copies of the genes being expressed. The glycoprotein being synthesized forms a macromolecular coating about 15 nm thick over the body and flagellum of the parasite. This coating functions only during the mammalian stage of the life cycle, since it is shed when the trypanosome enters the tsetse fly vector. *See* ***Glossina, Trypanosoma.***

antihemophilic factor   a protein participating in the cascade of reactions that result in blood coagulation (*q.v.*). A deficiency in AHF results in classic hemophilia (*q.v.*). AHF is encoded by a gene (HEMA), located at the end of the long arm of the X chromosome. The gene has been cloned and shown to contain 186,000 nucleotide pairs. It consists of 26 exons ranging in size from 69 to 3,106 bp and introns as large as 32 kb. The deduced amino acid sequence predicts a polypeptide containing 2,322 amino acids. *See* **Appendix C,** 1984, Gitschier *et al;* **cross-reacting material, von Willebrand disease.**

antimetabolite   in general, a molecule that functions as an antagonist or metabolic poison.

antimitotic agent   any compound that suppresses the mitotic activity of a population of cells.

antimongolism   a syndrome of congenital defects accompanying hypoploidy for chromosome 21. Such children carry a normal chromosome 21 and one that contains a large deficiency. From the standpoint of chromosome balance, mongolism and antimongolism represent reciprocal phenomena.

antimorph   **1.** enantiomer (*q.v.*). **2.** a mutant allele that acts in a direction opposite to the normal allele.

antimutagen   a compound (generally a purine nucleoside) that antagonizes the action of mutagenic agents (generally purines or purine derivatives) on bacteria. Some antimutagens reduce the spontaneous mutation rate.

anti-oncogenes   a class of genes that are involved in the negative regulation of normal growth. The loss of these genes leads to malignant growth. The *Rb* gene is an example of an anti-oncogene. *See* **retinoblastoma.**

antiparallel   **1.** the opposite strand orientations with which all nucleic acid duplexes (DNA-DNA, DNA-RNA, or RNA-RNA) associate; if one strand is oriented left to right 5′ to 3′, the complementary strand is oriented left to right 3′ to 5′ antiparallel to it. **2.** two segments of a polypeptide chain may lie antiparallel with one end N-terminal to C-terminal and the other end C-terminal to N-terminal, respectively. *See* **Appendix C,** 1961, Josse, Kaiser, and Kornberg.

antipodal   one of a group of three haploid nuclei formed during megasporogenesis. In corn, these divide mitotically and eventually form a group of 20 to 40 antipodal cells, which aid in nourishing the young embryo.

antirepressor   in lambda phage, the dimeric protein product of gene *cro* (control of *r*epressor and *o*ther things) that prevents the synthesis of repressor and inhibits the expression of early genes (those involved in replication of the phage genome), allowing the phage to enter the late stage of development leading to cell lysis. *See* ***cro* repressor.**

Antirrhinum majus   the snapdragon; a classic species for the study of multiple alleles.

antisense RNA (asRNA)   an RNA molecule with a nucleotide sequence complementary to a specified mRNA. Some bacteria generate asRNA as a mechanism for gene regulation. In the laboratory, asRNA is synthesized by splicing the gene under study in a reverse orientation to a viral promoter. The coding strand is now transcribed. Once isolated and purified, the asRNA can be injected into an egg or embryo, where it will combine with natural messages to form duplexes. These block either the further transcription of the message or its translation. Using this strategy, mutant phenotypes can be mimicked. *See* **viroid.**

antisense strand   *See* **coding strand.**

antiserum   a serum containing antibodies.

antisigma factor   a protein that is synthesized during infection of *Escherichia coli* by T4 phage and prevents recognition of initiation sites by the sigma factor of RNA polymerase.

antitermination factor   a protein that allows an RNA polymerase to ignore signals to stop translation at particular sites in DNA molecules.

antitoxin   an antibody that neutralizes a specific toxin.

anucleate   without a nucleus.

anucleolate   without a nucleolus.

anucleolate mutation   a mutation arising from the loss of the nucleolus organizer. Such mutations have been observed in *Drosophila, Chironomus,* and *Xenopus.* The study of such mutations led to the understanding of their role in the transcription of rRNA. *See* **Appendix C,** 1966, Wallace and Birnstiel; ***bobbed,* ribosome.**

AP endonuclease   any enzyme that cleaves DNA on the 5′ side of a site in a DNA molecule that lacks a purine of pyrimidine. Such sites result when DNA glycosylases (*q.v.*) remove chemically altered bases from the polymer. An AP endonuclease will then break the phosphodiester backbone, and this allows excision of the damaged region. *See* **cut-and-patch repair.**

aphasic lethal   a lethal mutation in which death may occur randomly throughout development.

apholate   an aziridine mutagen (*q.v.*) also used as an insect chemosterilant.

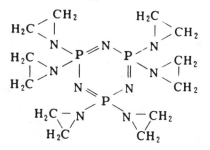

apical meristem   *See* **meristem.**

Apis mellifera   the domesticated honeybee. There are about 25 races recognized, but the ones most commonly used are the Italian, Caucasian, Carniolan, and German races. These are generally referred to as European bees, to distinguish them from African bees (*q.v.*). Bees reproduce by arrhenotokous parthenogenesis (*q.v.*). An alternative spelling, *A. mellifica,* also occurs in the literature.

aplasia   the failure of development of an organ.

apnea, drug-induced   a hereditary disease in man arising from a deficiency of the enzyme pseudocholinesterase.

apoenzyme   the protein portion of a holoenzyme (*q.v.*), which requires a specific coenzyme for its functioning.

apoferritin   *See* **ferritin.**

apoinducer   a protein that, when bound to DNA, activates transcription by RNA polymerase.

apomict   an organism produced by apomixis.

apomixis   reproduction in which meiosis and fertilization are altered, so that only one parent contributes genes to the offspring (e.g., agamospermy or thelytoky). *Contrast with* **amixis, amphimixis.**

apomorphic   a term applied to those characters of species that have evolved only within the taxonomic group in question. Plesiomorphic characters, on the other hand, are shared with other taxonomic groups as a consequence of their common ancestry. Thus, in mammals, the possession of hair would be an apomorphic character, whereas the possession of a backbone would be a pleisiomorphic character. *See* **cladogram.**

aporepressor a regulatory protein that, when bound to another molecule (corepressor, *q.v.*), undergoes an allosteric transformation that allows it to combine with an operator locus and inhibit transcription of genes in an operon. For example, in the histidine system of *E. coli,* excess histidine functions as *corepressor* by binding to an aporepressor to form a functional repressor (holorepressor); the holorepressor binds to the histidine operator and inhibits transcription of ten genes of this operon.

aposematic coloration warning coloration (*q.v.*).

apospory the development in plants of a diploid embryo sac by the somatic division of a nucellus or integument cell; a form of agamospermy (*q.v.*).

apostatic selection frequency-dependent selection for the rarer morph (e.g., in Batesian mimicry, the rarer the mimic, the greater its selective advantage). *See* **mimicry.**

a posteriori descriptive of biometric or statistical tests in which the comparisons of interest are unplanned and become evident only after experimental results are obtained.

apposition a process of increasing volume by adding new material to old at the periphery.

a priori descriptive of biometric or statistical tests in which the comparisons of interest are determined on theoretical grounds in advance of experimentation.

aptation any character currently subject to selection whether its origin can be ascribed to selective processes (adaptation) or to processes other than selection or selection for a different function (exaptation).

apterygote any of the primitive insects belonging to the cohort Apterygota, which contains the orders Thysanura (silverfish) and Diplura (campodeids and japigids). The species are wingless and possess abdominal styli. *See* **Appendix A.**

aptitude in microbial genetics, the specific physiological state of a lysogenic bacterium during which it can produce infectious bacteriophages when exposed to an inducing agent.

apyrene sperm *See* **sperm polymorphism.**

Arabidopsis thaliana an herb in which gene dosage and interaction have been studied extensively.

Arachnida a class of Arthropoda containing the spiders, scorpions, opilionids, and mites. *See* **Appendix A.**

Araldite trademark for a plastic commonly used for embedding tissues for electron microscopy.

Araenida the subclass of the arachnida containing the spiders.

arbovirus a virus that replicates in both an arthropod and a vertebrate host.

Archaebacteria a subkingdom of the Prokaryotae (*see* **Appendix A**), that is subdivided into two phyla. Archaebacteria that generate methane are placed in the phylum Methanocreatices, while nonmethanogenic archaebacteria are placed in a second, as yet unnamed, phylum. The archaebacteria are placed in a group separate from the rest of bacteria (the eubacteria) on the basis of a variety of biochemical characteristics (distinctive compounds in their cell walls and membranes, differences in rare bases found in their tRNAs, and distinctive structures of RNA polymerase subunits). The archaebacteria are thought to have been dominant organisms in the primeval biosphere, since its atmosphere was rich in carbon dioxide and included hydrogen but virtually no oxygen. The methanogenic bacteria are strictly anaerobic and derive their energy for growth by the reduction of CO_2 to CH_4. It appears that the archaebacteria evolved as a group before either the eubacteria or the eukaryotes. The eubacterial line of descent seems to have arisen from within the methanogen-halophile group of archaebacteria, while the eukaryote ancestor belonged to the sulfur-dependent thermophiles (*q.v.*). Some authorities have suggested that archaebacteria should be assigned to a kingdom of their own (the Urkingdom) and that the eubacteria and eukaryotes be treated as separate kingdoms derived from the ancestral Urkingdom.

Archaeopteryx a genus of Jurassic fossil birds showing teeth, three digits with claws on their wings, and a long feathered tail containing numerous vertebrae. The dinosaurian skeletal characteristics of these fossil birds demonstrate their reptilian ancestry.

Archaeozoa a name proposed by T. Cavalier-Smith for a kingdom to contain the most primitive surviving eukaryotes. These unicellular organisms lack mitochondria, they have 70 S ribosomes, and their small rRNAs show greater similarities in nucleotide sequence to the small RNAs of prokaryotes than do the small rRNAs of other eukaryotes. Archaeozoans may be derived directly from the ancient eukaryotic common ancestor of plants, fungi, and animals. *See* **classification,** *Giardia,* **microsporidia, Protozoa, ribosomes.**

Archean the earlier of the two eras making up the Precambrian eon. *See* **geologic time divisions.** Life began during the Archean, and the prokaryotes evolved.

archegonium the female sex organ of liverworts, mosses, ferns, and most gymnosperms.

archenteron the primitive digestive cavity of any metazoan embryo. It is formed by gastrulation.

area code hypothesis a theory proposing that as a cell passes through successive stages of differentiation, new patterns of molecules are displayed on its surface as a kind of area code that designates to which tissue or organ the cell should ultimately belong.

arg arginine. *See* **amino acid.**

arginine-urea cycle *See* **ornithine cycle.**

arithmetic mean an average; the number found by dividing the sum of a series by the number of items in the series.

arithmetic progression a series of numbers in which each number is larger (or smaller) than the number that precedes it by a constant value.

aromatic designating a chemical compound that contains a closed ring of carbon atoms, as found in benzene (*q.v.*). The aromatic amino acids phenylalanine and tyrosine serve as examples. *See* **amino acid.**

Arrhenius plot a plot of the logarithm of the growth rate versus the reciprocal of the absolute temperature. *See* **generation time.**

arrhenotokous parthenogenesis the phenomenon by which unfertilized eggs produce haploid males and fertilized eggs produce diploid females.

arrhenotoky arrhenotokous parthenogenesis.

arthritis an autoimmune disease of the joints involving an attack by the body's immune system upon the synovial membranes.

Arthropoda the largest phylum in the animal kingdom in number of species. It contains the Chelicerata and Mandibulata. *See* **Appendix A.**

Articulata a division of the invertebrates containing segmented, coelomate protostomes. *See* **Appendix A.**

artifact any structure that is not typical of the actual specimen, but that results from cytological processing, postmortem changes, etc.

artificial chromosomes genetically engineered chromosomes in *Saccharomyces.* They are generated from plasmids that have been constructed to contain a centromere, several origins of replication, and a pair of fused telomeres. Once inside a yeast cell, an endogenous enzyme cuts the fused telomeres and the circular chromosome becomes linear. Such artificial chromosomes segregate correctly in subsequent divisions even though they are shorter than the true yeast chromosomes.

artificial insemination (AI) the placement of sperm into a female reproductive tract or the mixing of male and female gametes by other than natural means. In AID, the sperm are provided by a donor other than the woman's husband; in AIH, the sperm are those of her husband. *See* **Appendix C,** 1769 (1780), Spallanzani.

artificial parthenogenesis induction of the development of an unfertilized egg by chemical or physical stimulation.

artificial selection the choosing by man of the genotypes contributing to the gene pool of succeeding generations of a given organism.

Ascaris megalocephala *Parascaris equorum* (*q.v.*) (the preferred nomenclature).

ascertainment bias *See* **proband method.**

Aschelminthes a phylum containing the pseudocoelomate species with an anterior mouth, posterior anus, and straight digestive tube. *See* **Appendix A.**

Ascobolus immersus an ascomycete fungus that is convenient for tetrad analysis (*q.v.*). *See* **Appendix A.**

ascogenous hypha a hypha that develops from the surface of an ascogonium after plasmogamy (*q.v.*). The hypha therefore contains nuclei of both mating types. After crozier formation (*q.v.*) such hyphae produce asci.

ascogonium in fungi, a female cell that receives haploid nuclei from an antheridium.

ascomycete a fungus of class Ascomycetes that produce ascospores. *See* **Appendix A.**

ascorbic acid vitamin C (*q.v.*). Scurvy is a deficiency disease resulting from inadequate ascorbic acid in the diet.

ascospore a meiospore (*q.v.*) contained within an ascus sac.

ascus a sac containing ascospores. The fact that all the products of a meiotic division are contained in an ascus in some fungi makes tetrad analysis (*q.v.*) possible.

asexual reproduction reproduction without sexual processes; vegetative propagation.

A-site-P-site model *See* **translation.**

asn asparagine. *See* **amino acid.**

asp aspartate; aspartic acid. *See* **amino acid.**

asparagine *See* **amino acid.**

Asparagus officinalis the asparagus.

aspartic acid *See* **amino acid.**

Aspergillus a genus of filamentous fungi belonging to the phylum Deuteromycota. *See* **Appendix A.** *A. flavus* is the source of aflotoxin (*q.v.*). *A. nidulans* is the species in which in 1952 Pontecorvo and Roper discovered parasexuality (*q.v.*) *See* **Appendix C.** *A. nidulans* has a haploid chromosome number of eight, and there are eight well-mapped linkage groups. Mitochondrial genes are also characterized.

asRNA antisense RNA (*q.v.*).

association the joint occurrence of two genetically determined characteristics in a population at a frequency that is greater than expected according to the product of their independent frequencies.

associative overdominance linkage of a neutral locus to a selectively maintained polymorphism that increases heterozygosity at the neutral locus, *Compare with* **hitchhiking.**

associative recognition requirement for the initiation of an immune response of the simultaneous recognition by T lymphocytes of the antigen in association with another structure, normally a cell surface alloantigen encoded within the major histocompatibility complex.

assortative mating sexual reproduction in which the pairing of male and female is not random, but involves a tendency for males of a particular kind to breed with females of a particular kind. If the two parents of each pair tend to be more (less) alike than is to be expected by chance, then positive (negative) assortative mating is occurring.

assortment the random distribution to the gametes of different combinations of chromosomes. Each 2N individual has a paternal and a maternal set of chromosomes forming N homologous pairs. At anaphase of the first meiotic division, one member of each chromosome pair passes to each pole, and thus the gametes will contain one chromosome of each type, but this chromosome may be of either paternal or maternal origin.

aster *See* **mitotic apparatus.**

asynapsis the failure of homologous chromosomes to pair during meiosis. *Contrast with* **desynapsis.**

atavism the reappearance of a character after several generations, the character being caused by a recessive gene or by complementary genes. The aberrant individual is sometimes termed a "throwback."

Atebrin a trade name for quinacrine (*q.v.*). Also spelled Atabrine.

ateliosis retarded growth resulting in a human of greatly reduced size but normal proportions. Such midgets generally show a marked deficiency in pituitary growth hormone. *See* **pituitary dwarfism.**

(A+T)/(G+C) ratio the ratio between the number of adenine-thymine pairs and the number of guanine-cytosine pairs in a given DNA sample.

atom the smallest particle of an element that is capable of undergoing a chemical reaction. *See* **chemical element.**

atomic mass the mass of a neutral atom of a nuclide, usually expressed in terms of atomic mass units.

atomic mass unit one-twelfth the weight of a ^{12}C atom; equivalent to 1.67×10^{-24} g.

atomic number the number of protons in the nucleus, or the number of positive charges on the nucleus; symbolized by Z. It also represents the number of orbital electrons surrounding the nucleus of a neutral atom.

atomic weight the weighted mean of the masses of the neutral atoms of an element expressed in atomic weight units.

ATP adenosine triphosphate. *See* **adenosine phosphate.**

atresia congenital absence of a normal passageway.

atrichia hairlessness. In the domestic dog, the condition is inherited as an autosomal dominant. Homozygotes are stillborn.

attached X chromosome monocentric elements containing two doses of the X-chromosome. *Drosophila* females carrying attached X chromosomes usually also have a Y chromosome. They produce XX and Y eggs and therefore generate patroclinous sons (which inherit their X from their father and their Y from their mother) and matroclinous daughters (which inherit their X's from their mother and their Y from their father). The terms *double X* and *compound X* are also used for aberrant chromosomes of this types. *See* **Appendix C,** 1922, Morgan; **detached X.**

attachment efficiency *See* **seeding efficiency.**

attachment point (ap) a hypothetical analog of the centromere on chloroplast DNA in *Chlamydomonas.* Chloroplast genes may be mapped with respect to the attachment point.

attenuation **1.** in physics, the loss in energy of an electromagnetic radiation as it passes through matter. **2.** in microbiology, the loss in virulence of a pathogenic organism as it is repeatedly subcultured or is let multiply on unnatural hosts. **3.** in immunology, the reduction in virulence of a substance to be used as an immunogen. Attenuation may result from aging, heating, drying, or chemically modifying the immunogen. **4.** in molecular genetics, a mechanism for regulating the expression of bacterial operons that encode enzymes involved in amino acid biosynthesis. *See* **attenuator.**

attenuator a nucleotide sequence that is located upstream of those bacterial operons which encode the enzymes that are involved in the synthesis of amino acids. The expression of such operons is switched on and off by controlling the transcription of the messages for these operons. The leader sequence of the tryptophan operon of *E. coli* illustrates how attenuators function. In drawing A. below, an RNA polymerase is moving along the coding strand of a bacterial DNA molecule. The

RNA molecule transcribed by the polymerase dangles behind it.

Near its 5′ end is a site for binding ribosomes. One has attached and is moving along the RNA, forming a polypeptide as it goes. How fast the ribosome moves is determined by the availability of amino acid–charged tRNAs. The detailed structure of the RNA transcript is shown in drawing B.

The blocks designated A and B can pair because their base sequences are complementary and so can C and D. However, B can also pair with C, so there are three hairpin loops that can form, A/B, B/C, C/D. However, only the A segment contains tryptophan codons (symbolized by Xs). When tryptophan is abundant, the ribosome moves without pause past A to B. As the polymerase transcribes C and D, these pair to form a termination hairpin (*q.v.*), and the RNA polymerase together with its transcript detach from the DNA strand. Therefore, the trp operon is silenced. However, when there is no tryptophan in the environment, the ribosome pauses at the trp codons. Since A is covered by the ribosome, B pairs with C, and now D cannot form a termination hairpin. Therefore, the polymerase continues to the operon and transcribes it. The transcript is later translated, and the enzymes that result catalyze the formation of tryptophan. Thus, enzymes required for making an essential amino acid are synthesized only when the amino acid is scarce.

att sites loci on a phage and the chromosome of its bacterial host where recombination integrates the phage into or excises it from the bacterial chromosome.

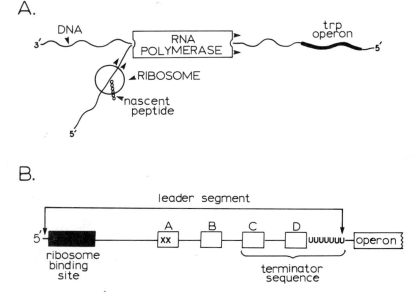

27

audiogenic seizure convulsions induced by sound. Certain strains of mice, rats, and rabbits are especially prone to such seizures.

aureomycin *See* **tetracycline.**

Australian one of the six primary biogeographic realms (*q.v.*), comprising Australia, the Celebes, New Guinea, Tasmania, New Zealand, and the oceanic islands of the South Pacific.

Australopithecine referring to early hominids, fossils of which have been found in Africa. According to some classifications, the genus *Australopithecus* contains four species: *A. afarensis, A. africanus, A. robustus,* and *A. boisei,* all of which lived between 4 and 1 million years ago. The famous fossil "Lucy" is a 40% complete skeleton found in 1977. She lived about 3 million years ago and belonged to *A. afarensis.*

autapomorphic character **1.** a derived character evolved from a plesiomorphic character state in the immediate ancestor of a single species **2.** uniquely derived characters shared by several synapomorphous taxa. For example, hair was an autapomorphy of the first mammalian species and is also a synapomorphy of all mammals.

autarchic genes in mosaic organisms, these genes are not prevented from manifesting their phenotypic effects by gene products diffusing from genetically different neighboring tissues, whereas *hyparchic* genes are so inhibited.

autocatalysis the promotion of a reaction by its product.

autochthonous pertaining to a species that has evolved within the region where it is native.

autofertilization *See* **thelytoky.**

autogamy that mode of reproduction in which the zygote is formed by fusion of two haploid nuclei from the same individual. In *Paramecium,* a process of self-fertilization resulting in homozygosis. In single individuals, the two micronuclei each undergo meiosis and seven of the eight resulting nuclei degenerate. The remaining haploid nucleus divides mitotically and the two identical nuclei fuse. The fusion nucleus gives rise to the micro- and macronuclei of the paramecium and its progeny. Autogamy is therefore a mode of nuclear reorganization that constitutes an extreme form of inbreeding. In *P. aurelia,* autogamy occurs spontaneously at regular intervals.

autogenous insect species in which females can produce eggs without first feeding. *See* **anautogenous insect.**

autogenous control regulation of gene expression by its own product either inhibiting (negative autogenous control) or enhancing (positive autogenous control) its activity. For example, in *E. coli,* AraC protein (the regulatory factor for the arabinose operon) controls its own synthesis by binding to the AraC promoter.

autograft the transplantation of a living piece of tissue from one site to another on the same animal.

autoimmune disease any pathological condition resulting from an individual's immune response to its own cells or tissues. *See* **arthritis, lupus erythematosus.**

autologous referring to a graft from one region to another on the same animal.

automimic a palatable individual that is an automatic mimic of members of the same species that are unpalatable to predators. *See* **automimicry.**

automimicry the phenomenon, seen for example in the Monarch butterfly (*q.v.*), in which the species has a polymorphism in terms of its palatability to predators. The polymorphism arises from the different food plants chosen by the ovipositing female. Most members of the species are rendered unpalatable because they feed as larvae upon plant species rich in substances toxic to birds. Those insects that feed on plants containing no toxins are palatable. However, such insects mimic perfectly the more abundant, unpalatable members of the species. *See* **mimicry.**

automixis fusion of nuclei or cells derived from the same parent to yield homozygous offspring. An example would be autogamy in paramecia or automictic parthenogenesis, as seen in certain species of Lepidoptera. *See* **autogamy, thelytoky.**

automutagen any mutagenic chemical formed as a metabolic product by an organism.

autonomous controlling element a controlling element (*q.v.*), apparently having both receptor and regulator functions, that enters a gene and makes it an unstable mutant.

autophagic vacuole an enlarged lysosome containing mitochondria and other cellular organelles in the process of being digested.

autophene a phenotype due to the genetic constitution of the cells showing it. Transplantation of such mutant cells to a wild-type host does not modify their mutant phenotype.

autopoiesis the ability of an organism to maintain itself through its own metabolic processes at the expense of carbon and energy sources. Cells are autopoietic; viruses and plasmids are not.

autopolyploid a polyploid that originates by the multiplication of one basic set of chromosomes.

autoradiograph a photographic picture showing the position of radioactive substances in tissues, obtained by coating a squash preparation or a section with a photographic emulsion in the dark, and subsequently developing the latent image produced by the decay radiations. In the case of colony hybridization (*q.v.*), a filter containing the radioactive chimeric vectors is taken to a dark room, placed in an x-ray film holder, and covered with a sheet of x-ray film. The film is then left to expose for several hours or a few days before it is processed. The position of the silver grains on the film marks the location of the colonies of interest. *Also see* **DNA fiber autoradiography.**

autoradiographic efficiency the number of activated silver grains (produced in a photographic emulsion coating a section) per 100 radioactive disintegrations occurring within the tissue section during the exposure interval.

autoradiography a technique for localizing radioactively labeled molecules by applying a photosensitive emulsion to the surface of a radioactive specimen. *See* **autoradiograph.**

autoregulation regulation of the synthesis of a gene product by the product itself. In the simplist autoregulated systems, excess gene product behaves as a repressor and binds to the operator locus of its own structural gene. *Contrast with* **end product inhbition.**

autoselection the process whereby a genetic element tends to increase in frequency by virtue of the nature of its transmission, even though it has no effect on the viability, fecundity, or fertility of the individual that bears it.

autosexing the use of sex-linked genes with obvious phenotypic effects to permit the identification by external inspection of the sex of immature organisms (larval silkworms or chicks, for example) before sexual dimorphic traits become obvious.

autosome a chromosome other than a sex chromosome. The genes residing on autosomes follow the mode of distribution of these chromosomes to the gametes during meiosis. This pattern (autosomal inheritance) differs from that of genes on the X or Y chromosomes, which show the sex-linked mode of inheritance. *See* **human pseudoautosomal region, sex linkage.**

autosyndesis *See* **allosyndesis.**

autotetraploid an autopolyploid with four similar genomes. If a given gene exists in two allelic forms A and a, then five genotypic classes can be formed in an autotetraploid: $AAAA$ (quadruplex), $AAAa$ (triplex), $AAaa$ (duplex), $Aaaa$ (simplex), and $aaaa$ (nulliplex).

autotrophs organisms that can build their own macromolecules from very simple, inorganic molecules, such as ammonia and carbon dioxide. Autotrophs include photosynthetic bacteria, protoctists, and plants that can convert visible light into chemical energy. In addition, chemoautotrophic bacteria can produce organic molecules from CO_2 in the absence of light. They use as energy sources for biosynthesis the oxidation of inorganic compounds such as molecular hydrogen, ammonia, and hydrogen sulfide. *Contrast with* **heterotrophs.**

autozygote an individual homozygous at a given locus whose two homologous genes are identical by descent, in that both are derived from the same gene in a common ancestor. *See* **allozygote.**

auxesis growth in size by increase in cell volume without cell division.

auxins a family of plant hormones that promote longitudinal growth and cell division. Natural auxins are indole derivatives biosynthesized from tryptophan. The most common auxin is indole acetic acid (*q.v.*) which is synthesized in all plants. *See* **antiauxin.**

auxocyte a cell whose nucleus is destined to enter meiotic prophase; a primary oocyte, primary spermatocyte, megasporocyte, or microsporocyte.

auxotroph a mutant microorganism that can be grown only upon minimal medium that has been supplemented with growth factors not required by wild-type strains.

Avena the genus to which the various species of oats belong. The most commonly cultivated oat is *A. sativa.*

Avena test a technique using the curvature of *Avena* coleoptiles as a bioassay for auxins.

average life in nuclear physics, the average of the individual lives of all the atoms of a particular radioactive substance. It is 1.443 times the radioactive half-life (*q.v.*).

avian leukosis *See* **leukemia.**

avian myeloblastosis virus an oncogenic RNA virus.

avidity the total combining power of an antibody with an antigen. It involves both the affinity of each binding site and the number of binding sites per antibody and antigen molecule. *Compare with* **affinity.**

Avogadro's number the number of atoms (6.025×10^{23}) in one gram atomic weight of an element; also the number of molecules in the gram molecular weight of a compound.

awn a stiff, bristlelike appendage occurring on the flowering glumes of grasses and cereals.

axenic growth of organisms of a given species in the complete absence of members of any other species.

axolotl the aquatic larval stage of salamanders of the genus *Ambystoma* that does not metamorphose but breeds while a larva. *See* **neoteny.**

axon the long process of a nerve cell, normally conducting impulses away from the nerve cell body.

axoneme a shaft of microtubules extending the length of a cilium, flagellum or pseudopod. Axonemes from all cilia and flagella (including sperm tails) contain the same "9 + 2" arrangement of microtubules. In the center of each axoneme are two singlet microtubules that run the length of the shaft. The central tubules are surrounded by a circle of doublet microtubules, each consisting of an A and B subfiber. Each A subfiber has longitudinally repeating pairs of armlike projections that contain dynein (*q.v.*). *See* **flagellum.**

axoplasm the cytoplasm contained in axons.

axopodia rigid, linear cellular projections composed mostly of microtubules found in species belonging to the Actinopoda. *See* **classification,** Protoctista.

5-azacytidine an analogue of cytidine in which a nitrogen atom is substituted for a carbon in the number 5 position of cytosine (*q.v.*). The analogue is incorporated into newly synthesized DNA, and such DNA is undermethylated. Since a reduction in the number of methyl groups attached to genes is associated with an increase in their transcriptional activities, 5-azacytidine can switch on certain genes. For example, patients given the drug may start making fetal hemoglobin, which implies that their gamma genes have been switched on. *See* **hemoglobin, hemoglobin genes.**

azaguanine a purine antagonist first synthesized in the laboratory and later shown to be identical to an antibiotic synthesized by *Streptomyces specta-*

balis. Azaguanine is incorporated into mRNA and causes errors in the translation.

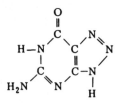

azaserine a glutamine analogue synthesized by various species of *Streptomyces.* Azaserine inhibits purine biosynthesis and produces chromosome aberrations. It is mutagenic and has antitumorgenic activity.

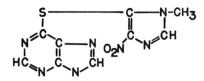

azathioprene a derivative of mercaptopurine (*q.v.*) that preferentially suppresses the primary antibody response and the rejection of allografts (*q.v.*).

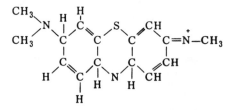

aziridine mutagen a mutagenic alkylating agent containing aziridinyl groups ($-N\begin{smallmatrix}CH_2\\ \\ CH_2\end{smallmatrix}$). *See* **apholate, ethylenimine, hemel, hempa, TEM, tepa, tetramine.**

azoospermia absence of motile sperm in the ejaculate.

Azotobacter a genus of free-living, rod-shaped, soil bacteria capable of nitrogen fixation. *See* **Appendix A,** Eubacteria.

azure B a basic dye used in cytochemistry. *See* **metachromasy.**

30

B

B₁, B₂, B₃, etc. the first, second, third, etc., backcross generations. The first backcross is made by mating an individual with one of its parents or with an individual of that identical genotype. The offspring produced belong to the B_1 generation. The second backcross is made by crossing B_1 individuals again with individuals of genotype identical to the parent referred to in the first backcross, etc.

Bacillus a genus of rod-shaped bacteria. *B. subtilis* is a Gram-positive, spore-forming, soil bacillus that grows readily in a chemically defined medium and undergoes genetic exchange by transformation and transduction (*q.v.*). While it has been studied in the most detail, considerable work has also been done on *B. cereus* and *B. megatherium.*

backbone in biochemistry, the supporting structure of atoms in a polymer from which the side chains project. In a polynucleotide strand, alternating sugar-phosphate molecules form such a backbone.

backcross a cross between an offspring and one of its parents or an individual genetically identical to one of its parents.

backcross parent that parent of a hybrid with which it is again crossed or with which it is repeatedly crossed. A backcross may involve individuals of genotype identical to the parent rather than the parent itself.

background constitutive synthesis the occasional transcription of genes in a repressed operon due to a momentary dissociation of the repressor that allows a molecule of RNA polymerase to bind to its promoter and initiate transcription. Sometimes called "sneak synthesis."

background genotype the genotype of the organism in addition to the genetic loci primarily responsible for the phenotype under study.

background radiation ionizing radiation arising from sources other than that under study. Background radiation due to cosmic rays and natural radioactivity is always present, and there may also be background radiation due to man-made contaminating radiation.

back mutation reverse mutation.

bacterial transformation *See* **transformation.**

bacteriochlorophyll *See* **chlorophyll.**

bacteriocins proteins synthesized by various bacterial species that are toxic when absorbed by bacteria belonging to sensitive strains. Resistance to and the ability to synthesize bacteriocins are controlled by plasmids. *Escherichia coli* strains produce bacteriocins called colicins (*q.v.*). Bacteriocins from *Pseudomonas aeruginosa* are called pycocins.

bacteriophage a virus whose host is a bacterium; commonly called *phage*. Below are listed some common bacteria and their viral parasites:

Escherichia coli	P_1, P_2, P_4, λ, Mu 1, N_4, ϕX174, R_{17}, T_1 through T_7.
Salmonella typhimurium	P_{22}
Corynebacterium diphtheriae	β
Bacillus subtilis	SP_{82}, ϕ 29
Shigella dysenteriae	P_1, P_2, P_4

Bacterial viruses show extreme variations in complexity. For example, the RNA phage R_{17} has a genome size of 1.1×10^6 daltons, whereas the DNA of the T_4 phage weighs 130×10^6 daltons. *See* **Appendix C,** 1915, Twort; 1917, d'Herelle; 1934, Schlesinger; 1934, Ellis and Delbrück; 1942, Luria and Anderson; 1945, Luria; 1949, Hershey and Rotman; 1952, Hershey and Chase; 1953, Visconti and Delbrück; 1966, Edgar and Wood; 1973, Fiers *et al.;* **filamentous phage, plaque, temperate phage, virulent phage.**

bacteroids intracellular, nitrogen-fixing symbionts found in the root nodules of leguminous plants. Bacteroids are derived from free-living species of *Rhizobium* (*q.v.*). *See* **leghemoglobin.**

bacteriophage packaging insertion of recombinant bacteriophage lambda DNA into *E. coli* for replication and encapsidation into plaque-forming bacteriophage particles.

bacteriostatic agent   a substance that prevents the growth of bacteria without killing them.

baculoviruses   a group of viruses that infect arthropods, especially insects. Baculoviruses utilize the synthetic machinery of the insect host cell to synthesize polyhedrin, a protein that coats the virus particle. By appropriate gene-splicing techniques baculoviruses have been engineered to synthesize foreign proteins, including the envelope protein of HIV (*q.v.*).

balanced lethal system   a strain of organisms bearing nonallelic recessive lethal genes, each in a different homologous chromosome. When interbred such organisms appear to breed true, because one-half of the progeny are homozygous for one or the other lethal gene and die prior to their detection. The surviving progeny, like their parents, are heterozygous for the lethal genes. *See* **Appendix C,** 1918, Muller; 1930, Cleland and Blakeslee.

balanced polymorphism   genetic polymorphism maintained in a population because the heterozygotes for the alleles under consideration have a higher adaptive value than either homozygote. *See* **Appendix C,** 1954, Allison.

balanced selection   selection favoring heterozygotes that produces a balanced polymorphism (*q.v.*).

balanced stock   a genetic stock that, though heterozygous, can be maintained generation after generation without selection. Such stocks may contain balanced lethal genes, or a recessive lethal gene, which kills hemizygous males, combined with a nonallelic recessive gene, which confers sterility on homozygous females. *See* **M5 technique.**

balanced translocation   synonymous with *reciprocal translocation. See* **translocation.**

Balbiani chromosome   a polytene chromosome. So called because such banded chromosomes were first discovered by E. G. Balbiani in *Chironomus* larvae in 1881.

Balbiani ring   a giant RNA puff present on a polytene chromosome of a salivary gland cell during a significant portion of larval development. The largest and most extensively studied Balbiani rings are on chromosome 4 of *Chironomus tentans*. See illustration. The transcription product of one of these puffs (BR 2) is a 75S RNA, which encodes the message for a giant polypeptide of the saliva. Balbiani rings contain thousands of DNA loops upon which mRNAs are being transcribed. These combine with proteins to form RNP granules (Balbiani

ring granules) that eventually pass into the cytoplasm through nuclear pores.

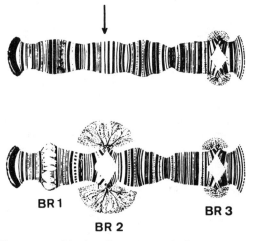

BR 1   BR 3

BR 2

Chromosome 4 in the salivary glands of *Chironomus tentans* with the BR 2 band in an unpuffed *(above)* and in a puffed stage *(below)*. The position of the BR 2 band is indicated by an arrow.

Bal 31 exonuclease   a nuclease that digests linear, double-stranded DNA fragments from both ends. The enzyme is used *in vitro* to shorten restriction fragments. The shortened segment can then be religated with a DNA ligase (*q.v.*) to generate deletion mutants. *See* **restriction endonucleases.**

Balzer freeze-fracture apparatus   *See* **freeze etching.**

Bam H1   *See* **restriction endonuclease.**

band   in chromosome studies, a vertical stripe on a polytene chromosome resulting from the specific association of a large number of homologous chromomeres at the same level in the somatically paired bundle of chromosomes.

BAP   6-benzylaminopurine (*q.v.*).

Bar   a sex-linked dominant mutation in *Drosophila melanogaster* which results in a reduction in the number of facets in the compound eye. This mutation, symbolized by *B*, is commonly used as a marker when constructing balancer X-chromosomes. The original mutation contained a tandem duplication of chromosome segment 16A. Unequal crossing over (*q.v.*) can cause this mutation to revert to wild type. The 16A duplication is now thought to be a transposon-induced rearrangement. Analysis of the *Bar* phenomenon led to the discovery of position effects (*q.v.*). *See* **Appendix C,** 1925, Sturtevant; **M5 technique, transposable elements.**

barley *See **Hordeum vulgare**.*

Barr body the condensed single X chromosome seen in the nuclei of somatic cells of female mammals. *See* **Appendix C,** 1949, Barr and Bertram; **dosage compensation, drumstick, late-replicating X chromosome, Lyon hypothesis, sex chromatin.**

basal body (granule) a structure generally composed of a ring of nine triplet microtubules surrounding a central cavity, found at the base of cilia. *See* **axoneme, centriole, flagellum.**

***Basc* chromosome** *See* **M5 technique.**

base analogue a purine or pyrimidine base that differs slightly in structure from the normal base. Some analogues may be incorporated into nucleic acids in place of the normal constituent. *See* **aminopurine, azaguanine, mercaptopurine.** Analogues of nucleosides behave similarly. *See* **5-bromodeoxyuridine.**

basement membrane a delicate acellular membrane that underlies most animal epithelia.

base pair a pair of hydrogen-bonded nitrogenous bases (one purine and one pyrimidine) that join the component strands of the DNA double helix. *See* **deoxyribonucleic acid.**

base-pairing rules the rule that adenine forms a base pair with thymine (or uracil) and guanine with cytosine in a double-stranded nucleic acid molecule.

base-pair ratio *See* $(A+T)/(G+C)$ **ratio.**

base-pair substitution a type of lesion in a DNA molecule that results in a mutation. There are two subtypes. In the case of *transitions* one purine is substituted by the other or one pyrimidine by the other, and so the purine-pyrimidine axis is preserved. In the case of *transversions,* a purine is substituted by a pyrimidine or vice versa, and the purine-pyrimidine axis is reversed. *See* **Appendix C,** 1959, Freese.

bases of nucleic acids the organic bases universally found in DNA and RNA. In a nucleotide sequence, a purine is often symbolized R, while a pyrimidine is symbolized by Y. The purines adenine and guanine occur in both DNA and RNA. The pyrimidine cytosine also occurs in both classes of nu-

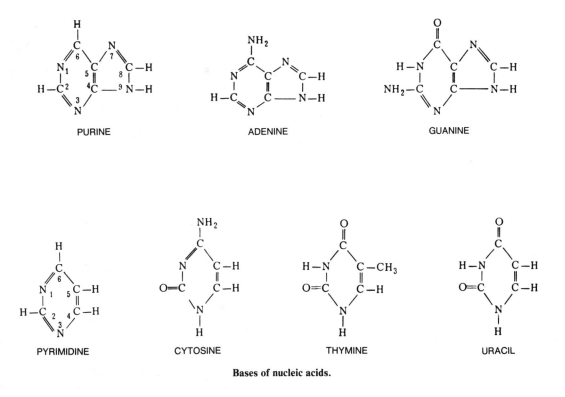

Bases of nucleic acids.

33

cleic acid. Thymine is found only in DNA, and uracil occurs only in RNA. *See* **rare bases.**

base stacking the orientation of adjacent base pairs with their planes parallel and with their surfaces nearly in contact, as occurs in double-stranded DNA molecules. Base stacking is caused by hydrophobic interactions between purine and pyrimidine bases, and results in maximum hydrogen bonding between complementary base pairs.

basic amino acids amino acids that have a net positive charge at neutral *p*H. Lysine and arginine bear positively charged side chains under most conditions.

basic dye any organic cation that binds to and stains negatively charged macromolecules, such as nucleic acids. *See* **azure B.**

basic number the lowest haploid chromosome number in a polyploid series (symbolized by x). The monoploid number.

basikaryotype the karyotype corresponding to the monoploid.

Basilarchia archippus the palatable Viceroy butterfly, which mimics the Monarch. *See* **mimicry.**

basophilic designating an acidic compound that readily stains with basic dyes.

Batesian mimicry *See* **mimicry.**

B cell B lymphocyte. *See* **lymphocyte.**

B chromosomes supernumerary chromosomes that are not duplicates of any of the members of the basic complement of the normal or "A" chromosomes. B chromosomes are devoid of structural genes. During meiosis, B chromosomes never pair with A chromosomes, and B's show an irregular and non-Mendelian pattern of inheritance. B chromosomes are extremely widespread among flowering plants and have been extensively studied in rye and maize. B chromosomes are believed to perpetuate and spread themselves in populations because they replicate faster than A chromosomes. *See* **Appendix C,** 1928, Randolph.

becquerel a unit of activity of a radioactive substance. 1 Bq = 1 disintegration per second.

bee dances circling and waggling movements (*Rundtanzen* and *Schwanzeltanzen*) performed by worker bees to give their hive mates information as to the location of a new source of food.

bees *See* ***Apis mellifera,*** **African bees.**

Beer-Lambert law the law that the absorption of light by a solution is a function of the concentration of solute. It yields a relation commonly used in photometry: $E = \log_{10} I_0/I = kcb$, where E = optical density; I_0 = the intensity of the incident monochromatic light; I = the intensity of the transmitted light; k = a constant determined by the solvent, wavelength, and temperature; c = the concentration of absorbing material in moles per liter; and b = the thickness in cm of the solution traversed by the light.

behavioral isolation a prezygotic (premating) isolating mechanism in which two allopatric species refuse to mate because of differences in courtship behavior; ethological isolation.

behavior genetics a branch of genetics that concerns the inheritance of forms of behavior such as courtship displays, nest building, etc., in lower animals and intelligence and personality traits in humans. Many traits that are of interest to behavioral scientists are quantitative characters (*q.v.*).

Bellevalia romana a lily used in cytology because it possesses a small number (N = 4) of large chromosomes.

Belling's hypothesis a hypothesis which assumes that crossing over does not require breakage and reunion. According to this theory the genes replicate first, and next the intergenic connections are formed between the adjacent newly synthesized genes. On rare occasions the interlinks form between genes, one of maternal and one of paternal origin. Crossovers arising by this mechanism would occur only between newly synthesized genes. Belling's hypothesis is a forerunner of the copy choice hypothesis (*q.v.*). *Contrast with* **breakage and reunion.**

Bence-Jones proteins proteins identified by the English physician Henry Bence-Jones in 1847. B-J proteins are excreted in the urine of patients suffering from malignancies of antibody-secreting cells (plasmacytomas or myelomas). B-J proteins consist of dimers of immunoglobulin chains. The proteins are synthesized by clones of identical cells, and therefore each patient produces identical peptides in sufficient quantities for the amino acid sequences to be determined. B-J proteins played a critical role in investigations of the chemical structure of the immunoglobulins (*q.v.*).

benign *See* **neoplasm.**

benign subtertian malaria *See* **malaria.**

34

benzene the simplest of the aromatic organic compounds.

Bergmann's rule the generalization that geographical races of a warm-blooded species possess smaller body sizes in the warmer parts of the range and larger body sizes in the colder parts of the range.

beta carotene *See* **carotenoids, retinene.**

beta chain one of the two polypeptides found in adult hemoglobin.

beta galactosidase an enzyme that breaks lactose into glucose and galactose. In *E. coli,* the enzyme is a tetramer of about 500,000 daltons encoded by the *lac Z* gene. *See* **homomeric protein, operon,** *lac* **operon, lactose.**

beta lactamase *See* **penicillinases.**

beta-2 microglobulin *See* **MHC molecules.**

beta particle a high-energy electron emitted from an atomic nucleus undergoing radioactive decay.

bicoid this *Drosophila* mutant belongs to the anterior class of maternal polarity genes. The bc^+ allele transcribes a message that is localized at the anterior pole of the oocyte during oogenesis. The sequences responsible for this localization reside in the trailer position of the mRNA. The protein product of *bc* is distributed in an exponential concentration gradient along the anteroposterior axis of the embryo. The type of differentiation target cells undergo is determined by their position in this gradient. *See* **Appendix C,** 1988, Macdonald and Struhl, Driever and Nüsslein-Volhard; **hunchback, maternal polarity mutants, trailer sequence.**

bidirectional genes a pair of open reading frames (*q.v.*), one on the plus strand and the other on the minus strand of the same DNA double helix and overlapping to a certain degree. *Compare with* **overlapping genes.**

bidirectional replication a mechanism of DNA replication involving two replication forks moving in opposite directions away from the same origin.

biennial designating a plant that requires two years to complete its life cycle, from seed germination to seed production and death. The plant develops vegetatively during the first growing season and flowers in the second.

bifunctional vector *See* **shuttle vector.**

bilateral symmetry a form of symmetry in which the body can be divided by a longitudinal plane into two parts that are mirror images of each other.

Bilateria animals with bilateral symmetry. *See* **classification.**

bilharziasis *See* **schistosomiasis.**

bilirubin an orange pigment formed as a breakdown product of the heme component of hemeproteins, especially the hemoglobin released during the normal destruction of erythrocytes by the reticuloendothelial system. Bilirubin released into the circulation by the reticuloendothelial system is taken up by the liver and excreted into the bile. The accumulation of bilirubin in plasma and tissues results in jaundice. *See* **Crigler-Najjar syndrome.**

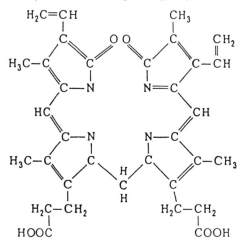

bimodal population a population in which the measurements of a given character are clustered around two values.

binary fission an amitotic, asexual division process by which a parent prokaryote cell splits transversely into daughter cells of approximately equal size.

Binet-Simon classification *See* **intelligence quotient classification.**

binomial distribution a probability function so named because the probabilities that an event will or will not occur, $n, n - 1, n - 2, \ldots, 0$ times are given by the successive coefficients in the binomial expansion $(a + b)^n$. Since a and b are the probabilities of occurrence and non-occurrence, respectively, their sum equals 1. The coefficients in a given binomial expansion can be found by referring to Pascal's pyramid (next page).

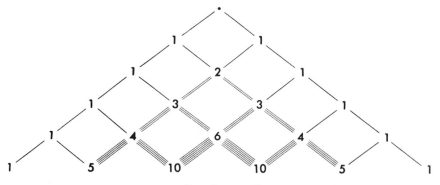

Pascal's pyramid.

Here each horizontal row consists of the coefficients in question for consecutive values of n. The expansions for n equaling 1, 2, 3, 4, or 5 are shown below:

$$(a + b)^1 = 1a + 1b$$
$$(a + b)^2 = 1a^2 + 2ab + 1b^2$$
$$(a + b)^3 = 1a^3 + 3a^2b + 3ab^2 + 1b^3$$
$$(a + b)^4 = 1a^4 + 4a^3b + 6a^2b^2 + 4ab^3 + 1b^4$$
$$(a + b)^5 = 1a^5 + 5a^4b + 10a^3b^2 + 10a^2b^3 + 5ab^4 + 1b^5$$

Note that each term of the triangle is obtained by adding together the numbers to the immediate left and right on the line above. Such a binomial distribution can be used for calculating the frequency of families in which a certain proportion of individuals show a given phenotype. If we ask, for example, what will be the distribution of girls and boys in families numbering four children and let the frequency of boys = a, and that of girls = b; then using the formula $(a + b)^4$, we conclude that the distribution would be 1/16 all boys; 4/16 3 boys and 1 girl; 6/16 2 boys and 2 girls; 4/16 1 boy and 3 girls; and 1/16 all girls.

binomial nomenclature   the current method of scientifically naming species of animals and plants. The name is in two parts: (1) the generic name designating the genus to which it belongs, and (2) the specific name peculiar to the new species. The capitalized generic name is written first; the specific second, usually with a lowercase letter; italic type is used (broader classifications are capitalized and written in ordinary roman letters). The name of the author responsible for naming and describing the species should follow (*Drosophila melanogaster* Meigen), but except in taxonomic publications the author's name is usually omitted.

bioassay   determination of the relative potency or effectiveness of a substance (e.g., a drug or a hormone) by comparing its effect on a group of test organisms or cells (e.g., tissue culture) using appropriate controls.

biochemical genetics   that branch of genetics that seeks to elucidate the chemical nature of hereditary determinants and their mechanisms of action during the life cycles of the organisms and their viruses. *See* **Appendix C,** 1909, Garrod; 1935, Beadle *et al.*

biocoenosis   a group of plant and animal species living together as a community in a particular habitat.

biogenesis   the production of a living cell from a parent cell. *Compare with* **biopoesis.**

biogenetic law   *See* **recapitulation.**

biogeographic realms   the divisions of the land masses of the world according to their distinctive floras and faunas. See map on facing page.

biogeography   the study of distributions of organisms over the earth and of the principles that govern these distributions.

biological clock   **1.** any mechanism that allows expression of specific genes at periodic intervals. **2.** any physiological factor that regulates body rhythms.

biological evolution   *See* **evolution.**

biological species   groups of naturally interbreeding populations that are reproductively isolated from other such species.

biolumniescence   the emission of light by living organisms.

biomass   the total weight of organic material in a particular sample, region, trophic level, etc. The dry biomass of the earth is estimated to be 3×10^{15} kg.

biome   a grouping of ecosystems (*q.v.*) into a larger group occupying a major terrestrial region (e.g., tropical rain forest biome, mixed conifer and deciduous forest biome).

biometry   the application of statistics to biological problems. *See* **Appendix C,** 1889, Galton.

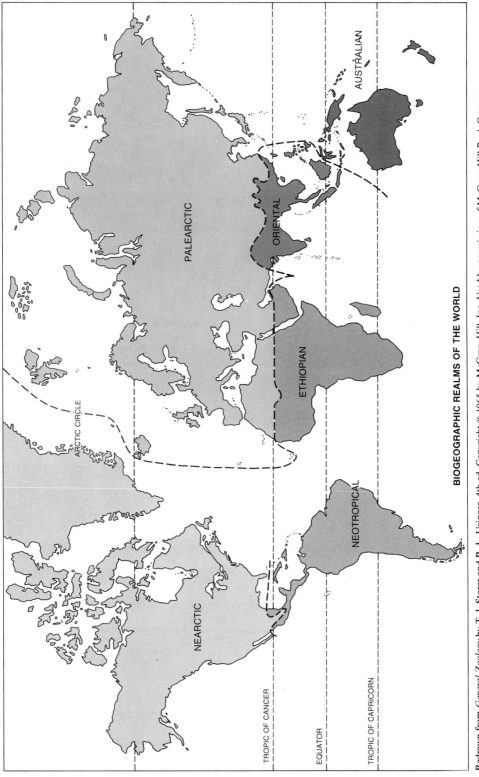

BIOGEOGRAPHIC REALMS OF THE WORLD

AUSTRALIAN

PALEARCTIC

ORIENTAL

ETHIOPIAN

NEARCTIC

NEOTROPICAL

ARCTIC CIRCLE

TROPIC OF CANCER

EQUATOR

TROPIC OF CAPRICORN

Redrawn from *General Zoology* by T. I. Storer and R. L. Usinger, 4th ed. Copyright © 1965 by McGraw-Hill, Inc. Used by permission of McGraw-Hill Book Company.

Biomphalaria glabrata *See* **schistosomiasis.**

biopoesis spontaneous generation (*q.v.*).

biosphere the surface of the earth where life resides.

biosynthesis the production of a chemical compound by a living organism.

biota a collective term to include all the organisms living in a given region.

biotechnology the collection of industrial processes that involve the use of biological systems. For some industries, these processes involve the use of genetically engineered microorganisms.

biotic potential *See* **reproductive potential.**

biotin a vitamin that functions as a cofactor in enzymes that catalyze carboxylation reactions. The antibiotic streptavidin (*q.v.*) has a very high affinity for this vitamin.

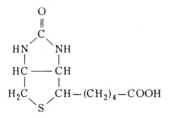

biotinylated DNA DNA probes labeled with biotin. Biotinylated deoxyuridine triphosphate is incorporated into the molecule by nick translation (*q.v.*). The probe is then hybridized to the specimen, such as denatured polytene chromosomes on a slide. The location of the biotin is visualized by complexing it with a streptavidin (*q.v.*) molecule that is attached to a color-generating agent. The technique is less time consuming than autoradiography and gives greater resolution. *See* **Appendix C,** 1981, Langer *et al.*

biotron a group of rooms designed for the control of environmental factors, singly and in combinations. Biotrons are used for producing uniform experimental organisms, and for providing controlled conditions for experiments.

biotype a physiologically distinct race within a species. If the biotype allows the race to occupy a particular environment, it is equivalent to an ecotype (*q.v.*).

biparental zygote **1.** the common state for nuclear genes in diploid zygotes to contain equal genetic contributions from male and female parents. **2.** the rare state for cytoplasmic genes in diploid zygotes

to contain DNA from both parents (e.g., chloroplast DNA in *Chlamydomonas*).

biparous producing two individuals at one birth.

birefringence *See* **anisotropy.**

birth defect **1.** any morphological abnormality present at birth (congenital); such abnormalities may have a genetic basis or they may be environmentally induced (*see* **phenocopy**). **2.** any biochemical or physiological abnormality present at birth; such abnormalities usually have a genetic basis and have been called "inborn errors of metabolism." *See* **Appendix C,** 1909, Garrod.

bisexual **1.** pertaining to a species made up of individuals of both sexes. **2.** pertaining to an animal having both ovaries and testes, or to a flower having both stamens and pistils.

Bison a genus including the American bison, *B. bison,* and the European bison, *B. bonasus,* the latter of which has been used in studies of the effects of inbreeding.

Biston betularia the peppered moth, a species showing industrial melanism (*q.v.*). A black form of the species, called *carbonaria,* spread through the industrial areas of England beginning about 1880. The melanic form is due to a dominant gene, and moths of this phenotype were less conspicuous in soot-darkened habitats. Thus, those individuals with genotypes that darkened their color tended to avoid bird predation and survived to reproduce. *See* **Appendix C,** 1956, Kettlewell.

bithorax a gene residing at 58.8 on the genetic map and within segment 89E of the salivary map of *Drosophila melanogaster.* The *bx* gene is one of a cluster of three genes that specify the type of differentiation that cells in the segments starting at the posterior portion of the second thoracic segment through the eighth abdominal segment will undergo. *See* **Appendix C,** 1978, Lewis; 1983, Bender *et al.; **Antennapedia,*** homeotic mutations, ***Polycomb,*** segment identity genes.

Bittner mouse milk virus *See* **mammary tumor agent.**

bivalent a pair of homologous, synapsed chromosomes. *See* **meiosis.**

Bkm sequences a satellite DNA containing repeats of the tetranucleotide sequences GATA and GACA that was first isolated from the banded krait. In this and many other snakes, the sequences are concentrated in the W chromosome. Bkm sequences also occur in the W chromosomes of birds. *See* **W, Z chromosomes.**

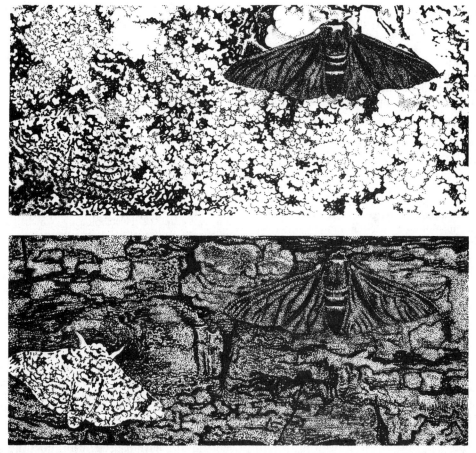

Biston betularia. The ancestral (left) and melanic (right) form of the peppered moth shown (in the upper drawing) resting upon lichen-covered bark in a nonpolluted woods and (in the lower drawing) resting upon soot-darkened bark.

blackwater fever   a name for malaria, with reference to the urinary excretion of heme. The malaria parasite uses only the globin fraction of hemoglobin for its metabolism and discards the heme, which is excreted, darkening the urine.

blast cell transformation   the differentiation, when antigenically stimulated, of a T lymphocyte to a larger, cytoplasm-rich lymphoblast.

blastema   a small protuberance composed of competent cells from which an animal organ or appendage begins its regeneration.

blastocyst   the mammalian embryo at the time of its implantation into the uterine wall.

blastoderm   the layer of cells in an insect embryo that completely surrounds an internal yolk mass. The cellular blastoderm develops from a syncytial blastoderm by the partitioning of the cleavage nuclei with membranes derived from infoldings of the oolemma.

blastodisc   a disc-shaped superficial layer of cells formed by the cleavage of a large yolky egg such as that of a bird or reptile. Mitosis within the blastodisc produces the embryo.

blastokinin   *See* **uteroglobin.**

blastomere   one of the cells into which the egg divides during cleavage. When blastomeres differ in size, the terms *macromere* and *micromere* are often used.

blastoporal lip   the dorsal rim of the amphibian blastopore, which functions as the organizer inducing the formation of the neural tube. *See* **chordamesoderm.**

blastula   an early embryonic stage in animals consisting of a hollow sphere of cells.

Blatella germanica   the German cockroach, the hemimetabolous insect for which the most genetic information is available.

blending inheritance 1. an obsolete theory of heredity proposing that certain traits of an offspring are an average of those of its parents because of the blending of their fluidlike germinal influences; hereditary characters transmitted in this way would not segregate in later generations. 2. a term incorrectly applied to codominant traits, to genes lacking dominance, or to additive gene action.

blepharoplast the basal granule of flagellates.

blocked reading frame *see* **reading frame.**

blood clotting a cascade of enzymatic reactions in blood plasma that produces strands of fibrin to stop bleeding. Fibrinogen, a protein found in the blood plasma, is acted upon by the enzyme thrombin. As a result a negatively charged peptide is split off the fibrinogen molecule, leaving monomeric fibrin, which is capable of rapid polymerization to produce a clot. Active thrombin is formed from an inactive precursor prothrombin, also found in blood plasma. The conversion of prothrombin into thrombin is a very complex process that requires a number of factors, including a lipoprotein factor liberated from rupturing blood platelets, a protein plasma thromboplastin component (PTC), factor VIII, calcium ions, and others. Factor VIII is a macromolecular complex composed of the antihemophilic factor and the von Willebrand factor. The vWF accounts for 99% of the mass of factor VIII. *See* **hemophilia A, von Willebrand disease.**

blood coagulation blood clotting (*q.v.*).

blood group a type in a system of classification of blood, based on the occurrence of agglutination of the red blood cells when bloods from incompatible groups are mixed. The classical human blood groupings were A, B, AB, and O. However, a multitude of more recently identified groups exist. *See* **Appendix C,** 1900, Landsteiner; 1925, Bernstein; 1951, Stormont *et al.;* **A, B antigens, Kell-Cellano antibodies, Kidd blood group, Lewis blood group, Lutheran blood group, MN, P,** and **Rh.**

blood group chimerism the phenomenon in which dizygotic twins exchange hematopoietic stem cells while *in utero* and continue to form blood cells of both types after birth. *See also* **radiation chimera.**

bloodline in domesticated animals, a line of direct ancestors.

blood plasma the straw-color fluid remaining when the suspended corpuscles have been removed from blood.

blood typing determination of antigens on red blood cells, usually for the purpose of matching donor and recipient for blood transfusion. Conventionally, only antigens of the ABO and Rh systems are typed for this purpose.

Bloom syndrome a type of dwarfism inherited as an autosomal recessive. Patients have decreased immunity and show sun sensitivity. This is one of the hereditary syndromes characterized by chromosome fragility. *See* **Fanconi anemia.**

blotting the general name given to methods by which electrophoretically or chromatographically resolved RNAs, DNAs, or proteins can be transferred from the support medium (e.g., gels) to an immobilizing paper or membrane matrix. Blotting can be performed by two major methods: (1) capillary blotting involves transfer of molecules by capillary action (e.g., Southern blotting, northern blotting, *q.v.*), and (2) electroblotting, which involves transfer of molecules by electrophoresis.

blue-green algae, blue-green bacteria *See* **Cyanobacteria.**

blunt end ligation the use of a DNA ligase (*q.v.*) to join blunt-ended restriction fragments. *Compare with* **cohesive end ligation.**

blunt ends *See* **restriction endonuclease.**

B lymphocyte a cell belonging to the class of lymphocytes that synthesize immunoglobulins. B lymphocytes mature within a microenvironment of bone marrow (in mammals) or within the bursa of Fabricius (in birds). At this time, the immunoglobulins synthesized by B lymphocytes are transferred to the cell surface. After the binding of an antigen molecule to a B lymphocyte, it goes through a cycle of mitotic divisions during which the immunoglobulins disappear from the cell surface. The plasma cells that result synthesize immunoglobulins and secrete them into the blood. However, some B lymphocytes do not differentiate into plasma cells, but retain membrane-bound immunoglobulins. These "memory" B lymphocytes function to respond to any subsequent encounter with the same antigen *See* **lymphocyte.**

bobbed a gene (*bb*) in *Drosophila melanogaster* producing a small bristle phenotype. The locus of *bb* is very near the centromere, and *bb* is the only gene known to have alleles on both the X and Y chromosomes. The plus allele of *bb* is the nucleolus organizer, and the various hypomorphic alleles may represent partial deletions of ribosomal RNA cistrons resulting from unequal crossing over (*q.v.*).

Bombardia lunata an ascomycete commonly used in tetrad analysis.

Bombay blood group a rare human variant of the ABO blood group system (first discovered in Bombay, India) that does not have A, B, or O antigens. Individuals homozygous for an autosomal recessive allele (*h/h*) cannot make the precursor H substance (*q.v.*) from which the A and B antigens are formed. This is a classical case of recessive epistasis in human genetics, because without the product of allele *H,* the products of the ABO locus cannot be formed. Bombay bloods appear to be group O when routinely tested by antibodies against the A or B antigens, but an individual with the Bombay phenotype may be carrying unexpressed genes for the A and/or B antigens. However, they make anti-H that is not found in individuals of groups A, B, or O. Therefore it is possible for a child of group A or B to be produced from parents that appear to be group O, if one of them is a Bombay phenotype and carries the genes for antigens A or B or both. *See* **A, B antigens.**

Bombus the genus containing the bumblebee.

Bombyx mori the commercial silkmoth, first domesticated for the purpose of silk production in China over 4,000 years ago. Next to *Drosophila melanogaster,* it is the insect whose genetics is best understood. *See* **Appendix C,** 1913, Tanaka; 1933, Hashimoto; **silk.**

bond energy the energy required to break a given chemical bond. For example, 58.6 kilogram calories per mol are required to break a carbon to carbon (C−C) bond.

border cell in *Drosophila* oogenesis, one of the migratory follicle cells that form the micropylar apparatus.

Bos the genus that includes the domestic cow, *B. taurus,* the Brahman, *B. indicus,* and the yak, *B. grunniens.* The haploid chromosome number for the domestic cow is 30, and about 30 genes have been assigned to 21 syntenic groups. *See* **cattle** for a listing of domestic breeds.

bottleneck effect fluctuations in gene frequencies occurring when a large population passes through a contracted stage and then expands again with an altered gene pool (usually one with reduced variability) as a consequence of genetic drift (*q.v.*).

botulism poisoning by an exotoxin (*q.v.*) synthesized by *Clostridium botulinum,* found in certain foods during decay.

bouquet configuration *See* **meiosis.**

bovine referring to members of the cattle family, especially to those of the domestic cattle species *Bos taurus.*

bovine achondroplasia hereditary chondrodystrophy seen in "bull-dog" calves of the Dexter breed.

The condition is inherited as an autosomal recessive.

bp abbreviation for "base pairs."

Bq becquerel (*q.v.*).

brachydactyly abnormal shortness of fingers or toes or both.

brachyury a short-tailed mutant phenotype in the mouse governed by a gene (*T*). It was through this mutant that the *T* complex was discovered.

Bracon hebetor *See* ***Microbracon hebetor*** (also called *Habrobracon juglandis*).

bradyauxesis *See* **heterauxesis.**

bradytelic used to refer to a lower-than-average rate of evolution.

bradytelic evolution *See* **evolutionary rate.**

Brahman a breed of humped domestic cattle *(Bos indicus).*

brain hormone prothoracicotropic hormone (*q.v.*).

Branchiostoma a genus of lancelets, commonly called *Amphioxus. Branchiostoma lanceolatum* is the sole living representative of the Cephalochordata. *See* **classification.**

branch migration *See* **Holliday model.**

Brdu 5-bromodeoxyuridine (*q.v.*)

breakage and reunion the classical and generally accepted model of crossing over by physical breakage and crossways reunion of broken chromatids during meiosis. *See* **Holliday model.**

breakage-reunion enzymes enzymes that use continuous stretches of DNA molecules, rather than preexisting termini, as substrates. The DNA duplex is broken and rejoined. The energy released by breakage is stored in a covalent enzyme-DNA intermediate and utilized in rejoining the molecules.

breakthrough an individual that escapes the deleterious action of its genotype. In a population of individuals homozygous for a given recessive lethal gene, almost all will die at a defined developmental stage. Those that develop past this stage are called "breakthroughs," or "escapers."

breathing in molecular genetics, the periodic, localized openings of a DNA duplex molecule to produce single-stranded "bubbles."

breed an artificial mating group derived from a common ancestor for genetic study and domestication.

breeding the controlled propagation of plants and animals.

breeding size the number of individuals in a population that are actually involved in reproduction during that generation.

breeding true to produce offspring of phenotype identical to the parents; said of homozygotes.

bridge-breakage-fusion-bridge cycle a cycle that begins with a dicentric chromosome forming a bridge as it is pulled toward both poles at once during anaphase. Such dicentric chromosomes may arise from an exchange within a paracentric inversion or may be radiation-induced. Once the dicentric breaks, the broken ends remain sticky, and these fuse subsequent to duplication. The result is another dicentric that breaks at anaphase, and so the cycle continues, with the chromosomes being broken anew at every mitosis (see diagram below). Since each subsequent break is likely to be at a different place than the previous ones, there will be a repeated regrouping of the genetic loci to produce duplications and deficiencies. *See* **Appendix C, 1938, McClintock; chromosome bridge.**

bridge migration synonymous with branch migration. *See* **Holliday model.**

bridging cross a mating made to transfer one or more genes between two reproductively isolated species by first transferring them to an intermediate species that is sexually compatible with the other two species.

brilliant cresyl blue a basic dye used in cytochemistry.

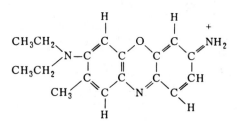

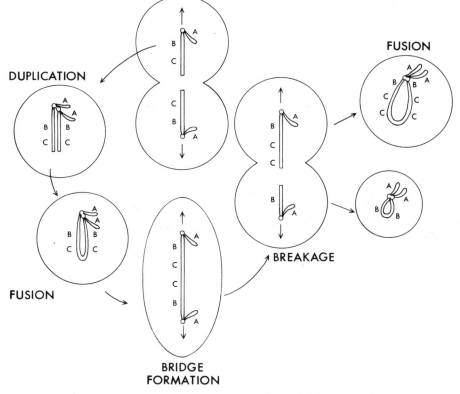

BREAKAGE

FUSION

DUPLICATION

FUSION

BREAKAGE

BRIDGE FORMATION

Bridge-breakage-fusion-bridge cycle.

42

bristle organ   each insect bristle is an organ consisting of four cells: the cell that secretes the bristle, the socket cell that secretes the ring that encloses the bristle, a sensory nerve cell whose process ends near the base of the bristle, and the sheath cell that surrounds the nerve axon. *See* **trichogen cell.**

broad bean   *Vicia faba* (*q.v.*). This is the European plant to which the term *bean* was originally applied.

broad heritability   the proportion of the total phenotypic variance (for a polygenic trait in a given population) that is attributed to the total genetic variance (including additive, dominance, epistatic, and other types of gene action); symbolized H^2. *See* **heritability.**

5-bromodeoxyuridine   a thymidine analogue that can be incorporated into DNA during its replication. This substitution profoundly affects that structure of the DNA. When both strands are substituted with BUDR, a chromatid stains less intensely than when only one strand is so substituted. Thus when cells are grown in the presence of BUDR for two replication cycles, the two sister chromatids stain differentially and therefore are called harlequin chromosomes. Consequently the BUDR labeling method can be used to detect sister chromatide exchanges. BUDR causes breakage in chromosomal regions rich in heterochromatin. Additional acronyms are *Budr* and *Brdu*. *See* **Appendix C,** 1972, Zakharov and Egolina.

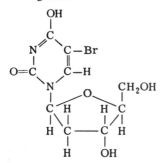

5-bromouracil   a mutagenically active pyrimidine analogue.

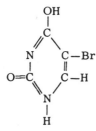

brood   the offspring from a single birth or from a single clutch of eggs.

broodiness   the tendency of female birds to incubate eggs.

Bryophyllum   a genus of succulent plants studied in terms of the genetic control of the photoperiodic flowering response. *See* **phytochrome.**

Bryophyta   the plant phylum containing mosses, liverworts, and hornworts. Bryophytes lack a vascular system. *See* **Appendix A.**

bubble   a structure formed in a duplex DNA at the site of initiation and strand separation during replication.

bud   1. a sibling cell produced by mitosis in yeast; one cell retains the cell wall of the parent cell, whereas the other produced by budding generates a new cell wall. 2. an underdeveloped plant shoot, consisting of a short stem bearing crowded, overlapping, immature leaves.

budding   1. in bacteria, yeast and plants, the process by which a bud (*q.v.*) is produced. 2. in enveloped viruses such as influenza virus and Sindbis virus, a mode of release from the host cell in which a portion of the cell membrane forms an envelope around the nucleocapsid. The envelope contains viral proteins, but no cellular proteins.

BUDR   5-bromodeoxyuridine (*q.v.*).

buffer   a compound that, in solution, tends to prevent or resist rapid changes in *p*H upon the addition of small quantities of acid or base.

buffering   the resistance of a system to change by outside forces.

Bufo   a genus of toads. Wild populations of species of this genus have been extensively studied by population geneticists.

bulb   a modified shoot consisting of a very much shortened underground stem enclosed by fleshy scalelike leaves. It serves as an organ of vegetative reproduction. The onion, daffodil, tulip, and hyacinth produce bulbs.

bull   the adult male of various animals including domesticated cattle, elephants, moose, or elk.

bull-dog calf   *See* **bovine achondroplasia.**

buoyant density   the equilibrium density at which a molecule under study comes to rest within a density gradient. *See* **centrifugation separation.**

Burkitt lymphoma   a monoclonal malignant proliferation of B lymphocytes primarily affecting the jaw and associated facial bones. The cancer is named after Denis Burkitt, who first described it in central African children in 1958. Most Burkitt tumors occurring in Africans contain Epstein-Barr virus (*q.v.*), and this virus is believed to be mosquito borne. Burkitt lymphomas from United States and European patients lack EBV. Burkitt lymphoma cells always contain a reciprocal translocation involving the long arm of chromosome 8 and chromosome 14, or less frequently 22 or 2. The break point on chromosome 8 is always near the *myc* oncogene (*q.v.*). The break point on the other chromosome is always near an immunoglobulin gene, namely, 14 (heavy chains), 22 (lambda light chains), or 2 (kappa light chains). In its translocated state *myc* is activated and the cancer ensues. *See* **immunoglobulin chains.**

bursa of Fabricius a saclike structure connected to the posterior alimentary canal in birds. The bursa is the major site where B lymphocytes become mature immunoglobulin (antibody)-secreting plasma cells. The equivalent organ in mammals has not been definitely identified. Most evidence suggests the bone marrow.

bursicon an insect hormone that appears in the blood after molting and is required for the tanning and hardening of new cuticle.

burst size the average number of bacteriophages released from a lysed host. *See* **Appendix C,** 1939, Ellis and Delbrück.

busulfan a mutagenic, alkylating agent.

$$CH_3-\underset{\underset{O}{\overset{\|}{\|}}}{\overset{\overset{O}{\|}}{S}}-O-(CH_2-CH_2)_2-O-\underset{\underset{O}{\overset{\|}{\|}}}{\overset{\overset{O}{\|}}{S}}-CH_3$$

C

C **1.** Celsius (also Centigrade), **2.** carbon. **3.** the haploid amount of DNA. **4.** cytosine or cytidine.

^{14}C a radioactive isotope of normal carbon (^{12}C) emitting a weak beta particle. The half-life of ^{14}C is 5,700 years.

CAAT box part of a conserved DNA sequence of about 75 bp upstream from the initiator for eukaryotic transcription; possibly involved in binding RNA polymerase II. *See* **Hogness box.**

cadherins a family of cell surface proteins that facilitate cell-to-cell adhesion during embryogenesis. There are a variety of cadherins, each specific for a certain cell type.

Caenobacter taenospiralis *See* **killer paramecia.**

Caenorhabditis elegans a small nematode whose developmental genetics has been extensively investigated. The worm is about 1 mm in length, and its life cycle, when reared at 20°C, is 3.5 days. Its transparent cuticle allows the visualization of every cell. The adult has 816 somatic cells, of which 302 are neurons. The complete lineage history and fate of every cell is known. *C. elegans* normally reproduces as a self-fertilizing hermaphrodite, which has two X chromosomes per cell, plus five pairs of autosomes. Loss of an X by meiotic nondisjunction leads to the production of males. These arise spontaneously among the progeny of hermaphrodites at a frequency of about 0.2%. The mating of hermaphrodites with males provides a method of genetic analysis, and over 500 genes have been defined and mapped. *See* **Appendix C,** 1974, Brenner; 1977, Sulston and Horvitz; 1981, Chalfe and Sulston; 1983, Greenwald *et al.;* **cell lineage, cell lineage mutants, *Panagrellus redivivus, Turbatrix aceti.***

caffeine an alkaloid stimulant found in tea and coffee. Caffeine is a purine analogue that is mutagenically active in microbial systems. Its structure is given opposite.

Cairns molecule *See* **theta replication.**

calciferol vitamin D (*q.v.*).

calcium, Ca an element universally found in small amounts in tissues. Atomic number 20;

atomic weight 40.08; valence 2 +; most abundant isotope ^{40}Ca; radioisotope ^{45}Ca, half-life 164d, radiation emitted—beta particles.

calico cat *See* **tortoiseshell cat.**

Calliphora erythrocephala a large fly in which polytene chromosomes occur in the ovarian nurse cells in certain inbred lines. The banding pattern of these giant chromosomes has been compared with those of pupal trichogen cells (*q.v.*).

callus the cluster of plant cells that results from tissue culturing a single plant cell.

calmodulin an intracellular calcium receptor protein that regulates a wide spectrum of enzymes and cellular functions, including the metabolism of cyclic nucleotides and glycogen. It also plays a role in fertilization and in the regulation of cell movement and cytoskeletal control, as well as in the synthesis and release of neurotransmitters and hormones. Calmodulin is a heat- and acid-stable, acidic protein with four calcium-binding sites. It is found in all eukaryotic cells and has a molecular weight of 16,700. It appears to be the commonest translator of the intracellular calcium message. *See* **second messenger.**

calyx the sterile, outer whorl of floral parts composed of sepals.

cambium the lateral meristem of vascular plants.

Cambrian the earliest period in the Paleozoic era. *See* **geologic time divisions.** Representatives of most animal phyla are present in Cambrian rocks. Algae and trilobites were plentiful.

Camelus the genus of camels including *C. bactri-*

Caffeine.

anus, the two-humped camel; and *C. dromedarius,* the one-humped camel, also called the dromedary.

cAMP *See* **cyclic AMP.**

Campbell model of λ integration a hypothesis that explains the mechanism of integration of phage lambda into the *E. coli* host chromosome. According to the model, linear lambda DNA is first circularized. Then prophage integration occurs as a physical breakage and reunion of phage and host DNA molecules precisely between the bacterial DNA site for phage attachment and a corresponding site in the phage DNA. *See* **Appendix C,** 1962, Campbell.

canalization the existence of developmental pathways that lead to a standard phenotype in spite of genetic or environmental disturbances.

canalized character a trait whose variability is restricted within narrow boundaries even when the organisms are subjected to disturbing environments or mutations.

canalizing selection elimination of genotypes that render developing individuals sensitive to environmental fluctuations.

cancer a class of diseases of animals characterized by uncontrolled cellular growth. *See* **leukemia, lymphoma, metastasis, neoplasm, oncogenic virus, sarcoma, teratoma.**

Canis familiaris the dog, the first domesticated animal. A favorite species for the study of behavior genetics. The haploid chromosome number is 39, and about 35 genes have been assigned to 26 linkage groups. *See* **dog** for a listing of breeds.

canonical sequence an archetypical amino acid sequence to which all variants are compared. A sequence that describes the nucleotides most often present in a DNA segment of interest. For example, in the Pribnow box and the Hogness box, the canonical sequences are T_{89} A_{89} T_{50} A_{65} A_{65} T_{100} and T_{83} A_{97} T_{93} A_{85} A_{63} A_{83} A_{50}, respectively. The subscript indicates the percent occurrence of the most frequently found base.

cap *See* **methylated cap.**

CAP catabolite activator protein (*q.v.*).

capacitation a process of physiological alterations whereby a sperm becomes capable of penetrating an egg as a consequence of exposure to one or more factors normally present in the female reproductive tract. It is theorized that a substance coating the sperm head must be removed by these female factors before the sperm can become fully functional for fertilization.

capon a castrated domestic fowl.

capped 5′ ends the 5′ ends of eukaryotic mRNAs containing methylated caps (*q.v.*).

capping 1. addition of a cap (*q.v.*) to mRNA molecules. 2. redistribution of cell surface structures to one region of the cell, usually mediated by cross-linkage of antigen-antibody complexes.

Capsicum a genus that includes red peppers and pimentos, *C. annum,* and green pepper, *C. frutescens.*

capsid the protein coat of a virus particle.

capsomere one of the subunits from which a virus shell is constructed. Capsomeres may contain several different polypeptide chains. The virus shell is formed by assembling capsomeres about the nucleic acid core in a precise geometrical pattern.

Carassius auratus the aquarium goldfish. A member of the carp family first described in China 2,300 years ago and bred for ornament since that time. *See* **Appendix C,** Osteichthyes, Cypriniformes.

carbohydrate a compound, having the general formula $C_xH_{2x}O_x$. Common examples of carbohydrates are glucose, cellulose, glycogen, and starches (*q.v.*).

carbon the third most abundant of the biologically important elements. Atomic number 6; atomic weight 12.01115; valence 4; most abundant isotope ^{12}C; radioisotope ^{14}C (*q.v.*).

carbonaria a melanic variety of the peppered moth, commonly found in soot-polluted woods surrounding industrial cities in Great Britain. It is an example of industrial melanism (*q.v.*).

3′ carbon atom end nucleic acids are conventionally written with the 3′ carbon of the pentose to the right. Transcription or translation from a nucleic acid proceeds from 5′ to 3′ carbon.

5′ carbon atom end nucleic acids are conventionally written with the end of the pentose containing the 5′ carbon to the left. *See* **deoxyribonucleic acid.**

carbon dioxide sensitivity *See* **sigma virus.**

Carboniferous the Paleozoic period that generated the great coal deposits. At this time the land was covered by extensive forests. Amphibians diversified, and the winged insects and reptiles arose. Cartilagenous fishes were the dominant marine vertebrates. In North America, where the stratigraphic record allows Carboniferous strata to be conveniently subdivided into upper and lower segments, the Carboniferous is replaced by the Pennsylvanian and Mississippian periods. *See* **geologic time divisions.**

carbonyl group a doubly bonded carbon-oxygen group ($C=O$). The secondary structure of a polypeptide chain involves hydrogen bonds between the carbonyl group of one residue (amino acid) and the imino (NH) group of the fourth residue down the chain. *See* **alpha helix.**

carboxyl group a chemical group (COOH) that is acidic because it can become negatively charged ($-C-O^-$) if a proton dissociates from its hydroxyl
$\parallel$
O
group.

carboxyl terminal C-terminus (*q.v.*).

carboxypeptidases two pancreatic enzymes (A and B) that hydrolyze protein chains beginning at the carboxyl terminal end of the chain and liberating amino acids one at a time. These enzymes are useful for amino acid sequence studies.

carcinogen a physical or chemical agent that induces cancer.

carcinogenic capable of inducing cancer.

carcinoma a cancer of epithelial tissues (e.g., skin cancer); *adenocarcinoma* is a cancer of gland epithelia.

carcinostasis inhibition of cancerous growth.

carnivore a meat-eating animal. Also applied to a few insectivorous plants.

carotenoids lipid-soluble pigments ranging in color from yellow to red. The carotenes whose structures appear below are plant carotenoids. Beta

carotene can be enzymatically hydrolyzed into two molecules of vitamin A (*q.v.*) and is therefore an important provitamin.

carpel the part of a flower that encloses the ovules and extends into a compound pistil.

carrier **1.** an individual heterozygous for a single recessive gene. **2.** a stable isotope of an element mixed with a radioisotope of that element to give a total quantity sufficient to allow chemical operations. **3.** an immunogenic molecule (e.g., a foreign protein) to which a hapten (*q.v.*) is coupled, thus rendering the hapten capable of inducing an immune response.

carrier-free radioisotope a radioisotope essentially undiluted with a stable contaminating isotope.

carrying capacity the size or density of a population that can be supported in stable equilibrium with the other biota of a community; symbolized K.

cartilage a skeletal connective tissue formed by groups of cells that secrete into the intercellular space a ground substance containing a protein, collagen, and a polysaccharide, chondroitin sulfuric acid.

Carya a genus that includes *C. ovata,* the shagbark hickory, and *C. pecan,* the pecan.

caryonide a lineage of paramecia that derive their macronuclei from a single macronuclear primordium. Such paramecia are generally immediate descendants of the exconjugants.

caryopsis a dry indehiscent multiple-seeded fruit derived from a compound ovary. The corn ear is an example.

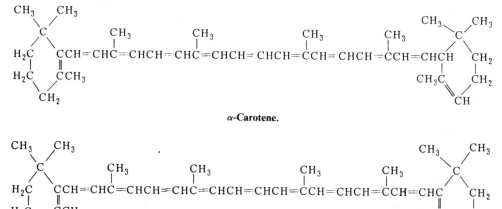

α-Carotene.

β-Carotene.

47

cassette mutagenesis a technique that involves removing from a gene a stretch of DNA flanked on either end by a restriction site (*q.v.*) and then inserting in its place a new DNA segment. This cassette can contain base substitutions or deletions at specific sites, and the phenotypic effects that result give insight into relative importance of specific subsegments of the region to the functioning of the gene or its product.

cassettes loci containing functionally related nucleotide sequences that lie in tandem and can be substituted for one another. The mating type reversals observed in yeast result from removing one cassette and replacing it by another containing a different nucleotide sequence.

caste a class of structurally and functionally specialized individuals within a colony of social insects.

cat any of a number of domesticated breeds of the species *Felis catus.* Popular breeds include: SHORT-HAIRED BREEDS: Domestic Shorthair, Siamese, Burmese, Abyssinian, Russian Blue, Havana Brown, Manx, and Rex. LONG-HAIRED BREEDS: Persian, Angora, and Himalayan.

catabolism metabolic breakdown of complex molecules to simpler products, often requiring catabolic enzymes and accompanied by the release of energy.

catabolite a compound generated by the breakdown of food molecules.

catabolite activating protein (CAP) a constitutively produced, dimeric, positive regulator protein in bacteria that, when bound to a promoter region and cAMP, facilitates transcription by RNA polymerase of certain catabolite-sensitive adjacent genes in inducible and glucose-sensitive operons (such as the *lac* operon of *E. coli*). Also known as cyclic AMP receptor protein (CRP) or catabolite gene activator (CGA) protein.

catabolite repression the reduction or cessation of synthesis of enzymes involved in catabolism of sugars such as lactose, arabinose, etc., when bacteria are grown in the presence of glucose. The enzyme adenyl cyclase is inhibited by glucose from converting ATP to cyclic adenosine monophosphate (cAMP); cAMP must complex with catabolite activator protein (CAP) in order for RNA polymerase to bind to promoters of genes responsible for enzymes capable of catabolizing sugars other than glucose. Therefore, in the presence of glucose, less CAP protein is available to facilitate the transcription of mRNAs for these enzymes.

catalase an enzyme that catalyzes the reaction of $H_2O_2 \rightarrow H_2O + \frac{1}{2}O_2$. Catalase is especially abundant in the liver, where it is contained in peroxisomes (*q.v.*). *See* **acatalasemia.**

catalyst a substance that increases the rate of a chemical reaction without being consumed. Enzymes are biological catalysts.

catarrhine pertaining to certain higher primates including humans, apes, and Old World (African and Asian) monkeys; characterized by close-set nostrils directed forward or downward.

catastrophism a geological theory proposing that the earth has been shaped by violent events of great magnitude (e.g., worldwide floods, near collisions with comets, etc.); the opposite of uniformitarianism (*q.v.*).

cat cry syndrome a syndrome of multiple congenital malformations in humans with a deficiency in the short arm of chromosome 5. Infants with this condition produce a peculiar cry that sounds like a cat mewing. Also known as the *cri du chat* syndrome.

category a rank in a taxonomic hierarchy to which one or more taxa may be assigned: e.g., phylum, class, order, family, genus, species.

catenane a structure made up of two or more interlocking rings.

catenate to convert two or more rings into a system of interlocking rings.

cathepsin any of certain proteolytic enzymes thought to reside in lysosomes (*q.v.*). Such enzymes are abundant, for example, in metamorphosing tadpoles during the resorption of the tail.

cathode the negative electrode to which positive ions are attracted. *Contrast with* **anode.**

cation a positively charged ion so named because it is attracted to the negatively charged cathode. *Contrast with* **anion.**

Cattanach's translocation a translocation in the mouse discovered by B. M. Cattanach. The aberration involves an X chromosome into which a segment of autosome 7 has been translocated. The insertion carries the wild-type alleles of three autosomal genes that control the color of the fur. Studies on mice heterozygous for Cattanach's translocation have shown that during X-chromosomal inactivation in somatic cells, the genes in the inserted autosomal segment are turned off sequentially in order of their distances from the X chromosomal element. Thus, the X inactivation spreads into the attached autosomal segment, but does not travel unabated to the end of the segment.

cattle any of a number of domesticated breeds of the species *Bos taurus.* Popular breeds include: BEEF CATTLE: Hereford, Shorthorn, Aberdeen-Angus, and Santa Gertrudis. DAIRY CATTLE: Holstein-Friesian, Jersey, Guernsey, Ayrshire, and Brown Swiss.

Cavia porcellus the domestic guinea pig, a species of rodent used as a laboratory animal. Numerous mutants are known affecting coat color and texture.

C banding a method for producing stained regions around centromeres. *See* **chromosome banding techniques.**

C^{13}/C^{12} ratio the ratio between the heavy, stable isotope of carbon and the normal isotope in a sample of interest. Since organisms take up C^{12} in preference to C^{13}, the ratio is used to determine whether or not the carbon in the specimen is of biological origin.

$CD4^+$ cells, $CD8^+$ cells *See* **T lymphocyte.**

cDNA complementary DNA produced from a RNA template by the action of RNA-dependent DNA polymerase (reverse transcriptase). If the RNA template has been processed to remove introns, the cDNA will not be identical to the gene from which the RNA was transcribed. Also called *copy DNA.*

cDNA clone a duplex DNA sequence complementary to an RNA molecule of interest, carried in a cloning vector.

cDNA library a collection of cDNA (*q.v.*) molecules, representative of all the various mRNA molecules produced by a specific type of cell of a given species, spliced into a corresponding collection of cloning vectors such as plasmids or lambda phages. Since not all genes are active in every cell, a cDNA library is usually much smaller than a gene library (*q.v.*). If it is known which type of cell makes the desired protein (e.g., only pancreatic cells make insulin), screening the cDNA library from such cells for the gene of interest is a much easier task than screening a gene library.

CD3 proteins *See* **T lymphocyte.**

Ceboidea the superfamily containing the monkeys of Central and South America.

cecidogen a gall-forming substance.

cell the smallest, membrane-bound protoplasmic body capable of independent reproduction. *See* **Appendix C,** 1665, Hooke.

cell affinity a property of eukaryotic cells of the same type to adhere to one another but not to those of a different type; this property is lost when the cell transforms to the cancerous state.

cell culture a term used to denote the growing cells *in vitro,* including the culturing of single cells. In cell cultures the cells are not organized into tissues. *See* **Appendix C,** 1940, Earle; 1956, Puck *et al.*

cell cycle the sequence of events between one mitotic division and another in a eukaryotic cell. Mitosis (M phase) is followed by a growth (G_1) phase, then by DNA synthesis (S phase), then by another growth (G_2) phase, and finally by another mitosis. In HeLa cells (*q.v.*), for example, the G_1, S, G_2, and M phases take 8.2, 6.2, 4.6, and 0.6 hours, respectively. *See* **Appendix C,** 1953, Howard and Pelc.

cell determination an event in embryogenesis that specifies the developmental pathway that a cell will follow.

cell differentiation the process whereby descendants of a single cell achieve and maintain specializations of structure and function. Differentiation presumably is the result of differential transcriptions.

cell division the process (binary fission in prokaryotes, mitosis in eukaryotes) by which two daughter cells are produced from one parent cell. *See* **Appendix C,** 1875, Strasburger.

cell-driven viral transformation a method for creating immortalized human antibody-producing cells *in vitro* without forming a hybridoma (*q.v.*). Normal B lymphocytes from an immunized donor are mixed with other cells infected with the Epstein-Barr virus (*q.v.*). The virus enters the B lymphocytes. The cells originally infected with the virus are experimentally destroyed and the virally transformed cells producing the antibody of interest are isolated. In cell-driven viral transformation, about 1 in 50 B lymphocytes is transformed, whereas with the cell hybridization technique only about 1 human cell in 10 million is transformed.

cell fractionation the separation of the various components of cells after homogenization of a tissue and differential centrifugation. Four fractions are generally obtained: (1) the nuclear fraction, (2) the mitochondrial fraction, (3) the microsomal fraction, and (4) the soluble fraction or cytosol. *See* **Appendix C,** 1946, Claude.

cell-free extract a fluid obtained by rupturing cells and removing the particulate material, membranes, and remaining intact cells. The extract contains most of the soluble molecules of the cell. The preparation of cell-free extracts in which proteins and nucleic acids are synthesized represent milestones in biochemical research. *See* **Appendix C,**

1955, Hoagland; 1961, Nirenberg and Matthaei; 1973, Roberts and Preston.

cell fusion   the experimental formation of a single hybrid cell with nuclei and cytoplasm from different somatic cells. The cells that are fused may come from tissue cultures derived from different species. Such fusions are facilitated by the adsorption of certain viruses by the cells. *See* **Sendai virus, polyethylene glycol, Zimmermann cell fusion.**

cell hybridization   the production of viable hybrid somatic cells following experimentally induced cell fusion (*q.v.*). In the case of interspecific hybrids, there is a selective elimination of chromosomes belonging to one species during subsequent mitoses. Eventually, cell lines can be produced containing a complete set of chromosomes from one species and a single chromosome from the other. By studying the new gene products synthesized by the hybrid cell line, genes residing in the single chromosome can be identified. *See* **Appendix C**, 1960, Barski *et al.;* **HAT medium, hybridoma, syntenic genes.**

cell interaction genes   a term sometimes used to refer to some genes in the I region of the mouse H-2 complex that influence the ability of various cellular components of the immune system to cooperate effectively in an immune response.

cell line   a heterogeneous group of cells derived from a primary culture (*q.v.*) at the time of the first transfer.

cell lineage   a pedigree of the cells produced from an ancestral cell by binary fission in prokaryotes or mitotic division in eukaryotes. *Caenorhabditis elegans* (*q.v.*) is the only multicellular eukaryote for which the complete pattern of cell divisions from single-celled zygote to mature adult has been elucidated. Cell lineage diagrams are available that detail each cell or nuclear division and the fate of each cell produced by a terminal division.

cell lineage mutants   mutations that affect the division of cells or the fates of their progeny cells. Cell lineage mutants generally fall into two broad classes. The first contains mutations that affect general cellular processes, such as cell division or DNA replication. Mutants perturbing the cell division cycle have been analyzed most extensively in *Saccharomyces cerevisiae*. The second class of mutations shows a striking specificity in their effects. For example, cell lineage mutants are known in *Caenorhabditis elegans* where particular cells are transformed to generate lineages or to adopt differentiated fates characteristic of cells normally found in different positions, at different times, or in the opposite sex. Some of these mutants result from trans-

formations in cell fates. For example, a particular cell "A" will adopt the fate of another cell "B," and this results in the loss of the cells normally generated by A and the duplication of cells normally generated by B. Such transformations resemble the homeotic mutations (*q.v.*) of *Drosophila*. In *Caenorhabditis,* mutations of this type are generally symbolized by *lin. See* **developmental control genes, heterochronic mutations.**

cell lysis   disruption of the cell membrane, allowing the dissolution of the cell and exposure of its contents to the environment. Examples: bacteria undergo *bacteriolysis,* red blood cells experience *hemolysis.*

cell-mediated immunity   immune responses produced by T lymphocytes rather than by immunoglobulins (humoral- or antibody-mediated immunity); abbreviated CMI.

cell-mediated lympholysis   the killing of "target" cells by activated T lymphocytes through direct cell-cell contact. Often used as an *in vitro* test of cell-mediated immunity.

cell plate   a semisolid structure formed by the coalescence of droplets that are laid down between the daughter nuclei following mitosis in plants. The cell plate is the precursor of the cell walls, and it is synthesized by the phragmoplast (*q.v.*).

cell strain   cells derived from a primary culture or cell line by the selection and cloning of cells having specific properties or markers. The properties or markers must persist during subsequent cultivation. *See* **in vitro marker, in vivo marker.**

cell theory   the theory that all animals and plants are made up of cells, and that growth and reproduction are due to division of cells. *See* **Appendix C**, 1838, Schleiden and Schwann; 1855, Virchow.

cellular immunity   immune responses carried out by active cells rather than by antibodies.

cellular transformation   *See* **transformation.**

cellulase   an enzyme that degrades cellulose to glucose.

cellulifugal   moving away from the center of the cell.

cellulose   a complex structural polysaccharide that makes up the greater part of the walls of plant cells. As shown on the facing page, cellulose is composed of a linear array of β-D-glucose molecules.

cell wall   a rigid structure secreted external to the plasma membrane. In plants it contains cellulose and lignin; in fungi it contains chitin; and in bacteria it contains peptidoglycans.

Cellulose.

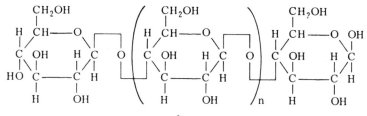

glucose

cen *See* **symbols used in human cytogenetics.**

cenospecies a group of species that, when intercrossed, produce partially fertile hybrids.

Cenozoic the most recent geologic era, occupying the last 65 million years and often called the age of mammals. *See* **geologic time divisions.**

center of origin an area from which a given taxonomic group of organisms has originated and spread.

centimorgan *See* **Morgan unit.**

central dogma the concept describing the functional interrelations between DNA, RNA, and protein; that is, DNA serves as a template for its own replication and for the transcription of RNA which, in turn, is translated into protein. Thus, the direction of the transmission of genetic information is DNA → RNA → protein. Retroviruses (*q.v.*) violate this central dogma during their reproduction.

centric fusion breakage in the very short arms of two acrocentric chromosomes, followed by fusion of the long parts into a single chromosome; the two small fragments are usually lost; also termed a *Robertsonian translocation* or *whole arm fusion*. *See* **Appendix C, 1911, Robertson.**

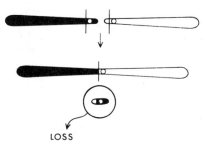

LOSS

centrifugal acting in a direction away from the center.

centrifugal selection *See* **disruptive selection.**

centrifugation separation any of various methods of separation dispersions by the application of centrifugal force. In the case of *density gradient equilibrium centrifugation,* a gradient of densities is established in a centrifuge tube by adding a high molecular weight salt such as cesium chloride. The mixture of molecules to be studied is layered in the surface of the gradient and then centrifuged until each molecule reaches the layer in the gradient with a buoyant density equal to its own. In the case of *density gradient zonal centrifugation* the macromolecules are characterized by their velocities of sedimentation through a preformed sucrose gradient. In this case the sedimentation velocity is determined by molecular size and shape.

centrifuge an apparatus used for the separation of substances by the application of centrifugal force generated by whirling at a high rate of rotation a vessel containing a fluid in which the substances are suspended. *See also* **ultracentrifuge.**

centriole a self-reproducing cellular organelle generally consisting of short cylinders containing nine groups of peripheral microtubules disposed about a central cavity. When a centriole reproduces, it forms a "daughter" centriole at right angles to itself. The daughter centriole grows outward away from the "mother" until it reaches its mature size. Centrioles are capable of movement and always come to lie at the polar regions of the spindle apparatus in dividing animal cells. The behavior of the centrioles is illustrated in the **meiosis** entry. An organelle that is ultrastructurally identical to the centriole forms the basal body of a cilium. Centrioles do not occur in the cells of higher plants. *See* **Appendix C, 1888, Boveri.**

centripetal acting in a direction toward the center.

centripetal selection *See* **stabilizing selection.**

centrolecithal egg one having centrally placed yolk. *See* **isolecithal egg, telolecithal egg.**

centromere a region of a chromosome to which spindle traction fibers attach during mitosis and meiosis. The position of the centromere determines whether the chromosome will appear as a rod, a J, or a V during its poleward migration at anaphase. In a very few species the traction fibers seem to attach along the length of the chromosome. Such chromosomes are said to be *polycentric* or to have a *diffuse centromere.* A replicated chromosome

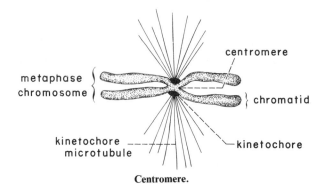

Centromere.

consists of two chromatids joined at the centromere region. Late in prophase, kinetochores develop on the two faces of the centromere that point toward the spindle poles. The microtubules of the traction fiber attach to the kinetochores as shown above.

In the older literature, the terms centromere and kinetochore are used synonymously. The centromere of metaphase chromosomes is narrower than the regions distal to it, and therefore it is called the *primary chromosomal constriction.* The centromere is generally bordered by heterochromatin containing repetitious DNA (*q.v.*) and it is late to replicate. *See* **Appendix C, 1903, Waldeyer;** *Luzula,* **meiosis, mitosis.**

centromere interference the inhibitory effect of the centromere upon crossing over in adjacent chromosomal regions.

centromere misdivision *See* **isochromosome.**

centromeric index the percentage of the total length of a chromosome encompassed by its shorter arm. For example, in human somatic cells during metaphase, chromosomes 1 and 13 have centromeric indexes of 48 and 17, respectively. Therefore, chromosome 1 is metacentric with its short arm occupying 48% of the total length of the chromosome, and chromosome 13 is acrocentric with a short arm that only occupies 17% of the total length.

centrosome a region of differentiated cytoplasm containing the centriole.

Cepaea a genus of land snails belonging to the family Helicidae. *C. hortensis* and *C. nemoralis* exhibit extensive variation in color and ornamentation of the shell with longitudinal bands. These species have been extensively studied in the field and in laboratory colonies by population geneticists.

cephalic designating the head or the anterior end of an animal.

cephalosporin an antibiotic with structural similarities to penicillin (*q.v.*). It has the advantage of not causing allergic reactions in patients that are allergic to penicillin and of being inert to penicillinases (*q.v.*).

Cephalosporium a genus of molds of importance because of the cephalosporin antibiotics they produce.

Cercopithecoidea a superfamily of primates containing the Old World (African and Asian) monkeys, baboons, macaques, colobines, etc. A sister group to the Hominoidea (*q.v.*). The divergence of the Cercopithecoidea and Hominoidea took place about 30 million years ago.

Cercopithecus aethiops the African green monkey. A catarrhine primate with a haploid chromosome number of 30. About 20 genes have been assigned to 9 different linkage groups.

cereal a cultivated grass whose seeds are used as food; for example, wheat, oats, barley, rye, maize, etc.

cerebroside a molecule composed of sphingosine, a fatty acid, and a sugar; abundant in the myelin sheaths of nerve cells.

certation competition for fertilization among elongating pollen tubes.

ceruloplasmin a blue, copper protein present among the α_2 globulins of the plasma. Approximately 95% of the circulating copper of human beings is bound to ceruloplasmin. Ceruloplasmin is made up of eight subunits, each of molecular weight 18,000.

cesium-137 a radioisotope of cesium with a half life of about 30 years generated during the explosion of certain nuclear weapons. It is one of the major sources of radiation contamination from fallout.

cesium chloride gradient centrifugation *See* **centrifugation separation.**

C genes genes that code for the constant region of immunoglobulin protein chains. *See* **immunoglobulin.**

chaeta a bristle, especially of an insect.

chaetotaxy the taxonomic study of the bristle pattern of insects.

Chagas disease a disease in man produced by *Trypanosoma cruzi* transmitted by bloodsucking bugs in the genera *Rhodnius* and *Triatoma*. Darwin is thought to have contracted Chagas disease while in South America, and as a result spent the remainder of his life as a semi-invalid.

chain reaction a biological, molecular, or atomic process in which some of the products of the process, or energies released by the process, are instrumental in the continuation or magnification of the process.

chain termination codon *See* **stop codon.**

chain terminator a molecule that stops the extension of a DNA chain during replication. *See* **2′,3′-dideoxyribonucleoside triphosphates.**

chalcones a group of pigments biogenetically related to anthocyans (*q.v.*). Chalcones give yellow to orange colors to the flowers of composites (*q.v.*).

Chambon's rule the generalization that the base sequences of all introns begin with GT and end with AG (except in tRNA genes).

character any detectable phenotypic property of an organism.

character displacement the exaggeration of species markers (visual clues, scents, mating calls, courtship rituals, etc.) or adaptations (anatomical, physiological, or behavioral) in sympatric populations relative to allopatric populations of related species. This phenomenon is attributed to the direct effects of natural selection intensifying allesthetic traits useful for species discrimination or for utilizing different parts of an ecological niche (thereby avoiding direct competition).

character states a suite of different expressions of a character in different organisms. These different states are said to be homologues. A character may have a minimum of two states (present/absent or primitive/derived) or have many states.

Chargaff's rules for the DNA of any species, the number of adenine residues equals the number of thymine residues; likewise, the number of guanines equals the number of cytosines; the number of purines (A + G) equals the number of pyrimidines (T + C). *See* **Appendix C,** 1950, Chargaff.

charged tRNA a transfer RNA molecule to which an amino acid is attached; also termed *aminoacylated tRNA*.

charon phages a set of 16 derivatives of bacteriophage lambda that are designed as cloning vectors. They were named by their originators (F. R. Blattner and 11 colleagues) after the old ferryman of Greek mythology who conveyed the spirits of the dead across the River Styx.

chase *See* **pulse-chase experiment.**

chasmogamous designating a plant in which fertilization takes place after the opening of the flower. *See* **cleistogamous.**

Chédiak-Steinbrinck-Higashi syndrome a defect in lysosome formation involving leukocytes and melanocytes; inherited as an autosomal recessive. Patients show decreased pigmentation of hair and eyes, abnormal susceptibility to infections, and tendencies to develop malignant lymphomas. A similar syndrome occurs in mice, cattle, and mink. In the mouse, a homologous mutation results in an impairment in the functioning of natural killer cells (*q.v.*). *See* **Aleutian mink.**

cheetah *See Acinonyx jubatus.*

chelating agent a compound made up of heterocyclic rings that forms a chelate with metal ions. Heme (*q.v.*) is an example of an iron chelate. The porphyrin ring in chlorophyll (*q.v.*) forms a magnesium chelate.

chelation the holding of a metal ion by two or more atoms of a chelating agent.

Chelicerata a subphylum of arthropods containing the species that have no antennae and possess pincerlike chelicerae as the first pair of appendages. *See* **classification.**

chemoautotrophy *See* **autotroph, methanogens.**

chemical elements Listed alphabetically by their symbols. *See* **periodic table.**

Ac	Actinium	Be	Beryllium
Ag	Silver	Bi	Bismuth
Al	Aluminum	Bk	Berkelium
Am	Americium	Br	Bromine
Ar	Argon	C	Carbon
As	Arsenic	Ca	Calcium
At	Astatine	Cd	Cadmium
Au	Gold	Ce	Cerium
B	Boron	Cf	Californium
Ba	Barium	Cl	Chlorine

Cm	Curium	O	Oxygen
Co	Cobalt	Os	Osmium
Cr	Chromium	P	Phosphorus
Cs	Cesium	Pa	Protactinium
Cu	Copper	Pb	Lead
Dy	Dysprosium	Pd	Palladium
Er	Erbium	Pm	Promethium
Es	Einsteinium	Po	Polonium
Eu	Europium	Pr	Praseodymium
F	Fluorine	Pt	Platinum
Fe	Iron	Pu	Plutonium
Fm	Fermium	Ra	Radium
Fr	Francium	Rb	Rubidium
Ga	Gallium	Re	Rhenium
Gd	Gadolinium	Rh	Rhodium
Ge	Germanium	Rn	Radon
H	Hydrogen	Ru	Ruthenium
He	Helium	S	Sulfur
Hf	Hafnium	Sb	Antimony
Hg	Mercury	Sc	Scandium
Ho	Holmium	Se	Selenium
I	Iodine	Si	Silicon
In	Indium	Sm	Samarium
Ir	Iridium	Sn	Tin
K	Potassium	Sr	Strontium
Kr	Krypton	Ta	Tantalum
La	Lanthanum	Tb	Terbium
Li	Lithium	Tc	Technetium
Lr	Lawrencium	Te	Tellurium
Lu	Lutetium	Th	Thorium
Md	Mendelevium	Ti	Titanium
Mg	Magnesium	Tl	Thallium
Mn	Manganese	Tm	Thulium
Mo	Molybdenum	U	Uranium
N	Nitrogen	V	Vanadium
Na	Sodium	W	Tungsten
Nb	Niobium	Xe	Xenon
Nd	Neodymium	Y	Yttrium
Ne	Neon	Yb	Ytterbium
Ni	Nickel	Zn	Zinc
No	Nobelium	Zr	Zirconium
Np	Neptunium		

chemiosmotic theory the concept that hydrogen ions are pumped across the inner mitochondrial membrane, or across the thylakoid membrane of chloroplasts, as a result of electrons passing through the electron transport chain (*q.v.*). The electrochemical gradient that results supplies the energy required for ATP formation.

chemostat an apparatus allowing the continuous cultivation of bacterial populations in a constant, competitive environment. Bacteria compete for a limiting nutrient in the medium. The medium is slowly added to the culture, and used medium plus bacteria are siphoned off at the same rate. The concentration of the limiting nutrient in the fresh medium determines the density of the steady-state population, and the rate at which the medium is pumped into the chemostat determines the bacterial growth rate. In chemostat experiments, environmental variables can be changed, one by one, to ascertain how these affect natural selection, or the environment can be held constant and the differential fitness of two mutations can be evaluated.

chemotaxis the attraction or repulsion of organisms by a diffusing substance.

chemotherapy the treatment of a disease with drugs of known chemical composition, which are assumed to be directly toxic to the pathogenic microorganisms.

chiasma (*plural* **chiasmata**) the cytological manifestation of crossing over; the cross-shaped points of junction between nonsister chromatids first seen in diplotene tetrads. *See* **Appendix C,** 1909, Janssens; 1929, Darlington.

chiasma interference the more frequent (in the case of negative chiasma interference) or less frequent (in the case of positive interference) occurrence of more than one chiasma in a bivalent segment than expected by chance.

chiasmata (*singular* **chiasma**) chromosomal sites where crossing over produces an exchange of homologous parts between nonsister chromatids. *See* **meiosis.**

chiasmatype theory the theory that crossing over between nonsister chromatids results in chiasma formation.

chicken *See* **Gallus domesticus.**

chimera an individual composed of a mixture of genetically different cells. In plant chimeras, the mixture may involve cells of identical nuclear genotypes, but containing different plastid types. In more recent definitions, chimeras are distinguished from mosaics (*q.v.*) by requiring that the genetically different cells of chimeras be derived from genetically different zygotes. *See also* **aggregation chimera, heterologous chimera, mericlinal chimera, periclinal chimera, radiation chimera.**

chimpanzee *See* **Pan.**

Chinchilla lanigera a rodent bred on commercial ranches for its pelt. Numerous coat color mutants are known.

Chironomus a genus of delicate, primitive, gnat-like flies that spend their larval stage in ponds and

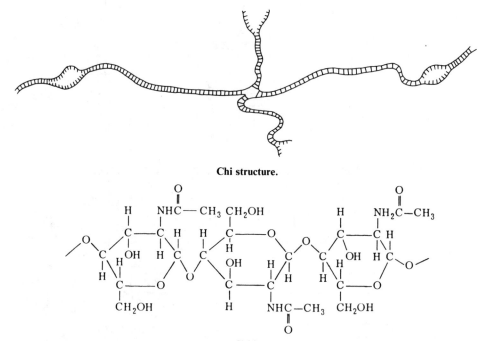

Chi structure.

Chitin.

slow streams. Nuclei from various larval tissues contain giant polytene chromosomes. The salivary gland chromosomes of *C. thummi* and *C. tentans* have been mapped, and the transcription processes going on in certain Balbiani rings (*q.v.*) have been studied extensively. See **Appendix C**, 1881, Balbiani; 1952, Beermann; 1960, Clever and Karlson.

chi sequence an octomeric sequence in *E. coli* DNA, occurring about once every 10 kb, acting as a "hotspot" for RecA-mediated genetic recombination.

chi-square test (x^2) a statistical procedure that enables the investigator to determine how closely an experimentally obtained set of values fits a given theoretical expectation. The relation between x^2 and probability is presented graphically on page 56. See **Appendix C**, 1900, Pearson.

chi structure a structure resembling the Greek letter χ, formed by cleaving a dimeric circle with a restriction endonuclease that cuts each DNA circle only once. The parental monomeric duplex DNA molecules remain connected by a region of heteroduplex DNA at the point where crossing over occurred. Thus, the identification of such chi structures provides evidence for cross-over events taking place between circular DNA molecules.

chitin a polymer of high molecular weight composed of N-acetylglucosamine residues joined together by beta glycosidic linkages between carbon atoms 1 and 4. Chitin is a component of the exoskeletons of arthropods.

Chlamydomonas reinhardi a species of green algae in which the interaction of nuclear and cytoplasmic genes has been extensively studied. The nuclear gene loci have been distributed among 18 linkage groups, and restriction maps are available for both chloroplast and mitochondrial DNAs. An alternative spelling, *C. reinhardtii*, occurs in the literature. See **Appendix A**, Protoctista, Chlorophyta; **Appendix C**, 1963, Sager and Ishida; 1970, Sager and Ramis.

chlorambucil a mutagenic, alkylating agent.

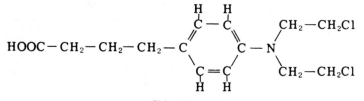

Chlorambucil.

55

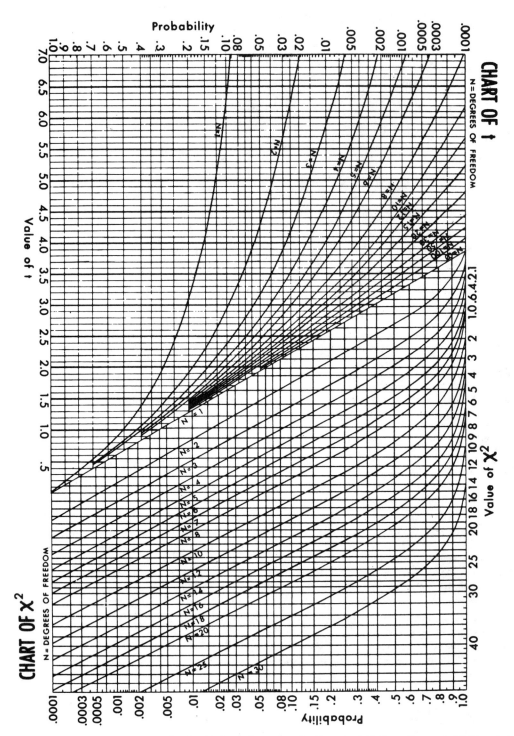

Charts giving the distribution of χ^2 and t. (Redrawn from *Genetics Notes* by J. F. Crow, Burgess Publishing Co.)

chloramphenicol an antibiotic produced by *Streptomyces venezuelae*. Chloramphenicol is a potent inhibitor of protein synthesis on the 70S ribosomes of prokaryotes. It attaches to the 50S ribosomal subunit and prevents the addition of an amino acid to the growing polypeptide chain. Chloroamphenicol does not bind to the 80S ribosomes of eukaryotes, but it does bind to the smaller ribosomes of the mitochondria present in eukaryotic cells. This is one of the lines of evidence that symbiotic prokaryotes were the ancestors of eukaryotic ribosomes. *See* **endosymbiont theory, ribosome, serial symbiosis theory, translation.**

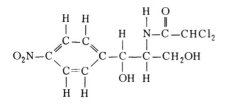

Chlorella a genus of green algae extensively used in studies of photosynthesis and its genetic control.

chlorenchyma tissue possessing chloroplasts.

chlorine an element universally found in small amounts in tissues. Atomic number 17; atomic weight 35.453; valence $1-$; most abundant isotope ^{35}Cl; radioisotopes ^{38}Cl, half-life 37 minutes, ^{39}Cl, half-life 55 minutes, radiation emitted—beta particles.

chlorolabe *See* **color blindness.**

chloromycetin chloramphenicol (*q.v.*).

chlorophyll a group of pigments that mediate photosynthesis. These include chlorophyll a and b, the green pigments found in the chloroplasts of plants. The structural formula for the chlorophyll a molecule and the appropriate dimensions of the porphyrin ring and phytol chain are illustrated. The chlorophyll b molecule differs from chlorophyll a in that the CH_3 indicated by the arrow is replaced by a CHO group. Other chlorophylls include chlorophyll c (found in the brown algae and some red algae), chlorophyll d (found in the red algae) and the bacteriochlorophylls (found in the green sulfur bacteria). *See* **Cyanobacteria.**

chloroplast the chlorophyll-containing, photosynthesizing organelle of plants. Chloroplasts are thought to be the descendants of endosymbiotic cyanobacteria. A typical chloroplast is shown on page 58. Each is surrounded by a double membrane and contains a system of internal thylakoid membranes. These form stacks of flattened discs called grana in which chlorophyll molecules are embedded. Chloroplasts contain DNA and can multiply. Replication of chloroplast DNA occurs throughout the cell cycle. Chloroplast DNAs are circular, like those of mitochondria, but many times larger. There are 40 to 80 DNA molecules per organelle. The DNA molecules form clusters in the stroma and are thought to attach to the inner membrane. The DNAs are devoid of histones. Chloroplasts contain 70S ribosomes and in this respect resemble bacterial ribosomes rather than those of the plant cytoplasm (*see* **ribosome**). The reproduction and functioning of chloroplasts is under the control of both nuclear genes and those of the organelle. Chloroplasts develop from protoplastids. These are small organelles surrounded by a double membrane. The inner one gives rise to a sparse internal membrane system from which the thylakoids develop. *See* **Appendix C**, 1837, von Mohl; 1883, Schimper; 1909, Correns and Bauer; 1951, Chiba; 1953, Finean *et al.*; 1962, Ris and Plaut; 1971, Manning and Richards; 1981, Steinbeck *et al.*; *Chlamydomonas reinhardi*, **chloroplast er.**

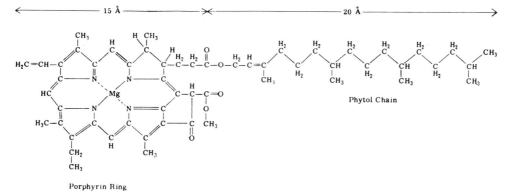

Chlorophyll.

57

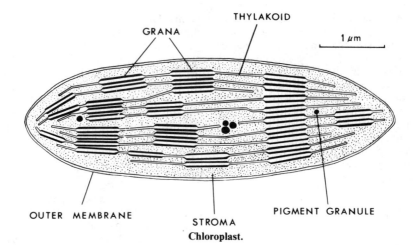

THYLAKOID

GRANA

1 μm

OUTER MEMBRANE

STROMA

PIGMENT GRANULE

Chloroplast.

chloroplast er chloroplasts embedded in a sheath of rough-surfaced endoplasmic reticulum. Structures of this sort are characteristic of organisms grouped within the Chromista (*q.v.*).

chlorosis failure of chlorophyll development.

CHO cell line A somatic cell line derived from Chinese hamster ovaries. The cells have a near diploid number, but over one-half of the chromosomes contain deletions, translocations, and other aberrations that have occurred during the evolution of the cell line. *See* **Cricetulus griseus.**

cholecystokinen a hormone secreted by the duodenum that causes gallbladder contraction.

cholesterol the quantitatively predominant steroid of man. A 27-carbon compound made up of a fused ring system. Cholesterol is a component of biomembranes and of the myelin sheaths that surround nerve axons. In insects, cholesterol serves as a precursor for ecdysone (*q.v.*). *See* **adrenal corticosteroid, familial hypercholesterolemia.**

cholinesterase *See* **acetylcholinesterase.**

chondriome all the mitochondria of the cell referred to collectively.

chondriosome a mitochondrion.

chondroitin sulfuric acid a mucopolysaccharide commonly found in cartilage. See figure on page 59.

chordamesoderm a layer of cells derived from the blastoporal lip that later in embryogenesis forms the mesoderm and notochord of the vertebrate embryo. The chordamesoderm acts as an organizer upon the overlying ectoderm to induce differentiation of neural structures.

Chordata the phylum of animals with a notochord, a hollow dorsal nerve cord, and gill slits at some developmental stage. *See* **Appendix A.**

chorea a nervous disorder, characterized by irregular and involuntary actions of the muscles of the extremities and face. *See* **Huntington disease.**

chorioallantoic grafting the grafting of pieces of avian or mammalian embryos upon the allantois of the chick embryo under the inner shell-membrane. The implant becomes vascularized from the allantoic circulation and continues development.

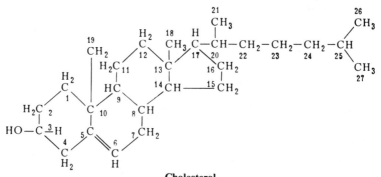

Cholesterol.

58

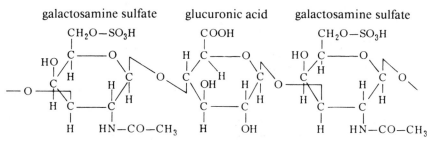

galactosamine sulfate glucuronic acid galactosamine sulfate

Chondroitin sulfuric acid.

chorion 1. the insect egg shell. 2. *see* **amnion.**

chorionic appendages the anterior, dorsal projections of the *Drosophila* egg shell that serve as breathing tubes when the egg is submerged.

chorionic gonadotropin (CG) a hormone produced by the placenta that continues to stimulate the production of progesterone by the corpus luteum and thus maintains the uterine wall in a glandular condition. This hormone is the one for which most pregnancy tests assay. For humans, it is abbreviated HCG.

chorionic villi sampling the harvesting of chorionic cells by introducing a catheter through the vagina and into the uterus until it touches the chorion, which surrounds the human embryo. Chorionic cells are the same genotype as that of the embryo, and therefore these can be used to detect enzymatic and karyotypic defects. Chorionic villi sampling can be performed between the 8th and 9th week of pregnancy. The technique has the advantage over amniocentesis (*q.v.*) of allowing fetal disorders to be detected earlier. Thus, when necessary, the pregnancy can be terminated at an earlier stage with less risk to the mother.

Christmas disease *See* **hemophilia.**

chromatid conversion a form of gene conversion (*q.v.*) rendered evident by identical sister-spore pairs in a fungal octad that exhibits a non-Mendelian ratio. For example, if an ordered octad of ascospores from a cross of + × m = (++) (++) (++) (mm), one chromatid of the m parental chromosome appears to have been converted to +. In *half-chromatid conversion,* an octad of (++) (++) (+m) (mm) indicates that one chromatid of the m parental chromosome was "half-converted." *See* **ordered tetrad.**

chromatid interference a deviation from the expected 1:2:1 ratio for the frequencies of 2-, 3-, and 4-strand double crossovers signaling a nonrandom participation of the chromatids of a tetrad in successive crossovers.

chromatids the two daughter strands of a dupli-

cated chromosome that are joined by a single centromere. Upon the division of the centromere, the sister chromatids become separate chromosomes. *See* **meiosis, mitosis.**

chromatin the complex of nucleic acids (DNA and RNA) and proteins (histones and nonhistones) comprising eukaryotic chromosomes. *See* **Appendix C, 1879, Flemming.**

chromatin-negative an individual (normally a male) whose cell nuclei lack sex chromatin.

chromatin-positive an individual (normally a female) whose cell nuclei contain sex chromatin. *See* **Barr body, sex chromatin.**

chromatograph the record produced by chromatography (*q.v.*). Generally applied to a filter paper sheet containing a grouping of spots that represent the compounds separated.

chromatography a technique used for separating and identifying the components from mixtures of molecules having similar chemical and physical properties. The population of different molecules is dissolved in an organic solvent miscible in water, and the solution is allowed to migrate through a stationary phase. Since the molecules migrate at slightly different rates, they are eventually separated. In *paper chromatography,* filter paper serves as the stationary phase. In *column chromatography (q.v.),* the stationary phase is packed into a cylinder. In *thin layer chromatography,* the stationary phase is a thin layer of absorbent silica gel or alumina spread on a flat glass plate. *See also* **affinity chromatography, counteracting chromatographic electrophoresis, gas chromatography, ion exchange column.** *See* **Appendix C, 1941, Martin and Synge.**

chromatophore submicroscopic particles isolated from photosynthetic bacteria that contain the photosynthetic pigments.

chromatosome a DNA protein complex consisting of a nucleosome (*q.v.*), the linker DNA segment, and its H1 histone. *See* **histones, nucleosome.**

Chromista the name proposed by T. Cavalier-

Smith for a kingdom to contain those eukaryotic species that show the following specific ultrastructural similarities. They all possess, at some time during their life cycles, cells that contain undulipodia covered with mastigonemes (*q.v.*) and/or chloroplast er (*q.v.*). Species belonging to the phyla Cryptophyta, Xanthophyta, Eustimatophyta, Bacillariophyta, Phaeophyta, Labyrinthulomycota, Hypochytridiomycota, and Oomycota fall into this group. *See* **Appendix A,** Protoctista; **classification, Protozoa.**

chromocenter a central aggregation of heterochromatic chromosomal elements in the *Drosophila* larval salivary-gland cell nucleus. The euchromatic chromosome arms extend from the chromocenter.

chromomere one of the serially aligned beads or granules of an eukaryotic chromosome, resulting from local coiling of a continuous DNA thread; best seen when most of the rest of the chromosome is relatively uncoiled as in the leptotene and zygotene stages of meiosis (*q.v.*). In polytene chromosomes (*q.v.*), the chromomeres lie in register and give the chromosome its banded appearance.

chromonema (*plural* **chromonemata**) the chromosome thread.

chromoneme the DNA thread of bacteria and their viruses.

chromoplast a carotenoid-containing plastid that colors ripe fruits and flowers.

chromosomal aberration an abnormal chromosomal complement resulting from the loss, duplication, or rearrangement of genetic material. *Intrachromosomal* or *homosomal* aberrations involve changes that occur in but one chromosome. Such aberrations include *deficiencies* and *duplications* that result in a reduction or increase in the number of loci borne by the chromosome. *Inversions* and *shifts* involve changes in the arrangement of the loci, but not in their number. In the case of an inversion a chromosomal segment has been deleted, turned through 180°, and reinserted at the same position on a chromosome, with the result that the gene sequence for the segment is reversed with respect to that of the rest of the chromosome. In the case of shift a chromosomal segment has been removed from its normal position and inserted (in the normal or reversed sequence) into another region of the same chromosome. *Interchromosomal* or *heterosomal* aberrations arise from situations in which non-homologous chromosomes are broken, and interchange occurs between the resulting fragments, producing a *translocation. See* **radiation-induced chromosomal aberrations.**

chromosomal mutation *See* **chromosomal aberration.**

chromosomal polymorphism the existence within a population of two or more different structural arrangements of chromosomal material (e.g., inversions, translocations, duplications, etc.).

chromosomal puff a localized swelling of a specific region of a polytene chromosome due to localized synthesis of DNA or RNA. Extremely large RNA puffs are called *Balbiani rings* (*q.v.*). *See* **Appendix C,** 1952, Beerman; 1959, Pelling; 1960, Clever and Karlson; 1961, Beerman; 1980, Gronemeyer and Pongs; **heat shock puffs.**

chromosomal RNA ribonucleic acid molecules associated with chromosomes during either division (e.g., primer, *q.v.*) or interphase (e.g., incomplete transcripts).

chromosomal sterility sterility from the lack of homology between the parental chromosomes in a hybrid.

chromosomal tubules microtubules of the spindle apparatus that originate at kinetochores of centromeres. The chromosomal tubules interpenetrate on the spindle with polar tubules (*q.v.*) and are hypothesized to slide by one another during anaphase as a consequence of making and breaking of cross bridges between them.

chromosome 1. in prokaryotes, the circular DNA molecule containing the entire set of genetic instructions essential for life of the cell. 2. in the eukaryotic nucleus, one of the threadlike structures consisting of chromatin (*q.v.*) and carrying genetic information arranged in a linear sequence. *See* **Appendix C,** 1883, Roux; 1888, Waldeyer.

chromosome arms the two major segments of a chromosome, whose length is determined by the position of the centromere. *See* **acrocentric, metacentric, submetacentric, telocentric.**

chromosome banding techniques there are four popular methods for staining human chromosomes. To produce *G-banding,* chromosomes are usually treated with trypsin and then stained with Giemsa. Most euchromatin stains lightly, and most heterochromatin stains darkly under these conditions. *C-bands* are produced by treating chromosomes with alkali and controlling the hydrolysis in a buffered salt solution. C-banding is particularly useful for staining and highlighting centromeres and polymorphic bands (especially those of meiotic chromosomes). With *Q-banding,* chromosomes are stained with a fluorochrome dye, usually quinacrine mustard or quinacrine dihydrochloride, and are viewed under untraviolet light. The bright bands correspond to the dark G-bands (with the exception of some of the polymorphic bands). Q-banding is especially useful for identifying the Y chromosome and polymorphisms that are not easily demonstrated by the G-banding procedure. *R-bands* are produced by treating chromosomes with heat in a phosphate buffer. They can then be stained with Giemsa to produce a pattern that is the reverse (hence the R in the term) of G-bands, thereby allowing the evaluation of terminal bands that are light after G-banding. Alternatively, chromosomes can be heated in buffer and then stained with acridine orange. When viewed under ultraviolet light, the bands appear in shades of red, orange, yellow, and green. They can also be photographed in color, but printed in black and white to reveal more distinctive R-bands. *See* **Appendix C,** 1970, Caspersson *et al.,* 1971; O'Riordan *et al.*

chromosome bridge a bridge formed between the separating groups of anaphase chromosomes because the two centromeres of a dicentric chromosome are being drawn to opposite poles. Such bridges may form as the result of single- or three-strand double-exchanges within the reverse loop of a paracentric inversion heterozygote (*see* **inversion**). They may also arise as radiation-induced chromosome aberrations (*q.v.*). Such chromosome bridges are always accompanied by acentric chromosome fragments. *See* **bridge-breakage-fusion-bridge cycle.**

chromosome condensation the process whereby eukaryotic chromosomes become shorter and thicker during prophase as a consequence of coiling and supercoiling of chromatic strands.

chromosome congression the movement of chromosomes to the spindle equator during mitosis.

chromosome diminution or elimination the elimination during embryogenesis of certain chromosomes from cells that form somatic tissues. Germ cells, however, retain these chromosomes.

chromosome jumping *See* **chromosome walking.**

chromosome loss failure of a chromosome to be included in a daughter nucleus during cell division. *See* **anaphase lag.**

chromosome map *See* **cytogenetic map, genetic map.**

chromosome polymorphism the presence in the same interbreeding population of one or more chromosomes in two or more alternative structural forms.

chromosome puff *See* **chromosomal puff.**

chromosome rearrangement a chromosomal aberration involving new juxtapositions of chromosomal segments; e.g., inversions, translocations.

chromosome scaffold when histones are removed from isolated metaphase chromosomes and these are centrifuged onto electron microscope grids, extremely long loops of DNA can be seen to project

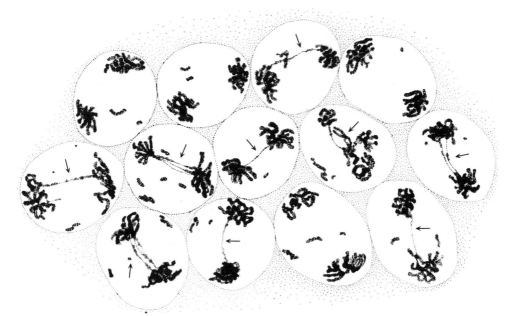

Chromosome bridges.

from an irregular mass (the scaffold) whose dimensions are similar to the original intact chromosome.

chromosome set   a group of chromosomes representing a genome (*q.v.*), consisting of one representative from each of the pairs characteristic of the somatic cells in a diploid species.

chromosome sorting   *See* **flow cytometry.**

chromosome substitution   replacement by a suitable crossing program of one or more chromosomes by homologous or homoeologous chromosomes from another source. This may be a different strain of the same species or a related species that allows hybridization. *See* **homoeologous chromosomes, homologous chromosomes.**

chromosome theory of heredity   the theory put forth by W. S. Sutton in 1902 that chromosomes are the carriers of genes and that their meiotic behavior is the basis for Mendel's laws (*q.v.*).

chromosome walking   the sequential isolation of clones carrying overlapping restriction fragments to span a segment of chromosome that is larger than can be carried in a phage or a cosmid vector. The technique is generally needed to isolate a locus of interest for which no probe is available but that is known to be linked to a gene that has been identified and cloned. This probe is used to screen a genome library. As a result, all fragments containing the marker gene can be selected and sequenced. The fragments are then aligned, and those segments farthest from the marker gene in both directions are subcloned for the next step. These probes are used to rescreen the genome library to select new collections of overlapping sequences. As the process is repeated, the nucleotide sequences of areas farther and farther away from the marker gene are identified, and eventually the locus of interest will be encountered. If a chromosomal aberration is available that shifts a particular gene that can serve as a molecular marker to another position on the chromosome or to another chromosome, then the chromosome walk can be shifted to another position in the genome. The use of chromosome aberrations in experiments of this type is referred to as *chromosome jumping.*

chromotrope   a substance capable of altering the color of a metachromatic dye (*q.v.*).

chronic exposure   radiation exposure of long duration. Applied to experimental conditions in which the organism is given either a continuous low level exposure or a fractionated dose.

chronocline   in paleontology, a character gradient in the time dimension.

chronospecies   a species that can be studied from its fossil remains through a defined period of time.

chrysalis   the pupa of a lepidopteran that makes no cocoon (*q.v.*).

chymotrypsin   a proteolytic enzyme from the pancreas that hydrolyzes peptide chains internally at peptide bonds on the carboxyl side of various amino acids, especially phenylalanine, tyrosine, and tryptophan.

Ci   abbreviation for curie (*q.v.*).

cilia (*singular,* **cilium**)   populations of thin, motile processes found covering the surface of ciliates or the free surface of the cells making up a ciliated epithelium. Each cilium arises from a basal granule in the superficial layer of cytoplasm. The movement of cilia propels ciliates through the liquid in which they live. The movement of cilia on a ciliated epithelium serves to propel a surface layer of mucus or fluid. *See* **axoneme.**

ciliate   a protozoan belonging to the phylum Ciliophora. *See* **Appendix A,** Protoctista.

circadian rhythm   a biological rhythm with a period of about 24 hours.

circular dichroism   the property of molecules to show differences in absorption between the clockwise and counterclockwise component vectors of a beam of circularly polarized light. Since helical molecules in solution often exhibit these properties, circular dichroism spectra have been used to study coiling changes of physiological significance, which various proteins can undergo. In the cases of chromatin fragments oriented in electric fields and exposed to circularly polarized ultraviolet light, measurements of circular dichroism allow conclusions to be drawn as to the way nucleosomes (*q.v.*) are stacked in 30 nm fibers.

circular linkage map   the linkage map characteristic of *Escherichia coli.* In preparation for genetic transfer during conjugation, the ring-shaped chromosome of an *Hfr* bacterium breaks in such a way that when the ring opens the *F* factor is left attached to the region of the chromosome destined to enter the *F⁻* cell last. Circular linkage maps have been constructed for several other bacteria and for certain viruses. *See also* **linkage map.**

circular overlap   the phenomenon in which a chain of continuous and intergrading populations of one species curves back until the terminal links overlap each other. Individuals from the terminal populations are then found to be reproductively isolated from each other; that is, they behave as if they belonged to separate species. A ring of races so formed is referred to as a *Rassenkreis.*

circumsporozoite protein   *See* **sporozoite.**

cis-acting locus   a genetic region affecting the activity of genes on that same DNA molecule. Cis-acting loci generally do not encode proteins, but rather serve as attachment sites for DNA-binding proteins. Enhancers, operators, and promoters are examples of cis-acting loci. *Contrast with* **trans-acting locus.**

cis dominance   the ability of a genetic locus to influence the expression of one or more adjacent loci in the same chromosome, as occurs in *lac* operator mutants of *E. coli.*

cisplatin   one of the most widely used antitumor drugs, especially effective for the management of testicular and ovarian cancers. When cisplatin binds to DNA, it loses two chloride ions and forms two platinum-nitrogen bonds with the N7 atoms of adjacent guanines on the same strand. This localized disruption of the double helix inhibits replication.

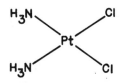

cisterna   a flattened, fluid-filled reservoir enclosed by a membrane. *See* **endoplasmic reticulum.**

cis, trans configurations   terminology that is currently used in the description of pseudoallelism. In the *cis* configuration both mutant recons are on one homologue and both wild-type recons are on the other ($a^1a^2/+ +$). The phenotype observed is wild type. In the *trans* configuration each homologue has a mutant and a nonmutant recon ($a^1 +/ + a^2$), and the mutant phenotype is observed. In the case of pseudoallelic genes, the terms *cis* and *trans* configurations correspond to the *coupling* and *repulsion* terminology used to refer to nonallelic genes. *See* **transvection effect.**

cis-trans test   a test used to determine whether two mutations of independent origin affecting the same character lie within the same or different cistrons. If the two mutants in the trans position yield the mutant phenotype, they are alleles. If they yield the wild phenotype, they represent mutations of different cistrons. However, different mutated alleles may represent cistrons with mutations at different sites. If these mutons are separable by crossing over, it is possible to construct a double mutant with the

mutant sites in the cis configuration ($m^1m^2/+ +$). Individuals of this genotype show the wild phenotype. *See* **pseudoalleles.**

cistron   originally the term referred to the DNA segment that specified the formation of a specific polypeptide chain (*see* **Appendix C,** 1955, Benzer). The definition was subsequently expanded to include the start and stop signals. In cases where a mRNA encodes two or more proteins, it is referred to as *polycistronic.* The proteins specified by a polycistron are often enzymes that function in the same metabolic pathway.

citric acid cycle (*also* **citrate cycle, Krebs cycle, tricarboxylic acid cycle)**   a cycle of enzyme-controlled reactions by which the acetyl-coenzyme A produced by the catabolism of fats, proteins, and carbohydrates is oxidized, and the energy released used to form ATP from ADP. The condensation of acetyl-coenzyme A with a four-carbon compound (oxaloacetic acid) yields the six-carbon compound citric acid for which the cycle is named. Through a further series of oxidative decarboxylations, citric acid is broken down to oxaloacetic acid, and so the cycle is completed. As a result, one molecule of activated acetate is converted to two CO_2 molecules, eight H atoms are taken up and oxidized to water, and twelve molecules of ATP are formed concurrently. The enzymes are localized in mitochondria (*q.v.*) *See* **glycolysis, cytochrome system.** (See diagram on page 64.)

citrullinuria   a hereditary disease in humans arising from a deficiency of the enzyme arginosuccinate synthetase. It is caused by a recessive gene on autosome 9. *See* **ornithine cycle.**

Citrus   a genus of fruit-bearing trees including *C. aurantium,* the bitter orange; *C. aurantifolia,* the lime; *C. paradisi,* the grapefruit; *C. limon,* the lemon; *C. reticulata,* the tangerine; *C. sinensis,* the sweet orange. The navel orange is a cultivar (*q.v.*) of the last-named species.

clade   **1.** in classification, any group of organisms that is defined by characters exclusive to all its members and that distinguish the group from all others. **2.** in evolutionary studies, a taxon or other group consisting of a single species and its descendents; a holophyletic group; a set of species representing a distinct branch on a phylogenetic tree.

cladistic evolution   splitting of a line of descent into two species. *Contrast with* **phyletic evolution.**

cladistics   a method of classification that attempts to reconstruct phylogenies utilizing only characters that are believed to be apomorphic (*q.v.*).

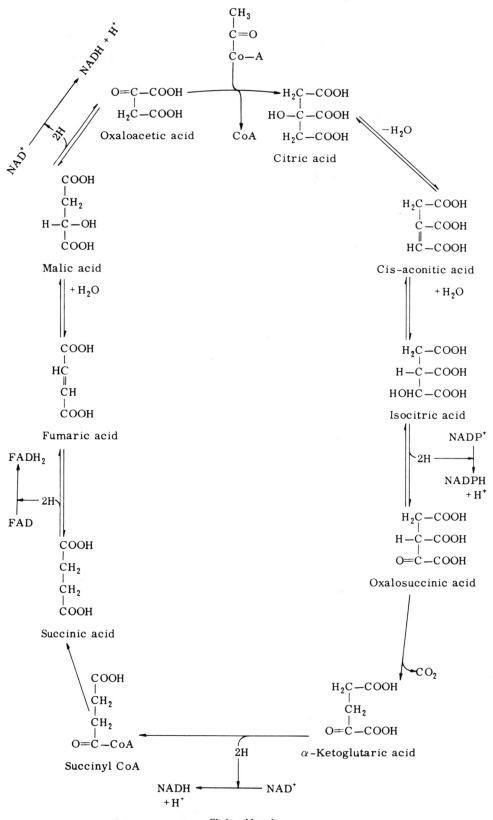

Citric acid cycle.

cladogenesis branching evolution; the splitting of a lineage into two or more lineages.

cladogram a branching diagram that displays the relationship between taxa in terms of their shared character states and attempts to represent the true evolutionary branchings of the lineage during its evolution from the ancestral taxon. In the accompanying cladogram, we start with ancestral species A, which is now extinct. It is characterized by six phenotypic characters (a–f) of taxonomic importance. The lineage undergoes speciation events (denoted by numbers 1 to 4) to produce five extant species, B–F. Each character (except f) undergoes a single change from the ancestral (plesiomorphic) state to a derived (apomorphic) state. The time and place of the change in the lineage is marked by letters carrying primes. Note that species E and F, which have the most recent common ancestor, share more apomorphic characters than species with more remote common ancestors. An ancestral character shared by several species is called symplesiomorphic, and a derived character shared by two or more species is called synapomorphic. Thus, character f is symplesiomorphic for the entire group, and character c is synapomorphic for species E and F.

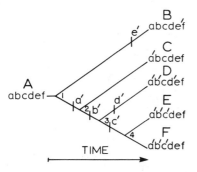

class a grouping used in the classification of organisms. It is a subdivision of a phylum, and it is in turn subdivided into orders.

classification 1. the process of grouping organisms on the basis of features they have in common, or on the basis of their ancestry, or both. 2. the resulting arrangement of living things into groups as a consequence of the above process. The traditional division of living organisms into two kingdoms (Plant and Animal) dates back to Linnaeus. This was supplanted by a system of five kingdoms (Bacteria, Protists, Fungi, Plants, and Animals) proposed by Whittaker. An updated variant of this system is used in **Appendix A**. More recently, a seven-kingdom system has been proposed by T. Cavalier-Smith. This system retains four of the five kingdoms but splits the protists into three separate kingdoms: Archaeozoa, Protozoa, and Chromista (all of which see).

class switching *See* **heavy chain class switching**.

clastogen a substance that causes chromosomal abnormalities.

clathrin *See* **receptor-mediated endocytosis**.

***ClB* technique** a technique used for detecting sex-linked lethal and viable mutations in *Drosophila melanogaster.* The name is derived from the X-chromosome used. It contains a *C*rossover suppressor (an inversion), a *l*ethal, and the dominant marker *B*ar eye.

cleavage the processes by which a dividing egg cell gives rise to all the cells of the organism. Cleavage in some species follows a definite pattern, where it is said to be determinate (permitting the tracing of *cell lineages*); in other species the pattern is lost after the first few cell divisions. *See* **contractile ring**.

cleavage furrow *See* **contractile ring**.

cleidoic egg an egg enclosed within a shell that is permeable only to gases.

cleistogamous designating a plant in which fertilization occurs within closed flowers, and therefore where self-pollination is obligatory. *See* **chasmogamous**.

cline a gradient of phenotypic and/or gene frequency change along a geographical transect of the population's range. *See* **isophene**.

clock mutants *Drosophila* mutants that show a disruption of the normal 24-hour circadian periodicity. Arrhythmic mutants are known, as well as mutants with short-periods (~19 hours) or long-periods (~28 hours). *See* **Appendix C**, 1971, Konopka and Benzer.

clomiphene any chemical that stimulates ovaries to release eggs.

clonal analysis the use of mosaics generated by genetic methods or surgical operations to investigate cell-autonomous and cell-nonautonomous developmental processes. Cell-autonomous genetic markers are used to label cells at an early stage in development, and the subsequent morphological fates of the offspring of these cells can then be followed. *See* **compartmentalization**.

clonal selection theory a hypothesis for explaining the specific nature of the immune response in which the diversity among various cells for the recognition of specific antigens exists prior to their exposure to the antigen. Subsequent exposure to a particular antigen causes the appropriate cells to undergo a clonal proliferation. *See* **Appendix C,** 1955, Jerne; 1959, Burnet.

clone **1.** a group of genetically identical cells or organisms all descended from a single common ancestral cell or organism by mitosis in eukaryotes or by binary fission in prokaryotes. **2.** genetically engineered replicas of DNA sequences.

cloned DNA any DNA fragment that passively replicates in the host organism after it has been joined to a cloning vector. Also called *passenger DNA.*

cloned library a collection of cloned DNA sequences representative of the genome of the organism under study.

cloning formation of clones or exact genetic replicas. *See* **gene cloning.**

cloning vector, cloning vehicle *See* **DNA vector, lambda cloning vector, plasmid cloning vector.**

clonotype the phenotype or homogenous product of a clone of cells.

close pollination the pollination of a flower with pollen of another flower on the same plant or of the same cone.

Clostridium the anaerobic, spore-forming, Gram-positive bacterial genus to which the species belong that cause tetanus, gangrene, and botulism.

club wheat *Triticum compactum* (N=21). *See* **wheat.**

clutch a nest of eggs.

C-meiosis colchicine-blocked meiosis. *See* **metaphase arrest.**

C-metaphase colchicine-blocked metaphase. *See* **metaphase arrest.**

C-mitosis colchicine-blocked mitosis. *See* **metaphase arrest.**

c-myc *See* ***myc.***

Cnidosporidia *See* **microsporidia.**

CNS abbreviation for central nervous system (the brain and spinal cord).

coadaptation the selection process that tends to accumulate harmoniously interacting genes in the gene pool of a population.

coated pit, vesicle *See* **receptor-mediated endocytosis.**

coat protein the structural protein making up the external covering of a virus.

cobalamin vitamin B_{12}, a coenzyme for a variety of enzymes. It is essential for deoxyribose synthesis. See figure on page 67.

cobalt a biological trace element. Atomic number 27; atomic weight 58.9332; valence 2, 3^+; most abundant isotope Co^{59}; radioisotope Co^{60}, half-life 5.2 years, radiation emitted—beta and gamma rays (used extensively as a gamma-ray source).

coconversion the concurrent correction of two sites during gene conversion.

cocoon **1.** a covering of silky strands secreted by the larva of many insects in which it develops as a pupa. Commercial silk is derived from the cocoon of *Bombyx.* **2.** a protective covering of mucus secreted about the eggs of some species.

code for the genetic code dictionary, *see* **amino acid, start codon, stop codon.**

code degeneracy *See* **degenerate code.**

coding strand the strand of a duplex DNA that has the same nucleotide sequence as mRNA except that T substitutes in DNA for U in RNA. The coding strand is also called the *sense* strand. The other strand, which is the actual template for mRNA synthesis, is the *anticoding* or *antisense* strand. *See* **Appendix C,** 1967, Taylor *et al.;* **transcription unit.**

coding triplet codon. *See* **amino acid.**

codominant designating genes when both alleles of a pair are fully expressed in the heterozygote. For example, the human being of AB blood group is showing the phenotypic effect of both I^A and I^B codominant genes.

codon (*also* **coding triplet**) the nucleotide triplet in messenger RNA that specifies the amino acid to be inserted in a specific position in the forming polypeptide during translation. A complementary codon resides in the cistron, specifying the mRNA in question. The codon designations are listed in the **amino acid** and **genetic code** entries.

coefficient of coincidence *See* **coincidence.**

coefficient of consanguinity the probability that two homologous genes drawn at random, one from each of the two parents, will be identical and therefore be homozygous in an offspring. The inbreeding coefficient of an individual is the same as the coefficient of consanguinity of its parents. *See* **Wright's inbreeding coefficient.**

coefficient of inbreeding Wright's inbreeding coefficient.

coefficient of kinship coefficient of consanguinity (*q.v.*).

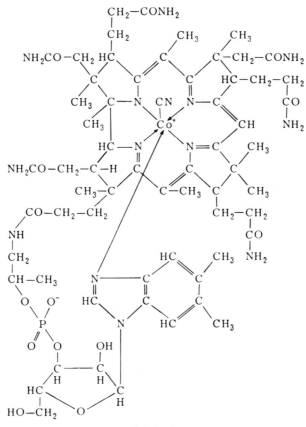

Cobalamin.

coefficient of parentage coefficient of consanguinity (*q.v.*).

coefficient of relationship (r) the proportion of alleles in any two individuals that are replicas inherited from a common ancestor.

coefficient of selection (s) *See* **selection coefficient.**

coelacanth *See* **living fossil.**

coelenterate an animal belonging to the phylum Cnidaria consisting mainly of marine organisms characterized by radial symmetry, a simple diploblastic body, a gastrovascular cavity, and usually showing metagenesis.

Coelomata a superphylum of animals having a coelom or body cavity formed in and surrounded by mesoderm. *See* **Appendix A.**

coenobium a colony of unicellular eukaryotes surrounded by a common membrane.

coenzyme an organic molecule that must be loosely associated with a given enzyme in order for it to function. Generally, the coenzyme acts as a donor or acceptor of groups of atoms that have been added or removed from the substrate. Well-known coenzymes include NAD, NADP, ATP, FAD, FMN, and coenzyme A. Some coenzymes are derivatives of vitamins.

coenzyme A a coenzyme made up of adenosine diphosphate, pantothenic acid, and mercaptoethylamine. The acetylated form of the coenzyme plays a central role in the citric acid cycle (*q.v.*). *See* formula on page 68.

coenzyme Q a molecule that functions as a hydrogen acceptor and donor in the electron transport chain (*q.v.*) Its formula is on page 68. *See* **cytochrome system (Q and QH₂).**

coevolution the evolution of one or more species in synchrony with another species as a consequence of their interdependence. Such a reciprocal adaptive evolution determines the patterns of host-plant utilization by insects. Also, parasites often evolve

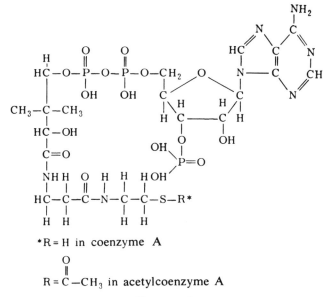

*R = H in coenzyme A

$$R = \overset{\displaystyle O}{\overset{\displaystyle \|}{C}} - CH_3 \text{ in acetylcoenzyme A}$$

Coenzyme A.

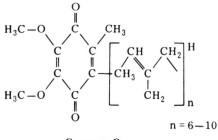

n = 6 − 10

Coenzyme Q.

and speciate in harmony with their hosts. *See* **Fahrenholz's rule.**

cofactor   a factor such as a coenzyme or a metallic ion required in addition to a protein enzyme for a given reaction to take place.

cognate tRNAs   transfer RNA molecules that can be recognized by aminoacyl-tRNA synthetase enzymes.

cohesive-end ligation   the use of DNA ligase to join double-stranded DNA molecules with complementary cohesive termini that base pair with one another and bring together 3'-OH and 5'-P termini. *Compare with* **blunt-end ligation.**

cohesive ends   *See* **restriction endonuclease, sticky ends.**

cohort   a group of individuals of similar age within a population.

coincidence, coefficient of   an experimental value

equal to the observed number of double crossovers divided by the expected number.

coincidental evolution   *See* **concerted evolution.**

cointegrate structure   the circular molecule formed by fusing two replicons, one possessing a transposon, the other lacking it. The cointegrate structure has two copies of the transposon located at both replicon junctions, oriented as direct repeats. The formation of a cointegrate structure is thought to be an obligatory intermediate in the transposition process. The donor molecule (containing the transposon) is nicked in opposite strands at the ends of the transposon by a site-specific enzyme. The recipient molecule is nicked at

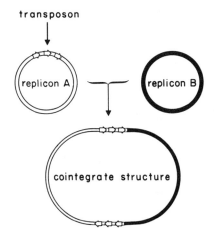

68

staggered sites. Donor and recipient strands are then ligated at the nicks. Each end of the transposon is connected to one of the single strands protruding from the target site, thereby generating two replication forks. When replication is completed, a cointegrate structure is formed, which contains two copies of the transposon oriented as direct repeats. An enzyme required in the formation of the cointegrate structure is called a *transposase*. The cointegrate can be separated into donor and recipient units each of which contains a copy of the transposon. This process is called *resolution* of the cointegrate, and it is accomplished by recombination between the transposon copies. The enzyme involved in resolution is called a *resolvase.*

coisogenic designating inbred strains of organisms that differ from one another only by a single gene as the result of mutation. *Contrast with* **congenic strains.**

coitus copulation; sexual intercourse in vertebrates.

Colcemid trademark for a colchicine derivative extensively used as a mitotic poison.

colchicine an alkaloid (see figure below) that inhibits the formation of the spindle and delays the division of centromeres. Colchicine is used to produce polyploid varieties of horticulturally important species. It is also used in medicine in the treatment of gout. Also used to stop mitosis at metaphase (when chromosomes are maximally condensed) for preparation of karyotypes. *See* **Appendix C, 1937, Blakeslee and Avery;** *Colchicum.*

Colchicum the genus of crocuses, including: *C. autumnale,* the autumn crocus, source of colchicine; *C. aureus,* the Dutch crocus; *C. sativus,* the Saffron crocus.

cold-sensitive mutant a gene that is defective at low temperature but functional at normal temperature.

coleoptile the first leaf formed during the germination of monocotyledons.

col factors bacterial plasmids that allow the cell to produce colicins (*q.v.*).

colicin any of a group of proteins produced by certain strains of *Escherichia coli* and related species that have bactericidal effects. Colicinogenic bacteria are immune to the lethal effects of their own colicins.

coliform designating a Gram-negative, lactose-fermenting rod related to *Escherichia coli.*

colinearity correspondence between the location of mutant sites within a bacterial cistron and the location of amino acid substitutions in its translational product. This correspondence is not complete in most eukaryotic genes because of the presence of nontranslated introns (*q.v.*) *See* **Appendix C, 1964, Sarabhai** *et al..*

coliphage a bacteriophage (*q.v.*) that parasitizes *Escherichia coli. See* **lambda phage**

collagen the most abundant of all proteins in mammals; the major fibrous element of skin, bone, tendon, cartilage, and teeth, representing one-quarter of the body's protein. It is the longest protein known and consists of a triple helix 3,000 Ångstroms long and 15 Ångstroms across. Five types of collagen are known, differing in amino acid sequence of the three polypeptide chains. In some molecules of collagen, the three chains are identical; in others, two of the chains are identical and the third contains a different amino acid sequence. The individual polypeptide chains are translated as longer precursors to which hydroxyl groups and sugars are attached. A triple helix is formed and secreted into the space between cells. Specific enzymes trim the ends of each helix. A mutation that produces exceptionally stretchable skin and loose-jointedness (Ehlers-Danlos syndrome) is near one end of the gene for the alpha-2-polypeptide chain and results in failure of that end to be trimmed off. *See* **cartilage.**

collagenase an enzyme that digests collagen.

collenchyma plant tissue composed of cells that fit

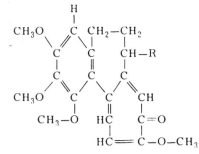

R = NHCOCH₃ for colchicine.

R = NHCH₃ for Colcemid.

Colchicine.

closely together and have thickened walls; in metazoans, undifferentiated mesenchyme cells lying in a gelatinous matrix.

Collinsia a genus of annual herbaceous plants containing about twenty species. The cytotaxonomy of the genus and the cytogenetics of interspecific hybrids have been extensively studied.

collision, inelastic the event occurring when a moving electron passes close to an atom. There is repulsion between the moving electron and an atomic electron sufficient to dislodge the latter from its nucleus.

colony in bacteriology, a contiguous group of single cells derived from a single ancestor and growing on a solid surface.

colony bank *See* **gene library.**

colony hybridization an *in situ* hybridization (*q.v.*) technique used to identify bacteria carrying chimeric vectors whose inserted DNA is homologous with the sequence in question. Colony hybridization is accomplished by transferring bacteria from a Petri plate to a nitrocellulose filter. The colonies on the filter are then lysed, and the liberated DNA is fixed to the filter by raising the temperature to 80°C. After hybridization with a labeled probe, the position of the colonies containing the sequence under study is determined by autoradiography (*q.v.*) *See* **Appendix C,** 1975, Grunstein and Hogness.

color blindness defective color vision in man due to an absence or reduced amount of one of the three visual pigments. The pigments *chlorolabe, erythrolabe,* and *cyanolabe* absorb green, red, and blue light, respectively. The pigments are made of three different opsins combined with vitamin A aldehyde. The green-blind individual suffers from *deuteranopia* and cannot make chlorolabe. The red-blind individual suffers from *protanopia* and cannot make erythrolabe. The blue-blind individual suffers *tritanopia* and cannot make cyanolabe.

Deuteranomaly, protanomaly, and *tritanomaly* are conditions caused by reduced amounts of the visual pigments in question, rather than their complete absence. The *protan* and *deutan* genes for red- and green-blindness, respectively, occupy different loci on the X chromosome. Blue blindness is quite rare and due to an autosomal gene. *See* **cone pigment genes.**

Columba livia the wild dove, ancestor to the domestic pigeon, *C. l. domestica. See* **Appendix A,** Aves, Columbiformes.

column chromatography the separation of organic compounds by percolating a liquid containing the compounds through a porous material in a cylinder. The porous material may be an ion exchange resin. *See* **chromatography.**

combinatorial association within a pool of immunoglobulin molecules, the association of molecules from any class of heavy chain with molecules from any type of light chain. Within a given immunoglobulin molecule, however, there is only one class of heavy and one type of light chain. *See* **immunoglobulin.**

combinatorial translocation for an immunoglobulin chain (heavy or light), the association of any variable region gene with any constant region gene within that same multigene family (*q.v.*). The two genes are brought together by interstitial translocation that may involve deletion of intervening genetic material. *See* **immunoglobulin.**

combining ability **1.** *general:* average performance of a strain in a series of crosses. **2.** *specific:* deviation of a strain's performance in a given cross from that predicted on the basis of its general combining ability.

comb shape a character influenced by two non-allelic gene pairs, *Rr* and *Pp,* in the chicken. One of the early examples of gene interaction (*q.v.*).

commaless genetic code successive codons that

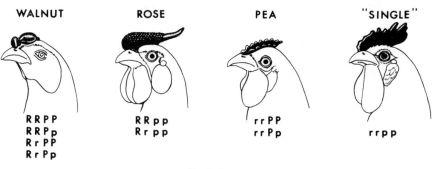

WALNUT   ROSE   PEA   "SINGLE"

RRPP   RRpp   rrPP
RRPp   Rrpp   rrPp   rrpp
RrPP
RrPp

Comb shape.

are contiguous and not separated by noncoding bases or groups of bases. Bacterial genes do not have introns (*q.v.*), and therefore the sequences of amino acids in polypeptides and of codons in the gene are colinear. Most eukaryotic genes contain coding regions called exons (specifying amino acids) interrupted by noncoding regions called introns, and in these situations the code is said to contain commas.

commensalism the association of organisms of different species without either receiving benefits essential or highly significant to survival. The species may live in the same shell or burrow or one may be attached to or live within the other. *See* **parasitism, symbiosis.**

community interacting populations of individuals belonging to different species and occupying a given region or a distinctive range of habitat conditions.

compartmentalization a phenomenon discovered in *Drosophila* by Garcia-Bellido during investigations of the distribution of genetically marked clones of cells during the development of imaginal discs. When such clones are analyzed it is found that they do not randomly overlap any area of the disc, but are confined to compartments and never cross the borders between compartments. A compartment contains all the descendants of a small number of founder cells called a *polyclone.* As development proceeds, large compartments are progressively split into smaller ones. The founder cells are related to each other by position, not ancestry; the progeny of the founder cells form the compartment under consideration, and no other cells contribute to it. The developmental pattern of each compartment is controlled by a selector gene. When selector genes mutate, the cells in a compartment may develop a pattern of cell types appropriate for another compartment. The homeotic mutants (*q.v.*) are examples of mutated selector genes. *See* **Appendix C,** 1973, Garcia-Bellido *et al.*

compatibility test any serological assay designed to detect whether blood or tissue from a prospective donor(s) can be transfused or transplanted without immunological rejection. *See* **cross matching, major histocompatibility complex.**

compensator genes sex-linked genes in *Drosophila* that when present in the female (in double dose) reduce the activity of her two doses of a given "primary sex-linked gene" so as to make the total phenotypic effect equivalent to that seen in the male (which has one dose of both the primary gene and the compensator gene).

competence 1. the state of a part of an embryo that enables it to react to a given morphogenetic stimulus by determination and subsequent differentiation in a given direction. 2. In a bacterium, competence refers to a stage of its life cycle during which the cell can naturally bind and internalize exogeneous DNA molecules, thereby allowing transformation (*q.v.*).

competition 1. the mutually exclusive use of the same limited resources (for example, food or a place to live, to hide, or to breed) by two or more organisms; 2. the mutually exclusive binding of two different molecules to the same site on a third molecule. For example, folic acid and aminopterin compete for combination sites on various folic acid-dependent enzymes.

competitive exclusion principle the assumption that two species with identical ecological requirements cannot coexist in the same niche in the same location indefinitely. One species will eventually supplant the other, unless they evolve adaptations that allow them to partition the niche and thereby reduce competition. Also known as *Gause's principle.*

complement a group of at least nine proteins (C1, C2, ... C9) normally found in vertebrate blood serum that can be activated immunologically (by antibodies of immunoglobulin classes IgG or IgM) or nonimmunologically (by bacterial lipopolysaccharides and other substances) through an alternate (properdin) pathway. Activation of the system involves sequential conversion of proenzymes to enzymes in a manner analogous to the formation of fibrin through the blood-clotting sequence. Some activated complement components enhance phagocytosis (opsonic activity), some make antigen-antibody-complement complexes "sticky" and cause them to become affixed to endothelial tissues or blood cells (serological adhesion or immune adherence), some cause release of vasoactive amines from blood basophils or tissue mast cells (anaphylotoxins), and some cause dissolution of bacterial cells (bacteriolysis). *See* **HLA complex.**

complementarity-determining region the segment of the variable region of an immunoglobulin or T cell receptor molecule that contains the amino acid residues that determine the specificity of binding to the antigen. *See* **paratope.**

complementary base sequence a sequence of polynucleotides related by the base-pairing rules. For example, in DNA a sequence A-G-T in one strand is complementary to T-C-A in the other strand. A given sequence defines the complementary sequence. *See* **deoxyribonucleic acid.**

complementary DNA *See* **cDNA.**

complementary factors   complementary genes.

complementary genes   nonallelic genes that complement one another. In the case of dominant complementarity, the dominant alleles of two or more genes are required for the expression of some trait. In the case of recessive complementarity, the dominant allele of either gene suppresses the expression of some trait (i.e., only the homozygous double recessive shows the trait).

complementary interaction   the production by two interacting genes of effects distinct from those produced by either one separately.

complementary RNA   *See* **cRNA.**

complementation   appearance of wild-type phenotype in an organism or cell containing two different mutations combined in a hybrid diploid or a heterokaryon. *See* **complementation test.**

complementation group   mutants lying within the same cistron; more properly called a *noncomplementation group.*

complementation map   a diagrammatic representation of the complementation pattern of a series of mutants occupying a short chromosomal segment. Mutually complementing mutants are drawn as nonoverlapping lines, and noncomplementing mutants are represented by overlapping, continuous lines. Complementation maps are generally linear, and the positions of mutants on the complementation and genetic maps usually agree. A complementation map is thought to show sites where lesions have been introduced into the polypeptides coded for by the DNA segment under study.

complementation test   the introduction of two mutant chromosomes into the same cell to see if the mutations in question occurred in the same gene. If the mutations are nonallelic, the genotype of the hybrid may be symbolized $(a +/+ b)$. The wild phenotype will be expressed, since each chromosome "makes up for" or "complements" the defect in the other. *See also* **allelic complementation, cis-trans test.**

complete dominance   *See* **dominance.**

complete linkage   a condition in which two genes on the same chromosome fail to be recombined and therefore are always transmitted together in the same gamete.

complete medium   in microbiology a minimal medium supplemented with nutrients (such as yeast extract, casein hydrolysate, etc.) upon which nutritional mutants can grow and reproduce.

complete metamorphosis   *See* **Holometabola.**

complete penetrance   the situation in which a dominant gene always produces a phenotypic effect or a recessive gene in the homozygous state always produces a detectable effect.

complete sex linkage   *See* **sex linkage.**

complexity   in molecular biology, the total length of different sequences of DNA present in a given preparation as determined from reassociation kinetics; usually expressed in base pairs, but the value may also be given in daltons or any other mass unit.

complex locus   a closely linked cluster of functionally related genes: e.g., the human hemoglobin gene complex or the *bithorax* locus in *Drosophila. See* **pseudoalleles.**

composite   a plant of the immense family Compositae, regarded as comprising the most highly developed flowering plants. It includes species such as the daisy and sunflower.

composite transposon   a DNA segment flanked on each end by insertion sequences (*q.v.*), either or both of which allow the entire element to transpose.

compound eye   the multifaceted kind of eye of insects. In *Drosophila,* each compound eye contains nearly 800 ommatidia.

compound X   *See* **attached X chromosomes.**

Compton effect   an attenuation process observed for x- or gamma-radiation. An incident photon interacts with an orbital electron of an atom to eject an electron and a scattered photon of less energy than the incident photon.

concanavalin A   a lectin (*q.v.*) derived from *Canavalia ensiformis,* the jack bean; abbreviated conA. The compound stimulates T lymphocytes to enter mitosis. *Compare with* **pokeweed mitogen.**

concatemer   the structure formed by concatenation of unit-sized components.

concatenation   linking of multiple subunits into a tandem series or chain, as occurs during replication of genomic subunits of phage lambda.

conception   *See* **syngamy.**

concordant   twins are said to be concordant for a given trait if both exhibit the trait.

conditional mutation   a mutation that exhibits wild phenotype under certain (permissive) environmental conditions, but exhibits a mutant phenotype under other (restrictive) conditions. Some bacterial mutants are conditional lethals that cannot grow above 45°C, but grow well at 37°C.

conditional probability   the probability of an event

that depends upon whether or not some other event has occurred previously or simultaneously. For example, the probability of a second crossover between two linked genes is usually greater as its distance from the first crossover increases. *See* **positive interference.**

conditioned dominance   referring to an allele that may or may not express itself depending upon environmental factors or the residual genotype. Thus, under different conditions the gene may behave as a dominant or as a recessive in heterozygotes.

cone   1. one of the elongate, unicellular photoreceptors in the vertebrate eye involved with vision in bright light and color recognition. **2.** ovule- or pollen-bearing scales, characteristic of conifers.

cone pigment genes   the genes that encode the pigments synthesized by the cone cells of the retina that are responsible for color vision. In humans, the blue cone pigment gene (BCP) resides on the long arm of chromosome 7. The red and blue pigments are encoded by genes near the tip of the long arm of the X chromosome. Since these genes show 98% sequence identity, they have arisen from a common ancestral gene. The green pigment gene is downstream of the red pigment gene. An upstream sequence, 4 kb from the red gene and 43 kb from the green gene, is essential for the activity of both. Deletions of this 580-bp element result in a rare form of X-linked colorblindness that is characterized by the absence of both red and green cone sensitivities. *See* **Appendix C,** 1986, Nathans *et al.;* **color blindness, rhodopsin.**

confusing coloration   a form of protective coloration that tends to confuse the predator by having a different appearance according to whether its possessor is at rest or in motion.

congenic strains   strains that differ from one another only with respect to a small chromosomal segment. Several congenic mouse strains differ only in the major or minor histocompatibility loci they contain. *Contrast with* **coisogenic.**

congenital   existing at birth. Congenital defects may or may not be of genetic origin.

congression   *See* **chromosome congression.**

congruence   in cladistics, congruent characters are shared features whose distribution among organisms fully corresponds to that in the same cladistic grouping. The most likely cladogram is the one that provides the maximum congruence between all of the characters involved.

conidium   an asexual haploid spore borne on an aerial hypha. In *Neurospora* two types of conidia

are found: oval *macroconidia,* which are multinucleate, and *microconidia,* which are smaller, spherical, and uninucleate. When incubated upon suitable medium, a conidium will germinate and form a new mycelium.

conifer   a tree or shrub that bears cones, including pines, firs, spruces, and sequoias.

conjugation   a temporary union of two single-celled organisms or hyphae with at least one of them receiving genetic material from the other. (1) In bacteria, the exchange is unidirectional with the "male" cell extruding all or a portion of one of its chromosomes into the recipient "female." (*See* **F factor**). (2) In *paramecium,* as shown in the illustration on page 74, entire nuclei are exchanged. (3) In fungi, conjugation also occurs between hyphae of opposite mating type to produce heterokaryons (*q.v.*).

conjugation tube   *See* **pilus.**

conjugon   a genetic element essential for bacterial conjugation. *See* **fertility factor.**

consanguinity   genetic relationship. Consanguineous individuals have at least one common ancestor in the preceding few generations.

consecutive sexuality   the phenomenon in which most individuals of a species experience a functional male phase when young, and later change through a transitional stage to a functional female phase. A situation common in some molluscs.

consensus sequence   synonymous with canonical sequence (*q.v.*).

conservative recombination   breakage and reunion of preexisting strands of DNA in the absence of DNA synthesis.

conservative replication   an obsolete model of DNA replication in which both old complementary polynucleotides are retained in one sibling cell, while the other gets the two newly synthesized strands. *Compare with* **semiconservative replication.**

conservative substitution   replacement of an amino acid in a polypeptide by one with similar characteristics; such substitutions are not likely to change the shape of the polypeptide chain, e.g., substituting one hydrophobic amino acid for another.

conserved sequence   a sequence of nucleotides in genetic material or of amino acids in a polypeptide chain that either has not changed or that has changed only slightly during an evolutionary period of time. Conserved sequences are thought to gen-

Conjugation.

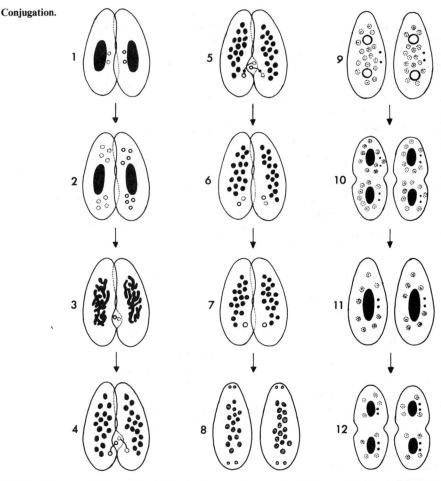

Nuclear changes accompany conjugation in *Paramecium aurelia*. (1) Two parental animals, each with one macronucleus and two diploid micronuclei. (2) Formation of eight haploid nuclei from the micronuclei of each conjugant. (3) Seven nuclei in each conjugant disappear, the remaining nucleus resides in the paroral cone; the macronucleus breaks up into fragments. (4–7) The nuclei in the paroral cones divide mitotically, forming "male" and "female" gamete nuclei. The female nuclei pass into the interior of the parental animals, while the male nuclei are transferred to the partners. Male and female haploid nuclei fuse. (8) Each fusion nucleus divides twice mitotically. (9) Two of the four nuclei so formed differentiate into macronuclear anlagen (white circles) while the other two become micronuclei. (10–12) Each micronucleus divides, and transverse fission of the exconjugants produces four animals (two of which are shown in 11). Fragments of the old macronucleus are gradually lost. In (12) tranverse fission begins in the animals seen previously in (11). (Redrawn from *Genetics of Paramecium aurelia* by G. H. Beale, 1954, published by Cambridge University Press.)

erally regulate vital functions and therefore have been selectively preserved during evolution.

conspecific belonging to the same species.

constant region *See* **immunoglobulin.**

constitutive enzyme an enzyme that is always produced irrespective of environmental conditions. *See* **Appendix C,** 1937, Karström.

constitutive gene a gene whose activity depends only on the efficiency of its promoter in binding RNA polymerase.

constitutive heterochromatin *See* **heterochromatin.**

constitutive mutation a mutation that results in an increased constitutive synthesis by a bacterium of several functionally related, inducible enzymes. Such a mutation either modifies an operator gene so that the repressor cannot combine with it or modifies the regulator gene so that the repressor is not formed. *See* **regulator gene.**

constriction an unspiralized region of a metaphase chromosome. Kinetochores and nucleolus organizers are located in such regions.

contact inhibition the cessation of cell movement on contact with another cell. It is often observed when freely growing cells, tissue cultured on a Petri plate, come into physical contact with each other. Cancer cells lose this property and tend to pile up in tissue culture to form multilayers called *foci.*

containment in microbiology, measures taken to diminish or prevent the infection of laboratory workers by the products of recombinant DNA techology and the escape of such products from the laboratory. *Biological containment* is accomplished by using genetically altered bacteria, phages, and plasmids that are unable to carry out certain essential functions (e.g., growth, DNA replication, DNA transfer, infection, and propagation) except under specific laboratory conditions. *Physical containment* is accomplished by design and use of special facilities and laboratory procedures such as limited access, safety hoods, aerosol control, protective clothing, pipeting aids, etc.

continental drift the concept that the world's continents once formed a part of a single mass and have since drifted into their present positions. The modern concept of plate tectonics (*q.v.*) has refined the theory of continental drift by placing the continents on larger sections of the earth's crust (tectonic plates) that are in motion. Continental drift began in the Proterozoic era; the continents collided to form a giant land mass named Pangaea in the late Permian period, and redispersed in the Mesozoic era. *See* **Appendix C,** 1912, Wegener; **geologic time divisions.**

continental island an island assumed to have been once connected to a neighboring continent. *Contrast with* **oceanic island.**

continuous distribution a collection of data that yield a continuous spectrum of values. For example, measurements such as height of plant or weight of fruit, carried out one or more decimal places. *See* **discontinuous distribution.**

continuous fibers the microtubules that connect the two poles of the mitotic apparatus, as distinct from traction fibers and astral fibers. *See* **mitotic apparatus.**

continuous variation the phenotypic variation exhibited by quantitative traits that vary by imperceptible degrees from one extreme to another. In human populations, phenotypes like body weight, height, and intelligence show continuous variation. *See* **discontinuous variation, quantitative inheritance.**

contractile ring a transitory organelle that, during late anaphase and telophase, assumes the form of a continuous equatorial annulus beneath the plasma membrane of the cleavage furrow. The ring is composed of an array of actin microfilaments aligned circumferentially along the equator of the cell. An interaction between cytoplasmic myosin and these actin molecules causes them to slide past one another, closing down the contractile ring, and producing the cleavage furrow.

control a standard of comparison; a test or experiment established as a check of other experiments, performed by maintaining identical conditions except for the one varied factor, whose causal significance can thus be inferred.

controlled pollination a common practice in plant hybridization of bagging the pistillate flowers to protect them from undesired pollen. When the pistillate flowers are in a receptive condition, they are dusted with pollen of a specified type.

controlling elements a class of genetic elements that renders target genes unstably hypermutable, as in the *Dissociation-Activator* system of corn (*q.v.*). They include receptors and regulators. The receptor element is a mobile genetic element that, when inserted into the target gene, causes it to become inactivated. The regulator gene maintains the mutational instability of the target gene, presumably by its capacity to release the receptor element from the target gene and thus return that locus to its normal function. *See* **transposable elements.**

controlling gene one that can switch cistrons on and off. *See* **regulator gene.**

convergence the evolution of unrelated species occupying similar adaptive zones, resulting in structures bearing a superficial resemblance (for example, the wings of birds and insects).

convergent evolution *See* **convergence.**

conversion *See* **gene conversion.**

Cooley anemia thalassemia (*q.v.*).

coordinated enzymes enzymes whose rates of production vary together. For example, the addition of lactose to the medium causes the coordinated induction of beta-galactosidase *and* beta-galactoside permease in *Escherichia coli.* Such enzymes are produced by cistrons of the same operon. *See* **regulator genes.**

Cope's "law of the unspecialized" a theory that

the evolutionary novelties associated with new major taxa are more likely to originate from a generalized, rather than from a specialized, member of an ancestral taxon.

Cope's rule the general tendency for animals to increase in body size during the course of phyletic evolution (*q.v.*).

copia elements transposable elements of *Drosophila* existing as a family of closely related base sequences that code for abundant mRNAs. There are usually between 20 and 60 copia elements per genome. The actual number depends on the *Drosophila* strain employed. The copia elements are widely dispersed among the chromosomes, and the sites occupied vary between strains. Each copia element ranges in size from 5–9 kb, and it carries direct terminal repeats about 280 bp long. Several *Drosophila* mutations have been found that result from the insertion of copia-like elements.

copolymer a polymeric molecule containing more than one kind of monomeric unit. For example, copolymers of uridylic and cytidylic acids (poly UC) are used as synthetic messengers.

copper a biological trace element. Atomic number 29; atomic weight 63.45; valance 1^+, 2^+; most abundant isotope ^{63}Cu; radioisotope ^{64}Cu, half-life 12.8 hours, radiations emitted—gamma rays, electrons, and positrons.

Coprinus radiatus a basidiomycete fungus that is of considerable genetic interest because of the existence of dikaryotic and diploid strains.

copy-choice hypothesis an explanation of genetic recombination based on the hypothesis that the new strand of DNA alternates between the paternal and maternal strands of DNA during its replication. *See* **Belling's hypothesis.**

copy DNA synonymous with complementary DNA. *See* **cDNA.**

copy error a mutation resulting from a mistake during DNA replication.

cordycepin 3′-deoxyadenosine, an inhibitor of the polyadenylation of RNA.

core 1. the region of a nuclear reactor containing the fissionable material. 2. synaptonemal complex.

core DNA the segment of DNA in a nucleosome (*q.v.*) that wraps around a histone octamer.

core granule RNP granules in the ommatidia of *Drosophila*. The xanthomatins and drosopterins are normally bound to these granules.

core particle a structural unit of eukaryotic chromosomes revealed by digestion with micrococcal nuclease to consist of a histone octamer and a 146 bp segment of DNA. *See* **nucleosome.**

corepressor in repressible genetic systems, the small effector molecule (usually an end product of a metabolic pathway) that inhibits the transcription of genes in an operon by binding to a regulator protein (aporepressor). Also called a *repressing metabolite.*

corm a swollen vertical underground stem base containing food material and bearing buds. It can function as an organ of vegetative reproduction. The crocus and gladiolus have corms.

corn *Zea mays,* the most valuable crop plant grown in the United States, it ranks along with wheat, rice, and potatoes, as one of the four most important crops in the world. Corn is generally classified in five commercial varieties on the basis of kernel morphology: (1) Dent corn (var. *indentata*), the most common variety of field corn; the kernel is indented, from drying and shrinkage of the starch in the summit of the grain. (2) Flint corn (var. *indurata*), the kernel is completely enclosed by a horny layer and the grain is therefore smooth and hard; flint corn is the fastest to mature. (3) Sweet corn (var. *saccharata*), which is grown for human consumption, is picked when it is filled with a milky fluid, before the grain hardens, (4) Popcorn (var. *everta*), the ear is covered with small kernels that are enclosed by a tough coat; when heated the contained moisture is turned to steam and the kernel explodes. (5) Flour corn (var. *amylacea*), the ear contains soft, starchy kernels; it requires a long growing season and is therefore grown primarily in the tropics. *See* **double cross, hybrid corn, *Zea mays.***

corpus allatum an endocrine organ in insects that synthesizes the allatum hormone (*q.v.*). In the larvae of cyclorrhaphous diptera, the corpus allatum forms part of the ring gland (*q.v.*).

corpus cardiacum an endocrine organ in insects consisting of a central bundle of axons enveloped by cortical cells. Axons from the corpus cardiacum enter the corpus allatum. Most axons associated with the corpus cardiacum have their cell bodies in the pars intercerebralis. The cortical cells and many of the axons contain numerous neurosecretory spheres (*q.v.*).

corpus luteum a mass of yellowish tissue that fills the cavity left after the rupture of the mature ovum from the mammalian ovary.

correction in a hybrid DNA sequence, the replacement (i.e., by excision and repair) of illegitimate nucleotide base pairs by bases that pair properly.

correlated response the change in one character occurring as an incidental consequence of the selection for a seemingly independent character. For example, reduced fertility may accompany selection for increased bristle number in *Drosophila. See* **pleiotropy.**

correlation the degree to which statistical variables vary together. It is measured by the *correlation coefficient (r),* which has a value from zero (no correlation) to -1 or $+1$ (perfect negative or positive correlation, respectively).

corridor a migration route allowing easy dispersal for certain species.

corticosterone one of a family of adrenal cortical hormones influencing glucose metabolism.

corticotropin *See* **adrenocorticotropic hormone.**

Corynebacterium diphtheriae the bacterium causing diphtheria. *See* **lysogenic conversion.**

COS cells a monkey cell line that has been transformed by an SV40 viral genome containing a defective origin of viral replication. When introduced into COS cells, recombinant RNAs containing the SV40 origin and a foreign gene should replicate many copies.

cosmic rays high-energy particulate and electromagnetic radiations originating outside the earth's atmosphere.

cosmid plasmid vectors designed for cloning large fragments of eukaryotic DNA. The term signifies that the vector is a plas*mid* into which phage lambda *cos* sites (*q.v.*) have been inserted. As a result the plasmid DNA can be packaged in a phage coat *in vitro. See* **Appendix C,** 1977, Collins and Holm.

cos sites *cohe*sive end sites, nucleotide sequences that are recognized for packaging a phage DNA molecule into its protein capsule.

cot the point (symbolized by $C_0t_{1/2}$) in a reannealing experiment where half of the DNA is present as double-stranded fragments; also called the *half reaction time.* If the DNA fragments contain only unique DNA sequences and are similar in length, then $C_0t_{1/2}$ varies directly with DNA complexity (*q.v.*). *See* **reassociation kinetics.**

cotransduction the simultaneous transduction of two or more genes because the transduced element contains more than one locus.

cotransformation 1. the simultaneous transfor-

mation of two or more bacterial genes; the genes cotransformed are inferred to be closely linked because transforming DNA fragments are usually small. Also called *double transformation.* **2.** in molecular biology, introduction of two physically unlinked sets of genes, one of which codes for a selectable marker, into a cell. This technique is useful in animal cells in which the isolation of cells transformed with a gene that does not code for a selectable marker has been problematic.

cotton *See* ***Gossypium.***

Coturnix coturnix japonica the Japanese quail, a small bird used as a laboratory animal.

cot value *See* **cot.**

cotyledon the leaf-forming part of the embryo in a seed. Cotyledons may function as storage organs from which the seedling draws food, or they may absorb and pass on to the seedling nutrients stored in the endosperm. Once the cotyledon is exposed to light, it develops chlorophyll and functions photosynthetically as the first leaf.

counteracting chromatographic electrophoresis a group of methods for purifying specific molecules from a mixture by the application of two counteracting forces; specifically, the chromatographic flow of a solute down a separation column vs. solute electrophoresis in the opposite direction.

countercurrent distribution apparatus an automated apparatus used for separating mixtures. The method takes advantage of differences in the solubilities of the components of the mixture in two immiscible solvents. An example of the usefulness of the technique is in the separation of different transfer RNA molecules.

counterselection a technique used in bacterial conjugation experiments to allow recovery of recombinant F$^-$ cells, while at the same time selecting against (preventing growth of) Hfr donor cells. For example, suppose the Hfr donor strain is susceptible to an antibiotic (such as streptomycin) and can synthesize histidine; the streptomycin locus must be so far from the origin of chromosome transfer that the mating pairs, which inevitably break apart, have separated before the *str* locus has been transferred. Suppose further that the recipient F$^-$ cell cannot make histidine (*his$^-$*) but is resistant to the antibiotic (*Str^r*). Only His$^+$*Str^r* recombinants can survive on a medium lacking histidine and containing streptomycin. The desired gene (*His$^+$* in this case) is called a *selected marker;* the gene that prevents growth of the male (*str^s* in this case) is called the *counterselective marker.*

coupled reactions chemical reactions having a common intermediate and therefore a means by which energy can be transferred from one to the other. In the following pair of enzyme catalyzed reactions, glucose-1-phosphate is the common intermediate that is formed in the first reaction and used up in the second:

$$ATP + glucose \rightarrow ADP + glucose\text{-}1\text{-}phosphate$$
$$glucose\text{-}1\text{-}phosphate + fructose \rightarrow sucrose + phosphate$$

A molecule of sucrose is synthesized from glucose and fructose at the expense of the energy stored in ATP and transferred by glucose-1-phosphate.

coupled transcription-translation a characteristic of prokaryotes wherein translation begins on mRNA molecules before they have been completely transcribed.

coupling, repulsion configurations when both nonallelic mutants are present on one homologue and the other homologous chromosome carries the plus alleles ($a\ b/ + +$), the genes are said to be in the *coupling* configuration. The *repulsion* configuration refers to a situation in which each homologue contains a mutant and a wild-type gene ($a + / + b$). *See* **cis, trans configurations.**

courtship ritual a characteristic genetically determined behavioral pattern involving the production and reception of an elaborate sequence of visual, auditory, and chemical stimuli by the male and female prior to mating. Such rituals are interpreted as ensuring that mating will occur only between individuals of opposite sex and the same species.

cousin the son or daughter of one's uncle or aunt. The children of siblings are *first* cousins. Children of *first* cousins are *second* cousins. Children of second cousins are *third* cousins, etc. The child of a first cousin is a *cousin once removed* of his father's or mother's first cousin.

covalent bond a valence bond formed by a shared electron between the atoms in a covalent compound.

covariance a statistic employed in the computation of the correlation coefficient between two variables; the covariance is the sum of $(x\text{-}\bar{x})\ (y\text{-}\bar{y})$ over all pairs of values for the variables x and y, where $\bar{x}$ is the mean of the x values and $\bar{y}$ is the mean of all y values.

cpDNA chloroplast DNA. Also abbreviated *ctDNA. See* **chloroplast.**

Craniata the subphylum of the Chordata containing animal species with a true skull. *See* **Appendix A.**

Crassostrea virginica *See* **Pelecypoda.**

Creeper a dominant autosomal gene in the chicken causing embryonic death when homozygous. Heterozygotes show malformations of the limbs.

Crepis a genus of weedy herbs much studied with respect to the evolution of karyotype (*q.v.*).

Cretaceous the most recent of the Mesozoic periods, during which the dinosaurs continued to diversify. Angiosperms, birds, and mammals persisted. At the end of the Cretaceous, there was a mass extinction involving 70% of all animal species. The dinosaurs became extinct. The continents formed from Pangaea were now widely separated. *See* **continental drift, geologic time divisions, impact theory.**

cretinism a stunting of bodily growth and mental development in humans due to a deficiency of thyroid hormones. Hereditary cretinism, which is often accompanied by goiter (*q.v.*) and deafness, consists of a group of metabolic disorders that results in a failure in the formation of sufficient thyroxine and triiodothyronine. The defects include the inability of the thyroid gland to accumulate sufficient iodine, to convert it into organically bound iodine, and to couple iodotyrosines to form iodothyronines. All hereditary defects in thyroid hormonogenesis are inherited as autosomal recessives. *See* **thyroid hormones.**

Cricetulus griseus the Chinese hamster. The rodent is a favorite for cytogenetic studies because of its small chromosome number (N = 11). A total of about 40 genetic loci have been assigned to specific chromosomes. *See* **CHO cell line.**

cri du chat syndrome *See* **cat cry syndrome.**

Crigler-Najjar syndrome a rare hereditary defect in bilirubin metabolism inherited as an autosomal recessive. Patients lack hepatic bilirubin UDP-glucuronyl transferase. This enzyme functions to conjugate bilirubin with glucuronic acid prior to biliary excretion. In the absence of the enzyme, excess bilirubin builds up in all tissues causing jaundice, brain damage, and death.

crisis period the time interval of a primary cell culture, following a number of cell divisions, during which most secondary progeny die even though culture conditions are adequate to initiate a new primary culture of low cell density from a fresh isolate. *See* **Hayflick limit, tissue culture.**

criss-cross inheritance referring to the passage of sex-linked traits from mother to son and from father to daughter.

cristae elaborate invaginations of the inner mitochondrial membrane.

CRM cross-reacting material (*q.v.*).

cRNA synthetic transcripts of a specific DNA molecule or fragment, made by an *in vitro* transcription system. This cRNA can be labeled with radioactive uracil and then used as a probe (*q.v.*).

Cro-Magnon man *Homo sapiens sapiens* living in the upper Pleistocene. Cro-Magnon replaced the Neanderthal (*q.v.*) throughout its range.

cro **repressor** a regulatory protein product of the *cro* gene of bacteriophage lambda of *E. coli*. The *cro* gene encodes a polypeptide containing 66 amino acids. These associate in pairs to form the active repressor. The *cro* dimer binds to six specific operator sites on the lambda chromosome. These operators all contain the same 17 base pair sequence and at least 14 of these are critical for *cro* binding. Studies of the three-dimensional structure of the dimer show that it fits into two adjacent grooves of the duplex DNA molecule. *See* **Appendix C,** 1981, Anderson *et al.; ***antirepressor, regulator gene.**

cross in higher organisms, a mating between genetically different individuals of opposite sex. In microorganisms, genetic crosses are often achieved by allowing individuals of different mating types to conjugate. In viruses, genetic crossing requires infecting the host cells with viral particles of different genotypes. The usual purpose of an experimental cross is to generate offspring with new combinations of parental genes. *See* **backcross, conjugation, dihybrid, E_1, F_1, I_1, monohybrid cross, P_1, parasexuality, test cross.**

cross-agglutination test one of a series of tests commonly employed in blood typing in which erythrocytes from a donor of unknown type are mixed with sera of known types.

crossbreeding outbreeding (*q.v.*).

cross-fertilization union of gametes that are produced by different individuals. *Compare with* **self-fertilization.**

cross hybridization (molecular) hybridization of a probe (*q.v.*) to a nucleotide sequence that is less than 100% complementary.

cross-induction the induction of vegetative phage replication in lysogenic bacteria in response to compounds transferred from UV-irradiated F^+ to non-irradiated F^- bacteria during conjugation.

crossing over the exchange of genetic material between homologous chromosomes. Crossing over is characterized by positive interference and by reciprocality; that is, wild-type and double mutants are produced simultaneously in the same tetrad from a *trans* heterozygote. *See* **Appendix C,** 1912, Morgan; 1913, Tanaka; 1931, Stern, Creighton and McClintock; 1964, Holliday; 1971, Howell and Stern; **intragenic recombination.**

crossing over within an inversion *See* **inversion.**

cross-linking formation of covalent bonds between a base in one strand of DNA and an opposite base in the complementary strand by mitotic poisons such as the antibiotic mitomycin C or the nitrite ion.

cross-matching. *See* **cross-agglutination test.**

crossopterygian a lobe-finned bony fish; one group of which was ancestral to the amphibians. *See* **living fossil.**

crossover fixation the spreading of a mutation in one member of a tandem gene cluster through the entire cluster as a consequence of unequal crossing over.

crossover region the segment of a chromosome lying between any two specified marker genes.

crossover suppressor a gene, or an inversion (*q.v.*), that prevents crossing over in a pair of chromosomes. *Gowen's crossover suppressor* gene on chromosome 3 of *Drosophila* prevents the formation of synaptonemal complexes.

crossover unit a 1% crossover value between a pair of linked genes.

cross-pollination the pollination of a flower with pollen from a flower of a different genotype.

cross-reacting material any nonfunctional protein reactive with antibodies directed against its functional counterpart. For example, some patients with classical hemophilia (*q.v.*) produce a CRM that reacts with anti-AHF serum, but this protein has lost its ability to take part in the blood-clotting process.

cross-reaction, serological union of an antibody with an antigen other than the one used to stimulate formation of that antibody; such cross-reactions usually involve antigens that are stereochemically similar or those that share antigenic determinants.

cross reactivation *See* **multiplicity reactivation.**

crown gall disease *See* ***Agrobacterium tumefaciens.***

crozier the hook formed by an ascogenous hypha

of *Neurospora* or related fungi previous to ascus development. The hook is formed when the tip cell of an ascogenous hypha grows back upon itself. Within the arched portion of the hypha, cell walls are subsequently laid down in such a way that three cells are formed. The terminal cell of the branchlet is uninucleate, the penultimate one is binucleate, and the antipenultimate one is uninucleate. Fusion of the haploid nuclei of different mating types in the penultimate cell occurs, and it enlarges to form the ascus in which meiosis occurs.

CRP cyclic AMP receptor protein. *See* **catabolite activating protein (CAP).**

cruciform structure a cross-shaped configuration of DNA produced by complementary inverted repeats pairing with one another on the same strand instead of with its normal partner on the other strand. *See* **palindrome.**

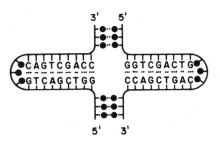

cryostat a device designed to provide low-temperature environments in which operations (like sectioning frozen tissues) may be carried out under controlled conditions.

cryptic coloration coloration patterns of organisms that render them inconspicuous against their natural backgrounds, hence less visible to potential predators.

cryptic gene a gene that has been silenced by a single nucleotide substitution, that is present at a high frequency in a population, and that can be reactivated by a single mutational event.

cryptic satellite a satellite DNA sequence that cannot be separated from the main-band DNA by density gradient ultracentrifugation. Cryptic satellite DNA can be isolated from the main-band DNA by its unique features (e.g., by the more rapid reannealing of the highly repeated segments that constitute the satellite).

cryptic species phenotypically similar species that never form hybrids in nature. *See* **sibling species.**

cryptoendomitosis somatic polyploidization taking place within an intact nuclear envelope. No stages comparable to the normal mitotic cycle are

observed, but the DNA content increases in multiples of the haploid value.

cryptogam a spore-bearing rather than a seed-bearing plant. In older taxonomy a member of the Cryptogamia, including the ferns, mosses, algae, and fungi. *See* **phanerogam.**

Cryptozoic a synonym for Precambrian (*q.v.*)

crystallins a family of structural proteins in the lens of the vertebrate eye. However, some crystallins play an enzymatic role in other tissues. For example, in reptiles and birds a form of crystallin is found in heart muscle, where it functions as a lactic dehydrogenase.

c-src a cellular gene, present in various vertebrates, that hybridizes with *src,* the oncogene of the Rous sarcoma virus (*q.v.*). The *c-src* genes code for *pp60c-src* proteins that resemble *pp60v-src* proteins in their enzymatic properties.

ctDNA chloroplast DNA. Also abbreviated *cpDNA. See* **chloroplast.**

C-terminus that end of the peptide chain that carries the free alpha carboxyl group of the last amino acid. By convention, the structural formula of a peptide chain is written with the C-terminus to the right. *See* **translation.**

"C"-type particles a group of RNA viruses with similar morphologies under the electron microscope, having a centrally placed, spherical RNA-containing nucleoid. These viruses are associated with many sarcomas and leukemias. The "C" refers to "cancer."

Cucumis a genus of nearly 40 species including several of considerable economic importance, such as the cucumber *(C. sativus)* and the muskmelon *(C. melo).* Considerable genetic information is available for both these species.

Cucurbita a genus of about 27 species, including 5 that are extensively cultivated: *C. pepo,* summer squash; *C. mixta,* cushaws; *C. moschata,* winter squash; *C. maxima,* Hubbard squash; and *C. ficifolia,* Malabar gourds. Most genetic information is available for *C. pepo* and *C. maxima.*

Culex pipiens the most widely distributed species of mosquito in the world. The genetics of insecticide resistance has been intensively studied in this species. Giant polytene chromosomes occur in the salivary gland and Malpighian tubule cells of larvae.

cull to pick out and discard inferior animals or plants from a breeding stock.

cultigen a plant that is known only under cultivation and whose place and method of origin is unknown.

cultivar a variety of plant produced through selective breeding by humans and maintained by cultivation.

curie the quantity of a radioactive nuclide disintegrating at the rate of 3.700×10^{10} atoms per second. Abbreviated Ci. 1 Ci = 3.7×10^{10} Bq.

cut a double-strand incision in a duplex DNA molecule. *Compare with* **nick**.

cut-and-patch repair repair of damaged DNA molecules by the enzymatic excision of the defective single-stranded segments and the subsequent synthesis of new segments. Using the complementary strand as a template, the correct bases are inserted and are interlinked by a DNA polymerase. A DNA ligase joins the two ends of the "patch" to the broken strand to complete the repair. *See* **AP endonuclease, repair synthesis**.

cuticle the chitinous, acellular outer covering of insects.

C value the amount of DNA (e.g., expressed in picograms per cell) comprising the haploid genome for a given species. *See* **Appendix C**, 1948, Boivin *et al.*

C value paradox the paradox that there is often no correlation between the C values of species and their evolutionary complexity. For example, the C values for mammals fall into a narrow range (between 2 and 3 pg). By contrast, the C values for amphibia vary from 1 to 100 pg. However, the minimum C values reported for species from each class of eukaryotes does increase with evolutionary complexity. It follows that in species with C values above the expected range, a large fraction of the DNA must have functions other than coding for proteins, such as acting as controlling elements, or else it is functionless, *selfish DNA* (*q.v.*).

CVS chorionic villi sampling (*q.v.*).

Cyanobacteria a phylum in the kingdom Eubacteria (*see* **Appendix A**). These bacteria are characterized by the thylakoids in which molecules of chlorophyll *a* are embedded. In the older literature, these bacteria were misclassified as blue-green algae and placed in the phylum Cyanophyta. The ancestors of present-day cyanobacteria were the dominant life form in the Proterozoic era, and the oxygen they generated from photosynthesis caused a transformation some two billion years ago of the earth's atmosphere from a reducing to an oxidizing one. The serial symbiosis theory (*q.v.*) derives chloroplasts from cyanobacteria. *See* **chlorophyll, stromatolites**.

cyanocobalamin cobalamin.

cyanogen bromide $Br-C\equiv N$, a reagent used for splitting polypeptides at methionine residues; commonly used in studies of protein structure and the determination of amino acid sequences.

cyanolabe *See* **color blindness**.

Cyanophyta *See* **Cyanobacteria**.

cyclically permuted sequences DNA sequences of the same length containing genes in the same linear order, but starting and ending at different positions, as in a circle. In T4 DNA, each phage contains a different cyclically permuted sequence that is also terminally redundant. Cyclic permutation is a property of a population of phage DNA molecules, whereas terminal redundancy is a property of an individual phage DNA molecule. *See* **headful mechanism, terminal redundancy**.

cyclical selection selection in one direction followed by selection in the opposite direction resulting from cyclical environmental fluctuations, such as seasonal temperature changes. If the generation time is short relative to the environmental cycle, different genotypes will be selected at different times, and the population will remain genetically inhomogeneous.

cyclic AMP adenosine monophosphate with the phosphate group bonded internally to form a cyclic molecule; generated from ATP by the enzyme adenylcyclase; abbreviated cAMP. Likewise, guanosine monophosphate (GMP) can become a cyclic molecule by a phosphodiester bond between 3′ and 5′ atoms. Cyclic AMP has been shown to function as an acrasin in slime molds and to be active in the regulation of gene expression in both prokaryotes and eukaryotes. In *Escherichia coli*, cyclic AMP is required for the transcription of certain operons. *See* **adenylcyclase, catabolite repression**.

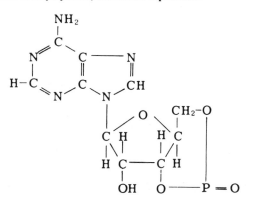

cycloheximide an inhibitor of translation on 80S ribosomes; an antibiotic synthesized by *Streptomyces griseus.*

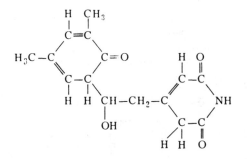

cyclophosphamide an immunosuppressive drug (*q.v.*).

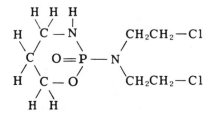

cyclorrhaphous diptera flies belonging to the suborder Cyclorrhapha, which contains the most highly developed flies. It includes the hover flies, the drosophilids, house flies, blow flies, etc.

cyclosis cytoplasmic streaming.

cyclotron *See* **accelerator.**

cys cysteine. *See* **amino acid.**

cysteine a sulfur-bearing amino acid found in biological proteins. It is important because of its ability to form a disulfide cross-link with another cysteine, either in the same or between different polypeptide chains. *See* **amino acid, cystine, insulin.**

cystic fibrosis (CF) the most common hereditary disease of Caucasians, CF is due to a recessive mutant gene on chromosome 7. In the United States, the frequency of homozygotes is 1/2,000, while heterozygotes make up about 5% of the population. The CF gene has been cloned and shown to be approximately 250 kb long. It contains 24 exons, which together encode a protein containing 1,480 amino acids. Some 70% of all CF patients carry the same mutation, a deletion of a specific codon that causes the CF protein to lack a phenylalanine at position 508. The CF protein is thought to control the transport of chloride ions through cell membranes. *See* **Appendix C, 1989, Tsui** *et al.*

cystine a derived amino acid formed by the oxi-

dation of two cysteine thiol side chains, which join to form a disulfide covalent bond. Such bonds play an important role in stabilizing the folded configurations of proteins. *See* **cysteine, insulin.**

cystoblast a mitotically active germarial cell that forms the interconnected cystocytes of the *Drosophila* ovary. *See* **nurse cells.**

cystocyte a daughter cell formed by the division of a cystoblast in the *Drosophila* germarium. *See* **nurse cells, pro-oocyte.**

cytidine *See* **nucleoside.**

cytidylic acid *See* **nucleotide.**

cytochalasin B a mold antimetabolite that prevents cells from undergoing cytokinesis. *See* **actin, contractile ring.**

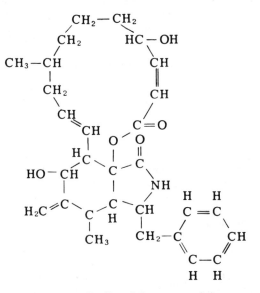

cytochromes a family of heme-containing proteins that function as electron donors and acceptors during the chains of reactions that occur during respiration and photosynthesis. Electron transport depends upon the continued oxidation and reduction of the iron atom contained in the center of the porphyrin prosthetic group (*see* **heme**). The first cytochrome is thought to have arisen about two billion years ago, and the genes that encode cytochromes have been modified slowly by base substitutions since then. The cytochromes were the first group of proteins for which amino acid sequence data allowed the construction of an evolutionary tree. *See* **Appendix C, 1963, Margoliash.**

cytochrome system a chain of coupled oxidation/reduction reactions that transports the electrons produced during the oxidations occurring in the cit-

$$\left\{\begin{matrix}NADH_2\\NAD\end{matrix}\right\}\xrightarrow[\;]{2e}\left(\begin{matrix}FADH_2\\FAD\end{matrix}\right)\xrightarrow[\;]{2e}\left(\begin{matrix}QH_2\\Q\end{matrix}\right)\xrightarrow[\;]{2e}\left(\begin{matrix}2Fe^{2+}\\CYT\ b\\2Fe^{3+}\end{matrix}\right)\xrightarrow[\;]{2e}\left(\begin{matrix}2Fe^{2+}\\CYT\ c\\2Fe^{3+}\end{matrix}\right)\xrightarrow[\;]{2e}\left(\begin{matrix}2Fe^{2+}\\CYT\ a\\2Fe^{3+}\end{matrix}\right)\xrightarrow[\;]{2e}\left(\begin{matrix}2Fe^{2+}\\CYT\ a_3\\2Fe^{3+}\end{matrix}\right)\xrightarrow[\;]{2e}\left(\begin{matrix}H_2O\\\tfrac{1}{2}O_2\end{matrix}\right)$$

electrons from citric acid cycle

Cytochrome system.

ric acid cycle (*q.v.*) to the final hydrogen and electron acceptor, oxygen, to form water. The molecules involved in this chain are NAD (*q.v.*), FAD (*q.v.*), coenzyme Q (*q.v.*), and cytochromes b, c, a, and a_3. The sequence of reactions is diagrammed above.

cytogenetic map a map showing the locations of genes on a chromosome.

cytogenetics the science that combines the methods and findings of cytology and genetics.

cytohet an eukaryotic cell containing two genetically different types of a specific organelle; the term is an abbreviation for *cyto*plasmically *het*erozygous. For example, in the single-celled alga *Chlamydomonas,* the frequency of rare cytohets (containing chloroplasts from both parents) can be greatly increased by treatment of one parent (mating type +) with ultraviolet light.

cytokinesis cytoplasmic division as opposed to karyokinesis (*q.v.*) *See* **cleavage, contractile ring.**

cytokinins a family of N-substituted derivatives of adenine (*q.v.*) synthesized mainly in the roots of higher plants. Cytokinins (also called kinins and phytokinins) promote cell division and the synthesis of RNA and protein. The first cytokinin to have its structure elucidated was kinetin (*see* **Appendix C,** 1956, Miller). The first cytokinin obtained from a plant was zeatin. It was isolated from maize kernels in 1964.

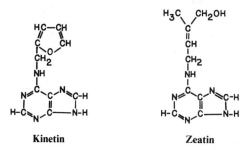

Kinetin   Zeatin

cytological hybridization synonymous with *in situ* hybridization (*q.v.*).

cytological map a diagrammatic representation of the physical location of genes at specific sites, generally on dipteran giant polytene chromosomes or on human mitotic chromosomes.

cytology the branch of biology dealing with the structure, function, and life history of the cell. *See* **Appendix C,** 1838, Schleiden and Schwann; 1855, Virchow; 1896, Wilson.

cytolysis the dissolution of cells.

cytophotometry quantitative studies of the localization within cells of various organic compounds using microspectrophotometry. Cytophotometric techniques are employed, for example, to determine changes in the DNA contents of cells throughout their life cycle. *See* **microspectrophotometer.**

cytoplasm the protoplasm exclusive of that within the nucleus (which is called nucleoplasm).

cytoplasmic inheritance non-Mendelian heredity involving replication and transmission of extrachromosomal genetic information found in organelles such as mitochondria and chloroplasts or in intracellular parasites such as viruses; also called *extranuclear inheritance. See* **Appendix C,** 1909, Correns and Bauer; **mtDNA lineages.**

cytoplasmic male sterility pollen abortion due to cytoplasmic factors that are maternally transmitted, but that act only in the absence of pollen-restoring genes. Such sterility can also be transmitted by grafting. This finding implicates a plant virus as the agent responsible for pollen abortion.

cytoplasmic matrix *See* **microtrabecular lattice.**

cytoplast the structural and functional unit of an eukaryotic cell formed by a lattice of cytoskeletal proteins to which are linked the nucleus and the cytoplasmic organelles.

cytosine *See* **bases of nucleic acids.**

cytosine deoxyriboside *See* **nucleoside.**

cytoskeleton an internal skeleton that gives the eukaryotic cell its ability to move, to assume a characteristic shape, to divide, to undergo pinocytosis, to arrange its organelles, and to transport them from one location to another. The cytoskeleton contains microtubules, microfilaments, and intermediate filaments.

cytosol the fluid portion of the cytoplasm exclusive of organelles; synonymous with hyaloplasm. *See* **cell fractionation.**

cytostatic referring to any agent that suppresses cell multiplication and growth.

cytotaxis the ordering and arranging of new cell structure under the influence of preexisting cell structure. The information controlling the three-dimensional architecture of the eukaryotic cell is thought to reside in the structure of the cytoplasmic ground substance. Evidence for this comes from microsurgical experiments on *Paramecium.* Cortical segments reimplanted with inverted polarity result in a changed pattern that is inherited through hundreds of generations. *See* **microtrabecular lattice.**

cytotoxic T lymphocyte a lymphocyte that binds to a foreign cell and kills it. Such lymphocytes recognize target cells on the basis of the antigenic properties of their class I histocompatibility molecules. *See* **helper T lymphocyte, T lymphocyte.**

D

d **1.** dextrorotatory. **2.** the dalton unit.

2,4D 2,4 dichlorophenoxyacetic acid (*q.v.*).

dalton a unit equal to the mass of the hydrogen atom (1.67×10^{-24} g). Abbreviated d.

Danaus plexippus the Monarch butterfly (*q.v.*); the model mimicked by the Viceroy. *See* **Batesian mimicry.**

dark-field microscope a microscope designed so that the entering center light rays are blacked out and the peripheral rays are directed against the object from the side. As a result, the object being viewed appears bright upon a dark background.

dark reactivation repair of mutagen-induced genetic damage by enzymes that do not require light photons for their action. *See* **photoreactivating enzyme.**

Darwinian evolution *See* **Darwinism.**

Darwinian fitness synonymous with adaptive value (*q.v.*).

Darwinian selection synonymous with natural selection (*q.v.*).

Darwinism the theory that the mechanism of biological evolution involves natural selection of adaptive variations. *See* **gradualism,** *Origin of Species.*

Darwin's finches *See* **Geospizinae.**

Datura stramonium the Jimson weed, a species in which classic studies on polysomy have been done. *See* **Appendix C,** 1920, 1922, Blakeslee *et al.*

dauermodification an environmentally induced phenotypic change in a cell that survives in the generative or vegetative descendants of the cell in the absence of the original stimulus. However, with time the trait weakens and eventually disappears.

daughter cells (nuclei) the two cells (nuclei) resulting from division of a single cell (nucleus). Preferably called *sibling* or *offspring cells* (nuclei).

day-neutral referring to a plant in which flowering is not controlled by photoperiod. *See* **phytochrome.**

DBM paper diazobenzyloxymethyl paper that binds all single-stranded DNA, RNA, and proteins by means of covalent linkages to the diazonium group; used in situations where nitrocellulose blotting is not technically feasible. *See* **Appendix C,** 1977, Alwine *et al.*

DEAE-cellulose diethylaminoethyl-cellulose, a substituted cellulose derivative used in bead form for chromatography of acidic or slightly basic proteins at *p*H values above their isoelectric point.

deamination the oxidative removal of NH_2 groups from amino acids to form ammonia.

decarboxylation the removal or loss of a carboxyl group from an organic compound and the formation of CO_2.

decay constant disintegration constant.

decay of variability the reduction of heterozygosity because of the loss and fixation of alleles at various loci accompanying genetic drift.

deciduous **1.** designating trees whose leaves fall off at the end of the growing season, as opposed to evergreen. **2.** designating teeth that are replaced by permanent teeth.

decoy protein *See* **sporozoite.**

dedifferentiation the loss of differentiation, as in the vertebrate limb stump during formation of a blastema.

defective virus a virus that is unable to reproduce in its host without the presence of another "helper" virus.

deficiency in cytogenetics, the loss of a microscopically visible segment of a chromosome. In a structural heterozygote (containing one normal and one deleted chromosome), the nondeleted chromosome forms an unpaired loop opposite the deleted segment when the chromosomes pair during meiosis. *See* **Appendix C,** 1917, Bridges.

deficiency loop in polytene chromosomes, deficiency loops allow one to determine the size of the segment missing. The illustration below shows a portion of the X chromosome from the nucleus of a salivary gland cell of a *Drosophila* larva structurally heterozygous for a deficiency. Note that

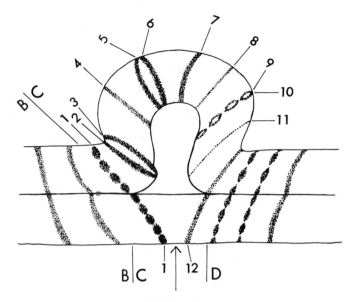

Deficiency loop.

bands C2–C11 are missing from the lower chromosome.

defined medium   a medium for growing cells, tissues, or multicellular organisms in which all the chemical components and their concentrations are known.

definitive host   the host in which a parasite attains sexual maturity.

deformylase   an enzyme in prokaryotes that removes the formyl group from the NH_2-terminal amino acid; fMet is never retained as the NH_2-terminal amino acid in functional polypeptides. *See* **start codon.**

degenerate code   one in which each different word is coded by a variety of symbols or groups of letters. The genetic code is said to be degenerate because more than one nucleotide triplet codes for the same amino acid (*q.v.*).

degrees of freedom   the number of items of data that are free to vary independently. In a set of quantitative data, for a specified value of the mean, only $(n\text{-}1)$ items are free to vary, since the value of the nth item is then determined by the values assumed by the others and by the mean.

dehiscent   designating fruit that opens when ripe to release seeds.

delayed dominance   *See* **dominance.**

delayed hypersensitivity   a cell-mediated immune response manifested by an inflammatory skin response 24–48 hours after exposure to antigen. *Compare with* **immediate hypersensitivity.**

delayed Mendelian segregation   *See* **Limnea peregra.**

deletion   the loss of a segment of the genetic material from a chromosome. The size of the deletion can vary from a single nucleotide to sections containing a number of genes. If the lost part is at the end of a chromosome, it is called a terminal deletion. Otherwise, it is called an intercalary deletion.

deletion mapping   1. the use of overlapping deletions to localize the position of an unknown gene on a chromosome or linkage map. 2. the establishment of gene order among several phage loci by a series of matings between point mutation and deletion mutants whose overlapping pattern is known. Recombinants cannot be produced by crossing a strain bearing a point mutant with another strain carrying a deletion in the region where the point mutant resides. *See* **Appendix C,** 1938, Slizynska; 1968, Davis and Davidson.

deletion method   a method of isolating specific messenger RNA molecules by hybridization with DNA molecules containing genetic deletions.

deletion-substitution particles   a specialized transducing phage in which deleted phage genes are substituted by bacterial genes.

delta chain   a component of hemoglobin A_2. *See* **hemoglobin.**

delta ray the track or path of an electron ejected from an atomic nucleus when an ionizing particle passes through a detection medium, especially through a photographic emulsion.

delta T50H the difference between the temperature at which DNA homoduplexes and DNA heteroduplexes undergo 50% dissociation. The statistic is often used to measure the genetic relationship between the nucleotide sequences of two or more species. A delta T50H value can be converted into an absolute time interval if the fossil record can provide an independent dating estimate. In primates, a delta T50H value of 1 equals about 11 million years. If repetitive sequences have been removed from the DNAs, then a delta T50H value of 1 represents about a 1% difference in single copy genes between the samples. *See* **DNA clock hypothesis, reassociation kinetics.**

deme a geographically localized population within a species.

denaturation the loss of the native configuration of a macromolecule resulting from heat treatment, extreme *p*H changes, chemical treatment, etc. Denaturation is usually accompanied by loss of biological activity. Denaturation of proteins often results in an unfolding of the polypeptide chains and renders the molecule less soluble. Denaturation of DNA leads to changes in many of its physical properties, including viscosity, light scattering, and optical density. This "melting" occurs over a narrow range of temperatures and represents the dissociation of the double helix into its complementary strands. The midpoint of this transition is called the melting temperature. *See* T_m.

denaturation map a map, obtained through electron microscopy using the Kleinschmidt spreading technique (*q.v.*), of a DNA molecule that shows the positions of denaturation loops. These are induced by heating the molecules to a temperature where segments held together by $A=T$ bonds detach while those regions held together by $G\equiv C$ base pairs remain double-stranded. Formaldehyde reacts irreversibly with bases that are not hydrogen bonded to prevent reannealing. Thus, after the addition of formaldehyde the DNA molecule retains its denaturation loops when cooled. Denaturation maps provide a unique way to distinguish different DNA molecules.

denatured DNA *See* **denaturation.**

denatured protein *See* **denaturation.**

dendrite one of the many short, branching cytoplasmic projections of a neuron. Dendrites synapse with and receive impulses from the axons of other neurons. These impulses are then conducted toward the perikaryon.

Denhardt's solution a solution consisting of Ficoll, polyvinylpyrrolidone, and bovine serum albumin, each at a concentration of 0.02% (w/v). Preincubation of nucleic acid-containing filters in this solution prevents non-specific binding of single-stranded DNA probes.

de novo 1. arising from an unknown source. 2. denoting synthesis of a specified molecule from very simple precursors, as opposed to the formation of the molecule by the addition or subtraction of a side chain to an already complex molecule.

***de novo* pathway** a process for synthesizing ribonucleoside monophosphates from phosphoribosylpyrophosphate, amino acids, CO_2, and NH_3, rather than from free bases, as in the salvage pathway.

densitometer an instrument used for measuring the light transmitted through an area of interest. Densitometers are used for scanning chromatograms and electropherograms and for measuring the blackening of photographic films.

density-dependent factor an ecological factor (e.g., food) that becomes increasingly important in limiting population growth as the population size increases.

density-dependent selection selection in which the values for relative fitness depend upon the density of the population.

density gradient equilibrium centrifugation *See* **centrifugation separation.**

density gradient zonal centrifugation *See* **centrifugation separation.**

density-independent factor an ecological factor (e.g., temperature) that is uncorrelated with variations in size of a population.

dent corn *See* **corn.**

deoxyadenylic, deoxycytidylic, deoxyguanylic acids *See* **nucleotide.**

deoxyribonuclease any enzyme that digests DNA to oligonucleotide fragments. *See* **endonuclease, exonuclease, restriction endonuclease.**

deoxyribonucleic acid DNA, the molecular basis of heredity. DNA consists of a polysugar-phosphate backbone from which the purines and pyrimidines project. The backbone is formed by bonds between the phosphate molecule and carbon 3 and carbon 5 of adjacent deoxyribose molecules. The nitrogenous base extends from carbon 1 of each sugar. Ac-

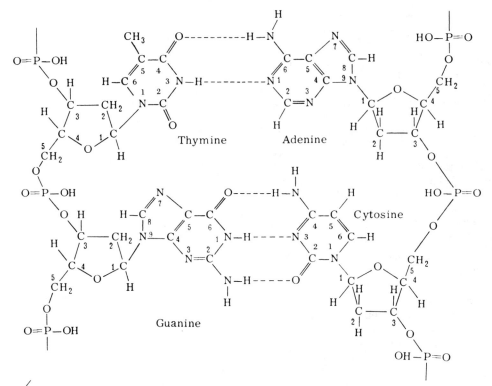

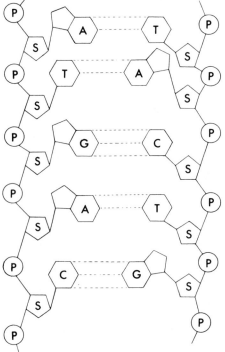

Deoxyribonucleic acid.

cording to the Watson-Crick model, DNA forms a double helix that is held together by hydrogen bonds between specific pairs of bases (thymine to adenine and cytosine to guanine). Each strand in the double helix is complementary to its partner strand in terms of its base sequence. The antiparallel strands form a right-handed helix that undergoes one complete revolution with each ten nucleotide pairs. DNA molecules are the largest biologically active molecules known, having molecular weights greater than 1×10^8 daltons. In the diagram to the left only five nucleotide pairs of the ladderlike DNA molecule are shown. "Uprights" of the ladder consist of alternating phosphate (P) and deoxyribose sugar (S) groups. The "cross rungs" consist of purine-pyrimidine base pairs that are held together by hydrogen bonds (represented here by dashed lines). A, T, G, and C represent adenine, thymine, guanine, and cytosine, respectively. Note that the AT pairs are held together less strongly than the GC pairs. In reality, the ladder is twisted into a right-handed double helix, and each nucleotide pair is rotated 36° with respect to its neighbor. A DNA molecule of molecular weight 2.5×10^7 d would be made up of approximately 40,000 nucleotide pairs. The type of DNA described above is the B form that occurs under hydrated conditions and is thought to be the principal biological conformation. The A form occurs under less hydrated conditions. Like the B form, it too is a right-handed double helix; however, it is more compact, with 11 base pairs per turn of the helix. The bases of the A form are tilted 20° away from perpendicular and displaced laterally in relation to the diad axis. The Z form of DNA is a left-handed double helix. It has 12 base pairs per turn of the helix, and presents a zigzag conformation (hence the symbolic designation). Unlike B DNA, Z DNA is antigenic. *See* **Appendix C**, 1953, Watson and Crick.

deoxyribonucleoside a molecule containing a purine or pyrimidine attached to deoxyribose.

deoxyribonucleotide a compound consisting of a purine or pyrimidine base bonded to deoxyribose, which in turn is bound to a phosphate group.

deoxyribose the sugar characterizing DNA.

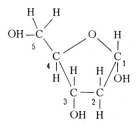

dependent differentiation differentiation of an embryonic tissue caused by a stimulus coming from other tissue and dependent upon that stimulus.

depolymerization the breakdown of an organic compound into two or more molecules of less complex structure.

derepression an increased synthesis of gene product accomplished by preventing the interaction of a repressor with the operator portion of the operon in question. In the case of inducible enzyme systems, the inducer derepresses the operon. A mutation of the regulatory gene that blocks synthesis of the repressor or a mutation of the operator gene that renders it insensitive to a normal repressor will also result in derepression. *See* **regulator gene.**

derived the more recent stages or conditions in an evolutionary lineage; the opposite of primitive.

dermatoglyphics the study of the patterns of the ridged skin of the palms, fingers, soles, and toes.

desmids green algae that exist as pairs of cells with their cytoplasms joined at an isthmus that contains a single shared nucleus. *See* **Appendix A**, Protoctista, Gamophyta; *Micrasterias thomasiana.*

desmin a 51,000 d cytoskeletal protein. Desmin molecules fall into the intermediate filament class and are found in glial and muscle cells.

desmosome an intercellular attachment device. It is a discontinuous button-like structure consisting of two dense plaques on the opposing cell surfaces, separated by an intercellular space about 25 mμ wide. On each symmetrical half-desmosome a thin layer of dense material coats the inner leaf of the cell membrane, and bundles of fine cytoplasmic filaments converge upon and terminate in this dense substance.

desynapsis the failure of homologous chromosomes that have synapsed normally during pachynema to remain paired during diplonema. Desynapsis is usually the result of a failure of chiasma formation. *Contrast with* **asynapsis.**

detached X an X chromosome formed by the detaching of the arms of an attached X chromosome (*q.v.*), generally through crossing over with the Y chromosome.

determinant in immunology, the portion of the antigen that is responsible for the specificity of the response and that is recognized by the binding sites of immunoglobulins and antigen-recognizing lymphocytes.

determinant cleavages a successive series of cleavages that follow a specific three-dimensional pattern such that with each division, cells are produced each of which can be shown to serve as the progenitor of a specific type of tissue. In developing mollusc eggs, for example, cell 4d, which is formed at the sixth cleavage, is always the progenitor of all primary mesodermal structures.

determinate inflorescence an inflorescence in which the first flowers to open are at the tip or inner part of the cluster, and the later ones are progressively lower or farther out.

determination the establishment of a single kind of histogenesis for a part of an embryo, which it will perform irrespective of its subsequent situations.

deutan *See* **color blindness.**

deuteranomaly *See* **color blindness.**

deuteranopia *See* **color blindness.**

deuterium *See* **hydrogen.**

deuteron the nucleus of a deuterium atom, containing one proton and one neutron.

Deuterostomia a subdivision of the Bilateria containing species in which the mouth does not arise from the blastopore. The coelom arises from the primitive gut. *See* **Appendix A, Protostomia.**

deuterotoky parthenogenesis in which both males and females are produced.

developer a chemical that serves as a source of reducing agents that will distinguish between exposed and unexposed silver halide and convert the exposed halide to metallic silver, thus producing an image on a photographic film.

development an orderly sequence of progressive changes resulting in an increased complexity of a biological system. *See* **differentiation, morphogenesis.**

developmental control genes genes that have as their primary functions the control of developmental decisions. Such genes that regulate cell fates during development have been extensively studied in *Caenorhabditis elegans* (*q.v.*). For example, in this nematode a specific set of cells can differentiate in two different ways to form the vulva or part of the uterus. The *lin-12* locus acts as a binary switch to affect the alternative cell fates. High activity of the locus specifies a ventral uterine precursor cell; low activity specifies an anchor cell that organizes the development of the vulva. *See* **Appendix C,** 1983, Greenwald *et al.;* **cell lineage mutants.**

developmental genetics the study of mutations that produce developmental abnormalities in order to gain understanding of how normal genes control growth, form, behavior, etc.

developmental homeostasis canalization (*q.v.*).

developmental homology anatomical similarity due to derivation from a common embryological source; e.g., the halteres of flies are developmentally homologous to the hind wings of moths.

deviation the departure of a quantity (derived from one or more observations) from its expected value (usually the mean of a series of quantities).

devolution regressive evolution.

Devonian the Paleozoic period during which cartilagenous and bony fishes evolved. On land the first seed-bearing plants arose, as well as amphibians and wingless insects. A mass extinction occurred late in the period. *See* **geologic time divisions.**

dex dextrorotatory. *See* **optical isomers.**

dextran a polysaccharide composed of a single repeating simple sugar.

dextrose glucose (*q.v.*).

d.f., D/F degrees of freedom (*q.v.*).

diabetes insipidus excessive excretion of normal urine; brought about because of inadequate output of pituitary antidiuretic hormone.

diabetes mellitus a hereditary metabolic disorder characterized by glucosuria and caused by the inability to manufacture insulin, often detected by excessive sugar in the urine.

diakinesis *See* **meiosis.**

diallelic referring to a polyploid in which two different alleles exist at a given locus. In a tetraploid, $A_1A_1A_2A_2$ and $A_1A_2A_2A_2$ would be examples.

dialysis the separation of molecules of differing size from a mixture by their differential diffusibility through a porous membrane. In the procedure knowns as *equilibrium dialysis,* soluble molecules of the same size are allowed to reach equivalent concentrations on either side of a semipermeable membrane. At equilibrium, if more molecules are detected on one side of the membrane, it indicates that they have become bound to some other larger molecules (e.g., repressor proteins, transport proteins, antibodies, etc.) present only on that side of the membrane, and thus are too large to pass through the pores of the membrane. This procedure is also used in immunology as a method of determining association constants for hapten-antibody reactions.

2,6-diaminopurine a mutagenically active purine analogue.

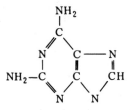

diapause a period of inactivity and suspension of growth in insects accompanied by a greatly decreased metabolism. In a given species, diapause usually takes place in a specific stage in the life cycle, and it often provides a means of surviving the winter.

diasteromer epimer (*q.v.*).

diauxy the adaptation of a microorganism to culture media containing two different sugars. The organism possesses constitutive enzymes for one of the sugars, which it utilizes immediately. Induced enzyme synthesis is required before the second sugar can be metabolized.

dicentric designating a chromosome or chromatid having two centromeres.

dichlorodiphenyltrichloroethane DDT; an insecticide to which many insect species have developed resistant races.

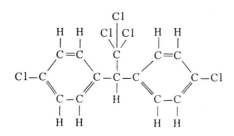

2,4-dichlorophenoxyacetic acid a phytohormone used as a weed killer.

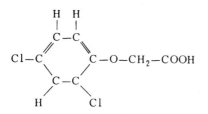

2,6-dichlorophenoxyacetic acid an antiauxin (*q.v.*).

dichogamous referring to flowers or hermaphroditic animals characterized by male and female sex organs that become mature at different times.

dichroism *See* **circular dichroism.**

dictyosome the Golgi apparatus in plants.

Dictyostelium discoideum *See* **Acrasiomycota.**

dictyotene stage a prolonged diplotene stage of meiosis seen in oocytes during vitellogenesis. The chromosomes that have already undergone crossing over may remain in this stage for months or even years in long-lived species.

2′,3′-dideoxynucleoside triphosphates analogues of normal 2′-deoxyribonucleoside triphosphates used in a modified "minus" technique for base sequencing of DNA molecules. Because these analogs have no oxygen at the 3′ position in the sugar, they act as specific chain-terminators (*q.v.*) for primed synthesis techniques (*see* **DNA sequencing techniques**). Nucleotides in which arabinose is substituted for deoxyribose also exhibit this chain-terminating effect.

differential affinity the failure of two partially homologous chromosomes to pair during meiosis when a third chromosome is present that is more completely homologous to one of the two. In its absence, however, pairing of the partially homologous chromosomes can occur. *See* **autosyndesis, homoeologous chromosomes.**

differential segment *See* **pairing segment.**

differential splicing *See* **alternative splicing.**

differentiation the complex of changes involved in the progressive diversification of the structure and functioning of the cells of an organism. For a given line of cells, differentiation results in a continual restriction of the types of transcription that each cell can undertake. *See* **development, morphogenesis.**

differentiation antigen a cell-surface antigen that is expressed only during a specific period of embryological differentiation.

diffuse centromere (kinetochore) *See* **centromere.**

diffusion the tendency for molecules because of their random heat motion to move in the direction of a lesser concentration, and so make the concentration uniform throughout the system.

digenetic descriptive of organisms of the subclass Digenea of the class Trematoda within the flatworm phylum Platyhelminthes. The term means "two beginnings," referring to a life cycle with alternation of generations, one parasitic and the other free-living. Digenea is the largest group of trematodes and the most important medically and economically. All members are endoparasitic with two or more hosts in the life cycle, the first host usually being a mollusc. The digenetic flukes include blood flukes and schistosomes that are generally considered to be the most serious helminthic human parasite. *See* **schistosomiasis.**

dihaploid a diploid cell, tissue, or organism having arisen from a haploid cell by chromosome doubling.

dihybrid a genotype characterized by heterozygosity at two loci. Mendel found that crosses between pure lines of peas that differed with respect to two unrelated traits produced genetically uniform F_1 dihybrid offspring. Intercrossing F_1 dihybrids produced parental and recombinant types in the F_2 population.

dihydrofolate reductase an enzyme essential for *de novo* thymidylate synthesis. It regenerates an intermediate (tetrahydrofolate) in thymidylate synthesis and is also essential for other biosynthetic events that depend on tetrahydrofolate, such as the synthesis of purines, histidine, and methionine. *See* **folic acid.**

dihydrouridine *See* **rare bases.**

2,5-dihydroxyphenylacetic acid homogentisic acid (*q.v.*).

dihydroxyphenylalanine (DOPA) a precursor of melanin (*q.v.*).

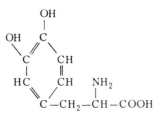

dimer a chemical entity consisting of an association of two monomeric subunits; e.g., the association of two polypeptide chains in a functional enzyme. If the two subunits are identical, they form a homodimer; if they are nonidentical, they form a heterodimer. Hexosaminidase (*q.v.*) is an example of a heterodimeric enzyme.

dimethylguanosine *See* **rare bases.**

dimethyl sulfate protection a method for identifying specific points of contact between a protein (such as RNA polymerase) and DNA based on the principle that, within an endonuclease-protected region (*see* **DNAase protection**), the adenines and guanines in the site of contact are not available to be methylated by exposure to dimethyl sulfate.

dimorphism the phenomenon of morphological differences that split a species into two groups, as in the sexual dimorphic traits distinguishing males from females.

dinitrophenol (DNP) a metabolic poison that prevents the uptake of inorganic phosphate and the production of energy-rich phosphorus compounds like ATP. DNP is a commonly used hapten in immunological experiments.

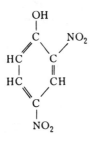

dioecious having male and female flowers or other reproductive organs on separate unisexual plants, or male and female reproductive organs in separate unisexual animals. *See* **flower.**

diploblastic having a body made of two cellular layers only (ectoderm and endoderm), as the coelenterates.

diplochromosome a chromosome arising from an abnormal duplication in which the centromere fails to divide and the daughter chromosomes fail to move apart. The resulting chromosome contains four chromatids.

Diplococcus pneumoniae the pneumonia bacterium. The classical studies on transformation were performed on this organism. *See* **Appendix C,** 1944, Avery *et al.;* 1964, Fox and Allen.

diplo-haplont an organism (such as an embryophyte) in which the products of meiosis form haploid gametophytes that produce gametes. Fertilization generates a diploid sporophyte in which meiosis takes place. Thus, diploid and haploid generations alternate. *Contrast with* **diplont, haplont.**

diploidy the chromosome state in which each type of chromosome (except the sex chromosomes) is represented twice (2N).

diplonema *See* **meiosis.**

diplont an organism (such as any multicellular animal) characterized by a life cycle in which the products of meiosis function as gametes. There is no haploid multicellular stage as in a diplo-haplont and haplont (*q.v.*).

diplophase the diploid phase of the life cycle between the formation of the zygote and the meiosis.

diplospory a type of apomixis in plants in which a diploid gametophyte is formed after mitotic divisions of the spore-forming cells.

diplotene *See* **meiosis.**

dipole a molecule carrying charges of opposite sign at opposite poles.

Diptera an insect order containing midges, mosquitoes, and flies. *See* **Appendix A,** Animalia, Arthropoda.

directional selection selection resulting in a shift in the population mean in the direction desired by the breeder or in the direction of greater adaptation by nature. For example, the breeder might select for a number of generations seeds from only the longest ear of corn in the population. *See* **disruptive selection.**

direct repeats identical or closely related DNA sequences present in two or more copies in the same orientation in the same molecule, although not necessarily adjacent.

DIS *Drosophila Information Service* (*q.v.*).

discoidal cleavage   cleavage occurring at the surface of an enormous yolk mass.

discontinuous distribution   a collection of data recorded as whole numbers, and thus not yielding a continuous spectrum of values; e.g., the number of leaves per plant in a population of plants. *See* **continuous distribution.**

discontinuous replication   *See* **replication of DNA.**

discontinuous variations   variations that fall into two or more non-overlapping classes.

discordant   twins are said to be discordant with respect to a trait if one shows the trait and the other does not.

disequilibrium   *See* **gametic disequilibrium, linkage disequilibrium.**

disintegration constant   the fraction of the number of atoms of a radioactive element that decay in unit time; d in the equation $N = N_o e^{-dt}$, where N_o is the initial number of atoms present, and N is the number of atoms present after some time, t.

disjunction   the moving apart of chromosomes during anaphase of mitotic or meiotic divisions.

dispersal mechanism   any means by which a species is aided in extending its range. For example, sticky seeds can cling to animals and be transported by them to new regions.

dispersive replication   an obsolete model of DNA replication in which parental and newly synthesized daughter molecules are interspersed in an essentially random fashion.

disruptive selection   the selection of divergent phenotypic extremes in a population until, after several generations of selection, two discontinuous strains are obtained. For example, the breeder might select for a number of generations seeds from the longest and the shortest ears of corn in a population. *See* **directional selection.**

disseminule   a plant part that gives rise to a new plant.

Dissociation-Activator system   *See* ***Activator-Dissociation*** system.

dissociation constant   (pK) a relation between proton-donor and proton-acceptor molecules in solution. It is represented by: $\log \dfrac{[pa]}{[pd]} = pH - pK$, where [pa] is the concentration of the molecule functioning as a proton acceptor, [pd] is the concentration of the molecule functioning as a proton donor, and pH is the hydrogen ion concentration of the aqueous solution. A single molecular species may be present, if the molecule is amphiprotic.

Thus, if the pK of a carboxyl group is 4 and the pH is 7, $\log \dfrac{[COO^-]}{[COOH]} = 7 - 4 = 3$. Thus, there will be 1000 ionized groups for each un-ionized carboxyl group in a neutral aqueous solution.

distal   situated away from the place of attachment. In the case of a chromosome, the part farthest from the centromere.

distributive pairing   the pairing of chromosomes at metaphase I of meiosis that leads to their proper distribution to daughter cells. Synaptonemal complexes play no role in this type of chromosomal association.

distylic species   a plant species composed of two types of individuals each characterized by a different flower morphology.

disulfide linkage   the sulfur-to-sulfur bonding of adjacent cysteine residues in or between protein molecules.

diurnal   **1.** pertaining to the daytime. **2.** recurring in the period of a day; daily.

divergence   in molecular biology, the percent difference between nucleotide sequences of two related DNA segments or between amino acid sequences of the two related polypeptide chains.

divergence node   the branching point in an evolutionary tree. The place where two lineages diverge from a common ancestor.

divergent transcription   the transcriptional orientation of different DNA segments in opposite directions from a central region.

diversifying selection   *See* **disruptive selection.**

diversity   in ecology, the number of species or other taxa in a particular ecological unit.

Division   *See* **Appendix A: Classification.**

dizygotic twins   *See* **twins.**

D loop   **1.** a displacement loop formed early in the replication of duplex DNA (either circular or linear) consisting of a single, unreplicated, parental strand on one side, and a double-stranded branch (composed of one parental strand paired with the leading strand, *q.v.*) on the other side. Because the leading strand displaces the unreplicated parental strand, the replication "bubble" or "eye" is called a displacement or D loop. **2.** A segment of mitochondrial DNA in which one strand is displaced by a short stretch of RNA base paired to the other complementary strand.

DNA   deoxyribonucleic acid (*q.v.*). *Also see* **promiscuous DNA.**

DNA adduct   *See* **adduct.**

DNA-agar technique   a technique for testing the degree of homology between nucleic acid molecules from different sources by allowing fragments of radioactive nucleic acid from one source to react with nonlabeled nucleic acids from another source trapped in an agar gel. This procedure binds to the gel radioactive polynucleotide fragments that are complementary to those trapped in the agar. *See* **Appendix C,** 1963, McCarty and Bolton; **hybrid duplex molecule.**

DNAase   symbol for deoxyribonuclease (*q.v.*).

DNAase protection   a method for estimating the size of a DNA region that is interacting with a protein (e.g., the size of the region occupied by RNA polymerase during transcription). After the protein is bound to DNA, an endonuclease is added that degrades most of the DNA outside the region of interaction to mono- and dinucleotides. *See* **dimethyl sulfate protection.**

DNA clock hypothesis   the postulation that, when averaged across the entire genome of a species, the rate of nucleotide substitutions in DNA remains constant. Hence the degree of divergence in nucleotide sequences between two species can be used to estimate their divergence node (*q.v.*). *See* **delta T50H.**

DNA clone   a DNA segment that has been inserted via a viral or plasmid vector into a host cell with the following consequences: the segment has replicated along with the vector to form many copies per cell, the cells have mutiplied into a clone, and the insert has been magnified accordingly.

DNA complexity   a measure of the amount of nonrepetitive DNA characteristic of a given DNA sample. In an experiment involving reassociation kinetics (*q.v.*), DNA complexity represents the combined length in nucleotide pairs of all unique DNA fragments. The DNA of evolutionarily advanced species is more complex than that of primitive species.

DNA-dependent RNA polymerase   RNA polymerase (*q.v.*). *Contrast with* **RNA-dependent DNA polymerase.**

DNA-driven hybridization reaction   a reaction involving the reassociation kinetics of complementary DNA strands when DNA is in great excess of a radioactive RNA tracer; employed in cot analysis to determine the repetition frequencies of the corresponding genome sequences. *See* **reassociation kinetics.**

DNA duplex   a DNA double helix. *See* **deoxyribonucleic acid.**

DNA fiber autoradiography   light microscopic autoradiography of tritiated thymidine-labeled DNA molecules attached to millipore filters. The technique was devised by Cairns (1963) for studying DNA replication in *E. coli* and later adapted by Huberman and Riggs (1968) for visualizing the multiple replicons of mammalian chromosomes.

DNA fingerprint technique   a technique that relies upon the presence of simple tandem-repetitive sequences that are scattered throughout the human genome. Although these regions show considerable-length polymorphisms, they share a common 10–15 base-pair core sequence. DNAs from different individual humans are enzymatically cleaved and separated by size on a gel. A hybridization probe containing the core sequence is then used to label those DNA fragments that contain complementary sequences. The pattern displayed on each gel is specific for a given individual. The technique has been used to establish family relationships in cases of disputed parentage. In violent crimes, blood, hair, semen, and other tissues from the assailant are often left at the scene. The DNA fingerprinting technique provides the forensic scientist with a means of identifying the assailant from a group of suspects. *See* **Appendix C,** 1985, Jeffries *et al.;* **alphoid sequences, fingerprinting technique, restriction fragment polymorphisms, VNTR locus.**

DNA glycolases   a family of enzymes, each of which recognizes a single type of altered base in DNA and catalyzes its hydrolytic removal from the sugar-phosphodiester backbone.

DNA grooves   the DNA double helix contains two grooves that run its length. The major groove is 12 Å wide, while the minor groove is 6 Å wide. The major groove is slightly deeper than the minor groove (8.5 Å *vs.* 7.5 Å). Each groove is lined by potential hydrogen-bond donor and acceptor atoms, and these interact with DNA-binding proteins that recognize specific DNA sequences. For example, endonucleases bind electrostatically to the minor groove of the double-helical DNA. *See* **deoxyribonuclease, helix-turn-helix motif.**

DNA gyrase   *See* **gyrase.**

DNA hybridization   a technique for selectively binding specific segments of single-stranded DNA or RNA by base pairing to complementary sequences on ssDNA molecules that are trapped on a nitrocellulose filter (*q.v.*). 1. DNA-DNA hybridization is commonly used to determine the degree of sequence identity between DNAs of different species. **2.** DNA-RNA hybridization is the method used to select those molecules that are complementary to a specific DNA from a heterogeneous population of RNAs. *See* **Appendix C,** 1960, Doty *et al.;* 1963, McCarty and Bolton; 1972, Kohne *et al.; in situ* **hybridization, reassociation kinetics.**

DNA library *See* **genomic library.**

DNA ligases enzymes that catalyze the formation of a phosphodiester bond between adjacent 3′-OH and 5′-P termini in DNA. DNA ligases function in DNA repair to seal single-stranded nicks between adjacent nucleotides in a duplex DNA chain. *See* **Appendix C,** 1966, Weiss and Richardson; **blunt-end ligation, cohesive-end ligation, cut-and-patch repair, replication of DNA.**

DNA looping a phenomenon that involves proteins that bind to specific sites on a DNA molecule while also binding to each other. The DNA loops that form as a result stimulate or inhibit the transcription of associated genes. Enhancer (*q.v.*) sequences may represent DNA segments involved in DNA looping.

DNA modification *See* **modification.**

dna mutations mutations of *E. coli* that influence DNA replication. The *dna A, dna B* and *dna C* mutations are defective in proteins that interact with replication origins. The *dna E, dna X,* and *dna Z* genes encode subunits of DNA polymerase III, and *dna G* encodes primase (*q.v.*). *See* **DNA polymerase, replicon.**

DNA polymerase an enzyme that catalyzes the formation of DNA from deoxyribonucleoside triphosphates, using single-stranded DNA as a template. Three different DNA polymerases (pol I, pol II, and pol III) have been isolated from *E. coli.* Pol III is the major enzyme responsible for cellular DNA replication in this bacterium. The other two enzymes function primarily in DNA repair. Eukaryotes contain many different DNA polymerases. They reside in different parts of the cell (nucleus, cytosol, or mitochondria) and have different functions (replication, repair, and recombination). *See* **polymerase chain reaction.**

DNA probe *See* **probe.**

DNA puff *See* **chromosomal puff.**

DNA relaxing enzyme *See* **topoisomerase.**

DNA repair any mechanism that restores the correct nucleotide sequence of a DNA molecule that has incurred one or more mutations, or that has had its nucleotides modified in some way (e.g., methylation). *See* **cut-and-patch repair, error-prone repair, mismatch repair, photoreactivting enzyme, proofreading, recombination repair, SOS response.**

DNA replication *See* **replication of DNA.**

DNA restriction enzyme any of the specific endonucleases (*q.v.*) present in many strains of *Escherichia coli* that recognize and degrade DNA from foreign sources. These nucleases are formed under the directions of genes called *restriction alleles.* Other genes called *modification alleles* determine the methylation pattern of the DNA within a cell. It is this pattern that determines whether or not the DNA is attacked by a restriction enzyme. *See* **modification methylases, restriction endonuclease.**

DNA-RNA hybrid a double helix consisting of one chain of DNA hydrogen bonded to a complementary chain of RNA. *See* **Appendix C,** 1961, Hall and Spiegelman.

DNase also symbolized *DNAase. See* **deoxyribonuclease.**

DNase protection *See* **DNAase protection.**

DNA sequencing techniques **1.** the method developed by F. Sanger and A. R. Coulson (1975) is known as the "plus and minus" method or the "primed synthesis" method. DNA is synthesized *in vitro* in such a way that it is radioactively labeled and the reaction terminates specifically at the position corresponding to a given base. After denaturation, fragments of different lengths are separated by electrophoresis and identified by autoradiography. In the "plus" protocol, only one kind of deoxyribonucleoside triphosphate (dNTP) is available for elongation of the ^{32}P-labeled primer. In the "minus" protocol, one of the four dNTPs is missing; alternatively, specific terminator base analogues (2′,3′-dideoxyribonucleoside triphosphates, *q.v.*) can be used instead of the "minus" technique. **2.** in the 1977 procedure of A. M. Maxam and W. Gilbert (the "chemical" method), single-stranded DNA (derived from double-stranded DNA and labeled at the 5′ end with ^{32}P) is subjected to several chemical (dimethyl sulfate–hydrazine) cleavage protocols that selectively make breaks on one side of a particular base; fragments are separated according to size by electrophoresis on acrylamide gels and identified by autoradiography.

DNA topoisomerase *See* **topoisomerase.**

DNA unwinding protein a protein that binds to single-stranded DNA and facilitates the unwinding of the DNA duplex during replication and recombination. *See* **gene 32 protein.**

DNA vector a replicon, such as a small plasmid or

a bacteriophage, that can be used in molecular cloning experiments to transfer foreign nucleic acids into a host organism in which they are capable of continued propagation. *See* **lambda cloning vector, plasmid cloning vector.**

DNP **1.** 2:4 dinitrophenol. **2.** DNA-protein complex.

dog any of a large number of domesticated breeds of the species *Canis familiaris.* Popular breeds include: TERRIERS: Welsh, Bedlington, Dandie Dinmont, West Highland White, Skye, Cairn, Scottish, Sealyham, Fox (Smooth), Fox (Wire), Schnauzer, Airedale, Irish, Kerry Blue, Bull, Manchester. POINTERS: German Shorthaired Pointer, Irish Setter, English Setter, Gordon Setter, Weimaraner, Pointer, Brittany Spaniel. COURSING HOUNDS: Irish Wolfhound, Scottish Deerhound, Greyhound, Whippet, Borzoi, Saluki, Afghan. TRAILING HOUNDS: Basenji, Bloodhound, Dachshund, Bassett, Beagle, Black and Tan Coonhound. MISCELLANEOUS HOUNDS: Otterhound, Norwegian Elkhound. FLUSHING SPANIELS: English Springer, English Cocker, American Cocker, Welsh Springer. RETRIEVERS: Golden Retriever, Labrador Retriever, Chesapeake Bay Retriever, Irish Water Spaniel, Curly-coated Retriever. SHEEP DOGS: Briard, Kuvasz, Shetland Sheepdog, Collie, Belgian Sheepdog. SLED DOGS: Siberian Huskie, Eskimo, Samoyed, Alaskan Malamute. GUARD DOGS: Bouvier de Flandres, Mastiff, Rottweiller, Boxer, Great Dane, Bull Mastiff, Schnauzer, German Shepherd, Dobermann Pinscher. MISCELLANEOUS WORKING DOGS: St. Bernard, Welsh Corgi (Cardigan), Welsh Corgi (Pembroke), Newfoundland, Great Pyrenees. TOYS: Maltese, Pug, Japanese Spaniel, English Toy Spaniel (King Charles), Pekingese, Pomeranian, Yorkshire Terrier, Griffon, Chihuahua, Papillon, Poodle (Toy), Mexican hairless. NONSPORTING BREEDS: Lhasa Apso, Poodle (Standard), Poodle (Miniature), Dalmation, Chow Chow, Keeshond, Schipperke, English Bulldog, French Bulldog, Boston Terrier.

Dollo's law the proposition that evolution along any specific lineage is essentially irreversible. For example, no modern mammal can de-evolve back to a form identical in all respects to the mammallike, reptilian ancestor from which it was derived.

domain **1.** a homology unit; i.e., any of the three or four homologous regions of an immunoglobulin heavy chain that apparently evolved by duplication and diverged by mutation. **2.** any discrete, continuous part of a polypeptide sequence that can be equated with a particular function. **3.** any region of a chromosome within which supercoiling is independent of other domains. **4.** an extensive region of DNA including an expressed gene that exhibits pronounced sensitivity to degradation by endonucleases.

domesticated species *See* **Appendix B.**

dominance referring to alleles that fully manifest their phenotype when present in the heterozygous, heterokaryotic, or heterogenotic state. The alleles whose phenotypic expressions are masked by dominant alleles are termed *recessive alleles.* Sometimes the dominant allele expresses itself late in development (e.g., Huntington disease, *q.v.*), in which case the allele is said to show *delayed dominance. See* **codominant, incomplete dominance, semidominance.**

dominance variance genetic variance for a polygenic trait in a given population attributed to the dominance effects of contributory genes.

dominant complementarity *See* **complementary genes.**

dominant gene *See* **recessive gene.**

donkey *Equus asinus,* a close relative of the horse. The female is referred to as a jennet, the male as a jack. *See* **horse-donkey hybrids.**

donor splicing site *See* **left splicing junction.**

DOPA dihydroxyphenylalanine (*q.v.*).

dosage compensation a mechanism that regulates the expression of sex-linked genes that differ in dose between females and males in species with an XX-XY method of sex determination. In *Drosophilia melanogaster,* dosage compensation is accomplished by raising the rate of transcription of genes on the single X chromosome of males to double that of genes on either X chromosome in females. In mammals, the compensation is made by inactivating at random one of the two X chromosomes in all somatic cells of the female. The inactivated X forms the Barr body or sex chromatin. In cases where multiple X chromosomes are present all but one are inactivated. *See* **Appendix C,** 1948, Muller; 1961, Lyon, Russell; **Fabry disease, glucose-6-phosphate dehydrogenase deficiency, Lesch-Nyhan syndrome, ocular albinism.**

dose **1.** gene dose—the number of times a given gene is present in the nucleus of a cell. **2.** radiation dose—the radiation delivered to a specific tissue area or to the whole body. Units for dose specifications are the gray, roentgen, rad, rep, and sievert.

dose-action curve dose-response curve (*q.v.*).

dose fractionation the administration of radiation in small doses at regular intervals.

dose-response curve the curve showing the relation between some biological response and the administered dose of radiation.

dosimeter an instrument used to detect and measure an accumulated dosage of radiation.

dot blot *See* **dot hybridization.**

dot hybridization a semiquantitative technique for evaluating the relative abundance of nucleic acid sequences in a mixture or the extent of similarity between homologous sequences. In this technique, multiple samples of cloned DNAs, identical in amount, are spotted on a single nitrocellulose filter in dots of uniform diameter. The filter is then hybridized with a radioactive probe (e.g., an RNA or DNA mixture) containing the corresponding sequences in unknown amounts. The extent of hybridization is estimated semiquantitatively by visual comparison to radioactive standards similarly spotted.

Dotted a gene, symbolized by *Dt* residing on chromosome 9 of maize, that influences the rate at which *a* mutates to *A. A* is on chromosome 3. *See* **Appendix C,** 1938, Rhoades.

double cross the technique used for producing hybrid seed for field corn. Four different inbred lines (A, B, C, and D) are used. A × B → AB hybrid and C × D → CD hybrid. The single-cross hybrids (AB and CD) are then crossed and the ABCD seed is used for the commercial crop.

double crossover *See* **double exchange.**

double diffusion technique synonymous with Ouchterlony technique (*q.v.*).

double exchange breakage and interchange occurring twice within a tetrad involving two, three, or four of the chromatids.

double fertilization in flowering plants the union of one sperm nucleus with the egg nucleus to form the diploid zygote and of the other sperm nucleus with the polar nuclei to form a triploid endosperm nucleus. *See* **pollen grain.**

double haploids plants that are completely homozygous at all gene loci, generated when haploid germ cells, grown in tissue culture, double their chromosome sets. *See* **anther culture.**

double helix the Watson-Crick model of DNA structure, involving plectonemic coiling (*q.v.*) of two hydrogen-bonded polynucleotide, antiparallel (*q.v.*) strands wound into a right-handed spiral configuration. *See* **deoxyribonucleic acid.**

double infection infection of a bacterium with two genetically different phages.

double-sieve mechanism a model that explains the rarity of misacylation of amino acids by proposing that an amino acid larger than the correct one is rarely activated because (1) it is too large to fit into the active site of the tRNA synthetase (first sieving), and (2) the hydrolytic site of the same synthetase is too small for the correct amino acid (second sieving). Thus, an amino acid smaller than the correct one can be removed by hydrolysis.

double transformation *See* **cotransformation.**

double X in *Drosophila melanogaster*, an acrocentric, double- length X chromosome arising as a radiation-induced aberration. Such double X chromosomes are superior to the ordinary metacentric, attached X chromosomes (*q.v.*) for most stockkeeping operations, since they do not break up by crossing over with the Y. *See also* **detached X.**

doubling dose that dose of ionizing radiation that doubles the spontaneous mutation rate of the species under study.

doubling time the average time taken for the cell number in a population to double. The doubling time will equal the generation time (*q.v.*) only if (1) every cell in the population is capable of forming two daughter cells, (2) every cell has the same average generation time, and (3) there is no lysis of cells. The doubling time is generally longer than the generation time.

Dowex trademark of a family of ion-exchange resins.

down promoter mutations promoter alterations that decrease the frequency with which transcription is initiated relative to wild type; promoters with this property are called low-level or weak promoters.

downstream *See* **transcription unit.**

Down syndrome a type of mental retardation due to trisomy of autosome 21. Since the eyelid openings of the patient are oblique and the inner corner of the eyelid may be covered by an epicanthic fold, the condition is often called mongolism. The frequency of such trisomic births increases with advancing maternal age. *See* **Appendix C,** 1959, Lejuene *et al.;* 1968, Henderson and Edwards; **translocation Down syndrome.**

DPN diphosphopyridine nucleotide. NAD is the preferred nomenclature.

drawing tube a device attached to a compound

microscope that permits the image and a drawing surface to be viewed concurrently, so that the object under study may be drawn accurately.

drift   genetic drift (*q.v.*).

dRNA   DNA-like RNA. RNA molecules that are not included in the rRNA and tRNA classes. Much of the dRNA is of high molecular weight, short half-life, and never leaves the nucleus.

Drosophila   a genus of flies containing about 900 described species. The most extensively studied of all genera from the standpoint of genetics and cytology. The genus is subdivided into eight subgenera: (1) *Hirtodrosophila,* (2) *Pholadoris,* (3) *Dorsilopha,* (4) *Phloridosa,* (5) *Siphlodora,* (6) *Sordophila,* (7) *Sophophora,* and (8) *Drosophila. D. melanogaster,* the multicellular organism for which the most genetic information is available, belongs in the subgenus *Sophophora. See* **Appendix C,** 1910, Morgan; 1926, Chetverikov; 1927, Muller.

***Drosophila* eye pigments**   the ommatidia of the dull red compound eyes of *Drosophila* contain two classes of pigments, one brown (the ommochromes) and one bright red (the drosopterins).

Studies of the precursor compounds isolated from eye color mutants played an important role in the development of the one gene–one enzyme concept (*see* **Appendix C,** 1935, Beadle and Ephrussi). An example of an ommochrome is xanthommatin. Hydroxykynurenine, a compound biosynthesized from tryptophan, serves as a precursor of xanthommatin. Flies lacking the plus allele of the *cinnabar* gene are unable to synthesize hydroxykynurenine, and therefore this is sometimes called the *cn*[+] substance. Drosopterins are pteridine derivatives. Sepiapterin is a precursor of drosopterin that accumulates in *sepia* mutants. It also gives the *Drosophilia* testis its yellow color. *See* **formylkynurenine.**

***Drosophila* Information Service**   a yearly bulletin that lists all publications concerning *Drosophila* that year, the stock lists of major laboratories, the addresses of all *Drosophila* workers, descriptions of new mutants and genetic techniques, research notes, and new teaching exercises. *See* **Appendix D.**

***Drosophila* salivary gland chromosomes**   the most extensively studied polytene chromosomes. During larval development, the cells of the salivary gland

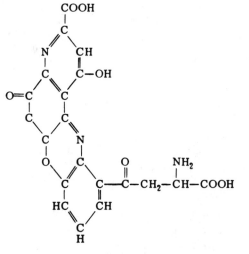

xanthommatin

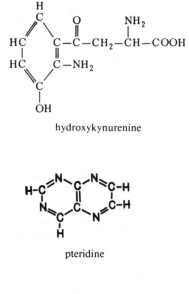

hydroxykynurenine

pteridine

drosopterin

sepiapteria

***Drosophila* eye pigments.**

98

undergo 9 or 10 cycles of endomitotic DNA replications to produce chromosomes that contain 1,000–2,000 times the haploid amount of DNA. The cytological map of the chromosomes of *D. melanogaster* contains slightly over 5,000 bands. It is divided into 102 divisions, distributed as shown below.

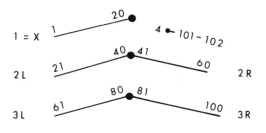

The solid circles represent the centromeres. Each division is subdivided into subdivisions lettered A–F, and the subdivisions contain varying numbers of bands. Genes have been localized within these bands by studying overlapping deficiencies and, more recently, by *in situ* hybridization with labeled probes. *See* **Appendix C**, 1933, Painter; 1935, Bridges; 1988, Sorsa; **biotinylated DNA, deficiency loop.**

drosopterins   *See* ***Drosophilia* eye pigments.**

drug-resistant plasmid   *See* **R plasmid.**

drumstick   a small protrusion from the nucleus of the human polymorphonuclear leukocyte, found in 3 to 5% of these cells in females, but not in males. *See* **Barr body.**

drupe   a simple, fleshy fruit, such as an olive, derived from a single carpel and usually single-seeded.

Dryopithecus   a genus of fossil primates from which the great apes and humans are thought to have diverged about 25 million years ago.

dsDNA   double-standed DNA.

D_1 trisomy syndrome   a well-defined set of congenital defects in man caused by the presence of an extra chromosome of the D group (13). Also called Patau syndrome. *See* **human mitotic chromosomes.**

dual recognition   an immunological model proposing that a T cell has two receptors, both of which must simultaneously bind specific molecules in order to activate the cell; one receptor binds to the antigen, the other binds to a self-molecule of the major histocompatibility system (*q.v.*); a form of *associative recognition* (*q.v.*)

Duchenne muscular dystrophy   a form of muscular dystrophy (*q.v.*) due to a gene on the distal portion of the short arm of the human X chromosome.

Duffy blood group system   the first human genetic locus to be assigned to a specific autosome. It resides near the centromere on the short arm of chromosome 1. Duffy antigenic determinants on erthyrocyte membranes serve as receptors for *Plasmodium vivax,* and individuals homozygous for the inactive allele Fy^- are therefore immune to tertian malaria. As a consequence, practically all Africans are Duffy negative. *See* **Appendix C**, 1968, Donahue *et al.;* 1976, Miller *et al.;* **Fy, malaria.**

duplex   *See* **autotetraploid.**

duplex DNA   DNA molecules as described in the Watson-Crick model; that is, with the two polynucleotide chains of opposite 3′-5′ polarity intertwined and annealed.

duplication   *See* **chromosomal aberration.**

durum wheat   *Triticum durum* (N = 14), an ancient wheat grown since Egyptian times. Used today in the manufacture of macaroni. *See* **wheat.**

dwarf   a short person with abnormal proportions, usually suffering from achondroplasia (*q.v.*). *See* **ateliosis.**

dyad   1. a pair of sister chromatids (*q.v.*). 2. the products resulting from disjunction of tetrads at the first meiotic division; dyads are contained in the nuclei of secondary gametocytes. *See* **meiosis.**

dynein   a complex protein, generally composed of two subunits, found in the armlike extensions of the A microtubules of the nine peripheral doublets of axonemes (*q.v.*). Dynein functions to couple a cycle of ATP binding and hydrolysis with a mechanical binding and release cycle to generate a sliding in relation to one another of the microtubule doublets in axonemes. In *Drosophila melanogaster,* the Y chromosomes appear to contain structural genes that encode the dyneins of sperm axonemes. Cytoplasmic dyneins also exist and are thought to function in the translocation of organelles. *See* **Appendix C**, 1963, Gibbons; 1982, Goldstein *et al.*

dysgenesis   *See* **hybrid dysgenesis.**

dysgenic   genetically deleterious.

dysploidy   the situation where the species in a genus have different diploid chromosome numbers, but the numbers do not represent a polyploid series.

dystrophin   *See* **muscular dystrophy.**

E

E_1, E_2, E_3 the first, second, and third generation of organisms following some experimental manipulation, such as irradiation with x rays.

early genes genes expressed during early development. In T_4 bacteriophage, those genetic elements that function in the period of phage infection before the start of DNA replication. *Compare with* **late genes.**

ecdysis 1. molting, the periodic shedding of the cuticle of arthropods. 2. in dinoflagellates, the shedding of the thecal wall layer.

ecdysones steriod hormones produced by the prothoracic gland (*q.v.*) of insects, required for molting and puparium formation; also called *molting hormones.* The structural formulas for α and β ecdysone are shown below. Beta ecdysone is sometimes called 20-hydroxy-ecdysone (20 HE). The precursor of α ecdysone is dietary cholesterol (*q.v.*). The α ecdysone synthesized by the prothoracic gland is converted in peripheral tissues, such as the fat body, to β ecdysone, which is the active molting hormone. The conversion involves the addition of a hydroxyl group at the position indicated by the arrow. The relative concentrations of ecdysone and allatum hormone (*q.v.*) determine whether juvenile or adult development will follow a given molt. Temperature-sensitive ecdysone mutants are known for *Drosophila melanogaster. See* **Appendix C**, 1965, Karlson *et al.;* **ring gland.**

echinoderm an animal with an external skeleton of calcareous plates and internal water-vascular system (starfish, sea urchins, sea lilies, sea cucumbers, etc.). *See* **Appendix A**, Echinodermata.

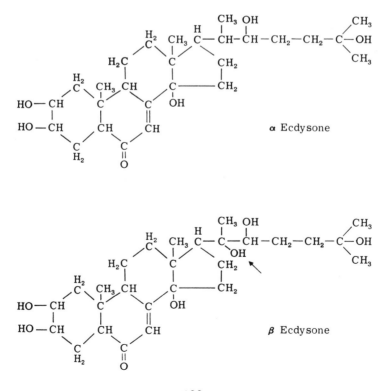

α Ecdysone

β Ecdysone

eclipsed antigens antigenic determinants of parasite origin that resemble antigenic determinants of their hosts to such a degree that they do not elicit the formation of antibodies by the host. The formation of eclipsed antigens by parasites is termed *molecular mimicry.*

eclipse period in virology, the time interval between infection and the first intracellular reappearance of infective phage particles.

eclosion the emergence of the adult insect from its pupal case.

ecodeme a deme (*q.v.*) associated with a specific habitat (a cypress swamp, for example).

ecogeographical divergence the evolution from a single ancestral species of two or more different species each in a different geographical area and each adapted to the local peculiarities of its habitat.

ecogeographic rules any of several generalizations concerning geographic variation within a species that correlate adaptations with climate or other environmental conditions: e.g., Allen's rule, Bergman's rule, Gloger's rule.

ecological isolation a premating (prezygotic) isolating mechanism in which members of different species seldom, if ever, meet because each species prefers to live (is adapted to) different habitats.

ecological niche the position occupied by a plant or animal in its community with reference both to its utilization of its environment and its required associations with other organisms.

ecology the study of the relationships between organisms and their environment.

ecophenotype a nongenetic phenotypic modification in response to environmental conditions.

Eco RI *See* **restriction endonuclease.**

ecosystem an assemblage of interacting populations of species grouped into communities in a local environment. Ecosystems vary greatly in size (e.g., a small pool *vs.* a giant reef). *See* **biome.**

ecotype race (within a species) genetically adapted to a certain environment. *See* **Appendix C,** 1948, Clausen *et al.*

ectoderm a germ layer forming the external covering of the embryo and neural tube (from which develop the brain, spinal cord, and nerves). Ectodermal derivatives include all nervous tissues, the epidermis (including cutaneous glands, hair, nails, horns, the lens of the eye, etc.), the epithelia of all sense organs, the nasal cavity sinuses, the anal canal, and the mouth (including the oral glands, and tooth enamel), and the hypophysis.

ectopic pairing nonspecific pairing of intercalary and proximal heterochromatic segments of *Drosophila* salivary chromosomes.

ectoplasm the superficial cytoplasm of ciliates.

ectotherm vertebrates such as fishes, amphibians, and reptiles that have little or no endogenous mechanisms for controlling their body temperature. Their temperature is determined by environmental conditions. *Contrast with* **endotherm.**

editing *See* **proofreading.**

EDTA an abbreviation for ethylene diaminetetracetic acid. A molecule capable of reacting with metallic ions and forming a stable, inert, water-soluble complex. EDTA is used to remove metals that occur in minute amounts even in distilled water.

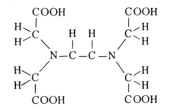

Edward syndrome a well-defined set of congenital defects in humans caused by the presence of an extra chromosome 18. Also called trisomy 18 syndrome or E_1 trisomy syndrome. *See* **human mitotic chromosomes.**

EEG electroencephalogram (*q.v.*).

effective fertility *See* **fertility.**

effective lethal phase the stage in development at which a given lethal gene generally causes the death of the organism carrying it.

effective population size the average number of individuals in a population that contribute genes to succeeding generations. If the population size shows a cyclical variation as a function of season of the year, predation, parasitism, and other factors, the effective population size is closer to the number of individuals observed during the period of maximal contraction.

effector **1.** a molecule that affects (positively or negatively) the function of a regulatory protein. **2.** an organ or cell that reacts to a nervous stimulus by doing chemical or mechanical work; for example, muscles, glands, electric organs, etc.

effector cell in immunology, a cell (usually a T

lymphocyte) that carries out cell-mediated cytotoxicity.

effector molecules   small molecules that combine with repressor molecules and activate or inactivate them with respect to their ability to combine with an operator gene. *See* **inducible system, repressible system, regulator gene.**

egg   a female gamete; an ovum. *See* **Appendix C,** 1651, Harvey; 1657, de Graff; 1827, von Baer.

egg chamber   the insect ovarian follicle. In *Drosophila,* it consists of a cluster of 16 interconnected cystocytes surrounded by a monolayer of follicle cells. The oocyte is the most posterior cystocyte. The remaining 15 cystocytes function as nurse cells (*q.v.*).

Ehlers-Danlos syndrome   a family of hereditary diseases characterized by overelasticity and brittleness of the skin and by excessive extensibility of the joints. The underlying defects involve blocks in the synthesis of collagen (*q.v.*). In type VI, for example, a hydroxylysine-deficient collagen is produced that is unable to form intermolecular cross-links. Both X-linked and autosomal genes are involved.

einkorn wheat   *Triticum monococcum* (N = 7), "one-grained" wheat, so called because it has a single seed per spikelet. A wheat in cultivation since stone age times. *See* **wheat.**

ejaculate   **1.** the process of semen release in higher vertebrates; also called *ejaculation.* **2.** the semen released in a given copulatory interaction or in a given artificially induced response.

elaioplast   an oil-rich plastid.

electroblotting   *See* **blotting.**

electrode   either terminal of an electrical apparatus.

electroencephalogram   the record of the rhythmical changes in the electrical potential of the brain.

electrofusion   *See* **Zimmermann cell fusion.**

electrolyte   a substance that when dissolved in water conducts an electric current.

electron   a negatively charged particle that is a constituent of every neutral atom. Its mass is 0.000549 atomic mass units.

electron carrier   an enzyme, such as a flavoprotein or cytochrome, that can gain and lose electrons reversibly.

electron-dense   said of a dense area seen on an electron micrograph, since the region has prevented electrons from passing through it. High electron density may mean that the area contains a high concentration of macromolecules or that it has bound the heavy metals (Os, Mn, Pb, U) used as fixatives and/or stains.

electron microscope   a magnifying system that uses beams of electrons focused in a vacuum by a series of magnetic lenses and that has a resolving power hundreds of times that of the best optical microscopes. Two main types of electron microscopes are available. In the *transmission electron microscope* (TEM), the image is formed by electrons that pass through the specimen. In the *scanning electron microscope* (SEM), image formation is based on electrons that are reflected back from the specimen. Thus, the TEM resembles a standard light microscope, which is generally used to look at tissue slices, and the SEM resembles a stereoscopic dissecting microscope, which is used to examine the surface properties of biological materials. *See* **Appendix C,** 1932, Knoll and Ruska.

electron microscope techniques   *See* **freeze-etching, freeze-fracture, Kleinschmidt spreading technique, negative staining.**

electron pair bond   convalent bond.

electron transport chain   a chain of molecules localized in mitochondria and acting as hydrogen and electron acceptors. The chain functions to funnel electrons from a given substrate to O_2. The energy released is used to phosphorylate ADP. *See* **cytochrome system.**

electron volt   a unit of energy equivalent to the amount of energy gained by an electron passing through a potential difference of one volt. Larger multiple units of the electron volt (eV) are frequently used: keV = kilo (thousand) eV; MeV = mega (million) eV; and GeV = giga (billion) eV.

electropherogram   a supporting medium containing a collection of molecules that have been separated by electrophoresis. The medium is generally in the form of a sheet, much longer and wider than it is thick, and it is often a gel such as agarose.

electrophoresis   the movement of the charged molecules in solution in an electrical field. The solution is generally held in a porous support medium such as filter paper, cellulose acetate (rayon), or a gel made of starch, agar, or polyacrylamide. Electrophoresis is generally used to separate molecules from a mixture, based upon differences in net electrical charge and also by size or geometry of the molecules, dependent upon the characteristics of the gel matrix. The SDS-PAGE technique is a method of separating proteins by exposing them to the anionic detergent sodium dodecyl sulfate (SDS) and polyacrylamide gel electrophoresis (PAGE). When SDS binds to proteins, it breaks all noncovalent interactions so that the molecules assume a random coil configuration, provided no disulfide bonds exist (the latter can be broken by treatment with mercaptoethanol). The distance moved per unit time by a random coil follows a mathematical formula involving the molecular weight of the molecule, from which the molecular weight can be calculated. *See* **Appendix C,** 1933, Tiselius; **pulsed field gradient gel electrophoresis, zonal electrophoresis.**

electroporation The application of electric pulses to animal cells or plant protoplasts to increase the permeability of their membranes. The technique is used to facilitate DNA uptake during transformation experiments.

electrostatic bond the force holding ions together in a crystal such as common salt (NaCl).

element a pure substance consisting of atoms of the same atomic number and that cannot be decomposed by ordinary chemical means. *See* **chemical elements, periodic table.**

Elodea canadensis a common pond weed that behaves as a facultative apomict (*q.v.*). *See* **Appendix C,** 1923, Santos.

elongation factors proteins that complex with ribosomes to promote elongation of polypeptide chains; they dissociate from the ribosome when translation is terminated. Elongation factor G (EF-G), also called *translocase,* is associated with the movement of the peptidyl tRNA from the "A" site to the "P" site of the ribosome. Elongation factor T (EF-T) is responsible for alignment of the AA⌐tRNA complex in the "A" site of the ribosome. *See* **translation.**

emasculation **1.** the removal of the anthers from a flower. **2.** castration.

EMB agar a complex bacterial growth medium containing two *p*H- sensitive dyes, eosin and methylene blue, and a sugar such as lactose. Lac⁺ cells preferentially metabolize lactose oxidatively, resulting in excretion of hydrogen ions; under low *p*H, the dyes become deep purple. Lac⁻ cells that use amino acids as a source of energy release NH_3, thus raising *p*H; this decolorizes the dyes and produces a white colony. These color differences occur for all metabolizable sugars and also for many other carbon sources.

Embden-Meyerhof-Parnas pathway glycolysis.

embryo an organism in early stages of development, especially before hatching from the egg when it is dependent upon its own yolk supply for nutrition. The developing human is designated as an embryo up to the beginning of the third month of pregnancy. Subsequently, when the body shape is essentially formed, the term *fetus* is used.

embryonic induction *See* **induction.**

embryo polarity mutants *See* **maternal polarity mutants.**

embryo sac the female gametophyte of angiosperms. It contains several haploid nuclei formed by the division of the haploid megaspore nucleus. *See* **synergid.**

embryo transfer (transplantation) artificial introduction of an early embryo into the oviduct or uterus of the biological mother or of a surrogate mother. For example, a woman with blocked Fallopian tubes may choose to have an egg fertilized *in vitro,* and the resulting embryo may be introduced into her uterus for implantation. In the case of a superior dairy or beef cow, many eggs can be simultaneously artificially induced (superovulation), flushed from the oviducts, fertilized *in vitro,* and each embryo transferred to a different surrogate mother for subsequent development.

emergent properties in a hierarchical system, properties that are manifest at higher levels because they are formed by measures or processes involving aggregations of lower-level units. For example, in an ecological hierarchy, populations have properties not expressed by an individual; a community has properties not expressed in a population, etc.

emmer wheat *Triticum dicoccum* (N = 14) (also called *starch wheat* or *two-grained spelt*), a wheat cultivated since Neolithic times. *See* **wheat.**

EMS ethyl methane sulfonate (*q.v.*).

enantiomers, enantiomorphs compounds showing mirror-image isomerism.

endemic native to a region or place; not introduced. In epidemiology, referring to a disease that occurs continually in a given geographical region.

endergonic reaction a reaction requiring energy from the outside before reactants will form products.

end labeling the attachment of a radioactive chemical (usually ^{32}P) to the 5′ or 3′ end of a DNA strand.

endocrine system a system of ductless glands that controls metabolism through the synthesis and release into the bloodstream of hormones (*q.v.*).

endocytosis the uptake by a cell of particles, fluids, or specific macromolecules by phagocytosis, pinocytosis, or receptor-mediated endocytosis, respectively.

endocytotic vesicles *See* **receptor-mediated endocytosis.**

endoderm the layer of cells lining the primitive gut (archenteron) in an early embryo, beginning in the gastrula stage. The endoderm forms the epithelial lining of the intestine and all of the outgrowths of the intestine (gill pouches and gills, the larynx, wind pipe and lungs, the tonsils, thyroid, and thymus glands, the liver, the gall bladder and bile duct, the pancreas, and the urinary bladder and adjacent parts of the urogenital system).

endogamy the selection of a mate from within a small kinship group. Inbreeding. *Contrast with* **exogamy.**

endogenote that portion of the original chromosome of a merozygote that is homologous to the exogenote (*q.v.*).

endogenous originating within the organism.

endogenous virus an inactive virus that is integrated into the chromosome of its host cell, and can therefore exhibit vertical transmission (*q.v.*).

endometrium the glandular lining of the uterus of mammals that undergoes cyclical growth and regression during sexual maturity.

endomitosis chromosomal replication within a cell nucleus that does not divide.

endomixis a process of self-fertilization in which the sperm and egg nuclei from one individual unite.

endonuclease an enzyme that breaks the internal phosphodiester bonds in a DNA molecule. Endonucleases of somatic tissue hydrolyze DNA by introducing double-strand breaks. Endonucleases isolated from cells in meiotic prophase produce single-strand breaks in the DNA with 5′hydroxy termini. Single-strand breaks are essential first steps in replication and recombination. *See* **Appendix C,** 1971, Howell and Stern; **restriction endonuclease.**

endoplasmic reticulum (*abbreviated* **er**) a system of cisternae in the cytoplasm of many cells. In places the er is continuous with the plasma membrane or the outer membrane of the nuclear envelope. If the outer surfaces of the er membranes are coated with ribosomes, the er is said to be rough-surfaced; otherwise it is said to be smooth-surfaced. *See* **Appendix C,** 1953, Porter; 1960, Siekevitz and Palade; 1965, Sabatini *et al.,* **chloroplast er, leader sequence peptide.**

endopolyploidy the occurrence in a diploid individual of cells containing 4C, 8C, 16C, 32C, etc., amounts of DNA in their nuclei. The nurse-cell nuclei in *Drosophila* egg chambers are good examples of endopolyploidy.

Endopterygota Holometabola (*q.v.*).

endorphins one of a group of mammalian peptide hormones liberated by proteolytic cleavage from a 29 kd prohormone called pro-opiocortin. Beta-endorphins are found in the intermediate lobe of the pituitary gland and are potent analgesics (pain suppressors). The name "endorphin" is a contraction of "endogenous morphine" because it is produced within the body (endogenously), attaches to the same neuron receptors that bind morphine, and produces similar physiological effects.

endoskeleton an internal skeleton, as in the case of the bony skeleton of vertebrates.

endosperm triploid nutritive cells surrounding and nourishing the embryo in seed plants.

endosymbiont theory the proposal that self-replicating eukaryotic organelles originated by fusion of formerly free-living prokaryotes with primitive nucleated cells. *See* **Appendix C,** 1972, Pigott and Carr; 1981, Margulis; **serial symbiosis theory.**

endotherm vertebrates that control their body temperature by endogenous mechanisms (e.g., birds and mammals). *Contrast with* **ectotherm.**

endotoxins complex lipopolysaccharide molecules that form an integral part of the cell wall of many Gram-negative bacteria and are released only when the integrity of the cell is disturbed. Endotoxins are capable of nonspecifically stimulating certain immune responses. *Compare with* **exotoxin.**

end product a compound that is the final product in a chain of metabolic reactions.

end product inhibition the situation in which the activity of an initial enzyme in a chain of enzymatic reactions is inhibited by the final product; feedback inhibition.

end product repression the situation in which the final product in a chain of metabolic reactions functions as a corepressor and shuts down the operon turning out the enzymes that control the reaction chain.

energy-rich bond a chemical bond that releases a large amount of energy upon its hydrolysis. The ATP molecule has an energy-rich phosphate bond, for example.

Engystomops pustulosus a tropic frog particularly suited to biochemical investigations of oogenesis because of the synchronous development of the egg masses.

enhancers sequences of nucleotides that potentiate the transcriptional activity of physically linked genes. The first enhancer to be discovered was a 72 bp, tandem repeat located near the replication origin of simian virus 40. Enhancers were subsequently found in the genomes of eukaryotic cells and in RNA viruses. Some enhancers are constitutively expressed in most cells, while others are tissue specific. Enhancers act by increasing the number of RNA polymerase II molecules transcribing the linked gene. An enhancer may be distant from the gene it enhances. The enhancer effect is mediated through sequence-specific DNA-binding proteins. These observations have led to the suggestions that, once the DNA-binding protein attaches to the enhancer element, it causes the intervening nucleotides to loop out to bring the enhancer into physical contact with the promoter of the gene it enhances. This loop structure then facilitates the attachment of polymerase molecules to the transcribing gene. *See* **Appendix C,** 1981, Banerji *et al.;* **DNA looping.**

enkephalins pentapeptides with opiate-like activity (*compare with* **endorphins**), first isolated in 1975 from pig brain. Met-enkephalin has the amino acid sequence Tyr-Gly-Gly-Phe-Met; Leu-enkephalin has the sequence Tyr-Gly-Gly-Phe-Leu. *See* **polyprotein.**

enol forms of nucleotides *See* **tautomeric shift.**

enrichment methods for auxotrophic mutants *See* **filtration enrichment, penicillin enrichment technique.**

enterovirus a member of a group of RNA-containing viruses (including poliomyelitis virus) inhabiting the human intestine.

entoderm endoderm (*q.v.*).

entomophilous designating flowers adapted for pollination by insects.

enucleate to remove the nucleus from a cell.

enucleated lacking a nucleus.

environment the complex of physical and biotic factors within which an organism exists.

environmental variance that portion of the phenotypic variance caused by differences in the environments to which the individuals in a population have been exposed.

enzymes protein catalysis. Enzymes differ from inorganic catalysts in their extreme specificity. They catalyze reactions involving only one or a few closely related compounds, and they are able to distinguish between stereoisomers. The main groups of enzymes are oxidoreductases, transferases, hydrolases, lyases, isomerases, and ligases. All chemical reactions have potential energy barriers. In order to pass this barrier, the reactant must be activated to reach a transition state from which the products of the reaction can be released. During an enzyme-catalyzed reaction, an enzyme-substrate complex is formed. The E-S complex has a lowered activation energy, and this allows the reaction to occur at body temperature. *See* **Appendix C,** 1878, Kuhne; 1926, Sumner; **abzymes, ribozyme.**

enzyme induction *See* **inducible enzyme.**

eobiogenesis the first instance of the generation of living matter from inorganic material.

Eocene an epoch in the Tertiary period. *See* **geologic time divisions.** Angiosperms and gymnosperms were the dominant plants, and representatives of all mammalian orders were present.

eon the most inclusive of the divisions of geologic time. *See* **geologic time divisions.**

Ephestia kühniella the Mediterranean meal moth. Numerous mutants affecting the color pattern of the forewing and eye pigmentation have been studied in this species. It is parasitized by *Habrobracon juglandis* (*q.v.*). *See* **Appendix C,** 1935, Kuhn and Butenandt.

epicanthus a fold of skin extending over the inner corner of the eye characteristic of members of the Mongoloid race.

epicotyl the portion of the seedling above the cotyledons that develops into the shoot and its derivatives.

epidemiology the study of disease epidemics, with an effort to tracing down the cause.

epidermal growth factor (EGF) a serum protein that stimulates cell division of a wide variety of cells in tissue culture and presumably can also do so *in vivo.* EGF binds to the plasma membrane of sensitive cells and thereby stimulates the enzymatic activity of a protein kinase located within the cell membrane. This kinase phosphorylates tyrosine residues in polypeptides as does the product of gene *src* of the Rous sarcoma virus. Both kinases are antigenically related but are not structurally identical. *See* **vinculin.**

epigamic serving to attract or stimulate members of the opposite sex during courtship.

epigamic selection *See* **sexual selection.**

epigenesis the concept that an organism develops by the new appearance of structures and functions, as opposed to the hypothesis that an organism develops by the unfolding and growth of entities already present in the egg at the beginning of development (preformation).

epigenetics the study of the mechanisms by which genes bring about their phenotypic effects.

epigenotype the series of interrelated developmental pathways through which the adult form is realized.

Epilobium hirsutum hairy willow-herb, the subject of classic studies on cytoplasmic inheritance.

epimers organic compounds that are partial isomers in that they differ from each other only in the three-dimensional positioning of atoms about a single asymmetric carbon atom.

epinephrine a hormone from the adrenal medulla that elevates blood glucose by mobilizing glycogen reserves.

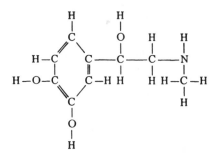

episome a class of genetic elements of which phage lambda and sex factor F are examples in *Escherichia coli.* Episomes may behave (1) as autonomous units replicating in the host independent of the bacterial chromosome, or (2) as integrated units attached to the bacterial chromosome and replicating with it. *Compare with* **plasmid.**

epistasis the nonreciprocal interaction of nonallelic genes. The situation in which one gene masks the expression of another. The recessive gene *apterous (ap)* in *Drosophila* produces wingless homozygotes. In such individuals, any other recessive gene affecting wing morphology will have its action masked. The *apterous* gene is said to be epistatic to a gene like *curled wing,* which is hypostatic to *ap. See* **Appendix C,** 1902–09, Bateson; **Bombay blood group.**

epistatic gene *See* **epistasis.**

epithelium a tissue that forms the surface of an organ or organism. For example, the outer skin is an epithelial tissue; the cells lining the gut and respiratory cavities are also epithelial cells.

epitope the antigenic determinant on an antigen to which the paratope on an antibody binds.

epizoite a nonparasitic sedentary protoctist or animal living attached to another animal.

epoch in geological time, a major subdivision of a geological period. *See* **geologic time divisions.**

eponym a term or phrase derived from the name of a person or place. Eponyms often honor the discoverer of a law or phenomenon, as in *Sewall Wright effect;* or the inventor of a technique, as in *Miller spread.* Genetic diseases are often named after the physicians who first described them (i.e., *Tay-Sachs disease*) or, more rarely, after the patient (i.e., *Christmas disease*). Geographic eponyms also occur in the genetic literature, as in *Bombay blood group.*

epoxide a family of chromosome-breaking, alkylating agents. Di(2,-3 epoxy)propyl ether is an example.

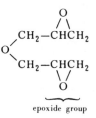

epoxide group

Epstein-Barr virus a DNA virus of the herpes group discovered in 1964 by M. A. Epstein and Y. M. Barr in cultures of Burkitt lymphoma cells. EBV is the cause of infectious mononucleosis, and it has an integration site on human chromosome 14. *See* **Burkitt lymphoma.**

equational division a division of each chromosome into equal longitudinal halves that are incorporated into two daughter nuclei. It is the type of division seen in mitosis.

equatorial plate *See* **mitosis.**

equilibrium centrifugation *See* **centrifugation separation.**

equilibrium dialysis *See* **dialysis.**

equilibrium population a population in which the allelic frequencies of its gene pool do not change through successive generations. An equilibrium can be established by counteracting evolutionary forces (e.g., a balance between selection and mutation pressures) or by the absence of evolutionary forces. *See* **Hardy-Weinberg law.**

equine referring to members of the horse family, especially the domestic horse *Equus caballus.*

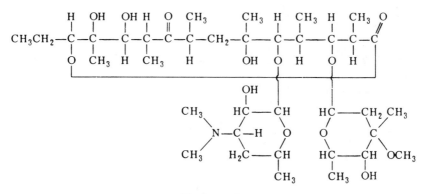

Erythromycin.

Equus the genus that contains two domesticated species: *E. caballus,* the horse (2n = 64), and *E. asinus,* the donkey (2n = 62). In the horse, about 20 genes have been assigned to 5 linkage groups. *See* **horse** for a listing of breeds. *Also see* **horse-donkey hybrids.**

er endoplasmic reticulum (*q.v.*).

era in geological time, a major division of a geological eon. *See* **geologic time divisions.**

ergastoplasm rough-surfaced endoplasmic reticulum.

ergosome polysome (*q.v.*).

error-prone repair *See* **SOS repair.**

erythroblastosis the discharge of nucleated red blood corpuscles from the blood-forming centers into the peripheral blood.

erythroblastosis fetalis a hemolytic disease of infants due to Rh incompatibilities between the fetus and its mother. *See* **Appendix C,** 1939, Levine and Stetson; **RhoGAM.**

erythrocyte the hemoglobin-containing cell found in the blood of vertebrates.

erythrolabe *See* **color blindness.**

erythromycin an antibiotic (shown above) produced by *Streptomyces erythieus.* Erythromycin interferes with prokaryotic protein synthesis by inhibiting translocase (*q.v.*).

escaper *See* **breakthrough.**

Escherichia coli the "colon bacillus," the organism about which the most molecular genetics is known. The distances between genes presented on its circular linkage map are measured in minutes, based on interrupted mating experiment (*q.v.*). *E. coli* is of preeminent importance in recombinant DNA research, since it serves as the host for a wide

variety of viral, plasmid, and cosmid cloning vectors. *See* **Appendix C,** 1946, Lederberg and Tatum; 1953, Hayes; 1956, 1958, Jacob and Wollman; 1961, Jacob and Monod, Nirenberg and Matthaei; 1963, Cairns; 1969, Beckwith *et al.;* 1972, Jackson *et al.;* 1973, Cohen *et al.*

essential amino acids amino acids required in the diet of a species because these molecules cannot be synthesized from other food materials; in contrast to nonessential amino acids that can be synthesized by normal members of the species. *See* **amino acid.**

established cell line a cell line (*q.v.*) that demonstrates the potential to be subcultured indefinitely *in vitro.* HeLa cells (*q.v.*) represent an established cell line.

estivate aestivate (*q.v.*).

estradiol a steroidal estrogen.

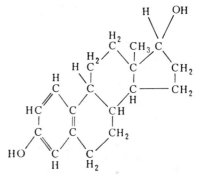

estrogen an ovarian hormone that prepares the mammalian uterus for implantation of an embryo; also responsible for development of secondary sexual characteristics in females.

estrous cycle a seasonal cycle of reproductive activity dependent upon endocrine factors. If the organism has one estrous period per year, it is called monestrous, if more than one, polyestrous.

estrus 1. the period of reproductive activity. 2. the estrous cycle.

ethidium bromide a compound used to separate covalent DNA circles from linear duplexes by density gradient centrifugation. Because more ethidium bromide is bound to a linear molecule than to a covalent circle, the linear molecules have a higher density at saturating concentrations of the chemical and can be separated by differential centrifugation. It is also used to locate DNA fragments in electrophoretic gels because of its fluorescence under ultraviolet light.

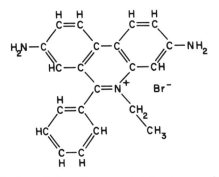

Ethiopian designating or pertaining to one of the six biogeographic realms (*q.v.*) of the world, comprising Africa, Iraq south of the Tropic of Cancer, Madagascar, and the adjacent islands.

ethological behavioral.

ethological isolation the failure of related species or semispecies to produce hybrid offspring because of differences in their mating behaviors.

ethology the scientific study of animal behavior, particularly under natural conditions.

ethylene dinitrilotetra-acetic acid *See* **EDTA.**

ethylenimine an aziridine mutagen (*q.v.*).

ethyl methane sulfonate one of the most commonly used mutagenic alkylating agents. The most common reaction of EMS is with guanine (*q.v.*), where it causes an ethyl group to be added to the number 7 nitrogen. Alkylation of guanine allows it to pair with thymine. Then, during replication, the

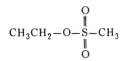

complementary strand receives thymine rather than cytosine. Thus, EMS causes base substitutions of the transition type.

etiolation a plant syndrome caused by suboptimal light, consisting of small, yellow leaves and abnormally long internodes.

etiology the study of causes, especially of disease.

E₁ trisomy syndrome a spectrum of congenital defects in man caused by the presence of an extra chromosome of the E group (18). *See* **Edward syndrome, human mitotic chromosomes.**

eubacteria a subkingdom of the Prokaryotae (*see* **Appendix A**) composed of bacteria that, unlike the archaebacteria, contain neuraminic acid (*q.v.*) in their cell walls. They also differ from archaebacteria in the composition of their tRNAs, rRNAs, and their RNA polymerases.

eucaryote *See* **eukaryote.**

Eucaryotes *See* **Eukaryotes.**

Euchlaena mexicana Mexican fodder grass (teosinte), a species postulated as the ancestor of maize.

euchromatic containing euchromatin.

euchromatin the chromatin that shows the staining behavior characteristic of the majority of the chromosomal complement. It is uncoiled during interphase and condenses during mitosis, reaching a maximum density at metaphase. In polytene chromosomes, the banded segments contain euchromatin. *See* **heterochromatin.**

Euentoma insects with genitalia, Malpighian tubes, and a tracheal system.

eugenics the improvement of humanity by altering its genetic composition by encouraging breeding of those presumed to have desirable genes (positive eugenics), and discouraging breeding of those presumed to have undesirable genes (negative eugenics).

Euglena a genus of unicellular flagellated protoctists often used in studes of molecular and developmental genetics. *See* **Appendix A**, Euglenophyta; **Appendix C**, 1971, Manning and Richards.

eukaryon the highly organized nucleus of an eukaryote.

eukaryote a member of the superkingdom Eukaryotes (*q.v.*).

Eukaryotes the superkingdom containing all organisms that are, or consist of, cells with true nuclei bounded by nuclear envelopes and that undergo meiosis. Cell division occurs by mitosis. Oxidative enzymes are packaged within mitochondria. The superkingdom contains four kingdoms: the Protoctista, the Fungi, the Animalia, and the Plantae. *See* **Appendix A,** Eukaryotes; **Appendix C,** 1937, Chatton. *Contrast with* **Prokaryotes.**

eumelanin   one of the pigment molecules found in the coat and pigmented retinal epithelium of mammals. It is derived from the metabolism of tyrosine and is normally black in color. Its precise coloration is affected by a large number of mutant genes. *See* **melanin.**

Eumetazoa   the subdivision of the animal kingdom containing organisms possessing organ systems, a mouth, and digestive cavity. *See* **Appendix A.**

euphenics   the amelioration of genotypic maladjustments brought about by efficacious treatment of the genetically defective individuals at some time in their life cycles.

euploid   a polyploid cell or organism whose chromosome number is an exact multiple of the basic number of the species from which it originated.

eupyrene sperm   *See* **sperm polymorphism.**

eusocial   a social system in which certain individuals incur obligate sterility, but enhance their fitness by aiding their collateral kin to rear their offspring. For example, sterile female worker bees may rear the offspring of their fertile sister queens. *See* **inclusive fitness.**

euthenics   the control of the physical, biological, and social environments for the improvement of humanity.

eV   electron volt.

evagination   an outpocketing.

eversporting   referring to a strain characterized by individuals that, instead of breeding true, produce variations of a specific sort in succeeding generations. Such strains generally contain mutable genes.

evocation   the morphogenetic effect produced by an evocator.

evocator   the morphogenically active chemical emitted by an organizer.

evolution   **1.** potentially reversible gene frequency changes within a population gene pool; microevolution. **2.** irreversible (*see* **Dollo's law**) genetic changes within a genealogical lineage producing anagenic or phyletic speciation. **3.** cladogenesis or splitting of one species into two. **4.** production of novel adaptive forms worthy of recognition as new taxa; e.g., the appearance of feathers on a reptilian

ancestor is considered to define a new taxon—Aves (birds). Items 2 through 4 are considered to be macroevolution. *See* **Appendix C:** 1831, 1837, Darwin; 1855, Wallace; 1858, Darwin and Wallace; 1859, Darwin; 1872, Gulick; 1937, Dobzhansky, Chatton; 1962, Zuckerkandl and Pauling; 1963, Mayr; 1974, Stebbins; 1967, Spiegelman *et al.;* 1968, Kimura; 1972, Kohne *et al.;* 1981, Margulis; 1983, Kimura and Ohta; **gradualism,** *Origin of Species,* **punctuated equilibrium.**

evolutionarily derived character   *See* **phylogenetic classification.**

evolutionarily primitive character   *See* **phylogenetic classification.**

evolutionary clock   *See* **DNA clock hypothesis, protein clock hypothesis.**

evolutionary rate   rapid evolution is called *tachytelic,* slow evolution is called *bradytelic,* and evolution at an average rate is called *horotelic.*

ewe   a female sheep. *See* **ram.**

exaptation   a character currently subject to selection, but whose origin can be ascribed to processes other than selection or to selection for a different function. *See* **aptation.**

exchange pairing   the type of pairing of homologous chromosomes that allows genetic crossing over to take place. Synaptonemal complexes (*q.v.*) play a critical role in exchange pairing.

excision   the enzymatic removal of a polynucleotide segment from a nucleic acid molecule.

excisionase   an enzyme required (in cooperation with an integrase) for deintegration of prophage from the chromosome of its bacterial host.

excision repair   *See* **cut-and-patch repair, repair synthesis.**

excitation   the reception of a quantum of energy by an atomic electron; an altered arrangement of planetary electrons in orbit about an atom resulting from absorption of electromagnetic energy.

exclusion principle   the principle according to which two species cannot coexist in the same locality if they have identical ecological requirements.

exclusion reaction   the healing reaction of a phage-infected bacterium that strengthens its envelope and prevents entry of additional phages.

exconjugant   **1.** ciliates (e.g., paramecia) that were partners in conjugation and therefore have exchanged genetic material. **2.** a female (F^-) recipient bacterial cell that has separated from a male (Hfr) donor partner after conjugation and therefore contains some of the donor's DNA.

exergonic reaction a reaction proceeding spontaneously and releasing energy to its surroundings.

exocytosis the discharge from a cell of materials by reverse endocytosis (*q.v.*).

exogamy the tendency of an individual to mate selectively with nonrelatives. *Contrast with* **endogamy.**

exogenic heredity transmission from generation to generation of information in the form of knowledge and various products of the human mind (i.e., books, laws, inventions, etc.).

exogenote the new chromosomal fragment donated to a merozygote (*q.v.*).

exogenous virus a virus that replicates vegetatively (productively in lytic cycle) and is not vertically transmitted in a gametic genome.

exon a portion of a split gene (*q.v.*) that is included in the transcript of a gene and survives processing of the RNA in the cell nucleus to become part of a spliced messenger of a structural RNA in the cell cytoplasm. Exons generally occupy three distinct regions of genes that encode proteins. The first, which is not translated into protein, signals the beginning of RNA transcription and contains sequences that direct the mRNA to the ribosomes for protein syntheisis. The exons in the second region contain the information that is translated into the amino acid sequence of the protein. Exons in the third region are transcribed into the part of the mRNA that contains the signals for the termination of translation and for the addition of a polyadenylate tail. *See* **Appendix C,** 1978, Gilbert; **intron, leader sequence, polyadenylation, posttranscriptional modification, terminator.**

exon shuffling the creation of new genes by bringing together, as exons of a single gene, several coding sequences that had previously specified different proteins or different domains of the same protein, through intron-mediated recombination. *See* **trans splicing.**

exonuclease an enzyme that digests DNA, beginning at the ends of the strands.

exonuclease III an enzyme from *E. coli* that attacks the DNA duplex at the 3′ end on each strand; used together with S1 nuclease (*q.v.*) to create deletions in cloned DNA molecules. *Compare with* **Bal 31 exonuclease.**

exonuclease IV an enzyme that specifically degrades single-stranded DNA. It initiates hydrolysis

at both the 3′ and 5′ ends to yield small oligonucleotides. This enzyme is active in the presence of EDTA.

exopterygota hemimetabola (*q.v.*).

exoskeleton a skeleton covering the outside of the body, characteristic of arthropods.

exotoxin a poison excreted into the surrounding medium by an organism (e.g., certain Gram-positive bacteria such as those causing diphtheria, tetanus, and botulism). Exotoxins are generally more potent and specific in their action than endotoxins (*q.v.*).

experimental error 1. the chance deviation of observed results from those expected according to a given hypothesis; also called random sampling error. 2. uncontrolled variation in an experiment. *See* **analysis of variance.**

explant an excised fragment of a tissue or an organ used to initiate an *in vitro* culture.

exponential growth phase that portion of the growth of a population characterized by an exponential increase in cell number with time. *See* **stationary phase.**

exponential survival curve a survival curve without a shoulder, or threshold region, and that plots as a straight line on semilog coordinates.

expression vector cloning vehicles designed to promote the expression of gene inserts. Typically, a restriction fragment carrying the regulatory sequences of a gene is ligated *in vitro* to a plasmid containing a restriction fragment possessing the gene but lacking its regulatory sequences. The plasmid with this new combination of DNA sequences is then cloned under circumstances that promote the expression of the gene under the control of the regulatory sequences.

expressivity the range of phenotypes expressed by a given genotype under any given set of environmental conditions or over a range of environmental conditions. For example, *Drosophila* homozygous for the recessive gene "eyeless" may have phenotypes varying from no eyes to completely normal eyes, but the usual condition is an eye noticeably smaller than normal.

expressor protein the product of a regulatory gene, necessary for the expression of one or more other genes under its positive transcriptional control.

extant living at the present time, as opposed to extinct.

extinction termination of an evolutionary lineage

without descendants. *See* **pseudoextinction, taxonomic extinction.**

extrachromosomal inheritance *See* **extranuclear inheritance.**

extragenic reversion a mutational change in a second gene that eliminates or suppresses the mutant phenotype of the first gene. *See* **suppressor mutation.**

extranuclear inheritance non-Mendelian heredity attributed to DNA in organelles such as mitochondria or chloroplasts; also called extrachromosomal inheritance, cytoplasmic inheritance, maternal inheritance.

extrapolation number in target theory (*q.v.*), the intercept of the extrapolated multitarget survival curve with the vertical logarithmic axis specifying the survival fraction. The extrapolation number gives the number of targets that each must be hit at least once to have a lethal effect on the biological system under study.

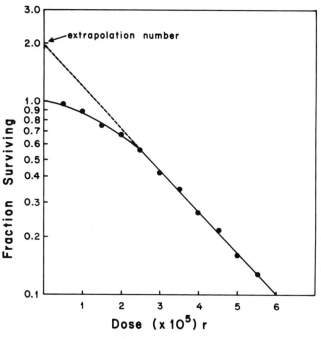

Extrapolation number.

F

F 1. Wright's inbreeding coefficient 2. fluorine. 3. Fahrenheit. 4. fertility factor.

F₁ first filial generation; the offspring resulting from first experimental crossing of the plants or animals. The parental generation with which the genetic experiment starts is referred to as P₁ (*q.v.*). *See* **Appendix C,** 1902–09, Bateson.

F₂ the progeny produced by intercrossing or self-fertilization of F₁ individuals.

Fab that fragment of a papain-digested immunoglobulin molecule that bears a single antigen-binding site and contains one intact light chain and a portion of one heavy chain. *See* **immunoglobulin.**

F(ab)₂ fragment that fragment of a pepsin-digested immunoglobulin molecule that contains portions of two heavy chains and two intact light chains and bears two antigen binding sites. *See* **immunoglobulin.**

Fabry disease a hereditary disease of glycosphingolipid metabolism in humans. The disease results from a mutation in the X-linked gene that codes for lysosomal *alpha galactosidase A* (*q.v.*). The enzyme defect causes the accumulation of glycosphingolipids in lysosomes of blood vessels and leads to vascular malfunctions. Fabry disease can be detected *in utero* by demonstration of a lack of alpha galactosidase A activity in amniocytes of XY fetuses. Heterozygous females can be documented by cloning cultured skin fibroblasts followed by the demonstration of two cell populations, one normal and one lacking enzyme activity (*see* **dosage compensation**). Hemizygotes survive to sexual maturity, so the mutation is transmitted by both heterozygotes and hemizygotes. Prevalence 1/40,000 males.

facet ommatidium (*q.v.*).

factor III *See* **blood clotting.**

factorial a continuing product of factors, calling for the multiplication together of all the integers from one to the given number. Factorial four means 1 × 2 × 3 × 4 = 24. The usual factorial symbol is an exclamation point (!) following the largest number in the series. Thus factorial four is written 4!

facultative having the capacity to live under more than one specific set of environmental conditions. For example, a facultative parasite need not live as a parasite (if another food source is available). A facultative apomict may reproduce sexually or asexually depending upon environmental conditions. *Compare with* **obligate.**

facultative heterochromatin *See* **heterochromatin.**

FAD flavin adenine dinucleotide, a coenzyme (*q.v.*).

Fagus sylvatica the European beech tree.

Fahrenholz's rule the hypothesis that in groups of permanent parasites the classification of the parasites corresponds directly with the natural relationships of their hosts. For example, closely related species of mammals are generally parasitized by closely related species of lice. The rule is based on the assumption that the intimate associations of parasites with their hosts necessitate that they evolve and speciate in harmony with their hosts. As a result of this coevolution, speciation and patterns of divergence in host taxa are paralleled by their parasites. An underlying assumption of Fahrenholz's rule is that there is no dispersal of parasites between unrelated hosts. *See* **resource tracking.**

falciparum malaria *See* **malaria.**

fallout the radioisotopes generated by nuclear weapons that are carried aloft and eventually fall to the earth's surface and contaminate it.

familial Down syndrome *See* **translocation Down syndrome.**

familial hypercholesterolemia (FHC) a human hereditary disease characterized by an elevation in the plasma concentration of low-density lipoproteins (LDL). FHC is inherited as an autosomal dominant at a locus on chromosome 14. The prevalence of heterozygotes is about 1/500 among American, European, and Japanese populations, and this makes FHC among the most common of hereditary diseases. Homozygotes are rare (1 per million in the U.S.A.). The mutant gene product is an LDL receptor. Receptor activity is absent in homozygotes and one-half normal in heterozygotes.

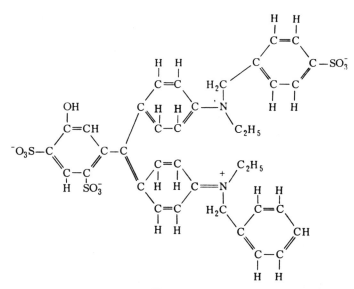

Fast green.

Deficient receptor-mediated endocytosis causes LDL to accumulate in the plasma. Cholesterol is deposited on arterial walls and atherosclerosis results. The disease is far more serious in homozygotes than in heterozygotes. *See* **Appendix C,** 1975, Goldstein and Brown; **plasma lipoproteins.**

family 1. in taxonomy, a cluster of related genera. 2. a set of parents (mother and father, sire and dam, etc.) together with their children (progeny, offspring) constitutes a *nuclear family;* an *extended family* could include half-sibs, aunts, uncles, grandparents, and/or other relatives.

family selection artificial selection of an individual(s) to participate in mating(s) based upon the merits of other members of the same family exclusive of parents and ancestors (e.g., full sibs or half sibs).

Fanconi anemia the first inherited disease in which hereditary chromosome fragility was established. The disease is characterized by a pronounced reduction in the number of erythrocytes, all types of white blood cells, and platelets in the circulating blood. Chromosome aberrations are common and usually involve nonhomologous chromosomes. In the Bloom syndrome (*q.v.*), most chromosome interchanges are between homologous chromosomes. The disease may result from a defect in the transport of enzymes functioning in DNA repair from cytoplasm to nucleus.

fast component 1. in reassociation kinetics, the first components to renature, containing highly repetitive DNA. 2. in electrophoresis, the molecules that move farthest in a given time from the origin.

fast green an acidic dye used in cytochemistry.

fat a glycerol ester of fatty acids. Glycerol tripalmitate may be taken as an example:

$$H_2-C-O-CO-(CH_2)_{14}CH_3$$
$$H-C-O-CO-(CH_2)_{14}CH_3$$
$$H_2-C-O-CO-(CH_2)_{14}CH_3$$

fat body the adipose tissue found in larval insects.

fate map a map of an embryo in an early stage of development that indicates the various regions whose prospective significance has been established by marking methods.

fatty acid an acid present in lipids, varying in carbon content from C_2 to C_{34}. Palmitic acid may be taken as an example: $CH_3(CH_2)_{14}COOH$.

fauna the animal life in a given region or period of time.

favism a hemolytic response to the consumption of beans produced by *Vicia faba* (*q.v.*). Various compounds in fava beans are enzymatically hydrolyzed to quinones, which generate oxygen radicals. Red blood cells deficient in glucose-6-phosphate dehydrogenase (*q.v.*) have a marked sensitivity to oxidating agents and lyse when oxygen radicals are abundant. The toxic properties of fava beans have been known for centuries in Mediterranean cultures where glucose-6-phosphate dehydrogenase deficiency (*q.v.*) is a common hereditary disease.

113

F⁺ cell a bacterial cell possessing a fertility (F) factor extrachromosomally in a plasmid. An F⁻ cell can donate the F factor to an F⁻ cell during conjugation. If the F factor integrates into the bacterial chromosome, the cell becomes an Hfr (*q.v.*), capable of transferring chromosomal genes. *See* **F factor.**

F⁻ cell a bacterial cell devoid of an F factor and that therefore acts only as a recipient ("female") in bacterial conjugation.

F$_c$ fragment that *c*rystallizable *f*ragment (hence the name) of a papain-digested immunoglobulin molecule that contains only portions of two heavy chains and no antibody binding sites. This fragment does, however, bind complement and is responsible for the binding of immunoglobulin to various types of cells in a non-antigen-specific manner. *See* **immunoglobulin.**

F$_c$ receptor a cell surface component of many cells of the immune system responsible for binding the F$_c$ portion of immunoglobulin molecules.

F-duction *See* **sexduction.**

fecundity potential fertility or the capability of repeated fertilization. Specifically, the term refers to the quantity of gametes, generally eggs, produced per individual over a defined period of time. *See* **fertility.**

feedback the influence of the result of a process upon the functioning of the process.

feedback inhibition end product inhibition (*q.v.*).

feeder cells irradiated cells, capable of metabolizing but not of dividing, that are added to culture media to help support the growth of unirradiated cells.

feline leukemia virus an oncogenic RNA virus.

Felis catus the domesticated cat. Its haploid chromosome number is 19, and about 40 genes have been assigned to 16 linkage groups. *See* **cat** for a listing of breeds.

female carrier in human pedigrees, a woman who is heterozygous for a recessive, X-chromosomal gene.

female gonadal dysgenesis Turner syndrome (*q.v.*).

female pronucleus the haploid nucleus of a female gamete, which functions in syngamy.

female-sterile mutation one of a class of mutations that cause female sterility generally because of a developmental block during oogenesis. Recessive female-sterile mutations are common in *Drosoph-*

ila melanogaster and *Bombyx mori.* Dominant female-steriles are much rarer.

female symbol ♀. The zodiac sign for Venus, the goddess of love and beauty in Roman mythology. The sign represents a looking glass.

F-episome *See* **fertility factor.**

F′-episome an F-episome carrying a genetically recognizable fragment of bacterial chromosome.

feral pertaining to formerly domesticated animals now living in a wild state.

fermentation an energy-yielding enzymatic breakdown of sugar molecules that takes place in bacteria and yeasts under anaerobic conditions.

ferritin an iron storage protein found in the liver and spleen, containing up to 20% of its weight in the form of iron. It consists of a protein component (apoferritin) and colloidal micelles of ferric hydroxide-ferric phosphate. Ferritin is often conjugated to proteins such as immunoglobulins, thus enabling their locations within tissues to be determined in electron micrographs due to the great electron scattering property of the iron atoms. *See* **Appendix C,** 1959, Singer.

ferritin-labeled antibodies *See* **ferritin.**

fertility the productivity of an individual or population in terms of generating viable offspring. The term is often used to refer to the number of offspring generated by a female during her reproductive period. In human genetics, the term *effective fertility* is used to refer to the mean number of offspring produced by individuals suffering from a hereditary disease as compared to the mean number of offspring produced by individuals free from the disease, but otherwise very similar. Effective fertility thus gives an indication of the selective disadvantage of the disease. *Compare with* **fecundity.**

fertility factor *See* **circular linkage map, F factor, F-prime factor, Hfr strain.**

fertility restorer a dominant nuclear gene in corn that nullifies the effect of a cytoplasmic male-sterility factor.

fertilization the union of two gametes to produce a zygote. *See* **Appendix C,** 1769, Spallanzani; 1875, Hertwig; **double fertilization, syngamy.**

fertilization cone a conical projection protruded from the surface of certain eggs at the point of contact with the fertilizing sperm.

fertilization membrane a membrane that grows outward from the point of contact of the egg and sperm and rapidly covers the surface of the egg.

fertilizin a substance secreted by the ovum of some species, that attracts sperm of the same species.

fetus *See* **embryo.**

Feulgen-positive stained by the Feulgen procedure and therefore containing DNA.

Feulgen procedure a cytochemical test that utilizes the Schiff reagent (*q.v.*) as a stain and is specific for DNA. *See* **Appendix C,** 1923, Feulgen and Rossenbeck.

F factor (fertility factor) a supernumerary sex chromosome, symbolized by F, that determines the sex of *E. coli.* In the presence of the F episome, the bacterium functions as a male. F is a circular DNA molecule made up of about 94,000 base pairs, about 2.5% the amount in the *E. coli* chromosome. About one third of the genes in the F chromosome are involved in the transfer of male genetic material to the female, including the production of the F-pilus, a hollow tube through which DNA is transferred during conjugation. *See* **circular linkage map, F-prime factor, Hfr strain.**

F′ factor *See* **F-prime factor.**

fibrin *See* **blood clotting.**

fibrinogen *See* **blood clotting.**

fibroblasts spindle-shaped cells responsible for the formation of extracellular fibers such as collagen (*q.v.*) in connective tissues.

fibroin the major protein component of silk (*q.v.*).

fibronectin a dimer made up of two similar protein subunits. Each has an Mr of 250,000, and the two are joined at one end by disulfide bonds. The proteins are modular in the sense that they are divided into a series of domains, each with specific binding properties. For example, there are different domains that bind specifically to actin, to collagen, and to certain receptor proteins embedded in the plasma membranes of cells. Fibronectin mediates the attachment of cells to collagenous substrates, participates in the organization of stress fibers, and facilitates cell-to-cell adhesions. The fibronectin gene contains a series of exons, and there is one-to-one correspondence of exons to the protein-binding domains. Fibronectins exist in a variety of isoforms, many of which result from alternative splicing (*q.v.*).

Ficoll a nonionic synthetic polymer of sucrose.

field *See* **prepattern.**

filaform thread-shaped.

filamentous phage a bacterial virus (e.g., M13, fd)

that specifically infects male (donor) cells and carries a single strand of DNA within a filamentous protein coat. A filamentous phage forms a double-stranded replicative form (*q.v.*) during its life cycle.

filial generations any generation following the parental generation. Symbolized F_1, F_2, etc.

filopodia very thin, fingerlike extensions of the plasma membrane; used by cells that move by amoeboid locomotion.

filter hybridization exposing DNA, denatured to single strands and immobilized on a nitrocellulose filter, to a solution of radioactively labeled RNA or DNA; only hybrid double-stranded molecules remain on the filter after washing. *Compare with* **liquid hybridization.**

filter route a migration path along which only a few species can easily disperse. *Compare with* **corridor.**

filtration enrichment a method for the isolation of nutritional mutants in fungal genetics. Mutagenized spores are placed upon a minimal medium. Normal spores germinate and send out an extensive mycelial network. These colonies are then filtered off, and the remaining germinated spores that show poor mycelial development are grown upon a supplemented medium, where each produces enough mycelia to allow further propagation and study.

finalism a philosophy that views evolution as being directed (by some rational force) toward an ultimate goal. *See* **teleology** (of which this is a special case); *see also* **orthogenesis.**

fine-structure genetic mapping the high-resolution analysis of intragenic recombination down to the nucleotide level.

fingerprint the pattern of the ridge skin of the distal surface of a finger.

fingerprinting technique in biochemistry, a method employed to determine differences in amino acid sequences between related proteins. The protein under study is enzymatically cleaved into a group of polypeptide fragments. These are separated in two dimensions using chromatography or electrophoresis. The result will be a two-dimensional array of spots, the "fingerprint." This is compared to the standard fingerprint. The difference in the position of one spot in the case of the Hb^S and Hb^A fingerprints led to the discovery that the normal and mutant hemoglobins differed in a single amino acid substitution. Similar methods have been used to determine the structure of nucleic acids. *See* **Appendix C,** 1957, Ingram; 1965, Sanger *et al.;* 1985, Jeffries *et al.;* **DNA fingerprint technique.**

first-arriver principle   a theory proposing that the first individuals to colonize a new environment or to become adapted to a specific niche acquire thereby a selective advantage over later arrivals, merely because they got there first; also known as "king-of-the-mountain" principle.

first cousin   *See* **cousin.**

first-division segregation ascus pattern   in ascomycetes, a 4-4 linear order of spore phenotypes within an ascus. This pattern indicates that a pair of alleles (e.g., those controlling spore pigmentation) separated at the first meiotic division, because no crossovers occurred between the locus and the centromere. *See* **ordered tetrad.**

first-order kinetics   the progression of an enzymatic reaction in which the rate at which the product is formed is proportional to the prevailing substrate concentration, with the result that the rate slows gradually, and the reaction never goes to completion. *See* **zero-order kinetics.**

fission   **1.** binary fission (*q.v.*). **2.** nuclear fission (*q.v.*).

fitness   the relative ability of an organism to survive and transmit its genes to the next generation.

fixation   the first step in making permanent preparations of tissues for microscopic study. The procedure aims at killing cells and preventing subsequent decay with the least distortion of structure. *See* **fixative.**

fixation, genetic   the status of a locus in which all members of a population are homozygous or hemizygous for a given allele; the frequency of the fixed allele is 1.0; all other alleles at that locus have been lost and therefore their frequencies are zero. *See* **monomorphic population.**

fixative   a solution used for the preparation of tissues for cytological or histological study. It precipitates the proteinaceous enzymes of tissues and so prevents autolysis, destroys bacteria that might produce decay of the tissue, and causes many of the cellular constituents to become insoluble.

fixing   in photography, the removal of the unchanged halide after the image is developed. An aqueous solution of sodium thiosulfate (hypo) is used.

flagellate   a protoctist belonging to the Zoomastigina or Euglenophyta. *See* **Appendix A.**

flagellin   the major protein of bacterial flagellae. Its molecular weight is about 20,000 d. Flagellin

molecules appear to be wound in a tight spiral within each flagellum.

flagellum (*plural,* **flagella**)   **1.** in bacteria, a whip-like motility appendage present on the surface of some species. Bacterial flagellae are composed of the protein flagellin *(q.v.)* and range in length from 2 to 20 μm. Bacteria having a single flagellum are called *monotrichous;* those with a tuft of flagella at one pole are called *lophotrichous;* and those with flagella covering the entire surface are called *peritrichous.* Antigens associated with flagella are called *H antigens. Compare with* **pilus. 2.** in eukaryotes, flagellum refers to a threadlike protoplasmic extension used to propel flagellates and sperm. Flagella have the same basic structure as cilia (*q.v.*), but are longer in proportion to the cell bearing them and are present in much smaller numbers (most sperm are monotrichous). *See* **axoneme, undulipodium.**

flanking DNA   nucleotide sequences on either side of the region under consideration. For example, the hallmarks of a transposon (*q.v.*) are (1) it is flanked by inverted repeats at each end, and (2) the inverted repeats are flanked by direct repeats.

flavin adenine dinucleotide (FAD)   a coenzyme composed of riboflavin phosphate and adenylic acid (formula on facing page). FAD forms the prosthetic group of enzymes such as d-amino acid oxidase and xanthine oxidase.

flavin mononucleotide   riboflavin phosphate, a coenzyme for a number of enzymes including l-amino acid oxidase and cytochrome c reductase.

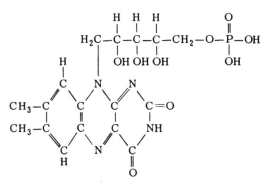

flavoprotein   a protein requiring FMN or FAD to function.

flint corn   *See* **corn.**

flora   the plant life in a given region or period of time.

floret   a small flower from an inflorescence, as in a grass panicle.

flour corn   *See* **corn.**

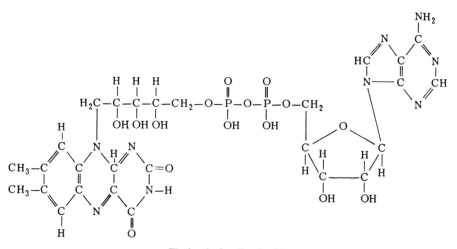

Flavin adenine dinucleotide.

flow cytometry a technology that utilizes an instrument in which particles in suspension and stained with a fluorescent dye are passed in single file through a narrow laser beam. The fluorescent signals emitted when the laser excites the dye are electronically amplified and transmitted to a computer. This is programmed to instruct the flow cytometer to sort the particles having specified properties into collecting vessels. Human mitotic chromosomes can be sorted to about 90% purity by this technique.

flower the specialized reproductive shoot of angiosperms (*q.v.*). Perfect flowers bear both pistils and stamens. Imperfect flowers bear either pistils or stamens. A plant bearing solely functional staminate or pistillate flowers is said to be *dioecious*. A plant bearing (1) imperfect flowers of both sexual types (corn is an example), or (2) perfect flowers (the garden pea), or (3) staminate, pistillate, and perfect flowers (the red maple, for example) is said to be *monoecious*.

fluctuation test a statistical analysis first used by Luria and Delbrück to prove that selected variants (*q.v.*), such as bacteriophage-resistant bacteria, are spontaneous mutants that arose prior to the exposure to the selective agent. They reasoned that if bacterial mutation is an event that is rare, discontinuous, and random, then there should be a marked fluctuation in the number of resistant variants present, at a given time, in a large number of independent cultures, each of which was grown from a small inoculum. The number of variants per sample would fluctuate because some cultures would contain large numbers of variants arising from the division of an early mutant, while in other cultures mutants occurring later during the growth

of the culture would produce clones of smaller size. Conversely, in separate samples taken from a single culture inoculated under identical conditions the variability in the number of mutants should be much less. However, if the agent used for selection induced the mutations, then the distribution of the number of mutants in any population of samples should be independent of the previous history of the culture. Since it was found that the variance was much larger when samples came from independent cultures than when they were taken from the same culture, it was concluded that spontaneous mutation was the source of the variants. *See* **Appendix C,** 1943, Luria and Delbrück.

fluid mosaic concept a model in which the cell membrane is considered to be a two-dimensional viscous solution consisting of a bilayer of highly oriented lipids. The layer is discontinuous, being interrupted by protein molecules that penetrate one or both layers. *See* **Appendix C,** 1972, Singer and Nicholson; **lipid bilayer model.**

fluke a common name for flatworms belonging to the class Trematoda. Flukes of medical importance are members of the order Digenea. These parasites have molluscs as intermediate hosts. *See* **schistosomiasis.**

fluourescein an orange-red compound that yields a bright-green fluorescence when exposed to ultra-

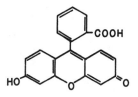

117

violet light. When conjugated to a specific antibody, this dye provides a means of localizing the antigen when the stained cell is viewed with a fluorescence microscope.

fluorescence   *See* **luminescence.**

fluorescence microscopy   the usual methods of microscopical examination are based on observing the specimen in the light transmitted or reflected by it. Fluorescence microscopical preparations are self-luminous. In most biological preparations the tissue sections are stained with a *fluorochrome,* a dye that emits light of longer wavelength when exposed to blue or ultraviolet light. The fluorescing parts of the stained object then appear bright against a dark background. The staining techniques are extremely sensitive and can often be used in living materials.

fluorescent antibody technique   a method for localizing a specific protein or other antigen in a cell by staining a section of the tissue with an antibody specific for that antigen. The antibody is tagged directly or indirectly by a fluorochrome for detection under a fluorescence microscope. *See* **immunofluorescence.**

fluorescent screen   a sheet of material coated with a substance such as calcium tungstate or zinc sulfide that will emit visible light when irradiated with ionizing radiation. Such screens are used in TV sets and as the viewing screens of electron microscopes.

fluorine   a biological trace element. Atomic number 9; atomic weight 18.9984; valence 1⁻; most abundant isotope ^{19}F.

fluorochrome   a fluorescent dye that can be conjugated to a compound that binds to a specific cell component. An example would be a fluorescein-labeled antibody of rodaminylphalloidin (*q.v.*).

flush end   synonymous with blunt end. *See* **restriction endonuclease.**

flying spot cytometer   an instrument used in cytometric DNA measurements. The heterogenous distribution of Feulgen stain within a nucleus leads to distributional errors when a single measurement is made of light transmission to estimate the amount of dye bound in the nucleus. Flying spot cytometers scan a defined microscopic area while making thousands of measurements with a minute measuring stop. The sum of these point absorbance measurements as determined by a built-in computer is proportional to the true absorbance of the specimen. *See* **Feulgen procedure, microspectrophotometer.**

F-mediated transduction   sexduction.

f-met-tRNA   the complex between N-formyl methionine and its transfer RNA.

FMN   flavine mononucleotide, a coenzyme.

foci   **1.** regions of growth of tumor cells appearing as raised clusters above a confluent monolayer of cells in tissue culture. **2.** opaque pocks appearing on the chorioallantoic membrane of a developing chick embryo that has been infected with certain viruses such as herpesviruses.

focus map   a fate map (*q.v.*) for regions of the *Drosophila* blastoderm determined to become adult structures, inferred from the frequencies of specific kinds of mosaics.

foldback DNA   single-stranded regions of DNA that have renatured by intrastrand reassociation between inverted repeats; hairpin DNA.

folic acid   the anti-pernicious-anemia vitamin. It is a compound made up of three components: a pteridine (*q.v.*), p-aminobenzoic acid, and glutamic acid.

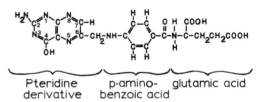

Pteridine   p-amino-   glutamic acid
derivative   benzoic acid

The active form of folic acid is tetrahydrofolate. This compound contains hydrogen atoms attached to nitrogens 5 and 8 and carbons 6 and 7. The enzyme dihydrofolate reductase catalyzes certain of these addition reactions. Tetrahydrofolate is an essential coenzyme in the biosynthesis of thymidylic acid. Thus, folic acid analogues like aminopterin and methotrexate block nucleic acid synthesis.

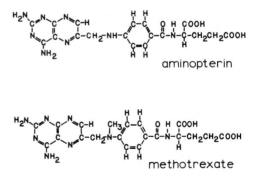

aminopterin

methotrexate

follicle-stimulating hormone   a glycoprotein hormone that stimulates the growth of ovarian follicles and estrogen secretion. It is produced by the adenohypophysis of vertebrates. Abbreviated FSH.

Følling disease *See* **phenylketonuria.**

footprinting a technique for identifying a segment of a DNA molecule that is bound to some protein of interest, on the principle that the phosphodiester bonds in the region covered by the protein are protected against attack by endonucleases. A control sample of pure DNA and one of protein-bound DNA are subjected to endonuclease attack. The resulting fragments are electrophoresed on a gel that separates them according to their lengths. For every bond position that is susceptible, a band is found on the control gel. The gel prepared from the protein-bound DNA will lack certain bands, and the missing bands identify the length of the site covered by the protein.

Forbes disease a hereditary glycogen storage disease in humans arising from a deficiency of the enzyme amylo-1,6-glucosidase. Inherited as an autosomal recessive. Prevalence 1/100,000.

formaldehyde CH_2O, a colorless gas readily soluble in water and having mutagenic properties. *See* **Appendix C,** 1946, Rapoport; **formalin.**

formalin an aqueous solution of formaldehyde (*q.v.*) commonly used as a fixative, which functions through cross linking protein molecules.

formamide a small organic molecule that combines with the free NH_2 groups of adenine and prevents the formation of A-T base pairs, thereby causing denaturation of double-stranded DNA. *See* **stringency.**

$$\overset{O}{\underset{H-C-NH_2}{\|}}$$

formylkynurenine the vermilion-plus substance (*q.v.*) in *Drosophila melanogaster*. *See* **Drosophila eye pigments.**

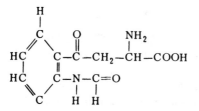

formyl methionine *See* **N-formyl methionine.**

forward mutation a change in a gene from wild-type (normal) allele to a mutant (abnormal) allele.

fossil any remains or traces of former life, including shells, bones, footprints, etc. exposed in rocks.

founder cells *See* **compartmentalization.**

founder effect the principle that when a small sample of a larger population establishes itself as a newly isolated entity, its gene pool carries only a fraction of the genetic diversity represented in the parental population. The evolutionary fates of the parental and derived populations are thus likely to be set along different pathways because the different evolutionary pressures in the different areas occupied by the two populations will be operating on different gene pools. *See* **Appendix C,** 1980, Templeton.

fox *See* **Vulpes vulpes.**

fowl *See* **poultry breeds.**

fowl achondroplasia hereditary chondrodystrophy affecting certain breeds of chickens (Scots, Dumpies, Japanese Bantams). The condition is inherited as an autosomal dominant. Homozygotes die as embryos.

fowl leukosis *See* **leukemia.**

F-pilus *See* **F factor.**

F-prime (F′) factor a bacterial episomal fertility (F) factor containing an additional portion of the bacterial genome. F-prime factors have been most extensively studied in *E. coli.*

fractionated dose the treatment of an organism by a series of short exposures to mutagenic radiations.

fraction collector an automated instrument that collects consecutive samples of fluids percolating through a column packed with porous material.

Fragaria chiloensis the majority of cultivated strawberries are derived from this species.

fragile chromosome site a nonstaining gap of variable width that usually involves both chromatids and is always at exactly the same point on a specific chromosome derived from an individual or kindred. Such fragile sites are inherited in a Mendelian codominant fashion and exhibit fragility as shown by the production of acentric fragments and chromosome deletions. In cultured human cells, fragile sites are expressed when the cells are deprived of folate and thymidine. Their expression is also enhanced by the addition of caffeine to the medium. Some fragile sites have been found at chromosomal bands where oncogenes have been mapped.

fragile X-associated mental retardation a moderate degree of mental retardation (IQs around 50) found in males carrying an X chromosome that has a fragile site at the interface of bands q27 and q28. The frequency of such hemizygotes is about 1.8 per 1,000. X-linked mental retardation accounts for about 25% of all mentally retarded males. *See* **Appendix C,** 1969, Lubs.

frame shift mutation *See* **reading frame shift.**

framework region the highly conserved, relatively invariant portion of the variable (V) region of an immunoglobulin chain, as distinguished from the hypervariable segments of the V region.

fraternal twins *See* **twins.**

free energy that component of the total energy of a system that can do work. *See* **thermodynamics, second law of.**

freemartin a mammalian intersex arising due to the masculinization of a female twin by hormones from its male sibling when the fetal circulations are continuous.

free radical an unstable and highly reactive molecule, bearing an atom with an unpaired electron, that nonspecifically attacks a variety of organic structures, including DNA. The interaction of ionizing radiation with water can generate hydroxyl and hydroperoxyl groups (free radicals that are potent oxidizing agents).

freeze-drying a method of dehydrating a cell or solution by rapidly freezing its moisture content to ice. The solid material is then dried in the frozen state under vacuum, so that ice sublimes directly to water vapor with a minimization of shrinkage. *See* **lyophilize.**

freeze-etching a technique for preparing biological material for electron microscopy. Live or fixed specimens are frozen in a liquid gas, such as freon or nitrogen, and then placed in a Balzer freeze-fracture apparatus. This is an instrument that allows frozen tissues to be sectioned in a vacuum. The exposed surface is allowed to sublime slightly (to etch), so that surface irregularities that reflect the type and distribution of cell constituents are accentuated. The surface is then replicated, and the replica is stripped away and viewed under the electron microscope. Preparations made in this way provide useful information concerning three-dimensional organization of protein particles embedded in the lipoidal membranes of cells.

freeze fracture a method for preparing samples for electron microscopy; frozen samples are fractured with a knife and the complementary surfaces are cast in metal. *See* **Appendix C,** 1961, Moor *et al.*

frequency-dependent fitness a phenomenon in which the adaptive value of a genotype varies with changes in allelic frequencies. For example, in Batesian mimicry, mimics have greater fitness when they are rare relative to their models. *See* **mimicry.**

frequency-dependent selection selection involving

frequency-dependent fitness (*q.v.*). *See* **Appendix C,** 1937; L'Héritier and Tiessier; 1951, Petit; **minority advantage.**

Freund's adjuvant a widely used adjuvant containing killed, dried mycobacteria suspended in the oil phase of a water-in-oil emulsion.

Friend leukemia virus a virus-inducing leukemia in mice and rats.

Fritillaria a genus of lilies. Species of this genus are widely used in cytogenetic investigations.

frizzle a feather mutation in domestic fowl. *FF* individuals are "extreme frizzle," with bristle feathers that wear off easily; whereas *Ff* individuals are "mild frizzle," with more normal curly feathers. Frizzle feather keratin shows a poorly ordered crystalline structure, and its amino acid composition is abnormal.

fructification 1. a reproductive organ or fruiting body. 2. the generation of fruit or spore-producing structures by plants.

fructose a six-carbon hexose sometimes called levulose. It is a component of sucrose.

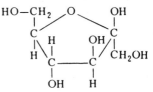

fructose intolerance a disorder of carbohydrate metabolism inherited as an autosomal recessive. Patients lack fructose-1,6-diphosphatase. Symptoms disappear if dietary fructose is restricted.

fruit the ripened ovary of the flower that encloses the seeds.

fruit fly *See* ***Drosophila.***

FSH follicle-stimulating hormone (*q.v.*).

F⁻ strain *Escherichia coli* behaving as recipients during unidirectional genetic transfer.

F⁺ strain *Escherichia coli* behaving as donors during unidirectional genetic transfer. *See* **F factor.**

F test *See* **analysis of variance.**

fundamentalism a conservative religious ideology that holds the origin and diversity of life is by divine creation, based upon a literal interpretation of the biblical account of Genesis.

fundamental theorem of natural selection a theorem developed by R. A. Fisher according to which the increase in fitness of a population at any given time is directly proportional to the genetic variance in fitness of its members.

Fungi the kingdom of fungi, eukaryotic animals that form spores and lack flagellae at all stages of their life cycle. *See* **classification, Eukaryotes.**

funiculus the plant stalk bearing on ovule.

fused protein a hybrid protein molecule produced when a gene of interest is inserted by recombinant DNA techniques into a recipient plasmid and displaces the stop codon for a plasmid gene. The fused protein begins at the amino end with a portion of the plasmid protein sequence and ends with the protein of interest. *Compare with* **polyprotein.** *See* **Appendix C,** 1970, Yourno *et al.*

fusidic acid an antibiotic that prevents translation by interfering with elongation factor G.

fusion gene a hybrid gene, composed of parts of two other genes, arising from deletion of a chromosomal segment between two linked genes or by unequal crossing over. Hemoglobin Lepore (*q.v.*) is an example of a fused gene.

Fy The symbol for alleles of the Duffy blood group system (*q.v.*). Fy^a and Fy^b are responsible for Duffy antigens a and b, respectively, while Fy^- homozygotes produce no antigen.

G

g gravity; employed in describing centrifugal forces. Thus, 2,000 × g refers to a sedimenting force 2,000 times that of gravity.

G guanine or guanosine.

galactose a six carbon sugar that forms a component of the disaccharide lactose and of various cerebrosides and mucoproteins. *See* **beta galactosidase.**

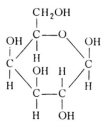

galactosemia a hereditary disease in humans inherited as an autosomal recessive due to a gene on the short arm of chromosome 9. Homozygotes suffer from a congenital deficiency of the enzyme galactosyl-1-phosphate uridyl-transferase, and galactose-1-phosphate accumulates in their tissues. They exhibit enlargement of the liver and spleen, cataracts, and mental retardation. Symptoms regress if galactose is removed from the diet. Prevalence 1/62,000. *See* **Appendix C,** 1971, Meril *et al.*

galactosidase *See* **alpha galactosidase, beta galactosidase.**

Galapagos finches *See* **Geospizinae.**

Galapagos Islands a cluster of 14 islands located in the Pacific Ocean 600 miles west of Ecuador. The species of plants and animals on these islands have been exhaustively studied by evolutionary biologists. *See* **Appendix C,** 1831, H.M.S. Beagle; **Geospizinae.**

gall an abnormal growth of plant tissues.

gallinaceous resembling domestic fowl.

Gallus domesticus the domesticated chicken, the bird for which the most genetic information is available. The haploid chromosome number is 39. The female is the heterogametic sex (ZW), whereas the male is the homogametic sex (ZZ). There are over 40 genes distributed among 10 linkage groups. For a listing of domesticated breeds, *see* **poultry breeds.**

Galton's apparatus an apparatus (illustrated on the facing page) consisting of a glass-faced case containing an upper reservoir where balls are stored. Below the reservoir are arranged row after row of equally spaced pegs that stand out from the wall, and below these are a series of vertical slots. The balls are allowed to fall one at a time through a central opening at the bottom of the reservoir. Since each ball after striking a peg has an equal probability of bouncing to the left or right, most will follow a zigzag course through the pegs and will eventually land in a central slot. The final distribution of balls in the slots will be a bell-shaped one. The apparatus demonstrates how the compounding of random events will generate a family of bell-shaped curves. *See* **Appendix C,** 1889, Galton.

gametangium an organ in which gametes are formed. *See* **antheridium, oogonium.**

gamete a haploid germ cell. *See* **Appendix C,** 1883, van Beneden.

game theory a mathematical theory dealing with the determination of optimum strategies where the policies adopted depend on the most likely behaviors of two or more competitors. Game theory is employed in mathematical models of species competition.

gametic disequilibrium the nonrandom distribution into the gametes in a randomly mating population of the alleles of genes occupying different loci. The nonrandom distribution may result from linkage of the loci in question or because the loci interact with respect to their effects on fitness. *See* **linkage disequilibrium.**

gametic meiosis *See* **meiosis.**

gametic number the haploid number of chromosomes (symbolized by N) characterizing a species.

gametoclonal variation the appearance of new traits in haploid plants that grow in tissue culture from anthers or other reproductive material rather

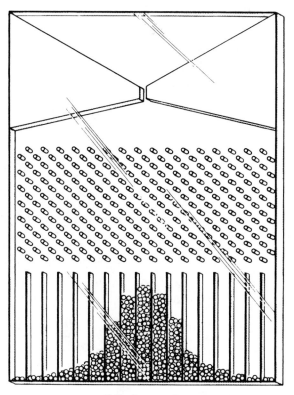

Galton's apparatus.

than from diploid body tissue as in somatoclonal variation (*q.v.*).

gametocyte   a cell that will form gametes through division; a spermatocyte or oocyte.

gametogamy   the fusion of gamete cells or nuclei.

gametogenesis   the formation of gametes.

gametophore   a branch bearing a gametangium or gametangia.

gametophyte   the haploid phase (of the life cycle of plants undergoing an alternation of generations) during which gametes are produced by mitosis. *See* **sporophyte.**

gamma chain   one of the two polypeptides found in fetal hemoglobin (*q.v.*).

gamma field   a field where growing plants may be exposed to chronic irradiation from a centrally placed multicurie ^{60}Co gamma-ray source.

gamma globulin   an antibody-containing protein fraction of the blood. *See* **Appendix C,** 1939, Tiselius and Kabat.

gamma ray   an electromagnetic radiation of short wavelength emitted from an atomic nucleus undergoing radioactive decay.

gamogony   a series of cell or nuclear divisions that eventually lead to the formation of gametes.

gamone   a compound produced by a gamete to facilitate fertilization. Chemotactic sperm attractants produced by eggs are examples.

gamont   the haploid adult form of those protoctists that have both haploid and diploid phases in their life cycles. Gamonts function in sexual reproduction; they undergo gametogony to produce diploid agamonts. Meiosis takes place in agamonts, and the haploid agametes that result disperse, undergo mitotic divisions, and differentiate into gamonts, completing the cycle.

gamontogamy   the aggregation of gamonts during sexual reproduction and the fusion of gamont nuclei to produce agamonts.

ganglion   a small nervous-tissue mass containing numerous cell bodies.

ganglioside   a family of complex lipids containing sphingosine, fatty acids, carbohydrates, and neuraminic acid. The Gm2 ganglioside that accumu-

123

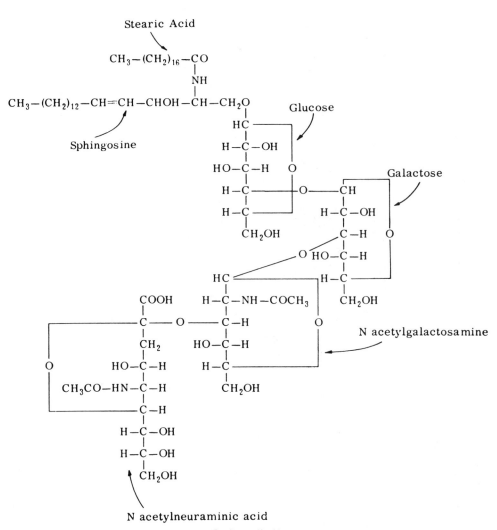

Stearic Acid

$CH_3-(CH_2)_{16}-CO$
|
NH
|
$CH_3-(CH_2)_{12}-CH=CH-CHOH-CH-CH_2O$

Sphingosine

Glucose

HC
|
H—C—OH
|
HO—C—H   O
|
H—C————O————CH
|
H—C   H—C—OH
|
CH$_2$OH

Galactose

C—H   O
|
O HO—C—H
|
H—C
|
CH$_2$OH

HC
|
H—C—NH—COCH$_3$ O
|
COOH   C—H
| |
C——O——C—H
| |
CH$_2$ HO—C—H
| |
HO—C—H   H—C
| |
CH$_3$CO—HN—C—H   CH$_2$OH
|
C—H
|
H—C—OH
|
H—C—OH
|
CH$_2$OH

N acetylgalactosamine

N acetylneuraminic acid

Gm2 ganglioside.

lates in the brain of patients with Tay-Sachs disease (*q.v.*) is shown above. *See* **Appendix C,** 1935, Klenk.

gap the position where one or more nucleotides are missing in a double-stranded polynucleotide containing one broken chain.

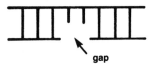

gap

gap genes *See* **zygotic segmentation mutants.**

gargoylism a term covering two genetically dis-

tinct hereditary diseases of connective tissue in man, Hunter syndrome (*q.v.*) and Hurler syndrome (*q.v.*).

Garrod disease *See* **alkaptonuria.**

gas chromatography a chromatographic technique in which an inert gas is used to sweep through a column the vapors of the materials to be separated.

gas-flow radiation counter a counter in which an appropriate atmosphere is maintained in the sensitive volume by allowing a suitable gas to flow slowly through it.

gastrin a hormone secreted by the stomach that causes secretion of digestive enzymes by other stomach cells.

Gastropoda the class of molluscs containing the snails. *See* **Appendix A.**

gastrula the stage of embryonic development when the gastrulation movements occur.

gastrulation the complex cell movements that carry those cells whose descendants will form the future internal organs from their largely superficial position in the blastula to approximately their definitive positions inside the animal embryo. Prior to gastrulation, the amphibian embryo relies on RNA molecules pre-loaded into the ooplasm during oogenesis. During gastrulation, however, newly synthesized nuclear gene products are required.

Gaucher cells large, lipid-filled cells scattered throughout the reticuloendothelial system of patients with Gaucher disease (*q.v.*).

Gaucher disease the most common hereditary disorder of glycolipid metabolism, inherited as an autosomal recessive. The disease is due to subnormal activity of lysosomal glucocerebrosidase, which cleaves glucose from cerebrosides. As a result, glucocerebrosides accumulate in lysosomes. Heterozygotes can be identified because their peripheral leukocytes show lower levels of lysosomal glucocerebrosidases. Homozygotes can be detected *in utero* by assays run on cells obtained by amniocentesis.

Gause's law *See* **competitive exclusion principle.**

Gaussian curve normal distribution.

G banding *See* **chromosome banding techniques.**

Geiger-Mueller (G-M) counter a sensitive gas-filled radiation-measuring device.

gel diffusion technique *See* **immunoelectrophoresis, Ouchterlony technique, Oudin technique.**

gemma (*plural,* **gemmae**) a multicellular, asexual reproductive structure, such as a bud or a plant fragment.

gemmules pangenes. *See* **pangenesis.**

gene a hereditary unit that, in the classical sense, occupies a specific position (locus) within the genome or chromosome; a unit that has one or more specific effects upon the phenotype of the organism; a unit that can mutate to various allelic forms; a unit that recombines with other such units. Three classes of genes are now recognized: (1) *structural genes* that are transcribed into mRNAs, which are then translated into polypeptide chains, (2) *structural genes* that are transcribed into rRNA or tRNA molecules which are used directly, and (3) *regulatory genes* that are not transcribed, but serve as recognition sites for enzymes and other proteins in-

volved in DNA replication and transcription. Seymour Benzer coined the following names for the gene in terms of its operational definitions. A *cistron* is the amount of genetic material that codes for one complete, mature tRNA, rRNA, or polypeptide chain. A cistron includes (if present) regions preceding and following the coding region (leader and trailer), as well as intervening sequences (introns) between coding segments (exons). A *muton* is the smallest unit that, when modified, changes the genetic code (corresponding to a nucleotide). A *recon* is the smallest unit that can experience recombination (corresponding to an adjacent pair of nucleotides in *cis* position). *See* **Appendix C,** 1909, Johannsen; 1933, Morgan; 1955, Benzer; 1961, Jacob and Monod; **operon, replicon.**

gene activation *See* **genetic induction.**

genealogy a record of the descent of a family, group, or person from an ancestor or ancestors; lineage; pedigree.

gene amplification any process by which specific DNA sequences are replicated to a disproportionately greater degree than their representation in the parent molecules. During development, some genes become amplified in specific tissues; e.g., ribosomal genes amplify and become active during oogenesis, especially in some amphibian oocytes (*see* **rDNA amplification, *Xenopus***). Genes encoding *Drosophila* chorion proteins are also amplified in ovarian follicle cells. Gene amplification can be induced by treating cultured cells with drugs like methotrexate (*q.v.*). *See* **Appendix C,** 1968, Gall, Brown and Dawid; 1978, Schimke *et al.*

gene bank *See* **genomic library.**

gene cloning creation of a line of genetically identical organisms, containing recombinant DNA molecules, which can be propagated and grown in bulk, thus amplifying the recombinant molecules.

gene cloning vehicle *See* **lambda cloning vector, plasmid cloning vector.**

gene cluster *See* **multigene family.**

gene conversion a situation in which the products of meiosis from an AA' individual are $3A$ and $1A'$ or $1A$ and $3A'$, not $2A$ and $2A'$ as is usually the case. Thus, one gets the impression that one A gene has been converted to an A' gene (or vice versa). *See* **Appendix C,** 1953, Lindegren.

gene dosage the number of times a given gene is present in the nucleus of a cell.

gene duplication the production of a tandem repeat of a DNA sequence by unequal crossing over or by an accident of replication.

gene expression the manifestation of the genetic material of an organism as a collection of specific traits.

gene family *See* **multigene family.**

gene flow the exchange of genes between different populations of the same species produced by migrants, and commonly resulting in simultaneous changes in gene frequencies at many loci in the recipient gene pool.

gene frequency the percentage of all alleles at a given locus in a population represented by a specific allele. Also referred to as *allelic frequency (q.v.).*

gene fusion the union by recombinant DNA techniques of two or more genes that code for different products, so that they are subject to control by the same regulatory systems.

gene insertion any technique that inserts into a cell a specific gene or genes from an outside source, including cell fusion, gene splicing, transduction, and transformation.

gene interaction interaction between different genes residing within the same genome in the production of a particular phenotype. Such interactions often occur when the products of the non-allelic genes under study function at steps in a sequence of reactions that result in compounds which generate the phenotype in question. These interactions can produce variations from the classical genetic ratios. An example would be the inheritance of aleurone color in *Zea mays*. In order for the corn kernel to possess colored aleurone, at least one *A* and one *C* gene must be present. Given *A* and *C* in the heterozygous or homozygous condition and, in addition, *R* in the heterozygous or homozygous condition, then a red pigment is produced. Purple pigment is synthesized if *P* is present in addition to *A, C,* and *R.* All four genes reside on different chromosomes. Thus, if a plant of genotype *AaCCRRPp* is self-pollinated, the offspring will contain the following aleurone classes: purple, red, and white in a 9:3:4 ratio. Here the 9:3:3:1 ratio has been converted into a 9:3:4 ratio, because the *P* gene cannot be expressed in the absence of *A.*

gene library *See* **genomic library.**

gene manipulation the formation of new combinations of genes *in vitro* by joining DNA fragments of interest to vectors so as to allow their incorporation into a host organism where they can be propagated. *See* **DNA vector, genetic engineering.**

gene mapping assignment of a locus to a specific chromosome and/or determining the sequence of genes and their relative distances from one another on a specific chromosome.

gene pair in a diploid cell, the two representative genes (either identical or nonidentical alleles) at a given locus on homologous chromosomes.

gene pool the total genetic information possessed by the reproductive members of a population of sexually reproducing organisms.

gene networking the concept that there exist functional networks of genes which program early development, and that genes which encode proteins with multiple conserved domains serve to cross-link such networks. Thus, a set of genes containing domain A and a set containing domain B are linked by genes containing both domains. The segmentation gene *paired (prd)* of *Drososphila* illustrates the theory. It contains a homeobox *(q.v.)* and a histidine-proline repeat domain. This *prd*-specific repeat occurs in at least 12 other genes, while the homeobox defines a second gene set. Presumably, the *prd* product can interact with products of genes containing only the homeobox sequence or the histidine-proline repeat, or both. The conserved domains are thought to serve as the sites to which the proteins bind to specific chromosomal regions to regulate neighboring genes. *See* **Appendix C,** 1986, Noll *et al.*

gene product for most genes, the polypeptide chain translated from an mRNA molecule, which in turn is transcribed from a gene; if the RNA transcript is not translated (e.g., rRNA, tRNA), the RNA molecule represents the gene product.

gene 32 protein the first DNA unwinding protein *(q.v.),* to be isolated. It is the product of gene 32 of phage T_4 and is essential for its replication. The protein has a molecular weight of 35,000 d and binds to a stretch of DNA about 10 nucleotides long. *See* **Appendix C,** 1970, Alberts and Frey.

generalized in evolution theory, an unspecialized condition or trait, usually considered to have a greater potential for evolving into a variety of alternative conditions than that possessed by a highly specialized one. Primitive traits tend to be generalized; derived or advanced traits tend to be more specialized.

generalized transduction *See* **transduction.**

generation time (Tg) the time required for a cell to complete one growth cycle. *See* **doubling time.**

gene redundancy the presence in a chromosome of many copies of a gene. For example, the nucleolus organizer of *Drosophila melanogaster* contains hundreds of duplicate copies of the cistrons that code for the 18S and 28S rRNA molecules.

gene splicing   *See* **recombinant DNA technology.**

gene substitution   the replacement of one gene by its allele, all the other genes (or all other relevant genes) remaining unchanged.

gene targeting   a technique for inserting into laboratory mice genetic loci modified in desired ways. Standard recombinant DNA techniques are used to introduce desired chemical changes into cloned DNA sequences of a chosen locus. The mutated sequence is then transferred into an embryo-derived, stem-cell genome, where it is allowed to undergo homologous recombination. Microinjection of mutant stem cells into mouse blastocysts is then performed to generate germ-line chimaeras. Interbreeding of heterozygous siblings eventually yields mice homozygous for the modified locus. *See* **Appendix C,** 1988, Mansour *et al.*

gene therapy   addition of a functional gene or group of genes to a cell by gene insertion (*q.v.*) to correct a hereditary disease.

genetic assimilation   the process by which a phenotypic character initially produced only in response to some environmental influence becomes, through a process of selection, taken over by the genotype, so that it is formed even in the absence of the environmental influence that at first had been necessary.

genetic background   all genes of the organism other than the one(s) under consideration; also known as the *residual genotype.*

genetic block   a block in a biochemical reaction generally due to a mutation that prevents the synthesis of an essential enzyme or results in the formation of a defective enzyme. If the defective enzyme has limited activity, the block may be a partial one, and the mutant is referred to as "leaky."

genetic burden   *See* **genetic load.**

genetic coadaptation   *See* **coadaptation.**

genetic code   the consecutive nucleotide triplets (codons) of DNA and RNA that specify the sequence of amino acids for protein synthesis. The code shown below is used by most organisms, but there are exceptions (*see* **universal code theory**). The mRNA nucleotide sequences are written 5′ to 3′ left to right, respectively, because that is the direction in which translation occurs. Thus, an mRNA segment specifying proline-tryptophan-methionine would be (5′) CCU-UGG-AUG (3′), whereas its complementary, antiparallel DNA template strand would be (3′) GGA-ACC-TAC (5′).

	SECOND BASE				
FIRST BASE	U	C	A	G	THIRD BASE
U	phe	ser	tyr	cys	U
	phe	ser	tyr	cys	C
	leu	ser	Ter	Ter	A
	leu	ser	Ter	trp	G
C	leu	pro	his	arg	U
	leu	pro	his	arg	C
	leu	pro	gln	arg	A
	leu	pro	gln	arg	G
A	ile	thr	asn	ser	U
	ile	thr	asn	ser	C
	ile	thr	lys	arg	A
	met and f met	thr	lys	arg	G
G	val	ala	asp	gly	U
	val	ala	asp	gly	C
	val	ala	glu	gly	A
	val	ala	glu	gly	G

See **amino acid** for symbols. Ter, terminator codon; f met, N-formyl methionine.

Genetic code.

127

The code is degenerate in that all amino acids, except methionine and tryptophan, are specified by more than one codon. Most of the degeneracy involves the third nucleotide at the 3' end of the codon (*see* **wobble hypothesis**). The code is read from a fixed starting point, in one direction, in groups of three consecutive nucleotides. The start codon is AUG, and in bacteria it specifies the insertion of N-formyl methionine (*q.v.*). When AUG occupies an internal position in the mRNA, it specifies methionine. *See* **Appendix C,** 1961, von Ehrenstein and Lipmann, Crick *et al.*, Nirenberg and Matthaei; 1966, Terzaghi *et al.*; 1967, Khorana; 1968, Holley *et al.*; 1979, Barrell *et al.*; 1985, Horowitz and Gorowski, Yamao; **transcription unit.**

genetic code dictionary *See* **amino acid, universal code theory.**

genetic colonization introduction of genetic material from a parasite into a host, thereby inducing the host to synthesize products that only the parasite can use. *See Agrobacterium tumefaciens,* **opine.**

genetic counseling the analysis of risk of producing genetically defective offspring within a family, and the presentation to family members of available options to avoid or ameliorate possible risks.

genetic death death of an individual without reproducing.

genetic detasseling a breeding technique used in the commercial production of corn seed. The breeding scheme produces pollen abortion with the result that the plants are no longer hermaphroditic and can only be cross-fertilized.

genetic differentiation the accumulation of differences in allelic frequencies between isolated or semi-isolated populations due to various evolutionary forces such as selection, genetic drift, gene flow, assortative mating, etc.

genetic dissection analysis of the genetic basis of a biological phenomenon through the study of mutations that affect that phenomenon. For example, spermatogenesis can be "genetically dissected" by inducing and then characterizing mutations that sterilize male *Drosophila.*

genetic distance 1. a measure of the numbers of allelic substitutions per locus that have occurred during the separate evolution of two populations or species. 2. the distance between linked genes in terms of recombination units or map units.

genetic divergence *See* **genetic differentiation.**

genetic drift the random fluctuations of gene frequencies due to sampling errors. While drift occurs in all populations, its effects are most evident in very small populations.

genetic engineering an all-inclusive term to cover all laboratory or industrial techniques used to alter the genetic machinery of organisms so that they can subsequently synthesize increased yields of compounds already in their repertoire, or form entirely new compounds, adapt to drastically changed environments, etc. Often, the techniques involve manipulating genes in ways that bypass normal sexual or asexual transmission. *See* **biotechnology, recombinant DNA technology.**

genetic equilibrium the situation reached in a population containing, as an example, the allelic genes *A* and *a,* where the frequencies of both alleles are maintained at the same values generation after generation. *See* **Hardy-Weinberg law.**

genetic fine structure *See* **fine-structure genetic mapping.**

genetic fitness the contribution to the next generation of a specified genotype in a population in relation to the contributions of all other genotypes of that same population.

genetic hitchhiking *See* **hitchhiking.**

genetic homeostasis the tendency of a population to equilibrate its genetic composition and to resist sudden changes.

genetic identity a measure of the proportion of genes that are identical in two populations.

genetic induction the process of gene activation by an inducer molecule, resulting in transcription of one or more structural genes. *See* **inducible system.**

genetic information the information contained in a sequence of nucleotide bases in a nucleic acid molecule. *See* **exon, intron.**

genetic load 1. the average number of lethal equivalents per individual in a population. 2. the relative difference between the actual mean fitness of a population and the mean fitness that would exist if the fittest genotype presently in the population were to become ubiquitous.

genetic map the linear arrangement of mutable sites on a chromosome as deduced from genetic recombination experiments. *See* **Appendix C,** 1913, Sturtevant.

genetic marker a gene, whose phenotypic expression is usually easily discerned, used to identify an individual or a cell that carries it, or as a probe to mark a nucleus, chromosome, or locus.

genetic polymorphism the long-term occurrence

in a population of two or more genotypes in frequencies that cannot be accounted for by recurrent mutation. Such polymorphism may be due to mutations which are (a) advantageous at certain times and under certain conditions and (b) disadvantageous under other circumstances, and which exist in habitats where situations (a) and (b) are encountered frequently. Genetic polymorphism may also result if genotypes heterozygous at numerous loci are generally superior to any homozygous genotype.

genetics the scientific study of heredity. *See* **Appendix C,** 1902–09, Bateson.

genetic surgery replacement of one or more genes of an organism with the aid of plasmid vectors, or the introduction of foreign genetic material into cells by microsyringes or micromanipulators.

genetic system the organization and method of transmission of the genetic material for a given species.

genetic variance the phenotypic variance of a trait in a population attributed to genetic heterogeneity.

genic balance a mechanism of sex determination, originally discovered in *Drosophila,* that depends upon the ratio of X chromosomes to sets of autosomes (A). Males develop when the X/A ratio is 0.5 or less; females develop when the X/A ratio is 1.0 or greater; an intersex develops when the ratio is between 0.5 and 1.0. *See* **Appendix C,** 1925, Bridges; **metafemales, metamales.**

genital disc the imaginal disc from which the reproductive duct system and the external genitalia are derived in *Drosophila.*

genome a term used to refer to all of the genes carried by a single gamete (i.e., by a single representative of each of all chromosome pairs).

genomic blotting *See* **Southern blotting.**

genomic exclusion an abnormal form of conjugation occurring in *Tetrahymena pyriformis* between cells with defective micronuclei and normal cells. The progeny are heterokaryons; each has an old macronucleus but a new diploid micronucleus derived from one meiotic product of the normal mate.

genomic formula a mathematical representation of the number of genomes (sets of genetic instructions) in a cell or organism. Examples: N (haploid gamete or monoploid somatic cell), 2N (diploid), 3N (triploid), 4N (tetraploid), 2N − 1 (monosomic), 2N + 1 (trisomic), 2N − 2 (nullisomic), etc.

genomic imprinting *See* **parental imprinting.**

genomic library a random collection of fragments of the DNA of a given species inserted into a corresponding collection of vectors (such as plasmids or lambda phages) and cloned in a suitable host. The collection must be large enough to include all the unique nucleotide sequences of the genome. *Compare with* **cDNA library.** *See* **Appendix C,** 1978, Maniatis *et al.*

genopathy a disease resulting from a genetic defect.

genophore the chromosome equivalent in viruses, prokaryotes, and certain organelles (e.g., the discrete, ringlike structure occurring in some algal chloroplasts). Genophores contain nucleic acids, but lack associated histones.

genotype the genetic constitution of an organism, as distinguished from its physical appearance (its phenotype). *See* **Appendix C,** 1909, Johannsen.

genotype-environment interaction an inference drawn from the observation that the phenotypic expression of a given genotype varies when measured under different environmental conditions.

genotype frequency the proportion of individuals in a population that possess a given genotype.

genotypic variance the magnitude of the phenotypic variance for a given trait in a population attributable to differences in genotype among individuals. *See* **heritability.**

genus (*plural* **genera**) a taxon that includes one or more species presumably related by descent from a common ancestor. *See* **hierarchy.**

geochronology a science that deals with the measurement of time in relation to the earth's evolution.

geographical isolate a population separated from the main body of the species by some geographical barrier.

geographic speciation the splitting of a parent species into two or more daughter species following geographic isolation of two or more parental populations; allopatric speciation.

geologic time divisions *See* page 130.

geometric mean the square root of the product of two numbers; more generally, the nth root of the product of a set of n positive numbers.

Geospizinae a subfamily of finches to which the 14 species found on the Galapagos Islands belong. The most marked distinction between the different species is in the size and shape of the bill, which is directly related to the bird's food habits. Charles

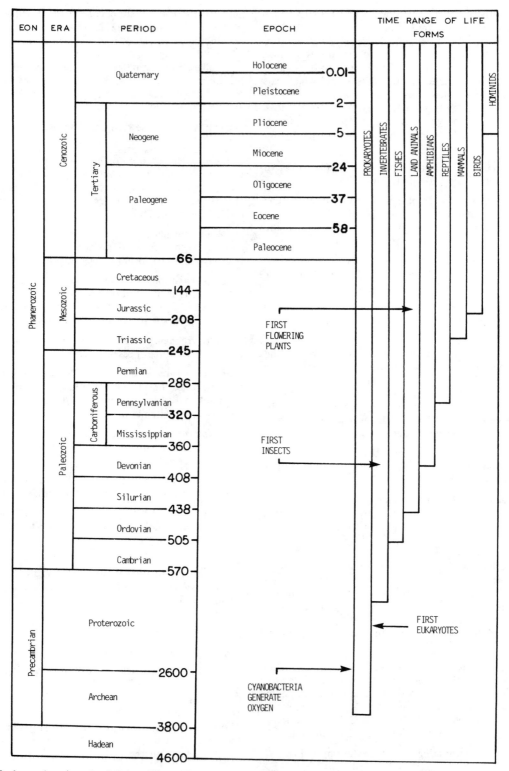

Each number gives the date in millions of years before the present. In the figure, the relative widths of each time division are not proportional to the absolute time spent in them.

Adapted from *The Earth Through Time*, third edition, by Harold L. Levin, W.B. Saunders Co. (1988).

Geologic time divisions.

Darwin was the first to suggest that the modern populations of these birds are the end product of an adaptive radiation from a single ancestral species. The evolutionary divergences resulted from adaptations that allowed different populations to utilize different food resources and so to avoid competition. *See* **Appendix C,** 1837, Darwin, **woodpecker finch.**

geotropism   the response of plant parts to the stimulus of gravity.

germarium   the anterior, sausage-shaped portion of the insect ovariole. It is in the germarium that cystocyte divisions occur and the clusters of cystocytes become enveloped by follicle cells. *See* **insect ovary types.**

germ cell   a sex cell or gamete; egg (ovum) or spermatozoan; a reproductive cell that fuses with one from the opposite sex in fertilization to form a single-celled zygote.

germinal cells   cells that produce gametes by meiosis: e.g., oocytes in females and spermatocytes in males.

germinal choice   the concept advocated by H. J. Muller of progressive human evolution by the voluntary choice of germ cells. Germ cells donated by individuals possessing recognized superior qualities would be frozen and stored in germ banks. In subsequent generations these would be available for couples who wished to utilize these rather than their own germ cells to generate a family. Such couples are referred to as "preadoptive" parents.

germinal mutations   genetic alterations occurring in cells destined to develop into germ cells.

germinal selection   1. selection by man of the germ cells to be used in producing a subsequent generation of a domesticated species. Such selection has been suggested for human beings. 2. selection during gametogenesis against induced mutations that retard the proliferation of the mutated cells. Such selection introduces errors in estimating the frequency of mutations induced in gonial cells.

germinal vesicle   the diploid nucleus of a primary oocyte during vitellogenesis. The nucleus is generally arrested in a postsynaptic stage of meiotic prophase.

germination inhibitor   any of the specific organic molecules present in seeds that block processes essential to germination and therefore are often the cause of dormancy.

germ line   pertaining to the cells from which gametes are derived. When referring to species, the cells of the germ line, unlike somatic cells, bridge the gaps between generations.

germ plasm   the hereditary material transmitted to offspring through the germ cells.

gerontology   the study of aging.

gestation period   in a viviparous animal, the time from conception to birth.

GH   growth hormone (*q.v.*).

Giardia lamblia   a species of protoctist that normally lives attached to the intestinal mucosa of its animal host. However, it can be cultured in the laboratory. The parasite has two nuclei and several flagella, but like prokaryotes it lacks mitochondria, Golgi material, and a rough-surfaced endoplasmic reticulum. Comparisons of the nucleotide sequences of the 16S-like RNAs of *Giardia* and several other eukaryotes show that *Giardia* represents the earliest diverging lineage in the eukaryotic line of descent yet encountered. *See* **Appendix A,** Protoctista, Zoomastigina; **Archaeozoa, ribosomes.**

gibberellin   one of a family of phytohormones of widespread distribution in plants. Many single-gene, dwarf mutants in *Pisum, Vicia,* and *Phaseolus* are cured by the application of gibberellins. Gibberellins also promote seed germination, the breaking of dormancy, and flowering. In maize, hybrids

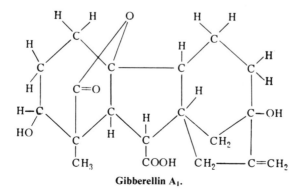

Gibberellin A$_1$.

131

contain higher concentrations of gibberellins than their homozygous parents. This suggests that heterosis (*q.v.*) has a phytohormonal basis.

gibbon *See Hylobates.*

gland an organ that synthesizes specific chemical compounds (secretions) that are passed to the outside of the gland.

gln glutamine. *See* **amino acid.**

globins a widespread group of respiratory proteins including the tetrameric hemoglobins of protochordates and various invertebrates, the monomeric myoglobins, and the leghemoglobins. In the case of myoglobins and the vertebrate hemoglobins, phylogenetic trees have been constructed showing that the globin sequences have diverged during evolution at a constant rate. From these data, the timing of the gene duplications that gave rise to the family of hemoglobins has been deduced. *See* **hemoglobin genes, myoglobin gene.**

globulin any of certain proteins that are insoluble in distilled water, but soluble in dilute aqueous salt solution. *See* **albumin, immunoglobulin.**

Gloger's rule an ecological rule stating that, in a geographically variable species, those populations living in the warmer, moister (more tropical) parts of the range tend to be of darker color than those living in the warmer, dryer (more desert) parts of the range; those living in cold, typically moist (more alpine or arctic) parts of the range tend to be colored very pale or white.

Glossina a genus of viviparous dipterans that serve as vectors of trypanosomes. *G. morsitans* is the African tsetse fly. *See* ***Trypanosoma.***

glu glutamate; glutamic acid. *See* **amino acid.**

glucagon a polypeptide hormone from the alpha cells of the pancreas that promotes the breakdown of liver glycogen and the consequent elevation of blood glucose.

glucocerebroside any of certain compounds related to sphingomyelins. They differ in that the phosphorylcholine seen in sphingomyelin (*q.v.*) is replaced by glucose. *See* **cerebroside.**

glucocorticoids steroid hormones produced by the adrenal cortex having effects on intermediary metabolism, such as stimulating glycogen deposition by the liver. Some glucocorticoids, such as cortisone, produce anti-inflammatory effects.

glucose a six-carbon sugar widely distributed in plants, animals, and microorganisms. It is a component of the disaccharide lactose and is present in

polysaccharides such as cellulose, starch, and glycogen.

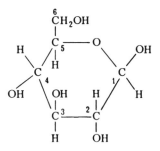

glucose-6-phosphate dehydrogenase (G6PD) an enzyme catalyzing the conversion of glucose-6-phosphate to 6-phosphogluconate. G6PD from human erythrocytes is a dimer consisting of identical subunits, each with a molecular weight of 55,000 d. Over 80 genetic variants of this enzyme are known. The four most common forms of the enzyme are

ENZYME	ACTIVITY	SOURCE
B	100%	widespread
A	90%	African blacks
A⁻	8%–20%	African blacks
M	0–7%	Mediterranean whites

glucose-6-phosphate dehydrogenase deficiency the most common disease-producing enzyme deficiency of humans. The gene encoding G6PD resides at the distal end of the long arm of the X chromosome. As a result of random X-chromosome inactivation (*see* **dosage compensation**), females heterozygous for the G6PD gene have some erythrocytes with the normal enzyme and some with the defective enzyme. The red cells of males with the A⁻ or M forms of the enzyme have a reduced life span, and exposure to antimalarial drugs such as primaquine results in life-threatening hemolysis. However, the possession of mutant proteins with reduced enzymatic activity appears to confer a resistance to malaria.

glucose-sensitive operons bacterial operons whose activity is inhibited by the presence of glucose. This lowers the level of cyclic AMP, thereby blocking a required positive control signal.

glucosylceramide lipidosis Gaucher disease (*q.v.*).

glume a chaffy bract, pairs of which enclose the base of grass spikelets.

glutamic acid *See* **amino acid.**

glutamine *See* **amino acid.**

glutathione a tripeptide containing glutamic acid, cysteine, and glycine and capable of being alternately oxidized and reduced. Glutathione plays an important role in cellular oxidations.

gly glycine. *See* **amino acid.**

glycerol a trihydric alcohol, that combines with fatty acids to form fats (*q.v.*).

$$
\begin{array}{c}
H \\
| \\
H-C-OH \\
| \\
H-C-OH \\
| \\
H-C-OH \\
| \\
H
\end{array}
$$

glycogen a soluble polysaccharide (see formula below) built up of numerous glucose molecules. Carbohydrate is stored as glycogen by vertebrates, especially in liver and muscles.

glycogenesis glycogen synthesis.

glycogenolysis the liberation of glucose from glycogen.

glycogenosis glycogen storage disease (*q.v.*).

glycogen storage disease any of a group of congenital and familial disorders characterized by the deposition of either abnormally large or abnormally small quantities of glycogen in the tissues. Anderson, Forbes, von Gierke, Hers, McArdles, and Pompe diseases are examples.

glycolipid a lipid containing carbohydrate.

glycolysis the sequential series of anaerobic reactions diagrammed on page 134 found in a wide variety of tissues that constitutes the principal route of carbohydrate breakdown and oxidation. The process starts with glycogen, glucose, or fructose and ends with pyruvic or lactic acids. The conversion of a molecule of glucose to two molecules of pyruvic acid generates two ATP molecules. Under aerobic conditions, the yield is eight ATP molecules. The pyruvic acid formed is broken down by way of the citric acid cycle (*q.v.*). *See* **pentosephosphate pathway.**

glycolytic participating in glycolysis.

glycoprotein a protein containing small amounts of carbohydrate (usually less than 4%). *See* **mucoprotein.**

glycoside a compound yielding a sugar upon enzymatic hydrolysis.

glycosidic bonds the bonds coupling the monosaccharide subunits of a polysaccharide.

glyoxylate cycle a series of metabolic reactions involved in photorespiration of plants, including oxidation of glycolate to glyoxylate by the enzyme glycolate oxidase within microbodies or peroxisomes (*q.v.*).

Glyptotendipes barbipes a midge possessing exceptionally large polytene chromosomes in the salivary glands of larvae and therefore a favorite for cytological studies.

gnotobiosis the rearing of laboratory animals in a germ-free state or containing only microorganisms known to the investigator.

gnotobiota the known microfauna and microflora of a laboratory animal in gnotobiosis (*q.v.*).

goiter a chronic enlargement of the thyroid gland that is due to hyperplasia, not neoplasia.

Golgi apparatus (or complex or material) a cell organelle identified in electron micrographs as a complex made up of closely packed broad cisternae and small vesicles. The Golgi apparatus is distinguished from the endoplasmic reticulum by the arrangement of the membranous vesicles and by the lack of ribosomes. The Golgi apparatus functions to collect and sequester substances synthesized by the endoplasmic reticulum. *See* **Appendix C,** 1954, Dalton and Felix.

gonad a gamete-producing organ of an animal; in the male, the testis; in the female, the ovary.

gonadotropic hormones pituitary hormones (such

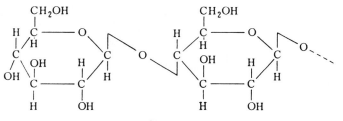

Glycogen.

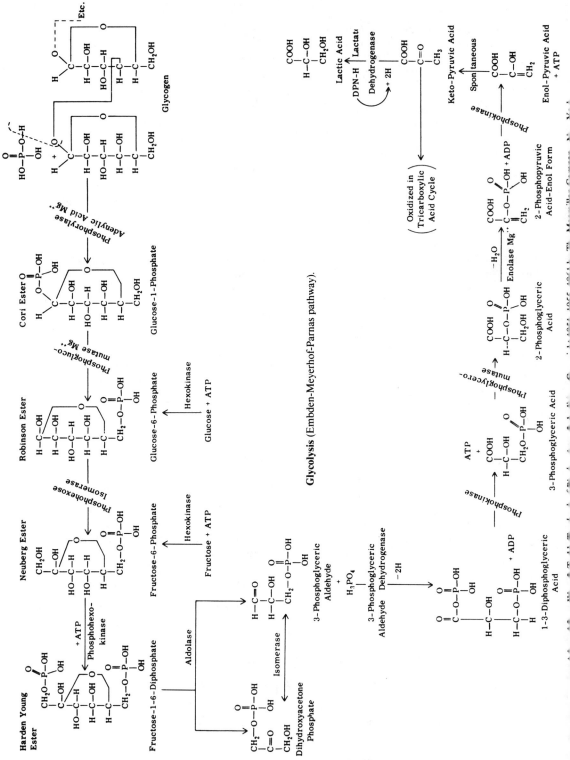

Glycolysis (Embden-Meyerhof-Parnas pathway).

as LH and FSH, *q.v.*) that stimulate the gonads; also called *gonadotropins.*

gonadotropin-releasing hormone (GnRH) a neurohormone from the hypothalamus that stimulates release of luteinizing hormone (LH) and follicle-stimulating hormone (FSH) from the pituitary gland.

gonochorism a sexual system in which each individual is either a male or a female. *Compare with* **hermaphrodite, monoecy.**

gonophore 1. in sessile coelenterates, the bud producing the reproductive medusae. **2.** in higher animals, any accessory sexual organ, such as an oviduct, or a sperm duct. **3.** in plants, a stalk that bears stamens and pistils.

gonomery the separate grouping of maternal and paternal chromosomes during the first few mitoses following fertilization as occurs in some insect embryos.

gonotocont an auxocyte (*q.v.*).

Gorilla gorilla the gorilla, a primate with a haploid chromosome number of 24. About 40 biochemical markers have been found to be distributed among 22 linkage groups. *See* **Hominoidea.**

Gossypium a genus composed of about 35 diploid and tetraploid species. Two tetraploid species *G. hirsutum* and *G. barbadense* are the source of most commercial cotton. The comparative genetics of these species has been intensively studied.

gout a hereditary disorder of purine metabolism characterized by increased amounts of uric acid in the blood and recurrent attacks of acute arthritis.

Gowen's crossover suppressor a recessive gene on the third chromosome of *Drosphila melanogaster* symbolized by *c(3)G.* Females homozygous for this gene are unable to form synaptonemal complexes (*q.v.*) in their oocyte nuclei and crossing over is abolished.

G6PD glucose-6-phosphate dehydrogenase (*q.v.*).

G6PD deficiency *See* **glucose-6-phosphate dehydrogenase deficiency.**

G1, G2 phases *See* **cell cycle.**

G proteins a family of membrane proteins that become activated only after binding guanosine triphosphate (GTP). Activated G proteins, in turn, activate an "amplifier" enzyme on the inner face of the membrane; the enzyme then converts precursor molecules into second messengers (*q.v.*). For example, an external signal molecule may bind to its cell-surface receptor and induce a conformational change in the receptor; this change is transmitted through the cell membrane to a G protein, making it able to bind GTP; binding of GTP causes another conformational change in the G protein that enables it to activate adenylate cyclase (the amplifier enzyme in this case) and thereby initiate the formation of cyclic adenosine monophosphate (the "second messenger" in this case).

Graafian follicle a fluid-filled spherical vesicle in the mammalian ovary, containing an oocyte attached to its wall. *See* **Appendix C,** 1657, de Graaf.

grade a stage of evolutionary advance. A level reached by one or more species in the development of a structure, physiological process, or behavioral character. Different species may reach the same grade because they share genes that respond in the same way to an environmental change. When two species that do not have a common ancestry reach the same grade, the term *convergence* is used to describe this evolutionary parallelism.

gradient a gradual change in some quantitative property of a system over a specific distance (e.g., a clinal gradient; density gradient).

gradualism a model explaining the mechanism of evolution that represents an updating of the original ideas set forth by Charles Darwin. According to this model, those individuals with hereditary traits that best adapt them to their habitat are most likely to survive and to transmit these adaptive genes to their offspring. As a result, with the passage of time the frequencies of beneficial genes rise in the population, and when the composition of the gene pool of the evolving population becomes sufficiently different from that of the original population, a new species will have arisen. Since a beneficial mutation must spread through an entire population to produce detectable evolutionary changes, speciation will be a gradual and continuous process. *Contrast with* **punctuated equilibrium.**

Graffi leukemia virus a virus that induces myeloid leukemia in mice and rats.

graft 1. a relatively small piece of plant or animal tissue implanted into an intact organism. **2.** the act of transferring a part of an organism from its normal position to another position in the same or another organism. *See* **autograft, heterograft, homograft, scion, stock, transplantation.**

graft hybrid a plant made up of two genetically distinct tissues due to fusion of host and donor tissues after grafting.

graft rejection a cell-mediated immune response to transplanted tissue that causes destruction of a graft. Rejection is evoked by the histocompatibility antigens (*q.v.*) of the foreign cells.

graft-versus-host reaction a syndrome arising when an allograft, containing immunocompetent cells, mounts an immune response against a host that is unable to reject it because the host is immunologically immature or immunologially compromised or suppressed (e.g., by radiation or drugs); synonymous with allogeneic disease, runt disease (*q.v.*).

gram atomic weight the quantity of an element that has a mass in grams numerically equal to its atomic weight.

gram equivalent weight the mass of an acid or a base that will release or neutralize one gram molecule (mole) of hydrogen ion. A 1 M solution of H_2SO_4 contains 2 gram equivalents. A thousandth of a gram equivalent weight is a millequivalent.

gramicidin S a cyclic antibiotic synthesized by *Bacillus brevis.* The molecule has the structure

$$\begin{bmatrix} d\text{-phe-pro-val-orn-leu} \\ leu\text{-orn-val-pro-d-phe} \end{bmatrix}$$

and it contains amino acids not usually found in proteins—namely, ornithine (orn) and d-phenylalanine (rather than the usual l-isomer). The synthesis of gramicidin S constitutes one of the best understood examples of a polypeptide that is not synthesized on a ribosome. Two enzymes are required (E_1 and E_2), which are bound together, forming a unit called gramicidin synthetase. One molecule each of proline, valine, ornithine, and leucine bind in that sequence to sulfhydryl groups of E_1. E_2 functions to isomerize l- to d-phenylalanine and to transfer it to the proline attached to E_1. The two identical polypeptides are then joined head to tail to form a decapeptide. This rare type of polypeptide synthesis is very uneconomical and cannot generate molecules greater than 20 amino acids long.

gram molecular weight the quantity of a compound that has a mass in grams numerically equal to its molecular weight. Gram molecular weight is often shortened to *gram-mole* or *mole.*

Gram staining procedure a staining technique perfected by the Danish physician Hans Christian Gram (1853–1938). It allows bacteria to be divided into two groups, Gram-positive (which stain deep purple) and Gram-negative (which stain light pink). The staining differences lie in the permeability properties of the cell walls of the two groups of bacteria.

grana (*singular* **granum**) long columns in chloroplasts made up of stacks of dense discs. Grana are sites where chlorophyll is located. Each disc contains a double layer of quantosomes. *See* **chloroplast.**

granulocytes white blood cells possessing distinct cytoplasmic granules and a multilobed nucleus; includes basophils, eosinophils, and neutrophils; also called *polymorphonuclear leukocytes.*

grass any species of monocotyledon belonging to the family Gramineae. Such species are characterized by leaves with narrow, spear-shaped blades, and flowers borne in spikelets of bracts.

gratuitous inducer a compound not found in nature that acts as an inducer although it cannot be metabolized. *See* **IPTG, ONPG.**

gravid an animal that is swollen from accumulated eggs or embryos.

gray a unit defining that energy absorbed from a dose of ionizing radiation equal to one joule per kilogram. Abbreviated Gy. 1 Gy = 100 rad.

grid 1. a network of uniformly spaced horizontal and vertical lines; 2. a specimen screen used in electron microscopy.

griseofulvin an antibiotic synthesized by certain *Penicillium* species that is used as a fungicide.

Gross leukemia virus a virus isolated by Gross that induces lymphoid leukemia in mice.

group selection natural selection acting upon a group of two or more individuals by which traits are selected that benefit the group rather than the individual. *See* **Hamilton's genetical theory of social behavior.**

group-transfer reactions chemical reactions involving the exchange of functional groups between molecules (excluding oxidations or reductions and excluding water as a participant). The enzymes that catalyze group-transfer reactions are called transferases or synthetases. For example, activation of an amino acid involves transfer of an adenosine monophosphate group from ATP to the COO^- group of the amino acid.

growth curve in microbiology, a curve showing the change in the number of cells in a growing culture as a function of time.

growth factor a specific substance that must be present in the growth medium to permit cell multiplication.

growth hormone (GH) a growth-stimulating hormone secreted by the adenohypophysis of vertebrates. Unlike many other polypeptide hormones, GH is highly species specific. Human growth hormone is a single polypeptide chain composed of 191 amino acids encoded by a gene on the long arm of chromosome 17. Limitations upon the supplies of human GH extracted from cadavers restricted its clinical use in the treatment of pituitary dwarfism (*q.v.*). These problems were overcome with the advent of recombinant DNA–derived human growth hormone. *See* **Appendix C,** 1979, Goeddel *et al.*

GSH   reduced glutathione.

GT-AG rule   intron junctions start with the dinucleotide GT and end with the dinucleotide AG, corresponding to the left and right ("donor and acceptor") splicing sites, respectively. Also called *Chambon's rule.*

GTP   guanosine triphosphate (*q.v.*).

guanine   *See* **bases of nucleic acids.**

guanine deoxyriboside   *See* **nucleoside.**

guanine-7-methyl transferase   *See* **methylated cap.**

guanosine   *See* **nucleoside.**

guanosine triphosphate   an energy-rich molecule (analogous to ATP) that is required for the synthesis of all peptide bonds during translation.

guanylic acid   *See* **nucleotide.**

guanylyl transferase   *See* **methylated cap.**

guinea pig   *See* ***Cavia porcellus.***

guppy   *See* ***Lebistes reticularis.***

Gy   abbreviation for gray (*q.v.*).

gymnosperm   a primitive plant having naked seeds (conifers, cycads, ginkgos, etc.).

gynander   synonymous   with   **gynandromorph** (*q.v.*).

gynandromorph   an individual made up of a mosaic of tissues of male and female genotypes. The fruit fly illustrated below is a bilateral gynandromorph with the right side female, the left side male. The zygote was + +/*w m.* Loss of the X chromosome containing the dominant (+) genes occurred at the first nuclear division. The cell with the single X chromosome containing the recessive marker genes gave rise to the male tissues. Therefore, the left eye is white and the left wing is miniature. Note the male abdominal pigmentation and the sex comb.

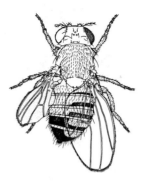

gynodioecy   a sexual dimorphism in plants having both bisexual and separate female individuals.

gynoecium   a collective term for all the carpels of a flower.

gynogenesis   reproduction by parthenogenesis requiring stimulation by a spermatozoan for the activation of the egg; synonymous with pseudogamy (*q.v.*).

gyrase   the colloquial name for a type II topoisomerase (*q.v.*) of *E. coli* that converts relaxed closed-circular, duplex DNA to a negatively superhelical form both *in vitro* and *in vivo.* This enzyme prepares the DNA for proteins that require unwinding of the duplex or single-stranded regions in order to participate in such processes as replication, transcription, repair, and recombination. Several drugs are known to inhibit gyrase, including adriamycin, naladixic acid, and novobiocin. *See* **Appendix C,** 1976, Gellert *et al.*

H

H **1.** symbol for broad heritability (also symbolized H^2); h^2 for narrow heritability. **2.** symbol for hydrogen.

H1, H2A, H2B, H3, H4 *See* **histones.**

^{3}H *See* **tritium.**

habitat the natural abode of an organism.

Habrobracon juglandis** See **Microbracon hebetor.

Hadean the geologic eon beginning with the origin of the earth about 4.5 billion years ago and ending with the formation of the earliest rocks about 3.8 billion years ago. *See* **geologic time divisions.**

Haemanthus katherinae the African blood lily. A favorite species used for the time-lapse photographic study of mitotic endosperm cells.

haemoglobin *See* **hemoglobin.**

hairpin loops any double-helical regions of DNA or RNA formed by base pairing between adjacent inverted complementary sequences on the same strand. *See* **palindrome, terminators.**

Haldane's rules the generalization that when one sex is absent, rare, or sterile, in the offspring of two different animal races or species, that sex is the heterogametic sex. Haldane's rule is known to apply for various species of mammals, birds, and insects. In *Drosophila* and *Mus,* the X and Y chromosomes interact during spermatogenesis, with the Y repressing the transcription of certain X-linked loci. Presumably, when the X and Y chromosomes are ⸀ ᴜm different species, such regulation does not take place and sterility results. Thus, Haldane's rule may be explained by the nonharmonious interaction of X- and Y-linked fertility genes in the hybrid.

half-chromatid conversion *See* **chromatid conversion.**

half-life, 1. biological the time required for the body to eliminate one-half of the dose of a given substance. This time is approximately the same for both stable and radioactive isotopes of any element. **2. radioactive** the time required for half the mass of a radioactive substance to decay into another substance. Each radionuclide has a unique half-life.

half-sib mating mating between half brother and half sister. Such individuals have one parent in common.

half-tetrad analysis recombinational analysis where two of the four chromatids of a given tetrad can be recovered, as in the case of attached X chromosomes in *Drosophila.*

half-value layer the thickness of a specified material that reduces the flux of radiation by one-half.

halide a fluoride, chloride, bromide, or iodide.

halogen fluorine (F), chlorine (Cl), bromine (Br), or iodine (I).

halophiles bacteria that require high concentrations of salt in order to survive. Examples are species belonging to *Halobacterium* and *Halococcus.* They are placed in the archaebacteria (*q.v.*) on the basis of the nucleotide sequences of their 16S rRNAs.

halteres paired club-shaped appendages that extend from the metathorax of Dipterans. They serve as gyroscopic sense organs adapted to perceive deviations from the plane of their vibration. Halteres are evolutionarily equivalent to the hind pair of wings in other insects. In *Drosophila,* certain homeotic mutations (*q.v.*) convert halteres into wings and vice versa.

Hamilton's genetical theory of social behavior a theory put forth by W. D. Hamilton to explain how altruism can evolve when it increases the fitness of relatives. The theory proposes that a social act is favored by natural selection if it increases the *inclusive fitness* of the performer. Inclusive fitness consists of the individual's own fitness as well as his effects on the fitness of any genetically related neighbors. The idea is that genetic alleles change in frequency in a population owing to effects on the reproduction of relatives of the individual in which the character is expressed, rather than on the personal reproduction of the individual. Therefore, Hamilton's theory is often referred to as *kin selection.* For example, a mutation that affected the behavior of a sterile worker bee so that she fed her fertile queen but starved herself would increase the inclusive fitness of that worker because, while her own fitness decreased, her actions increased the fitness of a close relative.

hamster   common laboratory rodent. *See **Cricetulus griseus, Mesocricetus auratus.***

hanging drop technique   a method for microscopic examination of organisms suspended in a drop on a special concave microscope slide.

Hansenula wingei   a yeast that has provided information concerning the genetic control of mating type (*q.v.*).

H antigens   **1.** histocompatibility antigens governed by histocompatibility genes (*q.v.*). **2.** flagellar protein antigens of motile Gram-negative enterobacteria.

haplo-   the prefix haplo-, when followed by a symbol designating a particular chromosome, indicates an individual whose somatic cells lack one member of the designated chromosome pair. Thus, in *Drosophila*, haplo-IV means a fly that is monosomic for chromosome 4.

haplodiploidy   a genetic system found in some animals (such as the honey bee) in which males develop from unfertilized eggs and are haploid, whereas the females develop from fertilized eggs and are diploid.

haplodiplomeiosis   *See **meiosis.***

haploid cell culture   *See **anther culture.***

haploidization   a phenomenon taking place during the parasexual cycle in certain fungi during which a diploid cell is transformed into a haploid cell by the progressive loss of one chromosome after another by nondisjunction.

haploid number   the gametic chromosome number, symbolized by N.

haploid parthenogenesis   the situation in which a haploid egg develops without fertilization, as in the honey bee.

haplont   an organism in which only the zygote is diploid (as in the algae, protozoa, and fungi). It immediately undergoes meiosis to give rise to the haplophase. *See **diplo-haplont, diplont.***

Haplopappus gracilis   a species of flowering plant showing the lowest number of chromosomes (N = 2), and therefore studied by cytologists. *See **Appendix A**, Dicotyledoneae, Asterales.*

haplophase   the haploid phase of the life cycle of an organism, lasting from meiosis to fertilization.

haplosis   the establishment of the gametic chromosome number by meiosis.

haplotype   the symbolic representation of a specific combination of linked alleles in a cluster of related genes. The term is a contraction of *haplo*id geno*type* and is often used to describe the combination of alleles of the major histocompatibility complex (*q.v.*) on one chromosome of a specific individual. *Compare with **phenogroup.***

hapten   an incomplete antigen; a substance that cannot induce antibody formation by itself, but can be made to do so by coupling it to a larger carrier molecule (e.g., a protein). *Complex haptens* can react with specific antibodies and yield a precipitate; *simple haptens* behave as monovalent substances that cannot form serological precipitates.

haptoglobin   a plasma glycoprotein that forms a stable complex with hemoglobin to aid the recycling of heme iron. In man this protein is encoded by a gene on chromosome 16.

Hardy-Weinberg law   the concept that both gene frequencies and genotype frequencies will remain constant from generation to generation in an infinitely large, interbreeding population in which mating is at random and there is no selection, migration, or mutation. In a situation where a single pair of alleles (*A* and *a*) is considered, the frequencies of germ cells carrying *A* and *a* are defined as p and q, respectively. At equilibrium the frequencies of the genotypic classes are p^2 (*AA*), 2 pq (*Aa*), and q^2 (*aa*). *See* graph (p. 140) and **Appendix C**, 1908, Hardy, Weinberg.

harlequin chromosomes   *See **5-bromodeoxyuridine.***

Harvey murine sarcoma virus   a virus carrying the oncogene *v-ras*[II] that is homologous to the cellular proto-oncogene *c-ras*[II], which resides on chromosome 11. *See **T24 oncogene.***

HAT medium   a tissue culture medium containing *h*ypoxanthine, *a*minopterin, and *t*hymidine. Mutant cells deficient in or lacking the enzymes thymidine kinase (TK⁻) and hypoxanthine-guanine-phosphoriboxyl transferase (HGPRT⁻) cannot grow in HAT medium because aminopterin blocks endogenous (*de novo*) synthesis of both purines and pyrimidines. Normal TK⁺ HGPRT⁺ cells can survive by utilizing the exogenous hypoxanthine and thymidine via the *salvage pathway* (*q.v.*) of nucleotide synthesis. HAT medium has been used to screen for hybridomas (*q.v.*) by mixing TK⁺ HGPRT⁻ myeloma cells with antigen-stimulated TK⁻ HGPRT⁺ spleen cells. The hybrid TK⁺ HGPRT⁺ clones that survive in HAT medium are then assayed for monoclonal antibodies specific to the immunizing antigen. *See **Appendix C**, 1964, Littlefield; 1967, Weiss and Green.*

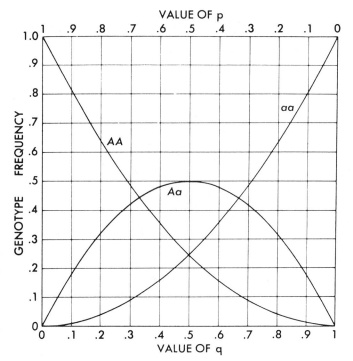

The relationships between the frequencies of genes *A* or *a* and the frequencies of genotypes *AA, Aa,* and *aa* as predicted by the **Hardy-Weinberg law.**

Hayflick limit   an experimental limit to the number of times a normal animal cell seems capable of dividing; mouse and human cells divide 30 to 50 times before they enter the "crisis period" (*q.v.*). *See* **Appendix C,** 1965, Hayflick.

Hb   hemoglobin. Hb^A symbolizes normal hemoglobin; Hb^F, fetal hemoglobin; Hb^S, sickle hemoglobin, etc.

HbO_2   oxyhemoglobin (*q.v.*).

H chain   *See* **immunoglobulin.**

H-2 complex   the major histocompatibility complex of the mouse lying in a segment of chromosome 17, which carries a number of polymorphic loci associated with various aspects of the immune system. It consists of four regions (K, I, S, D) that contain genes coding for classical transplantation antigens, Ia antigens and complement components, as well as immune response genes. The I region is further subdivided (*see* **I region**).

headful mechanism   a mechanism of packaging DNA in a phage head (e.g., T4) in which concatemeric DNA is cut, not at a specific position, but rather when the head is filled. This mechanism accounts for the observations of terminal redundancy and cyclic permutation in T4.

heat   in reproductive biology, that period of the sexual cycle when female mammals will permit coitus; estrus.

heat shock proteins   proteins synthesized in *Drosophila* cells within 15 minutes after a heat shock. These proteins are named in accordance with their molecular weights (in kd). After their synthesis in the cytoplasm, most heat shock proteins move to the nucleus and bind to chromatin. Here they protect the chromosomes from thermal damage in some unknown way. *See* **Appendix C,** 1974, Tissiers *et al.*; 1975, McKenzie *et al.*

heat shock puffs   a unique set of chromosomal puffs induced in *Drosophila* larvae exposed to ele-

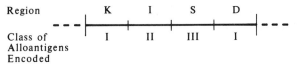

H-2 complex.

140

vated temperatures (for example, 40 minutes at 37°C). In *D. melanogaster,* there are nine heat-inducible puffs; in *D. hydei,* there are six. Heat shock results in the transcription of a specific set of mRNAs by the genes that form the puffs in polytene chromosomes. In culture, heat-shocked *Drosophila* cells also transcribe these same mRNAs into specific heat shock proteins. *See* **Appendix C,** 1962, Ritossa.

heat shock response the transcriptional activity that is induced at a small number of chromosomal loci following exposure of cells to a brief period of elevated temperature. At the same time, other loci that were active prior to the heat shock are switched off. This phenomenon appears to be universal, since it has been observed in *Drosophila, Tetrahymena,* sea urchin embryos, soy beans, chick fibroblasts, etc. Furthermore, heat shock proteins and the genes that encode them show a high degree of conservation during evolution.

heavy chain in a heteromultimeric protein, the polypeptide chain with the higher molecular weight (e.g., in an immunoglobulin molecule, the heavy chains are about twice the length and molecular weight of the light chains); abbreviated *H chains;* the smaller molecules are called *light (L) chains.* The heavy chain determines the class to which the immunoglobulin belongs.

heavy chain class switching the switching of a B lymphocyte from synthesis of one class of antibodies to another. For example, a B lymphocyte first synthesizes IgM, but it may later switch to the synthesis and secretion of IgG, and both antibodies will have the same antigen specificity. Thus, only the constant regions of the heavy chains differ between these two classes of antibodies. Class switching of this type involves both somatic recombination prior to transcription and processing of transcripts by eliminating some segments and splicing the fragments.

heavy isotope an atom, such as ^{15}N, that contains more neutrons than the more frequently occurring isotope, and thus is heavier.

heavy-metal stain one of the elements of high atomic weight, often used as stains in electron microscopy (U, Pb, Os, Mn).

heavy water nontechnical term for deuterium oxide.

HeLa cells An established cell line *(q.v.)* consisting of an aneuploid strain of human epithelial-like cells, maintained in tissue culture since 1951; originating from a specimen of tissue from a carcinoma of the cervix in a patient named *He*nrietta *La*cks. *See* **Appendix C,** 1951, Gey.

Helianthus annuus the sunflower.

helicase an enzyme that unwinds a DNA double helix molecule ahead of DNA polymerase III in *E. coli;* e.g., the rep protein.

heliotropism phototropism.

helix a curve on the surface of a cylinder or cone that cuts all the elements of the solid at a constant angle. Applied especially to the *circular helix* on a right circular cylinder. It resembles the thread of a bolt.

Helix a genus of gastropod containing common garden snails. Species from this genus are favorite experimental animals for the experimental analysis of molluscan development.

helix-destabilizing proteins any protein that binds to single-stranded regions of duplex DNA created by "breathing" *(q.v.),* and thereby causes unwinding of the helix; e.g., helicase *(q.v.);* also called *relaxation proteins* or *melting proteins.*

helix-turn-helix motif a term describing the three-dimensional structure of a segment that characterizes certain DNA-binding proteins. Regulatory proteins like the *cro* repressor *(q.v.)* are examples. The protein bends so that two successive alpha helices are held at right angles by a turn that contains four amino acids. One helix binds to the major groove of the DNA double helix in a region showing twofold rotational symmetry *(q.v.).*

Hellin's rule of multiple births a rule that allows calculation of the expected frequency of triplets, quadruplets, etc., from the observed frequency of twins. If the observed frequency of twins is n, then the expected frequency of triplets is n^2, of quadruplets is n^3, etc. This rule applies only to spontaneously occurring multiple births, not to cases of artificially induced twinning by the use of clomiphenes *(q.v.)* or other hormones resulting in multiple ovulations.

Helminthosporium maydis the corn leaf blight. A major effort is being made to produce corn strains resistant to this fungus, which produced a loss to the 1970 U.S. corn crop of approximately a billion dollars.

helper T lymphocytes a T lymphocyte that amplifies the activity of B lymphocytes, other T lymphocytes, and macrophages. Once a helper T lymphocyte recognizes an antigen, it divides and its progeny start to synthesize and secrete a variety of lymphokines *(q.v.).* These cause B lymphocytes to divide and differentiate into plasma cells, inactive precursor T cells to develop into cytotoxic T lymphocytes, and macrophages to be recruited and activated. *See* **cytotoxic T lymphocyte, T lymphocyte.**

helper virus a virus that, by its infection of a cell already infected by a defective virus, is able to supply something the defective virus lacks, thus enabling the latter to multiply.

hemagglutinins 1. antibodies involved in specific aggregation of red blood cells. 2. glycoproteins, formed on the surfaces of cells infected with certain viruses (e.g., influenza virus) or on the surfaces of enveloped viruses released from such cells, that can aggregate red blood cells of certain species. 3. lectins (*q.v.*).

hematopoiesis the formation of red blood cells.

hematopoietic pertaining to hematopoiesis.

heme an iron-containing porphyrin molecule that forms the oxygen-binding portion of hemoglobin. *See* **bilirubin, hemoglobin.**

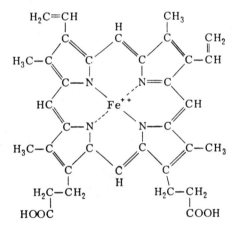

hemel an aziridine mutagen (*q.v.*).

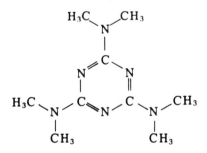

Hemimetabola a superorder of primitive insects having incomplete metamorphosis (mayflies, dragon flies, stoneflies, roaches, etc.). Synonymous with Exopterygota. *See* **Appendix A.**

hemimetabolous referring to those insects in which metamorphosis is simple and gradual. Each insect gradually acquires wings during a period of growth interrupted by several molts. Immature forms are called nymphs if terrestrial, naiads if aquatic. *Contrast with* **holometabolous.**

hemipteran a true bug; a member of the order Hemiptera. *See* **Appendix A.**

hemizygous gene a gene present in single dose. It may be a gene in a haploid organism, or a sex-linked gene in the heterogametic sex, or a gene in the appropriate chromosomal segment of a deficiency heterozygote.

hemocoel a body cavity of arthropods and molluscs that is an expanded part of the blood system. The hemocoel never communicates with the exterior, and never contains germ cells.

hemocyte an amoeboid blood cell of an insect. It is analogous to a mammalian leukocyte.

hemoglobin the oxygen-carrying pigment (molecular weight 64,500 d) of red blood cells. Hemoglobin is a conjugated protein composed of four separate chains of amino acids and four iron-containing ring compounds (heme groups). The protein chains of adult hemoglobin (HbA) occur as two pairs, a pair of alpha chains and a pair of beta chains. Each alpha chain contains 141 amino acids, and each beta chain contains 146 amino acids. Normal human adults also have a minor hemoglobin component (2%) called A_2. This hemoglobin has two alpha chains and two delta peptide chains. The delta chains have the same number of amino acids as the beta chains, and 95% of their amino acids are in sequences identical to those of the beta chains. The hemoglobin of the fetus (HbF) is made up of two alpha chains and two gamma chains. Each gamma chain also contains 146 amino acids. There are two types, G gamma and A gamma, which differ only in the presence of glycine or alanine, respectively, at position 136. The earliest embryonic hemoglobin tetramer consists of two zeta (alpha-like) and two epsilon (betalike) chains. Beginning about the eighth week of gestation, the zeta and epsilon chains are replaced by alpha and gamma chains, and just before birth the gamma chains begin to be replaced by beta chains. *See* **Appendix C,** 1960, Perutz *et al.*; 1961, Dintzis; 1962, Zuckerkandl and Pauling; **leghemoglobin.**

hemoglobin Bart's *See* **hemoglobin fusion genes.**

hemoglobin C a hemoglobin with an abnormal beta chain. Lysine is substituted for glutamic acid at position 6. *See* **hemoglobin S.**

hemoglobin Constant Spring a human hemoglobin possessing abnormal alpha chains that contain

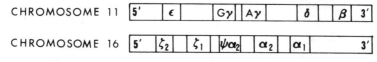

CHROMOSOME 11 | 5' | ε | Gγ | Aγ | δ | β | 3'

CHROMOSOME 16 | 5' | ζ₂ | ζ₁ | ψα₂ | α₂ | α₁ | 3'

Hemoglobin genes.

172 amino acids, rather than the normal number of 141. Hb Constant Spring seems to be the result of a nonsense mutation that converted a stop codon into a codon specifying an amino acid (glutamine, in this case). The 30 additional amino acids must be coded by adjacent base sequences that normally are not transcribed, or, if transcribed, not translated.

hemoglobin fusion genes abnormal hemoglobin genes arising as a result of unequal crossing over between genes sharing homologous nucleotide sequences. Examples are the fused gamma-beta gene that codes for the non-alpha chain of Hb Kenya and the variety of fused delta-beta genes that code for the non-alpha chains of the Hb Lepore (*q.v.*).

hemoglobin genes the genes coding for human hemoglobins are located on chromosomes 11 and 16. As diagrammed above, chromosome 11 contains the epsilon chain gene, two gamma chain genes (symbolized by Gγ and Aγ), a delta chain gene, and a beta chain gene. Chromosome 16 contains two zeta chain genes (symbolized ζ₂ and ζ₁) and two alpha chains (symbolized α₂ and α₁). Between ζ₁ and α₂ lies a DNA segment, with no coding function, that resembles the α₂ gene in about 75% of its nucleotide sequences. This is called the pseudogene $\psi_{\alpha 2}$. It is assumed that certain neighboring genes (Gγ and Aγ, δ and β, ζ₂ and ζ₁, and α₂ and α₁) arose during evolution as the results of duplications, and the same, of course, applies to the $\psi_{\alpha 2}$. *See* **Appendix C,** 1976, Efstratiadis *et al.*; 1979, Fritsch *et al.*; **myoglobin gene.**

hemoglobin H *See* **hemoglobin homotetramers, thalassemias.**

hemoglobin homotetramers abnormal hemoglobins made up of four identical polypeptides. Examples are Hb Bart's (only gamma chains) and Hb H (only beta chains).

hemoglobin Kenya *See* **hemoglobin fusion genes.**

hemoglobin Lepore an abnormal hemoglobin named after the Italian family in which it was first found. The protein contains a pair of normal alpha chains and a pair of abnormal chains, each chain made up of 146 amino acids. Each abnormal chain appears to be a hybrid molecule consisting of an N-terminal end containing amino acids in a sequence characteristic of the left end of the delta chain and a C-terminal end containing amino acids in a sequence characteristic of the right end of a beta chain. Hemoglobin Lepore is presumed to have arisen by unequal crossing over between a mispaired delta and beta cistron.

hemoglobin S a hemoglobin with an abnormal beta chain. Valine is substituted for glutamic acid at position 6. Thus, the mutations resulting in the hemoglobin S and C beta chains appear to involve the same codon. Unlike all other substitutions in the beta chain, this mutation affects the solubility of hemoglobin under conditions of low oxygen tension. As the oxygen concentration declines, Hb S polymerizes to form crystals that distort erythrocytes. These misshapen red blood cells are a diagnostic feature of sickle-cell anemia (*q.v.*). Hb C also causes sickling of erythrocytes. *See* **Appendix C,** 1949, Pauling; 1957, Ingram.

hemoglobinuria the excretion of hemoglobin in solution in the urine.

hemolysis the rupturing of blood cells.

hemolytic anemia *See* **anemia.**

hemophilia hereditary diseases characterized by defects in the blood-clotting mechanism. Hemophilia was the first human trait shown to display sex-linked inheritance (*See* **Appendix A,** 1820, Nasse). Hemophilia A (classical hemophilia) is due to a deficiency of functional antihemophilic factor (*q.v.*). Hemophilia B (Christmas disease) results from a deficiency of plasma thromboplastin component. Both clotting factors are proteins encoded by genes (HEMA and HEMB) that reside at a considerable distance from one another, on the X chromosome. The hemophilic population consists of hemizygotes for HEMA and for HEMB in a ratio of 4:1.

Hemophilus pertussis the bacterium responsible for whooping cough.

hemopoiesis *See* **hematopoiesis.**

hemopoietic hematopoietic.

hemopoietic histocompatibility (Hh) used in reference to transplantation of bone marrow in inbred strains of mice and their hybrids. As measured by the growth of the transplanted marrow cells in the spleen, a unique form of transplantation genetics seems to apply whereby hybrid recipients resist pa-

143

rental bone marrow grafts, but parental recipients accept hybrid bone marrow.

hempa an aziridine mutagen (*q.v.*).

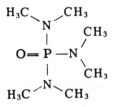

herb a plant with no persistent parts above ground, as distinguished from a shrub or tree.

herbage herbs collectively, especially the aerial portion.

herbarium a collection of dried and pressed plant specimens.

herbicide a chemical used to kill herbaceous plants.

herbivore a plant eater.

hereditary disease a pathological condition caused by a mutant gene. *See* **Appendix C**, 1966, McKusick.

heredity a familial phenomenon wherein biological traits appear to be transmitted from one generation to another. The science of genetics has shown that heredity results from the transmission of genes from parents to offspring. The genes interact with one another and with their environment to produce distinctive characteristics or phenotypes. Offspring therefore tend to resemble their parents or other close relatives rather than unrelated individuals who do not share as many of the same kinds of genes.

heritability an attribute of a quantitative trait in a population that expresses how much of the total phenotypic variation is due to genetic variation. In the broad sense, heritability is the degree to which a trait is genetically determined, and it is expressed as the ratio of the total genetic variance to the phenotypic variance (V_G/V_P). In the narrow sense, heritability is the degree to which a trait is transmitted from parents to offspring (i.e., breeding value), and it is expressed as the ratio of the additive genetic variance to the total phenotypic variance (V_A/V_P). The concept of additive genetic variance makes no assumption concerning the mode of gene action involved. Heritability estimates are most commonly made by (1) regression-correlation analyses of close relatives (e.g., parent-offspring, full sibs, half sibs), (2) experiments involving response to selection, and (3) analysis of variance components. Traits with high heritabilities respond readily to selection.

hermaphrodite an individual having both male and female reproductive organs. A *simultaneous hermaphrodite* has both types of sex organs throughout life. A *sequential hermaphrodite* may have the ovary first (protogyny), to be replaced by a testis later, or may develop the testis first (protandry), to be replaced later by an ovary. *See* **consecutive sexuality.**

herpes virus one of a group of animal viruses having a duplex DNA molecule within an icosahedral capsid. They range in size from 180 to 250 mμ and produce internuclear inclusions in host cells. Type 1 herpes simplex causes cold sores. Type 2 herpes simplex is associated with genital lesions and is sexually transmitted; both types 1 and 2 have been implicated in certain human cancers. The varicellazoster virus is the cause of chicken pox and shingles. The Epstein-Barr virus (EBV) and the cytomegalovirus (CMV) also belong to the herpes group.

Hers disease a hereditary glycogen storage disease in humans arising from a deficiency of the enzyme hepatic phosphorylase. It is inherited as an autosomal recessive trait with an incidence of 1/200,000.

het a partially heterozygous phage.

heterauxesis the relation of the growth rate of a part of a developing organism to the growth rate of the whole or of another part. If the organ in question grows more rapidly than the organism as a whole, it shows tachyauxesis; if less rapidly, bradyauxesis; if at the same rate, isauxesis. *See* **allometry.**

heteroalleles alternative forms of a gene that differ at nonidentical muton sites. Intragenic recombination between mutant heteroalleles can yield a functional cistron. *Compare with* **homoalleles.** *See* **Appendix C**, 1955, Pritchard; 1962, Henning and Yanofsky.

heterobrachial inversion pericentric inversion.

hetrocapsidic virus *See* **segmented genome.**

heterocaryon heterokaryon.

heterochromatin chromosomal material that, unlike euchromatin (*q.v.*), shows maximal condensation in nuclei during interphase. Chromosomal regions behaving in this way are said to show *positive heteropycnosis*. When entire chromosomes (like some Y chromosomes) behave this way, they are called *heterochromosomes*. In polytene chromosomes, heterochromatic regions adjacent to the centromeres of all chromosomes tend to adhere, forming a *chromocenter*. Heterochromatin is composed of repetitious DNA, is late to replicate, and is transcriptionally inactive. Such heterochromatic segments are often called *constitutive* to distinguish them from chromosomal segments or whole chromosomes that become dense and compact at different developmental stages. In such cases, one homologous chromosome may behave differently from the other. An example is the condensed, inactivated X chromosome characteristic of the diploid somatic cells of mammalian females. Such chromosomes are sometimes said to contain *facultative heterochromatin,* although there is no evidence that they contain a type of DNA different from euchromatin. *See* **Appendix C,** 1928, Heitz; 1936, Schultz; 1959, Lima-de-Faria; 1970, Pardue and Gall.

heterochromosome *See* **heterochromatin, B chromosomes.**

heterochronic mutations mutations that perturb the relative timing of developmental events. Certain of the cell lineage mutants (*q.v.*) of *Caenorhabditis elegans* are examples. In *lin-14* mutants, certain aspects of development occur precociously with respect to the molting cycle, while in *lin-4* certain stages of development are repeated at abnormally late times.

heterochrony a change during evolution in the onset of a developmental process so that the appearance or growth rate of a specific organ or other feature is altered in a descendant, relative to an ancestor. In the case of *acceleration,* the morphological feature appears earlier during the ontogeny of the descendant than it did in the ancestor. Conversely, *retardation* refers to the situation where the morphological feature appears later in the ontogeny of the descendant than it did during the ontogeny of the ancestor. *See* **heterauxesis, neoteny, paedomorphosis.**

heterocyclic referring to any organic compound forming a ring made up of carbon atoms and at least one atom other than carbon. Examples of heterocyclic amino acids are proline, hydroxyproline, tryptophan, and histidine (*see* **amino acid**). Purines and pyrimidines are heterocyclic compounds (*see* **bases of nucleic acids**). The porphyrin portion of heme is made up of heterocyclic rings (*see* **heme**).

heterodimer a protein made up of paired polypeptides that differ in their amino-acid sequences.

heteroduplex **1.** a DNA generated during genetic recombination by base pairing between complementary single strands from different parental duplex molecules. **2.** a double-stranded nucleic acid in which each chain has a different origin and therefore is not perfectly complementary; e.g., the hybrid molecule generated by *in vitro* annealing of eukaryotic mRNA with its DNA. *See* **Appendix C,** 1969, Westmoreland *et al. See* **R-loop mapping.**

heteroecious referring to a parasite requiring two or more hosts to complete its life cycle, as with some rust fungi and insects.

heterofertilization double fertilization of angiosperms in which the endosperm and egg are derived from genetically different sperm nuclei.

heterogametic sex the sex that produces gametes containing unlike sex chromosomes (e.g., male mammals produce X- and Y-bearing sperm, usually in equal proportions). Crossing over is often suppressed in the heterogametic sex. *See* **Appendix C,** 1912, Morgan; 1913, Tanaka; *compare with* **homogametic sex.**

heterogamy the alternation of bisexual reproduction with parthenogenetic reproduction.

heterogeneous nuclear RNA (hnRNA) the pool of extrachromosomal RNA molecules found in the nucleus, consisting of a heterogeneous mixture of primary transcripts, partly processed transcripts, discarded intron RNA, and small nuclear RNA. The term is often used to refer to the primary transcripts or to their modified products alone. *See* **spliceosome, Usn RNAs.**

heterogenetic antigens the same or similar (cross-reacting) antigens shared by several species (e.g., Forssman antigen, *q.v.*). Antibodies produced against one of these antigens will also react with the other antigens of the system even though these are derived from a different species; such antibodies are also called *heterophile antibodies. See* **heterophile antigen.**

heterogenote *See* **heterogenotic merozygote.**

heterogenotic merozygote a partially heterozygous bacterium carrying an exogenote containing alleles differing from those on the endogenote.

heterogony cyclical parthenogenesis, when one or more parthenogenetic generations alternate with an amphimictic one, usually in an annual cycle. Aphids, gall wasps, and rotifers are examples of animals undergoing heterogony.

heterograft heterologous graft (*q.v.*).

heterokaryon a somatic cell that contains nuclei derived from genetically different sources. The nuclei do not fuse, but divide individually and simultaneously to form new cells, as commonly occurs in fungal hyphae. *See* **interspecific heterokaryons.**

heterokaryon test a test for organelle mutations based upon the appearance of unexpected phenotypes in uninucleate cells derived from specially marked heterokaryons. For example, suppose that heterokaryons form between colonies thought to carry a mitochondrial mutant (A) and colonies carrying a known nuclear mutation (B). If uninucleate progeny cells or spores exhibiting both the A and the B phenotypes can be derived from the heterokaryons, then the A mutation is probably in an extranuclear gene, because recombination of nuclear genes does not occur in heterokaryons.

heterokaryosis the condition in which fungus hyphae contain haploid nuclei of different genotypes, as a result of nonsexual fusion of the different types of hyphae.

heterokaryotypic referring to an individual carrying a chromosomal aberration in the heterozygous condition.

heterologous 1. in immunology, referring to an antibody and an antigen that do not bind to one another; either may be said to be heterologous to the other. 2. in transplantation studies, referring to a graft originating from a donor belonging to a species different from the host's. 3. in nucleic acid studies, referring to a DNA of a different source from the rest. Thus, a rabbit hemoglobin gene used as a probe to detect a hemoglobin gene from a mouse gene library represents heterologous DNA.

heterologous chimera a chimera (*q.v.*) formed by cells or tissues from two different species.

heteromeric *See* **heteropolymeric protein.**

Heterometabola Hemimetabola (*q.v.*).

heteromixis in fungi, referring to the mating system where sexual reproduction involves the fusion of genetically different nuclei each from a different thallus. *Contrast with* **homomixis.** *See* **Appendix C,** 1904, Blakeslee.

heteromorphic bivalent a bivalent made up of chromosomes that are structurally different and consequently are only partly homologous (the XY bivalent is an example). *Contrast with* **homomorphic bivalent.**

heteromorphic chromosomes homologous chromosomes that differ morphologically.

heteromorphosis (*also* **homeosis**) the formation, whether in embryonic development or in regeneration, of an organ or appendage inappropriate to its site (for example, an antenna instead of a leg).

heteromultimeric protein *See* **heteropolymeric protein.**

heterophile antigen a substance that stimulates production in a vertebrate of antibodies capable of reacting with tissue components from other vertebrates or even from plants.

heteroplastic transplantation a transplantation between individuals of different species within the same genus.

heteroplastidy having two kinds of plastids, specifically, chloroplasts and starch-storing leukoplasts.

heteroploid referring in a given species to the chromosome number differing from the characteristic diploid number (or haploid number, if the species has a predominating haplophase).

heteropolymeric protein a protein made up of more than one kind of polypeptide (e.g., hemoglobin).

heteropycnosis referring to the appearance of chromosomes or chromosomal regions that have a coiling cycle out of phase with the rest of the genome. Positively heteropycnotic segments are more tightly coiled and negatively heteropycnotic segments are more loosely coiled than the rest of the chromosomal complement. *See* **allocycly, isopycnotic.**

heterosis the greater vigor in terms of growth, survival, and fertility of hybrids, usually from crosses between highly inbred lines. Heterosis is always associated with increased heterozygosity. *See* **gibberellin.**

heterosomal aberration *See* **chromosomal aberration.**

heterospory in plants, the existence (within the same species or within an individual organism) of two kinds of meiocytes (megasporocytes and microsporocytes) that produce two kinds of meiospores. *Compare with* **homospory.**

heterostyly a polymorphism of flowers that ensures cross-fertilization by producing flowers having stamens and styles of unequal lengths.

heterothallic fungus a fungal species producing a sexual spore that results from the fusion of genetically different nuclei which arose in different thalli. *Compare with* **homothallic fungus.**

heterotopic transplantation transplantation of tissue from one site to another on the same organism.

heterotrophs organisms that require complex organic molecules such as glucose, amino acids, etc., from which to obtain energy and to build macromolecules. *Contrast with* **autotrophs.**

heterozygosis heterozygosity.

heterozygosity the condition of having one or more pairs of dissimilar alleles.

heterozygote a diploid or polyploid individual that has inherited different alleles at one or more loci and therefore does not breed true. *See* **Appendix C**, 1902–09, Bateson; **homozygote.**

heterozygote advantage the situation where the heterozygote has a greater fitness than either homozygote. *See* **overdominance.**

HEXA *See* **Tay-Sachs disease.**

hexaploid a polyploid possessing six sets (6N) of chromosomes or genomes; e.g., bread wheat is thought to have originated by hybridizations involving three different species, each of which contributed two genomes to the allohexaploid. *See* **wheat.**

HEXB *See* **Sandhoff disease.**

hexosaminidase an enzyme functioning in the catabolism of gangliosides (*q.v.*). Hexosaminidase A is composed of alpha and beta subunits coded for by genes on human autosomes 15 and 5, respectively. Mutations at these loci result in Tay-Sachs and Sandhoff diseases (*q.v.*).

hexose monophosphate shunt *See* **pentose phosphate pathway.**

Hfr strain a strain of *Escherichia coli* that shows high frequencies of recombination (hence the abbreviation). In cells from such a strain, the F-factor is integrated into the bacterial chromosome. *See* **Appendix C**, 1953, Hayes; **circular linkage map.**

HGPRT hypoxanthine-guanine-phosphoribosyltransferase (*q.v.*); an enzyme involved in the *salvage pathway* (*q.v.*) of nucleotide synthesis. *See* **HAT medium.**

hibernate to be dormant during winter. Many mammals, reptiles, amphibians, and certain invertebrates hibernate. *See also* **aestivate.**

hierarchy an organization pattern involving groups within groups, as exemplified by the taxonomic hierarchy of organisms. *See* **classification.**

high-energy bond a covalent chemical bond (e.g., the terminal phosphodiester bond of adenosine triphosphate) that liberates at least 5 kcal/mole free energy upon hydrolysis.

high-energy phosphate compound a phosphorylated molecule that upon hydrolysis yields a large amount of free energy. *See* **phosphate bond energy, ATP.**

high frequency of recombination cell *See* **Hfr strain.**

highly repetitive DNA the fast component in reassociation kinetics, usually equated with satellite DNA. *See* **repetitive DNA.**

Himalayan a mutant allele at the albino locus (with known examples in the mouse, rat, rabbit, guinea pig, hamster, and cat) associated with a very lightly pigmented body and somewhat darker extremities. The form of tyrosinase encoded by this gene is temperature senstive and normally functions well only in the extremities where the body temperature is lower. Such animals raised in cold environments show darker pigmentation.

HindII *See* **restriction endonuclease.**

hinge region *See* **immunoglobulin.**

hinny *See* **horse-donkey hybrids.**

his histidine. *See* **amino acid.**

histidine *See* **amino acid.**

histidinemia a hereditary disease in man arising from a deficiency of the enzyme histidase.

histidine operon a polycistronic operon of *Salmonella typhimurium* containing nine genes involved in the synthesis of histidine.

histochemistry the study by specific staining methods of the distribution of particular molecules within sections of tissues. *See* **Appendix C**, 1825, Raspail.

histocompatibility antigen genetically encoded cell surface alloantigens that can cause the rejection of grafted tissues, cells, and tumors bearing them. *See* **Appendix C**, 1937, Gorer.

histocompatibility gene a gene belonging to the major histocompatibility (MHC) system or to any of numerous minor histocompatibility systems responsible for the production of histocompatibility antigens (*q.v.*) *See* **Appendix C**, 1948, Snell.

histocompatibility molecules genetically encoded, cell-surface alloantigens that can cause the rejection of grafted tissues, cells, and tumors bearing them. These cell-membrane glycoproteins are grouped into two classes. Class I molecules are found on the surfaces of all mammalian cells (except trophoblasts and spermatozoa). T lymphocytes (*q.v.*) of the CD8$^+$ subgroup recognize antigenic determinants of foreign class I histocompatibility molecules. Class II histocompatibility molecules are abundant on the surfaces of B lymphocytes (*q.v.*). T lymphocytes of the CD4$^+$ subgroup recognize antigenic determinants of foreign class II histocompatibility molecules. These T cells subsequently divide and secrete lymphokines (*q.v.*), which are important for B cell growth and differentiation.

HISTONE SYNONYMS	MOLECULAR WEIGHT (d)	TOTAL AMINO ACIDS	% LYSINE	% ARGININE	RELATIVE AMOUNT PER 200 bp DNA
H1 = I = F1	21,000	207	27	2	1
H2A = IIbl = F2A2	14,500	129	11	9	2
H2B = IIb2 = F2B	13,700	125	16	6	2
H3 = III = F3	15,300	135	10	15	2
H4 = IV = F2A1	11,300	102	10	14	2

Histones.

Class I histocompatibility molecules are heterodimers made up of heavy (alpha) and light (beta) polypeptide chains. The class I chains are encoded by genes residing in the right portion of the HLA complex (*q.v.*). The alpha chain contains regions showing sequence diversity, whereas the beta chain has an invariant amino acid composition. The class II histocompatibility molecules are also dimers composed of heavy alpha and light beta chains. These are encoded by genes in the left portion of the HLA complex. Most of the sequence diversity of class II histocompatibility molecules is localized within a segment of the beta chain. Class II histocompatibility dimers are associated with a third polypeptide chain that exhibits no polymorphism. *See* **Appendix C,** 1937, Gorer; 1987, Wiley *et al.*

histogenesis the development of histologically detectable differentiation.

histogenetic antigens or responses defined by means of cell-mediated immunity.

histogram a bar graph.

histoincompatibility intolerance to transplanted tissue.

histology the study of tissues.

histolysis tissue destruction.

histone genes in both the sea urchin and *Drosophila,* these genes are repetitive and clustered. In *Strongylocentrotus purpuratus* the genes for H1, H4, H2B, H3, and H2A lie in a linear sequence, each separated by a spacer. There are several hundred serial repeats of this five-membered unit transcribed as a polycistronic message. In *Drosophila,* there are about 110 copies of the histone genes, and these are localized in a four-band region in the left arm of chromosome 2. In man, the histone gene family resides on chromosome 7. Most histone genes lack introns (*q.v.*). *See* **Appendix C,** 1972, Pardue *et al.*; 1977, Old *et al.*; *Lyctechinus pictus.*

histones small DNA-binding proteins. They are rich in basic amino acids and are classified according to the relative amounts of lysine and arginine they contain (see above table).

Histones are conserved during evolution. For example, H4 of calf thymus and pea differ at only two sites. The nucleosome (*q.v.*) contains two molecules each of H2A, H2B, H3, and H4, arranged along the DNA in the order H2A, H2B, H4, H3, H3, H4, H2B, H2A. A single H1 histone molecule is bound to the DNA segment that lies between nucleosomal cores. The role of the H1 protein seems to be in packing nucleosomes into 30-nm fibers. Yeast histones lack the H1 subtype. *See* **Appendix C,** 1974, Kornberg; 1977, Leffak *et al.;* **chromatosome, nucleoprotein, nucleosome, ubiquitin.**

hitchhiking the spread of a neutral allele through a population because it is closely linked to a beneficial allele and therefore is carried along as the gene that is selected for increases in frequency.

HIV the *h*uman *i*mmunodeficiency *v*irus, a human RNA retrovirus known to cause AIDS. HIV is now referred to as HIV-1, to distinguish it from a related but distinct retrovirus, HIV-2, which has been isolated from West African patients with a clinical syndrome indistinguishable from HIV-induced AIDS. *See* **AIDS.**

HLA human leukocyte antigens concerned with the acceptance or rejection of tissue or organ grafts and transplants. These antigens are on the surface of most somatic cells except red blood cells, but are most easily studied on white blood cells (hence the name). *See* **Appendix C,** 1954, Dausset.

HLA complex the major histocompatibility gene complex of humans. The complex occupies a DNA segment about 3,500 kb long on the short arm of chromosome 6. The portion of the HLA complex closest to the telomere contains the genes that encode the class I histocompatibility molecules (HLA-B, -C, and -A). The portion closest to the centromere contains the genes encoding the class II histocompatibility molecules (DP, DQ, and DR). Genes encoding components of the complement (*q.v.*) system lie in the midregion of the complex. *See* **major histocompatibility complex (MHC).**

H locus in humans, a genetic locus that encodes a fucosyl transferase enzyme that is required during an early step in the biosynthesis of the antigens of the ABO blood group system. *See* **A, B antigens, Bombay blood group.**

hnRNA   heterogeneous nuclear RNA (*q.v.*).

hog   *See* **swine.**

Hogness box   a segment 19–27 bp upstream from the startpoint of eukaryotic structural genes to which RNA polymerase II binds. The segment is 7 bp long, and the nucleotides most commonly found are TATAAAA; named in honor of David Hogness. *See* **canonical sequence, Pribnow box, promoter.**

holandric   appearing only in males. Said of a character determined by a gene on the Y chromosome. *See* **hologynic.**

holism   a philosophy maintaining that the entirety is greater than the sum of its parts. In biology, a holist is one who believes that an organism cannot be explained by studying its component parts in isolation. *See* **reductionism.**

Holliday intermediate   *See* **Holliday model.**

Holliday model   a model that describes a series of breakage and reunion events occurring during crossing over between two homologous chromosomes. The diagram below illustrates this model. In (a) the duplex DNA molecule from two nonsister strands of a tetrad have aligned themselves in register so that the subsequent exchange does not delete or duplicate any genetic information. In (b) strands of the same polarity are nicked at equivalent positions. In (c) each broken chain has detached from its partner and paired with the unbroken chain in the opposite duplex. In (d) ligases have joined the broken strands to form an internal branch point. The branch is free to swivel to the right or left, and so it can change its position. This movement is called *branch migration*, and in (e) the branch is shown as having moved to the right. In (f) the molecule is drawn with its arms pulled apart, and in (g) the a b segment has been rotated 180° relative to the A B segment. The result is an x-shaped configuration, and in it one can readily see

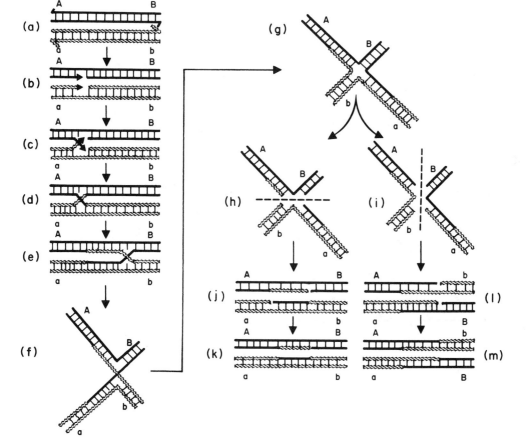

Holliday model.

149

the four sectors of single-stranded DNA in the branch region. To separate the structure into two duplexes, cuts must be made across the single strands at the branch region. The cuts can occur horizontally (h) or vertically (i). In the horizontal case, the separated duplexes will appear as shown in (j), and when the nicks are ligated, the duplexes will each contain a patch in one strand, as shown in (k). In the vertical case, the separated duplexes will appear as shown in (l), and when the nicks are ligased, the duplexes will contain splices in every strand, as shown in (m). It is only in this case that single crossover chromatids (Ab and aB) will be detected. Structures similar to the molecule drawn in (g) have been observed under the electron microscope and have been named *Holliday intermediates*. *See* **Appendix C**, 1964, Holliday; **chi structure.**

holoblastic cleavage   cleavage producing cells of approximately equivalent size.

Holocene   the epoch of the Quaternary period from the end of the Pleistocene to the present time. Neolithic to modern civilization. *See* **geologic time divisions.**

holocentric   referring to chromosomes with diffuse centromeres. *See* **centromere.**

holoenzyme   the functional complex formed by an apoenzyme (*q.v.*) and its appropriate coenzyme (*q.v.*).

hologynic   appearing only in females. Said of a trait passed from a P₁ female to all daughters occurring, for example, in the case of a gene linked to the W chromosome (*q.v.*). *See* **matrocliny, holandric.**

Holometabola   the superorder of insects containing species that pass through a complete metamorphosis.

holometabolous   referring to those insects in which larval and pupal stages are interposed between the embryo and the adult. *Contrast with* **hemimetabolous.**

holophyletic   an evolutionary lineage consisting of a species and all of its descendants.

holophytic nutrition   nutrition requiring only inorganic chemicals, as that of photosynthetic plants.

holorepressor   *See* **aporepressor.**

holotype   the single specimen selected for the description of a species.

holozoic nutrition   nutrition requiring complex organic foodstuffs, as that of organisms other than photosynthetic plants and protoctists.

homeobox   a sequence of about 180 base pairs near the 3′ end of certain homeotic genes (*q.v.*). The 60 amino acid peptide encoded by the homeobox appears to be a DNA-binding protein. *See* **Appendix C,** 1984, McGinnis *et al.*

homeologous chromosomes   *See* **homoeologous chromosomes.**

homeoplastic graft (*also* **homoeoplastic**)   a graft of tissue from one individual to another of the same species.

homeosis (*also* **homoeosis**)   heteromorphosis (*q.v.*).

homeostasis (*also* **homoeostasis**)   a fluctuation-free state. *See* **developmental homeostasis.**

homeotic mutations   those in which one developmental pattern is replaced by a different, but homologous one. The homeotic mutations of *Drosophila* cause an organ to differentiate abnormally and to form a homologous organ that is characteristic of an adjacent segment. Three such mutations are illustrated below: (A) A frontal view of the normal head. (B) The leg-like antenna of the mutation called *aristapedia*. (C) Leglike structures extending

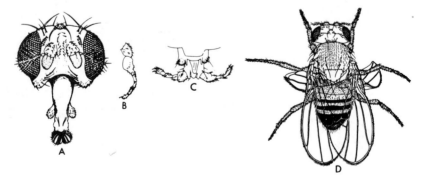

Homeotic mutations.

150

from the proboscis of the mutation called *proboscipedia*. (D) A *bithorax* male in which halteres are changed into winglike appendages. *See* **compartmentalization, segment identity genes.**

hominid a member of the family Hominidae including humans and related fossil species. Two genera are recognized: *Australopithecus* and *Homo.*

Hominoidea a superfamily of primates. Its living species include man, two species of chimpanzee, the gorilla, the orangutan, and nine species of gibbon. *See* **Appendix C, 1967, Sarich and Wilson; Cercopithecoidea.**

Homo a genus in which humans are placed. It contains two fossil species: *H. habilis,* which lived 2.3 to 1.5 million years ago, and *H. erectus,* which lived 1.5 to 0.3 million years ago. *Homo sapiens* is the most recent species, dating from about 300,000 years ago to present. It contains modern humans and Neanderthals.

homoalleles alternative forms of a gene that differ at the same muton site. Intragenic recombination between homoalleles is not possible. *Compare with* **heteroalleles.**

homoallelic referring to allelic mutant genes that have mutations at the same site (*q.v.*). A functional cistron cannot be generated by intragenic recombination between homoalleles. *Contrast with* **heteroallelic.**

homobrachial inversion paracentric inversion (*q.v.*).

homocaryon homokaryon (*q.v.*).

homocystinuria a hereditary disease in man arising from a deficiency of the enzyme serine dehydratase.

homodimer a protein made up of paired identical polypeptides.

homoeologous chromosomes chromosomes that are only partially homologous. Such chromosomes are derived from ancestral chromosomes that are believed to have been homologous. Evolutionary divergence has reduced the synaptic attraction of homoeologs. *See* **Appendix C, 1958, Okamoto; differential affinity, isoanisosyndetic alloploid.**

homoeosis homeosis. *See also* **homeoplastic graft, homeostasis, homeotic mutations.**

homoeotic homeotic (*q.v.*).

homogametic sex the sex that produces gametes all of which carry only one kind of sex chromosome; e.g., the eggs of female mammals carry only

an X chromosome. *Compare with* **heterogametic sex.**

homogamy the situation in which the male and female parts of a flower mature simultaneously.

homogenote a partial diploid (merozygote) bacterium in which the donor (exogenote) chromosomal segment carries the same alleles as the chromosome of the recipient (endogenote) cell.

homogentisic acid a compound derived from the metabolic breakdown of the amino acid tyrosine. *See* **alkaptonuria.**

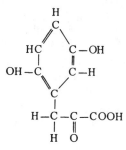

homograft homeoplastic graft (*q.v.*).

homoimmunity the resistance of a lysogenic bacterium (harboring a prophage, *q.v.*) to superinfection by phage of the same kind as that carried in the prophage state. The excess repressor molecules generated by the prophage bind to operators on the infecting DNA molecules and prevent their transcription.

homokaryon a dikaryotic mycelium in which both nuclei are of only one genotype.

homokaryotypic referring to an individual carrying a chromosomal aberration in the homozygous condition.

homolog 1. in classification, a character that defines a clade. 2. in evolution, homologs are characteristics that are similar in different species because they have been inherited from a common ancestor. 3. in cytology, *see* **homologous choromosomes.**

homologous referring to structures or processes in different organisms that show a fundamental similarity because of their having descended from a common ancestor. Homologous structures have the same evolutionary origin although their functions may differ widely: e.g., the flipper of a seal and the wing of a bat. *See* **analogous.**

homologous chromosomes chromosomes that pair during meiosis. Each homologue is a duplicate of one of the chromosomes contributed at syngamy by the mother or father. Homologous chromosomes contain the same linear sequence of genes and as a consequence each gene is present in duplicate.

homologue homolog (*q.v.*).

homology the state of being homologous. In molecular biology, the term is often misused when comparing sequences of nucleotides or amino acids from nucleic acids or proteins obtained from distantly related species. In such instances it is preferable to refer to sequence identities or similarities rather than "homologies."

homomeric protein referring to a protein made up of two or more identical polypeptide chains. An example would be beta galactosidase (*q.v.*), which is an aggregate of four identical polypeptides.

homomixis referring in fungi to the mating system in which sexual reproduction involves the fusion of genetically similar nuclei derived from one thallus.

homomorphic bivalent a bivalent made up of homologues of similar morphology. *See* **heteromorphic bivalent.**

homomultimer *See* **homopolymer.**

homoplasy parallel or convergent evolution; structural similarity in organisms not due directly to inheritance from a common ancestor or development from a common anlage.

homopolar bond covalent bond (*q.v.*).

homopolymer a polymer composed of identical monomeric units (poly U, for example).

homopolymer tails a segment containing several of the same kind of deoxyribonucleotides arranged in tandem at the 3′ end of a DNA strand. *See* **Appendix C,** 1972, Lobban and Kaiser; **terminal transferase.**

homosequential species species with identical karyotypes, as found in many species of *Drosophila* endemic to the Hawaiian Islands.

homosomal aberration *See* **chromosomal aberration.**

homospory in plants of both mating types or sexes, the production of meiospores of equivalent size. *Compare with* **isogamy.**

homothallic fungus a fungal species producing a sexual spore that results from the fusion of genetically different nuclei derived from the same thallus. *Compare with* **heterothallic fungus.**

homozygosity the condition of having identical alleles at one or more loci in homologous chromosome segments.

homozygote an individual or cell characterized by homozygosity. *See* **Appendix C,** 1902–09, Bateson.

homozygous having identical rather than different alleles in the corresponding loci of homologous chromosomes and therefore breeding true. *See* **heterozygosity.**

homunculus a miniature individual imagined by early biologists to be present in a sperm.

hopeful monster *See* **saltation.**

Hordeum vulgare cultivated barley. The haploid chromosome number is 7, and over a hundred mutations have been located in the 7 linkage groups.

horizontal classification a system of evolutionary classification that tends to unite transitional forms with their ancestors; the opposite of *vertical classification* (*q.v.*).

horizontal transmission *See* **vertical transmission.**

hormone an organic compound produced in one part of an organism and transported to other parts, where it exerts a profound effect. Mammalian hormones include ACTH, epinephrine, FSH, glucagon, GH, LH, insulin, intermedin, oxytocin, progesterone, prolactin, secretin, thyroxin, and vasopressin. *See also* **endocrine system.**

horotelic evolution *See* **evolutionary rate.**

horse any of a number of domesticated breeds of the species *Equus caballus.* Popular breeds include: DRAFT HORSES: Belgian, Clydesdale, Percheron, Shire, Suffolk. COACH HORSES: Cleveland Bay, French Coach, German Coach, Hackney. LIGHT HARNESS HORSES: American Trotter. SADDLE HORSES: American Saddle Horse, American Quarter Horse, Appaloosa, Arabian, Morgan, Palomino, Tennessee Walking Horse, Thoroughbred, Lippizzaner. PONIES: Hackney Pony, Shetland Pony, Welsh Pony.

horse bean broad bean (*q.v.*).

horse-donkey hybrids the horse female × donkey male cross produces a mule; the reciprocal cross produces a hinny. These hybrids have 63 chromosomes and are sterile. In such hybrids, mitochondria are of maternal origin. *See* **Appendix C,** 1974, Hutchison; ***Equus.***

host **1.** an organism infected by a parasite. **2.** the recipient of a graft.

host-cell reactivation cut and patch repair (*q.v.*) of UV-induced lesions in the DNAs of bacteriophages once they infect a host cell. Host-cell reactivation does not occur in the case of ssDNA or RNA viruses.

host-controlled restriction and modification *See* DNA restriction enzyme, restriction and modification model.

host range the spectrum of strains of a species of bacterium that can be infected by a given strain of phage. The first mutations to be identified in phage involved host range. *See* **Appendix C, 1945, Luria.**

host-range mutation a mutation of a phage that enables it to infect and lyse a previously resistant bacterium.

hot spot **1.** a site at which the frequencies of spontaneous mutation or recombination are greatly increased with respect to other sites in the same cistron. Examples are in the rII gene of phage T4 and in the *lacZ* and *trpE* genes of *E. coli.* **2.** a chromosomal site at which the frequencies of mutations are differentially increased in response to treatment with a specific mutagen.

housekeeping genes constitutive loci that are theoretically expressed in all cells in order to provide the maintenance activities required by all cells: e.g., genes coding for enzymes of glycolysis and the Kreb's cycle.

HPRT hypoxanthine-guanine-phosphoribosyl transferase (*q.v.*).

H substance a precursor polysaccharide for production of A and B antigens of the ABO blood group system. It is usually unmodified on group O cells, but is modified by the addition of different sugars to produce the A or B antigens. The molecule is specified by a gene on human chromosome 19. *See* **Bombay blood group.**

human chromosome band designations quinacrine and Giemsa-stained human metaphase chromosomes show characteristic banding patterns, and standard methods have been adopted to designate the specific patterns displayed by each chromosome. The X chromosome to the right illustrates the terminology. In the diagram, the dark bands represent those regions that fluoresce with quinacrine or are darkened by Giemsa. The short (p) arm and the longer (q) arm are each divided into two regions. In the case of longer autosomes, the q arm may be divided into three or four regions and the p arm into three regions. Within the major regions, the dark and light bands are numbered consecutively. To give an example of the methods used for assigning loci, the G6PD gene is placed at q28, meaning it is in band 8 of region 2 of the q arm. The color-blindness genes are both assigned to q27-qter. This means they reside somewhere between the beginning of q27 and the terminus of the long arm. *See* **human mitotic chromosomes.**

human cytogenetics *See* **symbols used in human cytogenetics, human mitotic chromosomes.**

human gene maps about 1,600 autosomal loci are known to exist, and about 30% of these have been assigned to one of the 22 autosomes. About 115 loci have been assigned to the X chromosome. Relatively few genes have been assigned to the Y chromosome; most of these control testis differentiation and spermatogenesis. The human mitochondrial chromosome is sometimes referred to as chromosome 25 or M. This circular chromosome is made up of 16,569 base pairs and contains about 40 genes.

human growth hormone *See* **growth hormone.**

human immunodeficiency virus *See* **AIDS, HIV.**

human mitotic chromosomes the mitotic chromosomes of man are generally grouped into seven classes (A–G) according to the following cytological criteria: Group A (chromosomes 1–3)—large chromosomes with approximately median centromeres. Group B (chromosomes 4–5)—large chromosomes with submedian centromeres. Group C (chromosomes 6–12 and the X chromosome)—medium-sized chromosomes with submedian centromeres. Group D (chromosomes 13–15)—medium-sized acrocentric chromosomes. Chromosome 13 has a prominent satellite on the short arm. Chromosome 14 has a small satellite on the short arm. Group E (chromosomes 16–18)—rather short chromosomes with approximately median (in chromosome 16) or submedian centromeres. Group F (chromosomes 19 and 20)—short chromosomes with approxi-

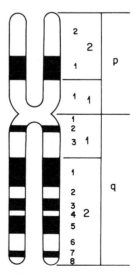

mately median centromeres. Group G (chromosomes 21, 22, and the Y chromosome)—very short acrocentric chromosomes. *See* **Appendix C,** 1956, Tjio and Levan; 1971, O'Riordan; 1981, Harper and Saunders; **human chromosome band designations, symbols used in human cytogenetics.**

human pseudoautosomal region a DNA segment in the distal part of the short arms of the human X and Y chromosomes that share homologous genes. These segments pair during meiosis, and obligatory crossing over takes place, so that genes in this region segregate like autosomal loci, rather than showing X or Y linkage. The gene *MIC2*, which encodes a molecule present on the surface of all human cells, has been shown to be pseudoautosomal. In the mouse, the steroid sulfatase gene (*Sts*) is pseudoautosomal. *See* **autosome.**

humoral immunity immunoglobulin (antibody)-mediated immune responses.

hunchback This *Drosophila* mutant belongs to the gap class of zygotic segmentation genes, and it interacts with a maternal polarity gene *bicoid (bc)*. The protein encoded by *bc* functions in development by binding to chromosomal sites upstream of the transcription start point of *hb*. When these sites are occupied, the *hb* gene is switched on, and its product stimulates the development of certain cells located in the head and thoracic segments. *See* **Appendix C,** 1989, Driever and Nüsslein-Volhard; **zygotic segmentation mutants, *bicoid*.**

Hunter syndrome an X-linked disorder of connective tissue in man associated with the storage of mucopolysaccharides.

Huntington disease a human disease characterized by irregular, spasmodic, involuntary movements of the limbs and facial muscles, mental deterioration, and death, usually within 20 years of the onset of the symptoms. These generally do not appear until the heterozygote is 35 to 40 years of age. It is caused by a dominant gene on chromosome 4, located near the telomere of the short arm.

In the earlier literature, the disease is called *Huntington's chorea. See* **parental imprinting.**

Hurler syndrome an autosomally linked disorder of connective tissue in man associated with the storage of mucopolysaccharides.

Hutchinson-Gilford syndrome *See* **progeria.**

HVL half-value layer (*q.v.*).

Hyalophora cecropia the giant cecropia moth; because of its large size a favorite experimental insect. *See* **Appendix C,** 1966, Röller *et al.*

hyaloplasm cytosol.

hyaluronic acid a mucopolysaccharide that is abundant in the jelly coats of eggs and in the ground substance of connective tissue. As shown below, hyaluronic acid is a polymer composed of glucosamine and glucuronic acid subunits.

hyaluronidase an enzyme that digests hyaluronic acid.

H-Y antigen an antigen detected by cell-mediated and humoral responses of homogametic individuals against heterogametic individuals of the same species, which are otherwise genetically identical. Antigenic responses of this sort have been demonstrated in mammals, birds, and amphibians. In mammals, the antigen is called H-Y because it acts as a *H*istocombatibility factor determined by the *Y*-chromosome. The location of the gene encoding the H-Y antigen is not known. However, the gene that induces synthesis of the H-Y antigen in humans is located on the Y-chromosome. A homologous locus, which suppresses H-Y production, lies on the distal end of the short arm of the X. Evidently, two doses of this gene are necessary for the complete suppression, since Turner syndrome (45, XO) females produce small amounts of the H-Y antigen. The H-Y locus is one of the areas that escapes X-chromosome inactivation (*q.v.*).

hybrid **1.** a heterozygote (e.g., a monohybrid is heterozygous at a single locus; a dihybrid is hetero-

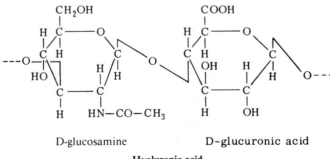

D-glucosamine   D-glucuronic acid

Hyaluronic acid.

154

zygous at two loci; etc.). **2.** an offspring from genetically dissimilar parents, even different species.

hybrid arrested translation   a method for identifying the cDNA corresponding to an mRNA that depends upon the ability of cDNA to hybridize with its mRNA and thereby to inhibit its translation in an *in vitro* system; the disappearance of the translation product from the system indicates the presence of the cDNA.

hybrid breakdown   the reduction in fitness of F_2 and/or backcross populations from fertile hybrids produced by intercrossing genetically disparate populations or species; a postzygotic reproductive isolating mechanism.

hybrid corn   commercial corn grown from seed produced by the "double cross" (*q.v.*) procedure. Such corn is characterized by its vigor and uniformity. *See* **Appendix C,** 1909, Shull.

hybrid DNA model   a model used to explain both crossing over and gene conversion by postulating that a short segment of heteroduplex (hybrid) DNA is produced from both parental DNAs in the neighborhood of a chiasma. *See* **Holliday model.**

hybrid duplex molecule   an experimentally reconstituted molecule containing a segment of single-stranded DNA hydrogen bonded to a second RNA or DNA molecule of complementary base sequence.

hybrid dysgenesis   a syndrome of correlated genetic abnormalities that occurs spontaneously when hybrids are formed between certain strains of *Drosophila.* The hybrids show germ line defects including chromosomal aberrations, high frequencies of lethal and visible mutations, and in most extreme cases, sterility. The cause of PM hybrid dysgenesis is a transposable element named P. Long established laboratory strains lack P elements. Those strains susceptible to P elements are called M strains. P elements are often present in strains established from newly captured wild flies. The cross P male $\times$ M female generates dysgenic F_1 individuals; the reciprocal cross does not. The P elements do not produce dysgenesis within P strains. P elements are dispersed over all the chromosome arms, and in P strains their transposition is repressed. When chromosomes carrying P elements are placed in M cytoplasm by suitable crosses, the P elements become derepressed and transpose at high rates, disrupting genetic loci and causing the dysgenic syndrome. *See* **Appendix C,** 1982, Bingham *et al.*; **P elements.**

hybrid inviability   a postmating (postzygotic) reproductive isolating mechanism in which hybrids

between disparate populations fail to survive to reproductive age.

hybridization   **1.** the mating of individuals belonging to genetically disparate populations or to different species. **2.** in Mendelian terms, the mating of any two unlike genotypes or phenotypes. **3.** the pairing of complementary RNA and DNA strands to produce an RNA-DNA hybrid, or the pairing of complementary DNA single strands to produce a DNA-DNA hybrid.

hybridization competition   a technique for distinguishing different mRNA molecules, using a variation of the basic filter hybridization (*q.v.*) technique. A specific DNA sequence trapped on a nitrocellulose filter is exposed to a tritiated RNA known to be complementary to that DNA. An unlabeled RNA of unknown specificity is added and, if complementary to the DNA, it will compete with the labeled molecules for hybridization to the DNA. Any diminution of labeling in the hybrid after equilibrium is reached is attributed to displacement by the unlabeled molecules.

hybridogenesis   a form of clonal reproduction in species hybrids whose gametes carry only the nuclear genome derived from one of the parental species. For example, a hybrid species of frog, *Rana esculenta,* is derived from crosses between *R. lessonae* and *R. ridibunda,* but only *ridibunda* chromosomes (and *lessonae* mitochondria) are found in gametes of *esculenta.*

hybridoma   a cell resulting from the fusion of an antibody-producing plasma cell (a B lymphocyte) and a myeloma (a bone-marrow cancer) cell. Such a hybrid cell produces a clone that can be maintained in tissue culture or as an animal tumor, and the clone may secrete only a single kind of antibody. Such monoclonal antibodies are used as probes in western blot experiments or in the histochemical localization of antigens of interest. *See* **Appendix C,** 1975, Köhler and Milstein; 1980, Olsson and Kaplan; **immunofluorescence.**

hybrid resistance   the phenomenon whereby tumors may grow more readily in homozygous recipients than in heterozygous recipients even though the tumor may be genetically histocompatible with both types of recipients.

hybrid sterility   the failure of hybrids between different species to produce viable offspring.

hybrid swarm   a continuous series of morphologically distinct hybrids resulting from hybridization of two species followed by crossing and backcrossing of subsequent generations.

hybrid vigor   heterosis (*q.v.*).

hybrid zone a geographical zone where hybrids between two geographical races are observed. *See* **Appendix C,** 1973, Hunt and Selander.

hydrocarbon an organic compound composed only of carbon and hydrogen atoms.

hydrogen the most abundant of the biologically important elements. Atomic number 1; atomic weight 1.00797; valence 1+, most abundant isotope ^{1}H, heavy isotope ^{2}H (deuterium); radioisotope ^{3}H (tritium, *q.v.*).

hydrogen bond the weak electrostatic attraction that exists between a hydrogen atom that is covalently bonded to an O or N atom and an atom containing an unshared electron pair. Hydrogen bonds are weaker than covalent bonds, but stronger than the van der Waals attractive forces between molecules. *See* **deoxyribonucleic acid.**

hydrogen ion concentration expressed as the logarithm of the reciprocal of the concentration of hydrogen ions in grams per liter of solution; abbreviated as pH. the scale runs from 0 to 14, with values above 7 basic; those below, acidic.

hydrogen peroxide H_2O_2. *See* **catalase, peroxisomes.**

hydrolases enzymes that catalyze the transfer of water between donor and receptor molecules. Proteolytic enzymes are a special class of hydrolases.

hydrolysis the splitting of a molecule into two or more smaller molecules with the addition of the elements of water.

hydroperoxyl radical HO_2, an oxidizing agent formed during the interaction of ionizing radiation with oxygenated water. *See* **free radical.**

hydrophilic water attracting; referring to molecules or functional groups in molecules that readily associate with water. The carboxyl, hydroxyl, and amino groups are hydrophilic.

hydrophobic water repelling; referring to molecules or functional groups in molecules (such as alkyl groups) that are poorly soluble in water. Populations of hydrophobic groups form the surface of water repellent membranes.

hydrophobic bonding the tendency of nonpolar groups to associate with each other in aqueous solution, thereby excluding water molecules.

hydrops fetalis *See* **thalassemias.**

hydroquinone *See* **quinone.**

hydroxyapatite a form of calcium phosphate that binds to double-stranded DNA.

hydroxykynurenine *See* ***Drosophila* eye pigments.**

hydroxylamine NH_2OH, a mutagen that converts the NH_2 of cytosine to NHOH, which pairs only with adenine.

5-hydroxymethyl cytosine a pyrimidine found instead of cytosine in the DNA of T-even coliphages. 5-Hydroxymethyl cytosine pairs with guanine. It is postulated that the phage-specific DNase, which breaks down the host DNA, attacks DNA molecules containing cytosine.

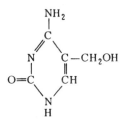

hydroxyurea a compound that inhibits semiconservative DNA replication, but not repair synthesis.

Hylobates a primate genus that contains nine species of gibbons. *H. concolor* is the best known from the genetic standpoint. Its haploid chromosome number is 26, and 20 genes have been assigned to 10 syntenic groups.

hymenopteran an insect belonging to the order Hymenoptera (which includes bees, ants, wasps, etc.). *See* **Appendix A.**

hyparchic genes *See* **autarchic genes.**

hyperammonemia a hereditary disease in man arising from a deficiency of the enzyme ornithine carbamoyl transferase.

hypercholesterolemia *See* **familial hypercholesterolemia.**

hyperchromic shift an increase in the absorbtion of ultraviolet light by a solution of DNA as these molecules are subjected to heat, alkaline conditions, etc. The shift is caused by the disruption of the hydrogen bonds of each DNA duplex to yield single-stranded structures.

hyperdontia the hereditary presence of one or more additional teeth.

hyperglycemia an increased glucose content in the blood.

hyperlipemia an increased concentration of neutral fat in the blood serum.

hypermorph a mutant gene whose effect is similar to, but greater than, that of the standard or wild-type gene.

hyperplasia an increase in amount of tissue produced by an increase in the number of cells. Hyperplasia often accompanies the regeneration of a damaged organ. *See* **hypertrophy.**

hyperploid referring to cells or individuals containing one or more chromosomes or chromosome segments in addition to the characteristic euploid number.

hyperprolinemia a hereditary disease in man arising from a deficiency of the enzyme proline oxidase.

hypersensitivity the characteristic of responding with clinical symptoms to allergens in amounts that are innocuous to most individuals. *See* **allergy.**

hypertension an increased blood pressure.

hypertrophy an increase in the size of a tissue or organ because of the increased volume of the component cells. *See* **hyperplasia.**

hypervariable (hv) sites amino acid positions within the variable region of an immunoglobulin light chain or heavy chain, exhibiting great variation among antibodies of different specificity; these noncontiguous sites are brought together in the active site where antigen is bound (a paratope) by complex folding of the polypeptide chain. *See* **immunoglobulin.**

hypha a filament of a fungus thallus.

hypo *See* fixing.

hypochromic anemia *See* **anemia.**

hypochromic shift reduction in the absorption of ultraviolet light as complementary single strands of DNA unite to form duplexes. *See* **hyperchromic shift.**

hypodontia the congenital absence of teeth.

hypoglycemia a decrease in sugar content of the blood serum.

hypomorph a mutant gene acting in the same direction as the normal allele, but with less effect.

hypophosphatasia a hereditary disease in man arising from a deficiency of the enzyme alkaline phosphatase.

hypophosphatemia a decreased concentration of inorganic phosphate in the blood serum.

hypophysis the pituitary gland.

hypoplasia an arrested development of an organ or part. The opposite of hyperplasia (*q.v.*).

hypoploid referring to cells or individuals containing one or more fewer chromosomes or chromosome segments than the characteristic euploid number.

hypostatic gene *See* **epistasis.**

hypothalamus the floor and sides of the vertebrate brain just behind the attachment of the cerebral hemispheres. The hypothalamus controls the secretion of antidiuretic hormone and oxytocin, and probably supplies them to the pituitary.

hypothyroidism a diminished production of thyroid hormone.

hypoxanthine 6-hydroxypurine. *See* **purine.**

hypoxanthine-guanine-phosphoribosyl transferase the enzyme that catalyzes the transfer of the phosphoribosyl moiety of 5-phosphoribosyl-1-pyrophosphate to the 9 position of hypoxanthine and guanine to form inosine monophosphate and guanosine monophosphate. Abbreviated HPRT or HGPRT. The Lesch-Nyhan syndrome (*q.v.*) is caused by deficiency of HPRT. *See* **Appendix C, 1987, Kuehn** *et al.*; **HAT medium.**

i the regulator gene of the lactose operon in *Escherichia coli. See* **regulator genes.**

I iodine.

I₁,I₂,I₃,etc. the first, second, third, etc., generations obtained by inbreeding.

*I*ᴬ,*I*ᴮ,*I*⁰ the allelic genes responsible for the ABO blood group system. *See* **A, B antigens.**

IAA indole acetic acid (*q.v.*).

Ia antigens alloantigens encoded by the Ia region of the mouse major histocompatibility complex (H-2). They are defined by serological methods and are found predominantly (but not exclusively) on B lymphocytes and macrophages.

icosahedron a regular geometric polyhedron composed of 20 equilateral triangular faces with 12 corners. The capsids of many spherical eukaryotic viruses and bacteriophages are icosahedral. *See* **Shope papilloma virus.**

ICSH interstitial cell-stimulating hormone. Identical to LH (*q.v.*).

identical twins *See* **twins.**

idiocy the most severe degree of mental retardation. An idiot reaches an intelligence level below that of a two-year-old child.

idiogram a diagrammatic representation of the karyotype (*q.v.*) of an organism.

idiotype antigenic determinants characteristic of a particular variable domain of a specific immunoglobulin or T cell receptor molecule. The idiotype is a unique attribute of a particular antibody from a specific individual. *Contrast with* **allotypes, isotypes.**

idling reaction production of ppGpp and pppGpp by ribosomes when an uncharged tRNA is present in the A site. *See* **translation.**

IF initiation factor (*q.v.*).

IFN interferon (*q.v.*).

Ig immunoglobulin (*q.v.*).

IgA human immunoglobulin A, found as a 160-kd monomer or as a 320-kd dimer in mucus and secretory fluids and on the surface of cell membranes.

IgD human immunoglobulin D, found as a 185-kd monomer on the surface of lymphocytes.

IgE human immunoglobulin E, found as a 200-kd monomer and involved in allergic reactions. It forms a complex with antigen and then binds to the surface of mast cells, triggering the release of histamine.

IgG human immunoglobulin G, found as a 150-kd monomer, which is the predominant molecule involved in secondary immune responses. It fixes complement and is the only immunoglobulin that crosses the placenta. *See* **Appendix C,** 1969, Edelman *et al.;* **immune response.**

IgM human immunoglobulin M, found as a 900-kd pentamer that is the predominant molecule involved in the primary immune response. It fixes serum complement and agglutinates effectively.

ile isoleucine. *See* **amino acid.**

imaginal discs inverted thickenings of epidermis containing mesodermal cells found in a holometabolous insect. During the pupal stage, the imaginal discs give rise to the adult organs, and most larval structures are destroyed. *See in vitro* **culturing of imaginal discs.**

imino forms of nucleotides *See* **tautomeric shifts.**

immediate hypersensitivity a type of hypersensitivity reaction that is mediated by antibodies and that occurs within minutes after exposure to the allergen or antigen in a previously sensitized individual. *Compare with* **delayed hypersensitivity.**

immortalizing genes genes carried by oncogenic viruses that confer upon cultured mammalian cells the ability to divide and grow indefinitely, thereby overcoming the Hayflick limit (*q.v.*).

immune competent cell a cell capable of producing antibody in response to an antigenic stimulus.

immune decoy protein *See* **sporozoite.**

immune globulins *See* **antibody.**

immune reaction the reaction between a specific antigen and antibody.

immune response the physiological response(s) stemming from activation of the immune system by antigens, including beneficial immunity to pathogenic microorganisms, as well as detrimental autoimmunity to self-antigens, allergies, and graft rejection. The cells mainly involved in an immune response are T and B lymphocytes and macrophages. T cells produce lymphokines (*q.v.*) that influence the activities of other host cells, whereas B cells mature to produce immunoglobulins (*q.v.*) or antibodies that react with antigens. Macrophages "process" the antigen into immunogenic units that stimulate B lymphocytes to differentiate into antibody-secreting plasma cells, and stimulate T cells to release lymphokines. Complement (*q.v.*) is a group of normal serum proteins that can aid immunity by becoming activated as a consequence of antigen-antibody interactions. The first contact with an antigen "sensitizes" the animal and results in a *primary immune response.* Subsequent contact of the sensitized animal with that same antigen results in a more rapid and elevated reaction, called the *secondary immune response* (also known as the "booster response" or the "anamnestic reaction"), which is most easily demonstrated by monitoring the level of circulating antibodies in the serum. The immune response can be transferred from a sensitized to an unsensitized animal via serum or cells. It is highly specific for the inciting antigen, and is normally directed only against foreign substances.

immune response (Ir) gene any gene that determines the ability of lymphocytes to mount an immune response to specific antigens. In the major histocompatibility complex of the mouse (the H-2 complex), the I region contains Ir genes and also codes for Ia (immune associated) antigens found on B cells and on some T cells and macrophages. In humans, the HLA D (DR) region is the homologue of the mouse H-2 I region. *See* **Appendix C,** 1948, Snell; 1963, Levine *et al.*; 1972, Benacerraf and McDevitt.

immunity 1. the state of being refractive to a specific disease, mediated by the immune system (T and B lymphocytes and their products—lymphokines and immunoglobulins, respectively). *Active immunity* develops when an individual makes an immune response to an antigen; *passive immunity* is acquired by receiving antibodies or immune cells from another individual. 2. the ability of a prophage to inhibit another phage of the same type from infecting a lysogenized cell (phage immunity). 3. the ability of a plasmid to inhibit the establishment of another plasmid of the same type in that

cell. 4. the ability of some transposons to prevent others of the same type from transposing to the same DNA molecule (transposon immunity). 5. phage-resistant bacteria are usually "immune" to specific phages because they lack the cell-surface receptors that define the host range of that phage.

immunity substance a cytoplasmic factor produced in lysogenic bacteria that prevents them from being infected by bacteriophages of the same type as their prophages and also prevents the vegetative replication of said prophages.

immunization administration of an antigen for the purpose of stimulating an immune response to it.

immunochemistry the study of the chemistry of immune responses.

immunocompetent (immune competent) cell a cell capable of carrying out its immune function when given the proper stimulus.

immunodominance within a complex immunogenic molecule, the ability of a specific component 1. to elicit the highest titer of antibodies during an immune response, or 2. to bind more antibodies from a given polyvalent antiserum than any other component of that same molecule. For example, in a glycoprotein antigen, a specific monosaccharide may be the most highly antigenic component of the entire molecule and therefore exhibits immunodominance over other components of the same molecule.

immunoelectrophoresis a technique that first separates a collection of different proteins by electrophoresis through a gel and then reacts them with a specific antiserum to generate a pattern of precipitin arcs. The proteins can thus be identified by their electrophoretic mobilities and their antigenic properties. *See* **Appendix C,** 1955, Grabar and Williams.

immunofluorescence a visual examination of the presence and the distribution of particular antigens on or in cells and tissues using antibodies that have been coupled with fluorescent molecules such as rhodamine and fluorescein. In the *direct method,* the fluorescent probe combines directly to the antigen of interest. In the *indirect method,* two antibodies are used in sequence. The first is the one specifically against the antigen under study. Subsequently, the tissue is incubated with a second antibody, prepared against the first antibody. The second antibody has been conjugated previously with a fluorescent dye, which renders the complex visible. The indirect method is often preferred because, if one wants to localize more than one antigen, only one fluorescently labeled antigen need be used, provided the first antibody in each case is from the same species of animal. The second, fluorescent antibody is generally commercially available. *See* **Appendix C**, 1941, Coons *et al.*

immunogen a substance that causes an immune response. Foreign proteins and glycoproteins generally make the most potent immunogens. *See* **antigen.**

immunogene any genic locus affecting an immunological characteristic; examples: immune response genes, immunoglobulin genes, genes of the major histocompatibility complex *(all of which see).*

immunogenetics studies using a combination of immunologic and genetic techniques, as in the investigation of genetic characters detectable only by immune reactions. *See* **Appendix C**, 1948, Snell; 1963, Levine *et al.*; 1972, Benacerraf and Mc-Devitt.

immunogenic capable of stimulating an immune response.

immunoglobulin an antibody secreted by mature lymphoid cells called plasma cells. Immunoglobulins are Y-shaped, tetrameric molecules consisting of two relatively long polypeptide chains called heavy (H) chains and two shorter polypeptide

chains called light (L) chains. Each arm of the Y-shaped structure has specific antigen-binding properties and is referred to as an antigen-binding fragment (Fab). The tail of the Y-structure is a crystallizable fragment (Fc). Five H chain classes of immunoglobulin are based upon their antigenic structures. Immunoglobulin class G (IgG) is the most common in serum and is associated with immunological "memory"; class IgM is the earliest to appear upon initial exposure to an antigen. Class IgA can be secreted across epithelial tissues and seems to be associated with resistance to infectious diseases of the respiratory and digestive tracts. The antibodies associated with immunological allergies belong to class IgE. Not much is known about the functions of IgD. Antibodies of classes IgG, IgD, and IgE have molecular weights ranging from 150,000 to 200,000d (7S); serum IgA is a 7S monomer, but secretory IgA is a dimer (11.4S); IgM is a pentamer (19S; 900,000d) of five 7S-like monomers.

In the case of IgG, each heavy chain consists of four "domains" of roughly equal size. The variable (V_H) domain at the amino (N-terminus) end contains different amino acid sequences from one immunoglobulin to another, even within the same H chain class. The other three domains have many regions of homology that suggest a common origin

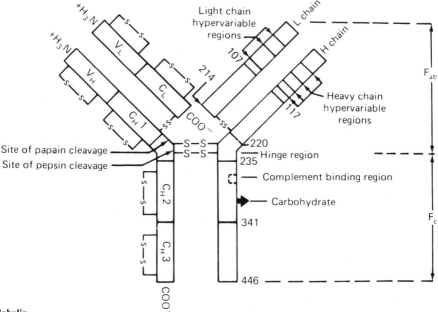

Immunoglobulin.
Diagram of a typical IgG molecule. Within each immunoglobulin molecule, the two L chains are identical and the two H chains are identical. Numbers represent approximate amino acid residues from the N terminus of the respective chain. From W. D. Stansfield, *Serology and Immunology,* Macmillan Publishing Co., Inc., 1981.

by gene duplication and diversification by mutation. These "constant" domains (C_H1, C_H2, C_H3) are essentially invariate within a given H chain class. An L chain is about half as long as an H chain. Its amino end has a variable region (V_L); its carboxyl end has a constant region (C_L). An Fab fragment consists of an L chain and an Fd segment of an H chain ($V_H + C_H1$). Within a tetrameric immunoglobulin molecule, the two L chains are identical and the two H chains are identical. The Fc fragment consists of carboxy-terminal halves of two H chains ($C_H2 + C_H3$). The region between C_H1 and C_H2 is linear rather than globular, and is called the "hinge region." Each mature antibody-producing plasma cell produces a single species of immunoglobulin, all of which contain identical L and H chains. *See* **Appendix C,** 1939, Tiselius and Kabat; 1959, Edelman; 1962, Porter; 1965, Hilschmann and Craig; 1969, Edelman *et al.*; **B lymphocyte, hybridoma, IgA–IgM, monoclonal antibody.**

immunoglobulin chains the components of the heteropolymeric immunoglobulin molecules. There are five groups of heavy chains, each characteristic of a specific class of immunoglobulin: gamma (IgG), mu (IgM), epsilon (IgE), alpha (IgA), and delta (IgD). The genes encoding all the heavy immunoglobulin chains are located on human chromosome 14. The constant region of each heavy chain makes up about 3/4 of the molecule, and the gene segments encoding the constant regions are arranged in the sequence mu, delta, gamma, epsilon, and alpha in both man and mouse. There are two groups of light chains: kappa chains, encoded by gene segments on human chromosome 2, and lambda chains, encoded by gene segments on chromosome 22. *See* **immunoglobulin genes.**

immunoglobulin genes genes encoding the light and heavy chains of the immunoglobulins. These genes are remarkable in that they are made up of segments that are shuffled as the B lymphocytes mature. The light chains contain segments that can be symbolized L-V, J, and C. The V or variable segment codes for the first 95 amino acids of the chain, whereas the C or constant segment codes for amino acids 108 to 214. The joining segment, J, codes for amino acids 96 to 107. L codes for a leader sequence 17–20 amino acids long. It functions in the transport of the molecule through the plasmalemma and is cleaved off the molecule in the process. There are about 300 L-V segments per light chain gene, and each of the V segments has a different base sequence. In the kappa gene, there are six J segments, each with a different base sequence, and one C segment. During differentiation of a given B lymphocyte stem cell, an immunoglobulin gene is assembled containing one L-V, one J, and one C segment, and this gene is transcribed by the lymphocyte and all of its progeny. The lambda gene also contains about 300 L-V segments, but each of the six J segments has its own adjacent C segment. The heavy chain gene is over 100,000 nucleotides long and contains a series of segments that can be symbolized L-V, D, J, C_μ, C_δ, $C_{\gamma3}$, $C_{\gamma1}$, $C_{\gamma2b}$, $C_{\gamma2a}$, C_ϵ, and C_α. There are about 300 L-V segments, 10–50 D segments, 4 J segments, and one each of the C segments. Each D segment codes for about 10 amino acids. During differentiation the segments are shuffled so that the variable region of a heavy chain is encoded by a segment that contains one L-V, one D, and one J segment. The gene also contains mu, delta, gamma, epsilon, and alpha subsegments, and which one of these is transcribed determines the class to which the antibody will belong. *See* **Appendix C,** 1965, Dreyer and Bennett; 1976, Hozumi and Tonegawa; **heavy chain class switching.**

immunological memory the capacity of the immune system to respond more rapidly and more vigorously to the second contact with a specific antigen than observed in the primary response to the first contact; the booster or anamnestic response.

immunological suppression a genetic or induced condition in which the ability of an individual's immune system to respond to most or all antigens is impaired. *See* **specific immune suppression.**

immunological surveillance theory the theory that the cell-mediated immune system evolved primarily to continuously monitor the body for spontaneously arising cancerous cells or those containing foreign pathogens and to destroy them.

immunological tolerance a state of nonreactivity toward a substance that would normally be expected to elicit an immune response. Tolerance to specific foreign antigens can be induced by the exposure of a bird or mammal to the foreign antigens during embryonic or neonatal (depending upon species) life. In adults, tolerance (usually of shorter duration) can be induced by using particular routes of administration for the antigens or administration of agents that are particularly effective against cells proliferating in response to antigen. Mechanisms may include actual deletion of potentially reactive lymphocytes or their "inactivation" by immunological suppression. *See* **Appendix C,** 1945, Owen; 1953, Billingham *et al.*

immunology the science dealing with immunity, serology, immunochemistry, immunogenetics, hypersensitivity, and immunopathology. *See* **Appendix C,** 1778, Jenner; 1900, Ehrlich; 1930, Landsteiner.

immunoselection a method for isolating cell-line variants lacking certain antigens, such as those of the major immunogene complex. By treating cells with a specific antiserum and complement, all cells die, except a few spontaneously arising variants. These do not express the corresponding antigen, and therefore they live and can be isolated. Many of these variants appear to be due to deletion mutations rather than to epigenetic changes or mitotic crossing-over. *Compare with* **antigenic conversion.**

immunosuppressive drugs compounds that block the immune response. *See* **azathioprene, mercaptopurine.**

impact theory a proposal, published in 1984 by Walter Alvarez and five colleagues, that the mass extinction of various groups of organisms that occured at the end of the Cretaceous (*q.v.*) resulted from the collision of the earth with an asteroid or comet. Rocks at the Cretaceous-Tertiary boundary have high iridium concentrations, and this iridium is postulated to have arisen from the pulverized asteroid.

impaternate offspring an offspring from parthenogenetic reproduction in which no male parent took part.

imperfect excision the release of a genetic element (e.g., an insertion sequence or prophage) from a DNA molecule in a way that either includes more than or less than the element itself.

imperfect flower *See* **flower.**

implant material artificially placed in an organism, such as a tissue graft, an electronic sensor, etc.

implantation **1.** attachment of a mammalian embryo to the uterine wall. **2.** the addition of tissue grafts to an organism without the removal of anything from it.

imprinting **1.** the imposition of a stable behavior pattern in a young animal by exposure, during a particular period in its development, to one of a restricted set of stimuli. **2.** *See* **parental imprinting.**

inactivation center a region of the mouse X chromosome that governs the degree to which translocated autosomal genes are inactivated when the associated X-linked genes are inactivated as the result of random X-inactivation. *See* **Cattanach's translocation, Lyon hypothesis.**

inactive X hypothesis Lyon hypothesis (*q.v.*).

Inarticulata a division of invertebrates containing the unsegmented, coelomate protostomes, such as sipunculids and molluscs. *See* **Appendix A.**

inborn error a genetically determined biochemical

disorder resulting in a metabolic defect that produces a metabolic block having pathological consequences. *See* **Appendix C,** 1909, Garrod.

inbred strain a group of organisms so highly inbred as to be genetically identical, except for sexual differences.

inbreeding the crossing of closely related plants or animals.

inbreeding coefficient *See* **Wright's inbreeding coefficient.**

inbreeding depression decreased vigor in terms of growth, survival, or fertility following one or more generations of inbreeding.

incapsidation the construction of a capsid around the genetic material of a virus.

inclusive fitness *See* **Hamilton's genetical theory of social behavior.**

incompatibility in immunology, genetic or antigenic differences between donor and recipient tissues that evoke an immunological rejection response.

incomplete dominance failure of a dominant phenotype to be fully expressed in an organism carrying a dominant and a recessive allele. The result is usually a phenotype that is intermediate between the homozygous dominant and the recessive forms. *See* **semidominance.**

incompletely linked genes genes on the same chromosome that can be recombined by crossing over.

incomplete metamorphosis *See* **Hemimetabola.**

incomplete sex linkage the rare phenomenon of a gene having loci on the homologous segments of both X and Y chromosomes.

incross mating between individuals from the same inbred line or variety, often of the same genotype.

independent assortment the random distribution to the gametes of genes located on different chromosomes. Thus, an individual of genotype *Aa Bb* will produce equal numbers of four types of gametes: *AB, Ab, aB,* and *ab. See* **Mendel's laws.**

independent probabilities in a group of events, the occurrence of any one event having no influence on the probability of any other event. For example, the orientation of one pair of homologous chromosomes on the first meiotic metaphase plate does not influence the orientation of any other pair of homologues. *See* **independent assortment.**

indeterminant inflorescence an inflorescence, such as a raceme (*q.v.*), in which the first flowers to open are at the base and are followed upward by progressively younger ones.

index case propositus (*q.v.*).

index fossil a fossil that appears only in rocks of a relatively limited geological age span.

indirect immunofluorescence microscopy. *See* **immunofluorescence.**

indole a tryptophan precursor in microorganisms.

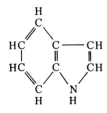

indoleacetic acid auxin, a phytohormone. *See* **auxins.**

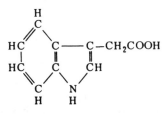

induced mutation a genetic alteration produced by exposure to a mutagen. *Compare with* **spontaneous mutation.**

inducer any of the small organic molecules that cause the cell to produce larger amounts of the enzymes involved in their metabolism. Inducers are a class of effector molecules (*q.v.*). *See* **gratuitous inducer, regulator genes.**

inducible enzyme an enzyme synthesized only in response to an inducer. *See* **adaptive enzyme, regulator gene.**

inducible system a regulatory system in which the product of a regulator gene (the repressor) is active and blocks transcription of the operon. The effector (called an inducer) inactivates the repressor and therefore allows mRNA synthesis to occur. Thus, transcription occurs only in the presence of effector molecules. *See* **regulator gene.**

induction 1. the determination of the developmental fate of one cell mass by another. The morphogenic effect is brought about by an evocator acting upon competent tissue. 2. the stimulation of a lysogenized bacterium to produce infective phage. 3. the stimulation of synthesis of a given enzyme in response to a specific inducer. *See* **Appendix C, 1918, Spemann.**

inductor any substance that carries out an induction similar to that performed by an organizer (*q.v.*).

industrial melanism the prevalence in industrial regions of black or dark brown forms of various animal species. The pigments involved are most often melanins. The most famous example involves the moth *Biston betularia* (*q.v.*). Studies on this species illustrated the importance of predation determining the direction of evolution. *See* **Appendix C, 1921, Goldschmidt; 1956, Kettlewell.**

inelastic collision *See* **collision.**

infectious nucleic acid purified viral nucleic acid capable of infecting a host cell and causing the subsequent production of viral progeny.

infectious transfer the rapid spread of extrachromosomal episomes (plus any integrated chromosomal genes) from donor to recipient cells in a bacterial population.

inflorescence 1. a flower cluster. 2. the arrangement and mode of development of the flowers on a floral axis. *See* **determinant inflorescence, indeterminant inflorescence.**

inheritance of acquired characteristics *See* **acquired characteristics, Lamarckism.**

initiation codon *See* **start codon.**

initiation factors proteins required for the initiation of protein synthesis. One (protein IF3) is required for the binding of the 30S particle to mRNA. A second (protein IF1) binds to f-met-tRNA and helps it attach to the 30S mRNA initiation complex. A third protein (IF2) is required, although its precise function is unclear. Initiation factors are symbolized IF in prokaryotes and eIF in eukaryotes, followed by a number. *See* **N-formyl-methionine, translation.**

initiator a molecule that initiates replication once it binds to a replicator. *See* **replicon.**

innervation the nerve supply to a particular organ.

inoculum a suspension of cells introduced into a nutrient medium to start a new culture.

inosine hypoxanthine riboside. *See* **rare bases.**

inquiline an animal that lives in the abode of another species.

insect ovary types three types of ovaries are found

among insects. The *panoistic* ovary appears to be the ancestral type. Here, all oogonia (except stem-line oogonia) are eventually transformed to oocytes. In *meroistic* ovaries, both oocytes and nurse cells (*q.v.*) are generated. These may be organized within the ovariole in two ways. In the *polytrophic meroistic* ovary, the nurse cells and oocytes alternate along the length of the ovariole. In the *telotrophic meroistic* ovary, the nurse cells are restricted to the germarium and are connected to oocytes in early stages of their development by cytoplasmic processes called nutritive chords. Panoistic ovaries are found in insects belonging to the more primitive orders (Archeognatha, Zygentoma, Ephemeroptera, Odonata, Plecoptera, Phasmida, Orthoptera, and Dictyoptera). Polytrophic meroistic ovaries occur in the Psocoptera, Phthiraptera, Hymenoptera, Trichoptera, Lepidoptera, and Diptera. Telotrophic ovaries occur in the Hemiptera, Coleoptera, Raphidioptera, and Megaloptera.

insertion the addition of one or more base pairs into a DNA molecule; a type of mutation commonly induced by acridine dyes or by mobile insertion sequences (*q.v.*).

insertional inactivation abolition of the functional properties of a gene product by insertion of a foreign DNA sequence into that gene's coding sequence; used in genetic engineering as a means of detecting when a foreign DNA sequence has become integrated into a plasmid or other recipient molecule of interest.

insertional mutagenesis alteration of a gene as a consequence of inserting unusual nucleotide sequences from such sources as transposons, viruses, transfection, or injection of DNA into fertilized eggs. Such mutations may partially or totally inactivate the gene product or may lead to altered levels of protein synthesis. *See* **insertional inactivation, insertion sequences, transgenic animals.**

insertional translocation *See* **translocation.**

insertion sequences transposable elements (*q.v.*) first detected as the cause of spontaneous mutations in *E. coli.* The majority of IS elements studied so far range in size from 0.7 to 1.8 kb. IS termini carry inverted repeats of about 10 to 40 bp, which are believed to serve as recognition sequences for a transposase (*q.v.*). The IS also contains a gene that encodes the transposase. *See* **Appendix C,** 1969, Shapiro.

insertion vector *See* **lambda cloning vehicle.**

in situ "in place"; in the natural or original position.

in situ **hybridization** a technique utilized to localize, within intact chromosomes, eukaryotic cells, or bacterial cells, nucleic acid segments complementary to specific labeled probes. To localize specific DNA sequences, specimens are treated so as to denature DNAs and to remove adhering RNAs and proteins. The DNA segments of interest are then detected via hybridization with labeled nucleic acid probes. The distribution of specific RNAs within intact cells or chromosomes can be localized by hybridization of squashed or sectioned specimens with an appropriate RNA or DNA probe. *See* **Appendix C,** 1969, Gall and Pardue; 1975, Grunstein and Hogness; 1981, Harper and Saunders.

instar the period between insect molts.

instinct an unlearned pattern of behavior.

instructive theory an early immunological theory in which it was believed that the specificity of antibody for antigen was conferred upon it by its initial contact with the antigen. This theory has been discarded in favor of the clonal selection theory (*q.v.*), in which specificity exists prior to contact with antigen.

insulin a polypeptide hormone produced by the beta cells in the islets of Langerhans. Insulin causes a fall in the sugar concentration of the blood, and its deficiency produces the symptoms of diabetes mellitus. Beef insulin was the first protein to have its amino acid sequence determined. This molecule (shown below) is made up of an A polypeptide (21

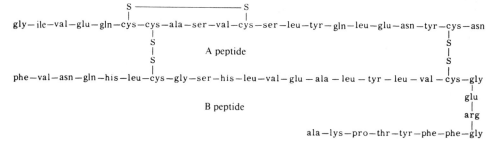

Insulin.

164

amino acids long) and a B peptide (containing 30 amino acids) joined by two disulfide bridges. In humans, insulin is encoded by a gene on the short arm of chromosome 11. *See* **Appendix C,** 1921, Banting and Best; 1952, Sanger *et al.*; 1982, Eli Lilly; **proinsulin.**

integrase an enzyme that catalyzes a site-specific recombination by which a prophage becomes integrated into or excised (deintegrated) from a bacterial chromosome; an excisionase enzyme is also required for the excision process.

integration efficiency the frequency with which a foreign DNA segment is incorporated into the genotype of a recipient bacterium, particularly with reference to transformation.

intelligence quotient (IQ) an individual can be assigned to a "mental age" group on the basis of performance on standardized intelligence tests. This mental age divided by the individual's chronological age and multiplied by 100 is the IQ.

intelligence quotient classification according to the Binet-Simon classification, intelligence quotients can be grouped as follows: genius, 140 and over; very superior, 120–139; superior, 110–119; average, 90–109; dull, 80–89; borderline, 70–79; mild retardation (moron), 50–69; moderate retardation (imbecile), 25–49; and severe retardation (idiot), 0–24.

interallelic complementation referring to the change in the properties of a multimeric protein as a consequence of the interaction of subunits coded by two different mutant alleles (in contrast to the protein consisting of subunits derived from a single mutant allele). The mixed protein (heteromultimer) may exhibit more activity (positive complementation) or less activity (negative complementation, *q.v.*) than the homomultimer. Also known as *intragenic complementation, allelic complementation.*

interbands the regions between bands in a polytene chromosome (*q.v.*). The DNA concentration in interbands is only a fraction of that in bands.

intercalary deletion *See* **deletion.**

intercalating agent a substance (e.g., acridine dyes) that inserts between base pairs in a DNA molecule, often disrupting the alignment and pairing of bases in the complementary strands. By causing addition or deletion of one or more base pairs during replication, a reading frame shift (*q.v.*) is often induced.

intercellular between cells.

interchange an exchange of segments between nonhomologous chromosomes resulting in translocations.

interchromosomal translocation *See* **translocation.**

intercistronic region the segment between the termination codon of one gene and the initiation codon of the next gene in a polycistronic transcription unit.

intercross mating of heterozygotes ($a/+$ $\times$ $a/+$).

interference *See* **positive interference.**

interference filter a filter used to produce a monochromatic light source.

interference microscope like the phase microscope, the interference microscope is used for observing transparent structures. However, with the interference microscope *quantitative* measurements of the relative retardation of light by various objects can be made. Such measurements can be used to determine the dry mass per unit area of specimen or the section thickness.

interferons a group of proteins that induce resistance to viral infections in cells. Interferons do not kill virus directly; rather, they induce host-cell enzymes that depress the transcription of viral genes and the translation of viral gene products. There are three major classes:

Type	IFN	Cell source
I	alpha	B lymphocytes and macrophages
II	beta	fibroblasts
III	gamma	T lymphocytes and natural killer cells

See **leader sequence peptide.**

intergenic suppression *See* **suppression.**

interkinesis the abbreviated interphase between the first and second meiotic division. No DNA replication occurs during interkinesis, unlike a premitotic interphase.

interleukins proteins secreted by mononuclear white blood cells that induce the growth and differentiation of lymphocytes. Interleukin-1 is secreted by mononuclear phagocytes, natural killer cells, and B lymphocytes, and it promotes the growth and differentiation of T lymphocytes and thymocytes. Interleukin-2 is secreted mainly by helper T lymphocytes following stimulation by interleukin-1 and binding of antigen to the T-cell receptor. The binding of IL-2 on T lymphocytes causes them to proliferate and to secrete lymphokines (*q.v.*). In humans, the gene encoding IL-2 resides on chromosome 4.

intermediary metabolism the chemical reactions in a cell that transform ingested nutrient molecules into the molecules needed for the growth of the cell.

intermediate filaments cytoplasmic filaments with diameters between 8 and 12 nm. They comprise a heterogeneous class of cytoskeletal proteins. In general, a given class of intermediate filaments is characteristic of a specific cell type. For example, keratin filaments are characteristic of epithelial cells, neurofilaments of neurons, vimentin filaments of fibroblasts, and desmin filaments of glial cells.

intermediate host a host essential to the completion of the life cycle of a parasite, but in which it does not become sexually mature.

intermedin a polypeptide hormone from the intermediate lobe of the pituitary gland that causes dispersion of melanin in melanophores. Also called melanocyte-stimulating hormone or MSH.

internal radiation the exposure to ionizing radiation from radioelements deposited in the body tissues.

interphase the period between succeeding mitoses.

interrupted mating experiment a genetic experiment in which the manner of gene transfer between conjugating bacteria is studied by withdrawing samples at various times and subjecting them to a strong shearing force in an electric blender. *See* **Appendix C**, 1956, Jacob and Wollman; **Waring blender.**

intersex a class of individuals of a bisexual species that have sexual characteristics intermediate between the male and female. *See* **Appendix C**, 1915, Goldschmidt.

interspecific heterokaryons cells containing nuclei from two different species produced by cell fusion (*q.v.*). *See* **Appendix C**, 1965, Harris and Watkins.

interstitial cells cells that lie between the testis tubules of vertebrates and secrete testosterone.

intervening sequence *See* **intron.**

intrachromosomal aberration *See* **translocation.**

intrachromosomal recombination sister chromatid exchange (*q.v.*).

intrachromosomal translocation *See* **translocation.**

intragenic complementation *See* **interallelic complementation.**

intragenic recombination recombination between mutons of a cistron. Such recombination is characterized by negative interference and by nonreciprocality (recovery of either wild-type or double-mutant recombinants, but not both from the same tetrad).

intragenic suppression *See* **suppression.**

intrasexual selection *See* **sexual selection.**

introgression *See* **introgressive hybridization.**

introgressive hybridization the incorporation of genes of one species into the gene pool of another. If the ranges of two species overlap and fertile hybrids are produced, they tend to backcross with the more abundant species. This process results in a population of individuals most of whom resemble the more abundant parents but who possess also some characters of the other parent species.

intromittent organ any male copulatory organ that implants sperm within the female.

intron in split genes (*q.v.*), a segment that is transcribed into nuclear RNA, but is subsequently removed from within the transcript and rapidly degraded. Most genes in the nuclei of eukaryotes contain introns and so do mitochondrial genes and some chloroplast genes. The number of introns per gene varies greatly, from one in the case of rRNA genes to more than 30 in the case of the yolk protein genes of *Xenopus*. Introns range in size from less than 100 to more than 10,000 nucleotides. There is little sequence homology among introns, but there are a few nucleotides at each end that are nearly the same in all introns. These boundary sequences participate in excision and splicing reactions. *See* **Appendix C**, 1977, Hogness *et al.*; 1978, Gilbert; **exon, posttranscriptional modification, splice junctions, spliceosome, transcription unit.**

intron intrusion the disruption of a preexisting gene by the insertion of an intron into a functional gene. Intron intrusion and the exon shuffling (*q.v.*) along with junctional sliding (*q.v.*), have been proposed as mechanisms for evolutionary diversification of genes.

intron-mediated recombination *See* **exon shuffling.**

intussusception 1. the growth of an organism by the conversion of nutrients into protoplasm. 2. the deposition of material between the microfibrils of a plant cell wall. 3. the increase in surface area of the plasmalemma by intercalation of new molecules between the existing molecules of the extending membrane.

in utero within the uterus.

inv *See* **symbols used in human cytogenetics.**

in vacuo in a vacuum.

invagination an inpocketing or folding in of a sheet of cells or a membrane.

inversion chromosome segments that have been turned through 180° with the result that the gene sequence for the segment is reversed with respect to that of the rest of the chromosome (see illustration below). Inversions may include or exclude the centromere. An inversion that includes the centromere is called *pericentric* or *heterobrachial,* whereas an inversion that excludes the centromere is called *paracentric* or *homobrachial.* Paracentric inversions are found more often in nature than pericentric inversions. A paracentric inversion heterozygote forms a reverse loop pairing configuration during pachynema. The various types of single and double exchanges that can occur within the loop are illustrated on page 168. Note that no monocentric, single-crossover chromatids are produced. For this reason, inversions give the impression of being *crossover suppressors,* and it was their action on crossing over that led to their discovery. *See* **Appendix C,** 1926, Sturtevant; 1933, McClintock; 1936, Sturtevant and Dobzhansky.

inversion heterozygote an organism in which one of the homologues has an inverted segment while the other has the normal gene sequence.

invertebrate an animal without a dorsal column of vertebrae; nonchordate metazoans.

inverted repeats (IR) two copies of the same DNA sequence orientated in opposite directions on the same molecule. IR sequences are found at opposite ends of a transposon (*q.v.*). *See* **palindrome.**

inverted terminal repeats short, related, or identical sequences oriented in opposite directions at the ends of some transposons (*q.v.*).

in vitro designating biological processes made to occur experimentally in isolation from the whole organism; literally "in glass," i.e., in the test tube.

Examples: tissue cultures, enzyme-substrate reactions.

in vitro **complementation** *See* **allelic complementation.**

in vitro **fertilization** experimental fertilization of an egg outside the female body. In humans, this is usually done because the woman's Fallopian tubes are blocked. The resulting embryo can then be inserted into the uterus for implantation. *See* **embryo transfer.**

in vitro **marker** a mutation induced in a tissue culture that allows subsequent phenotypic detection. Human *in vitro* markers include genes conferring resistance to various viruses, aminopterin, and purine analogues.

in vitro **mutagenesis** experiments in which segments of genomic DNA are treated with reagents that produce localized chemical changes in the molecule. The subsequent ability of the mutated molecules to function during replication, transcription, etc., is assayed either by using cell-free systems or *in vivo,* after splicing the fragment into an appropriate plasmid.

in vitro **packaging** the production of infectious particles from naked DNA by incapsidation of the DNA in question after supplying lambda phage packaging proteins and preheads.

in vitro **protein synthesis** the incorporation in a cell-free system of amino acids into polypeptide chains.

in vivo within the living organism.

in vivo **culturing of imaginal discs** the technique developed by Hadorn in which an imaginal disc is removed from a mature *Drosophila* larva, cut in half, and the half organ implanted into a young larva. Here regenerative growth occurs, and once the host larva has reached maturity the implant is removed once again, bisected, and one of the halves transplanted to a new host. By multiple repetition of this procedure, the cells are subjected to an abnormally long period of division and growth in a larval environment. If the regenerated disc is finally

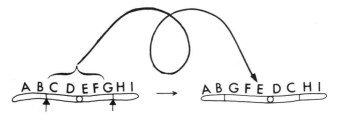

ABCDEFGHI → ABGFEDCHI

Inversion.

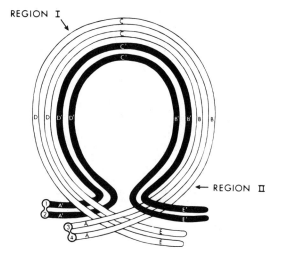

REGION I

REGION II

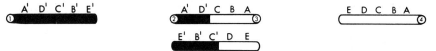

SINGLE EXCHANGE (REGION I, STRANDS 2 & 3)

A' D' C' B' E'
①

A' D' C B A
②   ③

E D C B A
④

E' B' C' D E

TWO STRAND DOUBLE EXCHANGE (REGION I, STRANDS 2&3; REGION II, STRANDS 2&3)

A' D' C B' E'
①

A' D' C B E'
②

E D C' B' A
③

E D C B A
④

THREE STRAND DOUBLE EXCHANGE (REGION I, STRANDS 2&3; REGION II, STRANDS 2&4)

A' D' C' B' E'
①

A' D' C B A
②   ③

E D C' B' A
④

E' B C D E

THREE STRAND DOUBLE EXCHANGE (REGION I, STRANDS 1&3; REGION II, STRANDS 2&3)

A' D C B E'
①

A' D' C' B' A
②   ③

E D C B A
④

E' B' C' D E

FOUR STRAND DOUBLE EXCHANGE (REGION I, STRANDS 1&4, REGION II, STRANDS 2&3)

A' D' C B A
①   ④

A' D' C' B' A
②   ③

E' B' C' D E

E' B C D E

Single and double exchanges within an inversion heterozgote.

allowed to undergo metamorphosis, it shows an abnormally high probability of producing structures characteristic of different discs. A regenerated genital disc may produce antennae, for example. Hadorn terms such differentiation *allotypic.* Since the allotypic organs appear in the offspring of cells that were previously determined to form genital structures, a change in determination must be postulated. This event is called *transdetermination. See* **Appendix C,** 1963, Hadorn.

in vivo marker a naturally occurring mutant mammalian gene that allows phenotypic detection of the tissue-cultured cells bearing it. Examples are the genes causing galactosemia and glucose-6-phosphate dehydrogenase deficiency in man, and the genes producing certain cell surface antigens in the mouse.

iodine a biological trace element. Atomic number 53; atomic weight 129.9044; valence 1^-; most abundant isotope ^{127}I; radioisotope ^{131}I; half-life, 8.0 days; radiations emitted—beta particles and gamma rays.

iojap a mutant nuclear gene in maize that induces changes in chloroplast characters. The mutant plastids behave autonomously thereafter.

ion exchange column a column packed with an ion exchange resin. *See* **chromatography, column chromatography.**

ion exchange resin a polymeric resin that has a higher affinity for some charged groups than it has for others. For example, resins with fixed cation groups will bind anions and thus can be used in column separation procedures. *See* **molecular sieve.**

ionic bond electrostatic bond (*q.v.*).

ionization any process by which a neutral atom or molecule acquires a positive or negative charge.

ionization chamber any instrument designed to measure the quantity of ionizing radiation in terms of the charge of electricity associated with ions produced within a defined volume.

ionization track the trail of ion pairs produced by an ionizing radiation during its passage through matter.

ionizing energy the average energy lost by an ionizing radiation in producing an ion pair in a given gas. The average ionizing energy for air is about 33 eV.

ionizing event the occurrence of any process in which an ion or group of ions is produced.

ionizing radiation electromagnetic or corpuscular radiation that produces ion pairs as it dissipates its energy in matter.

ionophores a class of antibiotics of bacterial origin that facilitate the movement of monovalent and divalent cations across biological membranes. Some of the major ionophores and the ions they transport are valinomycin (K^+, Rb^+), A 23187 (Ca^{++}, $2H^+$), nigericin (K^+, H^+), and gramicidin (H^+, Na^+, K^+, Rb^+).

ion pair the electron and positive atomic or molecular residue resulting from the interaction of ionizing radiation with the orbital electrons of atoms.

2ip 6-(γ,γ-dimethylallylamino) purine (*q.v.*).

IPTG isopropylthiogalactoside; a gratuitous inducer for the *E. coli lac* operon (*q.v.*). *See* **ONPG.**

IQ intelligence quotient (*q.v.*).

I region one of the central regions of the major histocompatibility complex (H-2) of the mouse. It contains genes coding for Ia antigens and controlling various immune responses. It has five subregions (A, B, J, E, C) and may be the equivalent of the D/DR region of the human major histocompatibility complex. *See* **HLA complex.**

IR inverted repeat (*q.v.*).

Ir gene *See* **immune response gene.**

iron a biological trace element. Atomic number 26; atomic weight 55.847; valence $2,3^+$; most abundant isotope ^{56}Fe; principal radioisotope ^{59}Fe, half-life 46 days, radiation emitted—beta particle.

isauxesis *See* **allometry, heterauxesis.**

IS element *See* **insertion sequences.**

islets of Langerhans clusters of hormone-secreting cells located in the pancreas of vertebrates. Two types of cells are found: alpha cells, which secrete glucagon (*q.v.*), and beta cells, which secrete insulin (*q.v.*).

isoacceptor transfer RNA one of a group of different tRNAs that accept the same amino acid but possess different anticodons. Higher organisms contain 2 to 4 isoacceptor tRNAs for certain amino acids. *See* **amino acid.**

isoagglutinin an antibody directed against antigenic sites on the red blood corpuscles of the same species and that causes agglutination.

isoagglutinogen an antigenic factor on the surface of cells that is capable of inducing the formation of homologous antibodies (isoagglutinins) in some members of the same species.

iso-allele an allele whose effect can only be distinguished from that of the normal allele by special tests. For example, two + alleles $+^1$ and $+^2$ may be indistinguishable (i.e., $+^1/+^1$, $+^2/+^2$, and $+^2/+^1$ individuals are phenotypically wild type). However, when compounded with a mutant allele *a*, $+^1$ and $+^2$ prove to be distinguishable (i.e., $a/+^1$ and $a/+^2$ individuals are observably different).

isoanisosyndetic alloploid an allopolyploid in which some chromosomes derived from both species are homoeologous and undergo a limited synapsis. *See* **isosyndetic alloploid.**

isoantibody an antibody formed in response to immunization with tissue constituents derived from an individual of the same species as the recipient.

isocapsidic viruses *See* **segmented genome.**

isochromatid break an aberration involving breaks in both sister chromatids at the same locus, followed by lateral fusion to produce a dicentric chromatid and an acentric fragment.

isochromosome a metacentric chromosome produced during mitosis or meiosis when the centromere splits transversely instead of longitudinally. The arms of such a chromosome are equal in length and genetically identical. However, the loci are positioned in reverse sequence in the two arms.

isocoding mutation a point mutation that alters the nucleotide sequence of a codon but, because of the degeneracy of the genetic code, does not change the amino acid that the codon specifies.

isoelectric point the *p*H at which the net positive and negative charge on a protein is zero.

isoenzymes isozymes (*q.v.*).

isofemale line a genetic lineage that began with a single inseminated female.

isoforms families of functionally related proteins that differ slightly in their amino acid sequences. Such proteins are encoded by genes that are now located at different chromosomal positions but are believed to be derived from a single ancestral gene. *See* **actin, actin genes, alternative splicing, fibronectin, myosin, myosin genes, tropomyosin, tubulin.**

isogamy that mode of sexual reproduction involving sex cells of similar size and morphology but opposite mating types. *See* **anisogamy.**

isogeneic referring to a graft involving genetically identical donor and host.

isogenic genetically identical (except possibly for sex); coming from the same individual or from a member of the same inbred strain.

isograft a tissue graft between two individuals of identical genotype.

isohemagglutinin isoagglutinin.

isoimmunization antibody formation in reaction to antigens of the same species.

isoionic point isoelectric point.

isolabeling labeling of both, or parts of both, daughter chromatids at the second metaphase after one replication in tritiated thymidine, as a result of sister chromatid exchange. In the absence of sister chromatid exchange, both daughter chromatids are labeled at metaphase I, but only one is labeled at metaphase II.

isolate a segment of a population within which assortative mating occurs.

isolating mechanism a cytological, anatomical, physiological, behavioral, or ecological difference, or a geographical barrier that prevents successful mating between two or more related groups of organisms. *See* **postzygotic isolation mechanism, prezygotic isolation mechanism.**

isolecithal egg one in which the yolk spheres are evenly distributed throughout the ooplasm. *See* **centrolecithal egg, telolecithal egg.**

isoleucine *See* **amino acid.**

isologous synonymous with isogeneic (*q.v.*).

isologous cell line cell lines derived from identical twins or from highly inbred animals.

isomerases a heterogeneous group of enzymes that catalyze the transfer of groups within molecules to yield isomeric forms. An example would be *racemase,* which interconverts D-lactic acid and L-lactic acid.

isomers compounds with the same molecular formula but with different three-dimensional molecular shapes or orientations in space.

isometry isauxesis.

isonymous marriage marriage between persons with the same surname. Isonymous marriages are used as indications of consanguinity in population genetics.

isophene a line on a map which connects points of equal expression of a character that varies clinally.

isopropylthiogalactoside a gratuitous inducer of the *lac* operon (*q.v.*).

isopycnic having the same density; used to refer to cell constituents having similar buoyant densities. *See* **centrifugation separation.**

isopycnotic referring to chromosomal regions or entire chromosomes that are not heteropycnotic, that is, are the same in appearance as the majority of the chromosomes. *See* **heteropycnosis.**

isoschizomers two or more restriction endonucleases (*q.v.*) isolated from different sources that cleave DNA within the same target sequences.

isosyndetic alloploid an allopolyploid where synapsis is restricted to the homologues derived from one species. *See* **isoanisosyndetic alloploid.**

isotonic solution a solution having the same osmotic pressure as another solution with which it is compared (usually blood or protoplasm).

isotope one of the several forms of a chemical element. Different isotopes have the same number of protons and electrons, but differ in the number of neutrons contained in the atomic nucleus. Hence they have identical chemical properties, but differ in atomic weights. *See* **Appendix C**, 1942, Schoenheimer; **radioactive isotope.**

isotopic dilution analysis a method of chemical analysis for a component of a mixture. The method is based on the addition to the mixture of a known amount of labeled component of known specific activity, followed by isolation of a quantity of the component and measurement of the specific activity of that sample.

isotopically enriched material material in which the relative amount of one or more isotopes of a constituent has been increased.

isotropic *See* **anisotropy.**

isotype exclusion synthesis of only kappa or lambda light chains by a given plasma cell as a consequence of allelic exclusion (*q.v.*). *See* **immunoglobulin.**

isotypes antigenic determinants shared by all individuals of a given species, but absent in individuals of other species. *Compare with* **allotypes, idiotypes.**

isozymes multiple forms of a single enzyme. While isozymes of a given enzyme catalyze the same reaction, they differ in properties such as the pH or substrate concentration at which they function best. Isozymes are complex proteins made up of paired polypeptide subunits. The lactic dehydrogenases, for example, are tetramers made up of two polypeptide units, A and B. Five isozymes exist and can be symbolized as follows: AAAA, AAAB, AABB, ABBB, and BBBB. Isozymes often have different isoelectric points and therefore can be separated by electrophoresis. The different monomers of which isozymes like lactic dehydrogenase are composed are specified by different gene loci. The term *allozyme* is used to refer to variant proteins produced by allelic forms of the same locus. *See* **allozymes.**

iteroparity repeated periods of reproduction during the life of an individual. *Compare with* **semelparity.**

IVS intervening sequence. *See* **intron.**

J

Japanese quail   *See Coturnix coturnix japonica.*

jarovization   synonym for vernalization (*q.v.*).

Java man   an extinct subspecies of primitive man known from fossils obtained in central Java. Now classified as *Homo erectus erectus,* but formerly referred to as *Pithecanthropus erectus.*

J chain   a small protein of about 15,000 daltons that holds the monomeric units of a multimeric immunoglobulin together, as occurs in the classes IgM and IgA.

J genes   a tandem series of four or five homologous nucleotide sequences coding for part of the hypervariable regions of light and heavy chains of mouse or human immunoglobulins; so named because they help *join* one of the genes for the variable region upstream to one of the genes for the constant region downstream and therefore are an important part of the mechanism generating antibody diversity.

JH   juvenile hormone. *See* **allatum hormones.**

Jordan's rule   an ecological principle stating that the ranges of closely related species or subspecies are generally adjacent and separated by a barrier of some sort.

jumping genes   mobile or "nomadic" genetic entities such as insertion elements and transposons.

junctional complex   a term used in electron microscopy to refer to any specialized region of intercellular adhesion, such as a desmosome (*q.v.*).

junctional sliding   a term descriptive of the fact that the location of intron-exon junctions is not constant within members of a gene family, such as the serine proteases. Some variation in length of such gene products can be attributed to extension or contraction of exons at the intron junctions.

junk DNA   *See* **selfish DNA.**

Jurassic   the middle period in the Mesozoic era, during which the dinosaurs became the dominant land vertebrates. Angiosperms and birds first appeared and archaic mammals persisted. Ammonites underwent a great diversification. The fragments formed from Pangaea began to separate. *See* **continental drift, geologic time divisions.**

juvenile hormone   *See* **allatum hormones.**

K

K **1.** degrees Kelvin. *See* **temperature. 2.** Cretaceous. **3.** potassium. **4.** the gene in *Paramecium aurelia* required for the maintenance of kappa. **5.** carrying capacity (*q.v.*).

kairomone a trans-specific chemical messenger the adaptive benefit of which falls on the recipient rather than the emitter. Kairomones are commonly nonadaptive to the transmitter. For example, a secretion that attracts a male to the female of the same species may also attract a predator. *See* **allomone.**

Kalanchoe a genus of succulent plants studied in terms of the genetic control of photoperiodic flowering response. *See* **phytochrome.**

kanamycin an antibiotic that binds to the 70S ribosomes of bacteria and causes misreading of the mRNA.

K and r selection theory *See* **r and K selection theory.**

K antigens *See* **O antigens.**

kappa symbiont *See* **killer paramecia.**

karyogamy the fusion of nuclei, usually of the two gametes in fertilization; syngamy.

karyokinesis nuclear division as opposed to cytokinesis (*q.v.*).

karyolymph nucleoplasm (*q.v.*).

karyon nucleus.

karyoplasm nucleoplasm.

karyosome a Feulgen-positive body seen in the nucleus of the *Drosophila* oocyte during stages 3–13. During stages 3–5, it contains synaptonemal complexes.

karyosphere the condensed Feulgen-positive mass seen in the anterior, dorsal portion of the mature primary oocyte of *Drosophila melanogaster*. This mass of DNA is not surrounded by a nuclear envelope. The tetrads subsequently emerge from the karyosphere and enter metaphase of the first meiotic division. The karyosphere stage of oogenesis is the most radiation-sensitive one.

karyotheca nuclear envelope.

karyotype the number, size, and morphology of the chromosome set of a cell, individual, or species. The term is often used for photomicrographs of chromosome preparations; e.g., the metaphase chromosomes arranged in a standard sequence; an idiogram (*q.v.*). *See* **human mitotic chromosomes.**

kb, kbp *See* **kilobase.**

KB cells a strain of cultured cells derived in 1954 by H. Eagle from a human epidermoid carcinoma of the nasopharynx.

K cells killer cells that mediate antibody-dependent cellular cytotoxicity (ADCC). These cells and natural killer (NK) cells have many similar properties, and may belong to the same cell lineage (lymphocyte or monocyte). Neither K nor NK cells have surface markers characteristic of either T cells (sheep red blood cell receptors) or B cells (endogenous surface immunoglobulins). Both K and NK cells possess Fc receptors for class Ig immunoglobulins and thus may acquire membrane-bound antibodies that react with target cells bearing the corresponding antigens. K cells cannot exhibit cytotoxicity without their bound antibodies; NK cells are not so restricted. Prior contact with the antigen is required by the host in order to arm its K cells with antibodies effective in ADCC.

kd kilodalton.

kDNA kinetoplast DNA.

Kell-Cellano antibodies antibodies against the red-cell antigens specified by the *K* gene, named for the first patient known to produce them. *See* **blood groups.**

kelp the largest of the marine brown algae.

keratin the major component of hair, nails, horns, etc. There are many different types of keratins, encoded by a large family of genes. Epidermal cells synthesize a sequence of different keratins as they mature.

keto forms of nucleotides *See* **tautomeric shift.**

keV *See* **electron volt.**

Kidd blood group a blood group defined by a human red cell antigen encoded by the *Jk* gene on the short arm of chromosome 2.

killer paramecia paramecia that secrete into the medium particles that kill other paramecia. The killer trait is due to kappa particles, which reside in the cytoplasm of those strains of *Paramecium aurelia* syngen 2 that carry the dominant *K* gene. Later it was found that kappa particles were symbiotic bacteria and that the particles with killing activity were defective DNA phages. The lysogenic, symbiotic bacterium has been named *Caenobacter taenospiralis,* and it is but one of many kinds of bacterial endosymbionts that occur in over 50% of the *P. aurelia* collected in nature. *See* **Appendix C,** 1938, Sonneborn; *Paramecium aurelia.*

killer particle in yeast, a double-stranded RNA plasmid containing ten genes for replication and several others for synthesis of a killer substance similar to bacterial colicin (*q.v.*). It is the only known plasmid that does not contain DNA.

kilobase a unit of length for nucleic acids consisting of 1,000 nucleotides; abbreviated kb, or kbp for kilobase pairs (DNA).

kilovolt a unit of electrical potential equal to 1000 volts, symbolized by **kV.**

kinase an enzyme that adds a phosphate group to its substrate.

kindred a group of human beings each of which is related, genetically or by marriage, to every other member of the group.

kinetic complexity *See* **DNA complexity.**

kinetin *See* **cytokinins.**

kinetochore *See* **centromere**

kinetoplast a highly specialized mitochondrion associated with the kinetosome of trypanosomes. Kinetoplast DNA is the only DNA known in nature that is in the form of a network consisting of interlocked circles. There are about 50 maxicircles and 5,000 minicircles per kinetoplast network. The maxicircles contain the genes essential for mitochondrial biogenesis. Unlike the maxicircles, the minicircles are not transcribed, and their function is unknown.

kinetosome a self-duplicating organelle homologous to the centriole. Kinetosomes reside at the base of cilia and flagella and are responsible for their formation. *See* **Appendix C,** 1976, Dippell.

kinety a row of interconnected kinetosomes on the surface of a ciliate.

kingdom systems *See* **classification.**

king-of-the-mountain principle *See* **first-arriver principle.**

kinin *See* **cytokinins.**

kin selection *See* **Hamilton's genetical theory of social behavior.**

Kjeldahl method a technique often used for the quantitative estimation of the nitrogen content of biological material.

Kleinschmidt spreading technique A procedure developed by A. K. Kleinschmidt that allows DNA molecules to be viewed under the electron microscope. The DNA is mounted in a positively charged protein film formed on the surface of an aqueous solution. The film of protein serves to hold the DNA in a relaxed but extended configuration and allows the sample to be transferred to a hydrophobic electron microscope grid when it is touched to the surface film. *See* **denaturation map.**

Klenow fragment a portion of bacterial DNA polymerase I derived by proteolytic cleavage; it lacks the 5′ to 3′ exonuclease activity of the intact enzyme.

Klinefelter syndrome a genetic disease that produces sterile males with small testes lacking sperm. The condition, which is sometimes associated with mental retardation, is due to the XXY AA karyotype. *See* **Appendix C,** 1959, Jacobs and Strong.

Km the Michaelis constant (*q.v.*).

knife breaker a mechanical apparatus that provides a method for breaking strips of plate glass first into squares and then into triangles. These are used as knives against which plastic-embedded tissues are cut into ultrathin sections for observation under the electron microscope. *See* **ultramicrotome.**

knob in cytogenetics, a heavily staining enlarged chromomere that may serve as a landmark, allowing certain chromosomes to be identified readily in the nucleus. In maize, knobbed chromatids preferentially enter the outer cells of a linear set of four megaspores during megasporogenesis and are therefore more likely to be included in the egg nucleus (*see* **meiotic drive**); genetic markers close to a knob tend to appear more frequently in gametes than those far from a knob.

Kornberg enzyme the DNA polymerase isolated from *E. coli* in 1959 by a group led by A. Kornberg; now called *DNA polymerase I;* it functions mainly in repair synthesis (*q.v.*).

Krebs' cycle citric-acid cycle (*q.v.*).

Krebs'-Henseleit cycle   ornithine cycle (*q.v.*).

K strategy   a type of life cycle relying on finely tuned adaptation to local conditions rather than on high reproductive rate. *See* **r and K selection theory.**

Kupffer cells   liver macrophages.

kurtosis   the property of a statistical distribution that produces a steeper or shallower curve than a normal distribution (*q.v.*) with the same parameters.

kuru   a chronic, progressive, degenerative disorder of the central nervous system found in the Fore natives living in a restricted area of New Guinea. The disease was at one time thought to be genetically determined, but it is now believed to be caused by a prion (*q.v.*).

kV   kilovolt.

kwashiorkor   a severe nutritional disorder due to a deficiency of certain amino acids (especially lysine). Kwashiorkor occurs in humans that subsist on a diet of cereal proteins deficient in lysine.

L

l **1.** line **2.** levorotatory. **3.** liter.

label, electron dense   *See* **ferritin.**

label, heavy   a heavy isotopic element introduced into a molecule to facilitate its separation from otherwise identical molecules containing the more common isotope.

label, radioactive   a radioactive atom introduced into a molecule to facilitate the study of its metabolic transformations.

labeled compound   a compound containing a radioactive label. The compound or its breakdown products may be followed through a series of biological reactions by detecting its radioactivity.

***lac* operon**   in *E. coli,* a DNA segment about 6000 bp long containing an operator gene and the structural genes *lac Z, lac Y,* and *lac A.* The structural genes code for β-galactosidase, β-galactoside permease, and β-galactoside transacetylase, respectively. The three structural genes are transcribed into a single mRNA from a promotor lying to the left of the operator. Whether or not this mRNA is transcribed depends upon whether or not a repressor protein is bound to the operator. The repressor protein is encoded by *lac I,* a gene lying to the left of the *lac* promotor. β-Galactosidase (*q.v.*) catalyzes the hydrolysis of lactose (*q.v.*) into glucose and galactose. After glucose and galactose are produced, a side reaction occurs forming allolactose. This is the inducer that switches on the *lac* operon. It does so by binding to the repressor and inactivating it. *See* **Appendix C,** Beckwith *et al.*; **IPTG, ONPG.**

***lac* repressor**   the first product of a regulator gene (*q.v.*) to be isolated. It is an acidic allosteric protein of molecular weight 152,000 daltons. It is a tetramer made up of four polypeptides, each containing 347 amino acids. A single *E. coli* contains only 10 to 20 copies of the *lac* repressor. *See* **Appendix C,** 1966, Gilbert and Muller-Hill.

lactamase   *See* **penicillin.**

lactic dehydrogenase   *See* **isozymes.**

lactogenic hormone   a protein hormone secreted by the anterior lobe of the pituitary that stimulates milk production in mammals and broodiness in birds.

lactose   4-(β-D-galactoside)-D-glucose. A disaccharide made up of two hexoses joined by a β-galactoside linkage. It is split into galatose and glucose by the enzyme β-galactosidase. Lactose differs from allolactose in that in lactose the galactose and glucose moieties are joined by a 1-4 linkage, whereas in allolactose the linkage is 1-6. As its name implies, lactose is abundant in the milk of mammals. *See lac* operon.

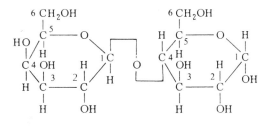

lagging   delayed movement from the equator to the poles at anaphase of a chromosome so that it becomes excluded from the daughter nuclei.

lagging strand   the discontinuously synthesized strand of DNA containing ligated Okazaki fragments (*q.v.*). *See* **leading strand, replication of DNA.**

lag growth phase   a period of time in the growth of a population during which little or no increase in the number of organisms occurs. The lag period precedes the exponential growth phase (*q.v.*).

lag load   a measure of the distance of a species from its local adaptive peak. The greater the lag load of a species, the more selective pressure is applied to the species, and hence the more rapid the rate of evolution it is likely to be experiencing. Also called *evolutionary lag. See* **Red Queen hypothesis.**

Lamarckism   a historically important, but no longer credited, theory that species can change gradually into new species by the willful striving of organisms to meet their own needs, together with the cumulative effects of use and disuse of body parts. All such acquired characteristics were thought to become part of the individual's heredity and as such could be transmitted to their offspring; otherwise known as the inheritance of acquired characteristics. *See* **Appendix C,** 1809, Lamarck.

lambda (λ) microliter (μl). The volume contained in a cube of side 1 mm. One μl of water weighs 1 mg.

lambda cloning vector a lambda phage that is genetically engineered to serve as a receptor for foreign DNA fragments in recombinant DNA experiments. Vectors that have a single target site at which foreign DNA is inserted are called *insertion vectors,* those having a pair of sites that span a DNA segment that can be exchanged with a foreign DNA fragment are called *replacement* or *substitution vectors.*

lambda d *gal* (λd*gal*) a lambda (λ) phage carrying a gene for galactose fermentation (*gal*) and also defective (d) for some phage function (usually lacking genes for making tails).

lambda phage a double-stranded DNA virus that infects *E. coli.* Once inside the host cell, the lambda genome can enter a lysogenic cycle or a lytic cycle of replication. Which pathway is chosen depends upon an intricate balance of host and viral factors. The complete nucleotide sequence of the lambda chromosome is known together with the position of most of its genes. *See* **Appendix C,** 1950, Lederberg; 1961, Meselson and Weigle; 1965, Rothman; 1966, Ptashne; 1969, Westmoreland *et al.*; 1974 Murray and Murray; **lambda cloning vector, lysogenic conversion, lysogenic cycle.**

lamellipodia extensive, lamellar cellular projections involved in attachment of eukaryotic cells such as fibroblasts to solid surfaces. Lamellipodia mark the forward edges of moving cells such as macrophages and are also called *ruffled edges.*

lampbrush chromosome a chromosome characteristic of the primary oocytes of vertebrates. Those found in the oocytes of salamanders are the largest known chromosomes, and they are generally stud-

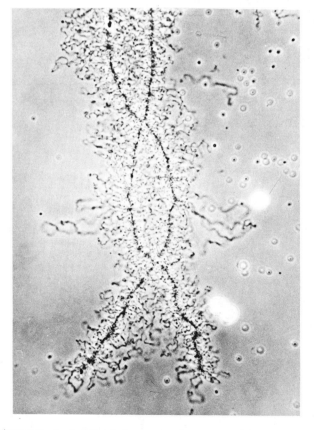

Phase contrast photomicrograph of a **lampbrush chromosome** from a *Notophthalmus viridescens* oocyte. Courtesy of J. G. Gall.

ied in diplonema. They have a rather fuzzy appearance when viewed at low magnification with a light microscope, and this is due to hundreds of paired loops that extend laterally from the main axis of each chromosome. The loops are active in RNA transcription. *See* **Appendix A,** 1882, Flemming; 1958, Callan and MacGregor; 1963, Gall; 1968, Davidson *et al.*; 1977, Old *et al.*

Lampyris a genus of beetles commonly called fireflies. The light signals sent by male fireflies serve as a visual, reproductive isolating mechanism (*q.v.*).

Laplacian curve. *See* **normal distribution.**

large angle x-ray diffraction a technique for the analysis of the small distances between individual atoms. *See* **x-ray crystallography, small angle x-ray diffraction.**

Laron dwarfism *See* **pituitary dwarfism.**

larva the preadult form in which some animals hatch from the egg. A larva is capable of feeding, though usually on a diet different from the adult, and is usually incapable of sexual reproduction.

larviparous depositing larvae rather than eggs. Fertilized eggs develop internally up to the larval stage. The female then lays these larvae. Some blowflies, for example, are larviparous.

laser an electronic device that generates and amplifies light waves coherent in frequency and phase in a narrow and extremely intense beam of light. The word is an acronym for *l*ight *a*mplification by *s*timulated *e*mission of *r*adiation. Microlaser beams are sometimes used to perform microcautery in experimental studies of cell division or morphogenesis.

laser microprobe a technique that uses a laser beam focused by a microscope to vaporize a minute tissue area. The vapor is then analyzed spectrographically.

late genes genes expressed late in the life cycle. In T_4 bacteriophage, those genes responsible for making capsid proteins, lysozyme, and other proteins after the initiation of phage DNA replication. *Compare with* **early genes.**

latent image the pattern of changes occurring in a surface of silver halide crystals when they absorb photons (from light) or ionizing particles. The pattern may be developed into a photographic image by chemical treatment.

latent period 1. the period during an infection when the causative agent cannot be detected by conventional techniques (e.g., the time from entry of phage DNA into a host cell until the release of infective phage progeny); prepatent period. *Compare with* **patent period. 2.** the time between infection and development of disease symptoms; period of incubation. *See* **Appendix C,** 1939, Ellis and Delbrück.

lateral element *See* **synaptonemal complex.**

late-replicating X chromosome in the mammalian somatic cell nucleus, all X chromosomes but one coil up into a condensed mass (the Barr body or sex chromatin body) and do not function in transcription. Such X chromosomes complete their replication later than the functional X and the autosomes. *See* **Barr body, Lyon hypothesis, sex chromatin.**

Latimeria *See* **living fossil.**

Latin square (*also* **Roman square**) a set of symbols arranged in a checkerboard in such a fashion that no symbol appears twice in any row or column. The Latin square, long a mathematical curiosity, was discovered to be useful for subdividing plots of land for agricultural experiments so that treatments could be tested even though the field had soil conditions that might vary in an unknown fashion in different areas. The technique required that the field be subdivided by a grid into subplots and the differing treatments be performed at consecutive intervals to plants from different subplots. Thus, if the subplots are A, B, C, and D, the experiments could be run as illustrated.

treatment	*day 1*	*day 2*	*day 3*	*day 4*
1	A	B	C	D
2	B	C	D	A
3	C	D	A	B
4	D	A	B	C

latitudinal parallel to the equator.

lattice a structure composed of elements arranged in a geometrical pattern with spaces between them.

lawn a continuous layer of bacteria on the surface of an agar plate.

law of parsimony *See* **Occam's razor.**

lazy maize a maize mutant characterized by a stalk that grows flat on the ground like a vine.

L cell *See* **mouse L cells.**

L chain *See* **immunoglobulin.**

LD50 the radiation dose required to kill half of a population of organisms within a specified time. Synonymous with median lethal dose.

LDH   lactate dehydrogenase. *See* **isozyme.**

leader protein   *See* **leader sequence peptide.**

leader sequence   the nontranslated segment of mRNA from its 5′ end to the start codon. The leader may contain regulatory signals, such as an attenuator (*q.v.*) or ribosome binding sites. *See* **exon, trailer sequence.**

leader sequence peptide   a sequence of 16 to 20 amino acids at the N-terminus of a eukaryotic protein that determines its ultimate destination. Proteins that are made and function in the cytosol lack leader sequences. Proteins destined for specific organelles require signal sequences appropriate for each organelle. The leader sequence for a protein destined to enter the endoplasmic reticulum always contains hydrophobic amino acids that become embedded in the lipid bilayer membrane, and it functions to guide the nascent protein to a receptor protein that marks the position of a pore in the membrane. Once the protein passes into the cysternal lumen through the pore, the leader segment is cleaved from the protein. For example, the leader sequence peptide of an interferon protein allows the cell to secrete the interferon, but is removed from the mature molecule during the secretion process. All of the leader sequences of mitochondrial proteins investigated thus far are strikingly basic, but otherwise have no similarities. The leader sequence peptide is also known as the signal peptide (*q.v.*). *See* **receptor-mediated translocation.**

leading strand   the DNA strand synthesized with few or no interruptions; as opposed to the *lagging strand,* which is produced by ligation of Okazaki fragments (*q.v.*). The leading strand is synthesized 5′ to 3′ toward the replication fork, whereas the lagging strand is synthesized 5′ to 3′ away from the replication fork. *See* **replication of DNA.**

leaky gene   hypomorphic gene

leaky protein   a mutant protein that has a subnormal degree of biological activity.

least squares, method of   a method of estimation based on the minimization of sums of squares. *See* **line of best fit.**

Lebistes reticularis   the guppy, a well-known tropical aquarium fish. The genetic control of sexuality has been extensively studied in this species.

lectins   proteins capable of agglutinating certain cells, especially erythrocytes, by binding to specific carbohydrate receptors on the surfaces of these cells. This agglutination behavior resembles antibody-antigen reaction but is not a true immunological reaction. Those lectins extracted from plant seeds have been called phytohemagglutinins. However, lectins have also been isolated from sources as varied as bacteria, snails, horseshoe crabs, and eels. *See* **concanavalin A, phytohemagglutinin, pokeweed mitogen.**

left splicing junction   the boundary between the left (5′) end of an intron and the right (3′) end of an adjacent exon in mRNA; also termed the donor splicing site.

leghemoglobin   an oxygen-binding protein found in the root nodules of leguminous plants. Leghemoglobin is structurally and functionally related to the myoglobins and hemoglobins of vertebrates. The heme portion of leghemoglobin is synthesized by the *Rhizobium* bacteroids, while the protein is encoded by the plant genome. Within the root nodule, bacteroids are bathed in a solution of leghemoglobin that serves to supply them with oxygen.

lek   a specific location where males of a species (especially birds) aggregate for courtship displays to females.

leptonema   *See* **meiosis.**

leptotene stage   *See* **meiosis.**

Lesch-Nyhan syndrome   the most common human hereditary defect in purine metabolism, due to a recessive gene on the long arm of the X chromosome. Hemizygotes lack hypoxanthine-guanine-phosphoribosyl transferase (*q.v.*). The disease is characterized by excessive production of uric acid, developmental and mental retardation, and death before sexual maturity. As a result of random X-chromosome inactivation (*see* **dosage compensation**), female heterozygotes are mosaics. About 60% of their cloned fibroblasts show HGPRT activity, whereas 40% have no detectable activity. The disease is transmitted via heterozygous mothers. Prevalence is 1/10,000 males.

LET   linear energy transfer (*q.v.*).

lethal equivalent value   the average number of recessive deleterious genes in the heterozygous condition carried by each member of a population of diploid organisms times the mean probability that each gene will cause premature death when homozygous. Thus a genetic burden of eight recessive semilethals each of which, when homozygous, produced only a 50% probability of premature death would be classified as a burden of four "lethal equivalents."

lethal mutation   a mutation that results in the premature death of the organism carrying it. Dominant lethals kill heterozygotes, whereas recessive lethals kill only homozygotes. *See* **Appendix C,**

1905, Cuénot; 1910, Castle and Little; 1912, Morgan; **aphasic lethal, monophasic lethal, polyphasic lethal.**

leu leucine. *See* **amino acid.**

leucine *See* **amino acid.**

leucine zipper a region in DNA-binding proteins spanning approximately 30 amino acids that contains a periodic repeat of leucines every seven residues. The region containing the repeat forms an alpha helix, with the leucines aligned along one face of the helix. Such helices tend to form stable dimers with the helices aligned in parallel. Leucine zippers occur in a number of transcriptional regulators. *See* **Appendix C,** 1988, Landschulz et al.; **helix-turn-helix motif, zinc finger proteins.**

leucocyte a variant spelling of leukocyte (*q.v.*).

leukemia a generally fatal disease characterized by an overproduction of white blood cells, or a relative overproduction of immature white cells. Leukemia is a common disease in the cow, dog, cat, mouse, guinea pig, and chicken, as well as man. Many virus-induced leukemias are known in the mouse and chicken. *See* **retroviruses.**

leukocyte a white blood cell.

leukopenia a decrease in the number of white blood corpuscles.

leukoplasts colorless plastids of tubers, endosperm, and cotyledons.

leukosis proliferation of leukocyte-forming tissue; leukemia (*q.v.*).

leukoviruses *See* **retroviruses.**

lev levorotatory. *See* **optical isomers.**

levulose fructose.

Lewis blood group a blood group determined by an antigen specified by the *Le* gene on human chromosome 19. The antigens are present in body fluids and are absorbed to red blood cells only secondarily. *See* **blood group, Lutheran blood group.**

Lex A repressor *See* **SOS response.**

Leydig epidermal cell a cell found in the epidermis and having the characteristics of a macrophage.

L forms bacteria that have lost their cell walls. *See* **protoplast.**

LH luteinizing hormone (*q.v.*).

library *See* **genomic library.**

lichen a composite plant consisting of a fungus and an alga living in symbiosis.

life cycle the series of developmental changes undergone by an organism from fertilization to reproduction and death.

life history strategies evolutionary adaptations in a biological lineage involving the timing of reproduction, fecundity, longevity, etc.

ligand a molecule that will bind to a complementary site on a given structure. For example, oxygen is a ligand for hemoglobin and a substrate of an enzyme molecule is a specific ligand of that molecule.

ligases enzymes that form C-C, C-S, C-O, and C-N bonds by condensation reactions coupled to ATP cleavage. *See* **DNA ligase.**

ligation formation of a phosphodiester bond to join adjacent nucleotides in the same nucleic acid chain (DNA or RNA).

light chain *See* **heavy chain.**

light repair *See* **dark reactivation, photoreactivation.**

Lilium the genus containing *L. longiflorum,* the Easter lily and *L. tigrinum,* the tiger lily, favorite species for cytological and biochemical studies of meiosis.

limited chromosome a chromosome that occurs only in nuclei of the cells of the germ line and never in somatic nuclei. *See* **chromosome diminution.**

Limnea peregra a freshwater snail upon which were first performed classical studies on the inheritance of the direction of the coiling of the shell. The trait showed delayed Mendelian segregation, since the phenotype of the snail is determined by the genotype of the maternal parent. *See* **Appendix C,** 1923, Boycott and Diver, Sturtevant.

line a homozygous, pure-breeding group of individuals that are phenotypically distinctive from other members of the same species.

lineage a linear evolutionary sequence from an ancestral species through all intermediate species to a particular descendant species.

linear accelerator *See* **accelerator.**

linear energy transfer the energy, in electron volts, dissipated per micron of tissue traversed by a particular type of ionizing particle.

linear regression regression line (*q.v.*).

linear tetrad a group of four meiotic products aligned linearly in such a way that sister products remain adjacent to one another. Ascospores show this order because the confines of the fungal ascus prevent nuclei from sliding past one another. *See* **ascus.**

line of best fit   a straight line that constitutes the best moving average for a linear group of observed points. This requires that the sum of the squares of the deviations of the observed points from the moving average be a minimum.

linkage   the greater association in inheritance of two or more nonallelic genes than is to be expected from independent assortment. Genes are linked because they reside on the same chromosome. *See* **Appendix A,** 1906, Bateson and Punnett; 1913, Sturtevant; 1915, Haldane *et al.*; 1951, Mohr.

linkage disequilibrium   the nonrandom distribution into the gametes of a population of the alleles of genes that reside on the same chromosome. The simplest situation would involve a pair of alleles at each of two loci. If there is random association between the alleles, then the frequency of each gamete type in a randomly mating population would be equal to the product of the frequencies of the alleles it contains. The rate of approach to such a random association or equilibrium is reduced by linkage and hence linkage is said to generate a disequilibrium. *See* **gametic disequilibrium.**

linkage group   the group of genes having their loci on the same chromosome. *See* **Appendix C,** 1919, Morgan.

linkage map   a chromosome map showing the relative positions of the known genes on the chromosomes of a given species.

linked genes   *See* **linkage.**

linker DNA   **1.** a short, synthetic DNA duplex containing the recognition site for a specific restriction endonuclease. Such a linker may be connected to ends of a DNA fragment prepared by cleavage with some other enzyme. **2.** a segment of DNA to which histone H1 is bound. Such linkers connect the adjacent nucleosomes of a chromosome.

linking number   the number of times that the two strands of a closed-circular, double-helical molecule cross each other. The *twisting number* (T) of a relaxed closed-circular DNA is the total number of base pairs in the molecule divided by the number of base pairs per turn of the helix. For relaxed DNA in the normal B form, L is the number of base pairs in the molecule divided by ten. The *writhing number* (W) is the number of times the axis of a DNA molecule crosses itself by supercoiling. The linking number (L) is determined by the formula: $L = W + T$. For a relaxed molecule, $W = 0$, and $L = T$. The linking number of a closed DNA molecule

cannot be changed except by breaking and rejoining of strands. The utility of the linking number is that it is related to the actual enzymatic breakage and rejoining events by which changes are made in the topology of DNA. Any changes in the linking number must be by whole integers. Molecules of DNA that are identical except for their linking numbers are called *topological isomers.*

Linnean system of binary nomenclature   *See* **Appendix C,** 1735, Linné.

lipase   an enzyme that breaks down fats to glycerol and fatty acids.

lipid   any of a group of biochemicals which are variably soluble in organic solvents like alcohol and barely soluble in water (fats, oils, waxes, phospholipids, sterols, carotenoids, etc.).

lipid bilayer model   a model for the structure of cell membranes based upon the hydrophobic properties of interacting phospholiplids. The polar head groups face outward to the solvent, whereas the hydrophobic tails face inward. Proteins (p) are embedded in the bilayer, sometimes being exposed on the outer surface, sometimes on the inner surface, and sometimes penetrating both surfaces. The proteins exposed on the outer surfaces of cells commonly serve as distinctive antigenic markers. Membrane proteins may serve a variety of functions such as communication, energy transduction, and transport of specific molecules across the membrane. *See* **fluid mosaic concept.**

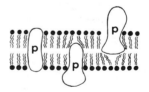

lipopolysaccharide   the active component of bacterial endotoxins associated with the cell walls of many Gram-negative species; a B-cell mitogen in some animal species.

liposomes   single- or multilaminar vesicles (made from lecithins and other lipids) that can function as artificial membrane systems for delivery of polynucleotides and genetic materials to cells, with a resultant stabilized expression of new genes in recipient cells.

lipovitellin   lipoprotein of molecular weight 400,000 found in amphibian yolk platelets. *See* **phosphovitin.**

liquid-holding recovery   a special form of dark reactivation (*q.v.*) in which repair of ultraviolet damage to DNA is enhanced by delaying bacterial growth and DNA replication through post-irradiation incubation of cells in a warm, nutrient-free buffer for several hours before plating them on nutrient agar.

liquid hybridization formation of double helical nucleic acid chains (DNA with DNA, DNA with RNA, or RNA with RNA) from complementary single strands in solution. *Compare with* **filter hybridization.**

liquid scintillation counter an electronic instrument for measuring radioisotopes dissolved in a solvent containing a fluorescent chemical that emits a flash of light (a scintillation) when struck by an ionizing particle or photon of electromagnetic radiation. The flash is captured by a photomultiplier tube, transformed into an electric pulse, amplified, routed through a scaler, and counted.

liter standard unit of capacity in metric system, equal to 1 cubic decimeter.

litter animals of one multiple birth.

littoral pertaining to the shore.

Liturgosa a genus of mantid. The evolution of karyotype has been extensively studied in this genus.

living fossil literally referring an organism that belongs to a group recognized on the basis of fossils and only subsequently found to be extant. The classic example of such a living fossil would be *Latimeria chalumnae,* the sole living representative of an ancient group of bony fishes called coelacanths. These belong to the Crossopterygii and are allies of the lung fishes and near the ancestry of the amphibians. The first fossil Permian *Coelacanthus* was named in 1839, while the first specimen of *Latimeria* was captured in 1938. The term is also used to refer to a living organism that is regarded as morphologically similar to a hypothetical ancestral missing link, even though the group has little or no fossil record. *Peripatus* is such an example, since it and other onycophorans have characteristics of both arthropods and annelids. Finally, living fossil is used to refer to the end member of a clade that has survived a long time and undergone little morphological change. The externally shelled cephalopod, *Nautilus,* would be an example of a living fossil by this criterion. *See* **Appendix A.**

local population a group of conspecific individuals together in an area within much most of them find their mates; synonymous with *deme,* and *Mendelian population. See* **subpopulations.**

locus (*plural,* **loci**) the position that a gene occu-

pies in a chromosome or within a segment of genomic DNA.

Locusta migratoria the Old World "plague" locust.

logarithmic phase the growth stage during which organisms are doubling their number each time a specified period of time elapses.

long-day plant a plant in which the flowering period is initiated and accelerated by a daily exposure to light exceeding 12 hours.

long period interspersion a genomic pattern in which long segments of moderately repetitive and nonrepetitive DNA sequences alternate.

long terminal repeats (LTRs) domains of several hundred base pairs at the ends of retroviral DNAs. LTRs may provide functions fundamental to the expression of most eukaryotic genes (e.g., promotion, initiation, and polyadenylation of transcripts).

lophophore a circular crescentic, or double spirally coiled ridge bearing ciliated tentacles found in species belonging to the Tentaculata. *See* **Appendix A,** Animalia, Coelomata.

low-energy phosphate compound a phosphorylated compound yielding relatively little energy upon hydrolysis.

LSD lysergic acid diethylamide (*q.v.*).

LTH lactogenic hormone (*q.v.*).

Lucilia cuprina the Australian blowfly, a major pest of the sheep industry in Australia. Good polytene chromosome preparations can be obtained from pupal trichogen cells (*q.v.*), and extensive cytogenetic studies have been made of chromosome rearrangements. The concept of pest control through the introduction into the field of chromosomally altered strains whose descendants manifest sterility has been tested in this species.

Lucké virus a virus causing renal cancer in frogs.

Lucy *See* **Australopithecine.**

Ludwig effect a generalization offered by W. Ludwig in 1950 that a species tends to be more diversified (polymorphic) both morphologically and chromosomally in the center of an old established range than at its margins.

luminescence light emission that cannot be attributed to the temperature of the emitting body. The light energy emitted may result from a chemical reaction going on within the emitter or it may be initiated by the flow of some form of energy into the body from the outside. The slow oxidation of phosphorus at room temperature is an example of the former, while luminescence resulting from electron bombardment of gaseous atoms in a mercury vapor lamp is an example of the latter. *Fluorescence* is defined as a luminescence emission that continues after the source of exciting energy is shut off. This afterglow is temperature independent. In the case of *phosphorescence,* the afterglow duration becomes shorter with increasing temperature.

lupus erythematosus a connective tissue disorder characterized by autoantibody production against cellular components that are abundant and highly conserved. Among the antibodies produced by patients with lupus are those that recognize U1 RNA. *See* **autoimmune disease, Usn RNAs.**

luteinizing hormone a glycoprotein hormone stimulating ovulation, growth of the corpus luteum and secretion of estrogen. LH is secreted by the adenohypophysis of vertebrates. Identical to ICSH.

luteotropin lactogenic hormone (*q.v.*).

Lutheran blood group a blood group determined by a red cell antigen specified by the *Lu* gene on human chromosome 19. Autosomal linkage in man was first demonstrated between *Lu* and *Le. See* **blood group, Lewis blood group.**

luxuriance a high degree of vegetative development, often seen in species hybrids; a special feature of heterosis (*q.v.*).

luxury genes genes coding for specialized (rather than "household") functions. Their products are usually synthesized in large amounts only in particular cell types (e.g., hemoglobin in erythrocytes; immunoglobulins in plasma cells).

Luzula a genus of plants containing the wood rushes. Many species in this genus have diffuse centromeres. *See* **centromere.**

lyases enzymes that catalyze the addition of groups to double bonds or the reverse.

Lycopersicon esculentum the cultivated tomato. The haploid chromosome number is 12. There are extensive linkage maps for each chromosome and also cytological maps for pachytene chromosomes.

Lyctechinus pictus a sea urchin often employed in studies of histone genes and their mRNAs. During oogenesis and egg maturation, large reservoirs of histone mRNAs are produced both by this species and by *Strongylocentrotus purpuratus* (*q.v.*). *See* **histone genes.**

Lymantria dispar the Gypsy moth. Classical studies on sex determination were performed on this species. *See* **Appendix C,** 1915, Goldschmidt.

lymphatic tissue tissues in which lymphocytes are produced and/or matured, including the thymus, spleen, lymph nodes and vessels, and the bursa of Fabricius (*q.v.*).

lymphoblast the larger, cytoplasm-rich cell that differentiates from antigenically stimulated T lymphocytes.

lymphocyte a spherical cell about 10 μm in diameter found in the lymph nodes, spleen, thymus, bone marrow, and blood. Lymphocytes, which are the most numerous cells in the body, are divided into two classes: B cells, which produce antibodies, and T cells, which are responsible for a variety of immunological reactions, including graft rejections. *See* **Appendix C,** 1962, Miller, Good *et al.,* Warner *et al.;* **B lymphocyte, T lymphocyte, immunoglobulin.**

lymphokines a heterogeneous group of glycoproteins (molecular weights 10,000–200,000) released from T lymphocytes after contact with a cognate antigen. Lymphokines affect other cells of the host rather than reacting directly with antigens. Various lymphokines serve several major functions: (1) recruitment of uncommited T cells; (2) retention of T cells and macrophages at the site of reaction with antigen; (3) amplification of "recruited" T cells; (4) activation of the retained cells to release lymphokines; and (5) cytotoxic effects against cells bearing foreign antigens (including foreign tissue grafts and cancer cells). Examples of lymphokines produced by T cells following *in vitro* antigenic stimulation are as follows: migration inhibition factor (MIF) prevents migration of macrophages; lymphotoxins (LT) kill target cells; mitogenic factor (MF) stimulates lymphocyte division; interleukin 2 (IL2) is required for T helper activity; interferons (IFN) promote antiviral immunity; chemotactic factors attract neutrophils, eosinophils, and basophils; macrophage activation factor (MAF) stimulates macrophages. *See* **histocompatibility molecules.**

lymphoma cancer of lymphatic tissue.

Lyon hypothesis the hypothesis that dosage compensation (*q.v.*) in mammals is accomplished by the random inactivation of one of the two X chromosomes in the somatic cells of females. *See* **Appendix C,** 1961, Lyon and Russell.

Lyonization a term used to characterize heterozygous females that behave phenotypically as if they were carrying an X-linked recessive in the hemizygous condition. An example would be a hemophilic mother that has produced nonhemophilic sons. It is assumed, according to the Lyon hypothesis, that occasionally a particular tissue comes to be made up entirely of cells that contain inactivated X chromosomes carrying the normal allele. The

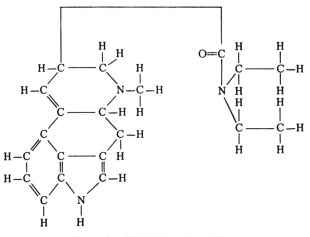

Lysergic acid diethylamide (LSD).

phenotype of such a female heterozygote would then resemble that of mutant males.

lyophilize   to render soluble by freeze drying.

lys   lysine. *See* **amino acid.**

lysate   a population of phage particles released from host cells via the lytic cycle.

Lysenkoism   a school of pseudoscience that flourished in the Soviet Union between 1932 and 1965. Its doctrines were advanced by T. D. Lysenko (1898–1976), who did not accept the gene concept and believed in the inheritance of acquired characteristics.

lysergic acid diethylamide (LSD)   a serotonin antagonist (shown above) that induces a schizophrenic-like state in humans.

lysine   *See* **amino acid.**

lysis   *See* **cell lysis.**

lysochrome   a compound that colors lipids by dissolving in them. *See* **Nile blue, Sudan black B.**

lysogen   a lysogenic bacterium.

lysogenic bacterium   one carrying a temperate virus in the prophage state.

lysogenic conversion   the change in phenotype of a bacterium (in terms of its morphology or synthetic properties) accompanying lysogeny (*q.v.*). Lysogenic cells exhibit immunity to superinfection by the same phage as that in the prophage state. Toxin production by *Corynebacterium diphtheriae* only occurs in strains that are lysogenic for phage beta.

lysogenic cycle   a method of temperate phage reproduction in which the phage genome is integrated into the host chromosome as a prophage and rep-

licates in synchrony with the host chromosome. Under special circumstances (e.g., when growth conditions for the host are poor), the phage may leave the host chromosome (deintegration or excision) and enter the vegetative state or the lytic cycle that produces progeny phage. *See* **Appendix C, 1950, Lwoff.**

lysogenic immunity   a phenomenon in which a prophage prevents another phage of the same type from becoming established in the same cell.

lysogenic repressor   a phage protein responsible for maintenance of a prophage and lysogenic immunity.

lysogenic response   the response following infection of a nonlysogenic bacterium with a temperate phage. The infecting phage does not multiply but rather behaves as a prophage. *See* **lytic response.**

lysogenic virus   a virus that can become a prophage.

lysogenization   the experimental production of a lysogenic strain of bacteria by exposing sensitive bacteria to a temperate phage.

lysogenized bacterium   a bacterium harboring an experimentally introduced, temperate phage.

lysogeny   the phenomenon in which genetic material of a virus and its bacterial host are integrated.

lysosomal diseases   hereditary diseases in man arising from a deficiency of enzymes located in lysosomes. Fabry, Gaucher, Niemann-Pick, Pompe, and Wolman diseases arise from lysosomal enzyme deficiencies.

lysosome   a membrane-enclosed intracellular vesicle that acts as the primary component for intracellular digestion in all eukaryotes. Lysosomes are known to contain at least 50 acid hydrolases, including phosphatases, glycosidases, proteases, sulfatases, lipases, and nucleases. Collectively, they can hydrolyze all classes of macromolecules. *See* **Appendix C,** 1955, de Duve *et al.*

lysozyme an enzyme digesting mucopolysaccharides. Lysozymes having a bacteriolytic action have been isolated from diverse sources (tears, egg white, etc.). An important lysozyme is the enzyme synthesized under the direction of a phage that digests the cell wall of the host from within and thus allows the escape of the phage progeny. Lysozyme was the first enzyme whose three-dimensional structure was determined by x-ray crystallography. *See* **Appendix C,** 1967, Blake *et al.*

lytic cycle the vegetative life cycle of a virulent phage, by which progeny phage are produced and the host is lysed. Temperate phage (*q.v.*) have the option of becoming a prophage (usually when growth conditions for its host cell are good) or entering the lytic cycle when growth conditions for its host are poor.

lytic response lysis following infection of a bacterium by a virulent phage, as opposed to lysogenic response (*q.v.*).

lytic virus a virus whose intracellular multiplication leads to lysis of the host cell.

M

M molar solution.

M13 a single-stranded bacteriophage cloning vehicle, with a closed circular DNA genome of approximately 6.5 kb. The major advantage of using M13 for cloning is that the phage particles released from infected cells contain single-stranded DNA that is homologous to only one of the two complementary strands of the cloned DNA, and therefore it can be used as a template for DNA sequencing analysis.

Ma megaannum (*q.v.*).

Macaca mulatta the rhesus monkey. The catarrhine primate most used in the laboratory and the original source of the Rh antigen. Its haploid chromosome number is 21, and about 30 genes have been located on 14 linkage groups. *See* **Rh factor.**

macaroni wheat *Triticum durum* (N = 14).

macha wheat *Triticum macha* (N = 21).

macroconidia *See* **conidia.**

macroevolution a large evolutionary pattern usually viewed through the perspective of geologic time, such as the evolution of the horse from *Eohippus* to *Equus.* Macroevolution includes changes in taxonomic categories above the species level, and events that result in the origin of a new higher taxon. *See* **microevolution.**

macromolecule a molecule of relative molecular mass ranging from a few thousand to hundreds of millions (proteins, nucleic acids, polysaccharides, etc.).

macromutation *See* **evolution.**

macronucleus the larger of the two types of nuclei in the ciliates; the "vegetative" nucleus. Macronuclei contain many copies of each gene and are transcriptionally active. *See* **Appendix C,** 1969, Ammermann; **micronucleus.**

macrophage a large, phagocytic, mononuclear leukocyte found in tissues, but derived from blood monocytes. Macrophages are called histiocytes in connective tissues, Kupffer's cells in the liver, Leydig cells in the skin, microglial cells in the nervous system, and alveolar macrophages in the lung. To stimulate an immune response, most antigens must be "processed" by macrophages and presented on their surfaces to lymphocytes in association with self-Ia molecules. *See* **major immunogen complex.**

macrophage activation factor a lymphokine (*q.v.*) that activates macrophages.

macroscopic visible to the unaided eye.

macula adherans desmosome (*q.v.*).

magnesium an element universally found in small amounts in tissues. Atomic number 12; atomic weight 24.312; valence 2^+; most abundant isotope ^{24}Mg, radioisotope ^{28}Mg, half-life 21 hrs, radiation emitted—beta particle.

mainband DNA the major DNA band obtained by density gradient equilibrium centrifugation of the DNA of an organism.

maize *See* ***Zea mays.***

major gene a gene with pronounced phenotypic effects, in contrast to its modifiers; **oligogene.** *Compare with* **polygene.**

major histocompatibility complex (MHC) a large cluster of genes on human chromosome 6 and on mouse chromosome 17. The MHC controls many activities of immune cells, including the transplantation rejection process and the killing of virus-infected cells by specific killer T lymphocytes. The MHC is part of a larger *major immunogene complex (q.v.)* with more diverse functions. In different mammals different symbols have been assigned to the MHC; for example: chicken (B), dog (DLA), guinea pig (GPLA), human (HLA), mouse (H-2), and rat (Rt-1). *See* **Appendix C,** 1948, Gorer *et al.*; 1953, Snell; **HLA complex.**

major immunogene complex (MIC) a genetic region containing loci coding for lymphocyte surface antigens (e.g., Ia), histocompatibility (H) antigens, immune response (Ir) gene products, and proteins of the complement system. The genes specifying immunoglobulins assort independently of the MIC, but the plasma cells responsible for their production are under the control of the MIC.

malaria the single most critical infectious disease of mankind. It affects about 200 million people annually and causes death in about 2 million. The disease is caused by protozoans of the genus *Plasmodium* transmitted to man by female mosquitoes of the genus *Anopheles.* The most dangerous form of malaria is subtertian malaria caused by *P. falciparum.* Benign tertian malaria and quartan malaria are caused by *P. vivax* and *P. malariae,* respectively. *Plasmodium knowlesi,* the species causing malaria in monkeys, and *P. berghei,* a species infecting rodents, are often used in laboratory investigations. *See* **Appendix C,** 1954, Allison; 1983, Godson *et al.*; **Duffy blood group system, glucose-6-phosphate dehydrogenase deficiency, merozoite, sickle-cell trait.**

male gametophyte *See* **pollen grain.**

male pronucleus the generative nucleus of a male gamete.

male symbol ♂; the zodiac sign for Mars, the Roman god of war. The sign represents a shield and spear.

maleuric acid $C_5H_6N_2O_4$, a mitotic poison.

malignancy a cancerous growth.

Maloney leukemia virus a virus of mice producing lymphocytic leukemia. The virus can be transmitted from an infected mother to newborn progeny through her milk. *See* **retrovirus.**

Malpighian tubule the excretory tubule of insects that opens into the anterior part of the hind gut.

Malthusian having to do with the theory advanced by the English social economist, T. R. Malthus, in his *An Essay on the Principle of Population,* published in 1798. According to this theory the world's population tends to increase faster than the food supply, and poverty and misery are inevitable unless this trend is checked by war, famine, etc.

Malthusian parameter the rate at which a population with a given age distribution and birth and death rate will increase.

mammary tumor agent a milk-borne virus that induces mammary cancer in mice of the appropriate genotype. *See* **Appendix C,** 1936, Bittner; **retrovirus.**

man *See Homo,* **human mitotic chromosomes.**

Mandibulata a subphylum of arthropods containing those species possessing antennae and a pair of mandibles. *See* **Appendix A,** Animalia, Arthropoda.

manganese a biological trace element. Atomic number 25; atomic weight 54.9380; valences 2^+, 3^+, 4^+, 6^+, 7^+; most abundant isotope ^{55}Mn; radioisotope ^{57}Mn, half life 7 days, radiation emitted—beta particles.

manic depressive psychosis a mental disorder characterized by periods of over-elated hyperactivity alternating with depression, anxiety, and even stupor. The condition is associated with a dominant gene on the short arm of the X chromosome. *See* **Appendix C,** 1972, Mendlewicz *et al.*

mantid an insect belonging to the orthopteran family Mantidae of the Dictyoptera. *See* **Appendix A.**

manifesting heterozygote a female heterozygous for a sex-linked recessive mutant gene who expresses the same phenotype as a male hemizygous for the mutation. This rare phenomenon results from the situation where, by chance, most somatic cells critical to the expression of the mutant phenotype contain an inactivated X chromosome carrying the normal allele of the gene. *See* **Lyonization.**

map *See* **genetic map.**

map distance the distance between genes expressed as map units or centiMorgans (cM).

mapping function a mathematical formula developed by J.B.S. Haldane that relates map distances to recombination frequencies. The function is represented graphically below. It demonstrates that no matter how far apart two genes are on a chromosome, one never observes a recombination value greater than 50%. It also shows that the relation between recombination frequencies and map distances is linear for genes that recombine with frequencies less than 10%.

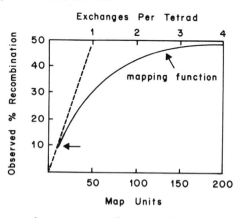

map unit a measure of genetic distance between two linked genes corresponding to a recombination frequency of 1% or 1 centiMorgan (cM). *See* **Morgan unit.**

mare   a female horse.

Marfan syndrome   a human hereditary disorder of connective tissue inherited as an autosomal dominant. The disease is characterized by excessively long fingers and toes and is often called arachnodactyly ("spider fingers") for this reason. Prevalence 1/20,000.

marker   **1.** a gene with a known location on a chromosome and a clear-cut phenotype, used as a point of reference when mapping a new mutant. **2.** antigenic markers serve to distinguish cell types **3.** marker DNAs, RNAs, and proteins are fragments of known sizes and/or properties that are used to calibrate an electrophoretic gel.

marker rescue   the phenomenon found when bacteria are mixedly infected with two genetically marked phages, of which only one type is irradiated. The progeny phages are predominately of the unirradiated type. However, some recombinants are found that contain genes from the irradiated parent. Such genes have been "rescued" by recombination.

marsupial   any of a group of primitive mammals, females of which have an abdominal pouch in which the young are carried and nourished (kangaroos, opossums, etc.). *See* **Appendix A.**

masked mRNA   messenger RNA that can be stored in large quantities because it is in some way inactivated and protected against digestion by nucleases (perhaps by being associated with proteins). Sea urchin eggs store mRNA in this masked form.

mass action, law of   the principle that the rate of a chemical reaction is proportional to the concentrations of the reacting substances, with each concentration raised to a power equal to the relative number of molecules of each species participating in the reaction.

mass extinction   *See* **Cretaceous, impact theory.**

mass number   the number of protons and neutrons in the nucleus of an atom (symbol, A).

mass spectrograph   an instrument for analyzing a substance in terms of the ratios of mass to charge of its components. It produces a focused mass spectrum of lines on a photographic plate.

mass unit   *See* **atomic mass unit.**

mast cells   tissue cells, thought to be the counterparts of blood basophils, that are abundant in lymph nodes, spleen, and bone marrow, in connective tissues, and in the skin. However, blood and lymph do not contain mast cells. The cytoplasm of mast cells contains granules rich in heparin, sero-tonin, and histamine. Mast cells possess receptors for IgE molecules. When antigen becomes attached to mast-cell-bound antibodies, these vasoactive amines are released and initiate an allergic reaction.

mastigonemes   slender, hollow, protein structures that extend laterally from some undulipodia and give them a hairy appearance. Mastigonemes serve to reverse the direction of thrust of the cell during swimming. *See* **Chromista.**

mate killers   *Paramecium aurelia* carrying mu particles. Such ciliates kill or injure sensitives with which they conjugate. The presence of mu in a paramecium protects it from the action of other mate killers.

maternal effect   a nonlasting influence of the genotype or phenotype of the mother upon the phenotype of the immediate offspring (i.e., phenotypic effects of stored material in the egg or maternally derived mRNAs transcribed by the F_1).

maternal inheritance   phenotypic differences controlled by cytoplasmic genetic factors (e.g., in mitochondria, long-lived mRNAs, viruses, or chloroplasts) derived solely from the maternal parent; also known as uniparental heredity, cytoplasmic inheritance, or extranuclear heredity.

maternal polarity mutants   mutations in *Drosophila melanogaster* that are maternally inherited and affect the development of embryos in a polarized manner. These genes are transcribed during oogenesis, and they imprint an anteroposterior pattern on the egg. The genes are grouped phenotypically into *anterior genes* that effect head and thoracic structures, *posterior genes* that define the abdominal region, and *terminal genes* that control the development of the acron and telson. *See* **Appendix C,** 1987, Nüsslein-Volhard *et al.*; 1988, Driever and Nüsslein-Volhard; *bicoid,* **zygotic segmentation mutants.**

mating   the union in pairs of individuals of opposite sexual type to accomplish sexual reproduction.

mating type   many species of microorganisms can be subdivided into groups (or mating types) on the basis of their mating behavior; only individuals of different mating types will undergo conjugation. Individuals from a given mating type possess on their surfaces proteins that will bind to complementary proteins or polysaccharides found only on the coats of individuals of opposite mating type. *See Paramecium aurelia.*

matrilinear inheritance   the transmission of cytoplasmic particles only in the female line.

matroclinous inheritance   *See* **matrocliny.**

matrocliny inheritance in which the offspring resembles the female parent more closely than the male. In *Drosophila,* the daughters produced by attached-X females are matroclinous in terms of their sex-linked genes. *See* **hologynic.**

maturase a protein encoded by an exon-intron combination that helps catalyze the excision of the intron from its own primary transcript. A maturase is probably not an enzyme that catalyzes intron removal and exon splicing, but rather a factor that modifies the specificity of a preexisting splicing enzyme.

maturation divisions a series of nuclear divisions in which the chromosome complement of the nuclei is reduced from diploid to haploid number. *See* **meiosis.**

Maxam-Gilbert method *See* **DNA sequencing techniques.**

maxicells bacterial cells that have degraded chromosomal DNAs because they have been heavily irradiated with ultraviolet light. Replication and transcription in such cells is inhibited because of the inability of the damaged DNA to act as template. However, if the maxicells contain multiple copies of a plasmid, the plasmid molecules that did not receive UV hits continue to replicate and transcribe gene products. Thus, maxicells provide a means of analyzing plasmid-encoded functions, while the products encoded by the host genome are reduced to a minimum. *See* **minicells.**

maximum permissible dose the greatest amount of ionizing radiation safety standards permit to be received by a person.

maze an experimental device consisting of a network of paths through which a test animal must find its way.

maze-learning ability the speed at which an animal learns to find its way through a maze without entering blind alleys.

McArdle disease a hereditary glycogen storage disease in humans with a frequency of 1/500,000. It arises from a deficiency of an enzyme, skeletal muscle glycogen phosphorylase, which is encoded by a gene on chromosome 11.

M chromosome the human mitochondrial chromosome. *See* **human gene maps.**

mean the sum of an array of quantities divided by the number of quantities in the group.

mean free path the average distance a subatomic particle travels between collisions.

mean square variance.

mechanical isolation reproductive isolation because of the incompatibility of male and female genitalia.

mechanistic philosophy the point of view that holds life to be mechanically determined and explicable by the laws of physics and chemistry. *Contrast with* **vitalism.**

medaka *See Oryzias latipes.*

medial complex *See* **synaptonemal complex.**

median the middle value in a group of numbers arranged in order of size.

median lethal dose the dose of radiation required to kill 50% of the individuals in a large group of organisms within a specified period. Synonymous with LD50.

medium the nutritive substance provided for the growth of a given organism in the laboratory.

megaannum one million years before the present; symbolized Ma.

megakaryocyte a large cell with a multilobed nucleus present in the bone marrow, but not circulating in the blood. Megakaryocytes bud off platelets (*q.v.*).

megasporangium a spore sac containing megaspores.

megaspore in angiosperms, one of four haploid cells formed from a megasporocyte during meiosis. Three of the four megaspores degenerate. The remaining megaspore divides to produce the female gametophyte or embryo sac (*q.v.*).

megaspore mother cell the diploid megasporocyte in an ovule that forms haploid megaspores by meiotic division.

megasporocyte megaspore mother cell.

megasporogenesis the production of megaspores.

meiocyte an auxocyte (*q.v.*).

meiosis in most sexually reproducing organisms, the doubling of the gametic chromosome number, which accompanies syngamy, is compensated for by a halving of the resulting zygotic chromosome number at some other point during the life cycle. These changes are brought about by a single chromosomal duplication followed by two successive nuclear divisions. The entire process is called meiosis, and it occurs during animal gametogenesis or sporogenesis in plants. The prophase stage is much longer in meiosis than in mitosis. It is generally di-

Meiosis.

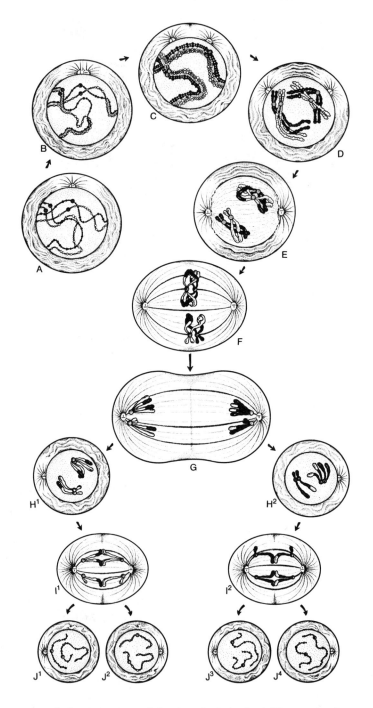

(A) The leptotene stage during spermatogenesis in a hypothetical animal. The organism has two pairs of chromosomes (one metacentric and one submetacentric). The maternal and paternal homologous chromosomes are drawn in different shades. The centromeres are represented by open circles. Note that the chromosomes are oriented with both of their

190

vided into five consecutive stages: *leptonema, zygonema, pachynema, diplonema,* and *diakinesis.*

During the leptotene stage the chromosomes appear as thin threads with clearly defined chromomeres. The chromosomes are often all oriented with one or both of their ends in contact with one region of the nuclear membrane, forming the so-called *bouquet configuration.* Although each chromosome appears single, it is actually made up of two chromatids. However, the doubleness of the chromosome does not become obvious until pachynema. The DNA replication that doubles the diploid value occurs before leptonema.

In diploid somatic cells, the 2N chromosomes are present as N pairs, and each chromosome is a replica of one contributed by the male or female parent at fertilization. In the somatic nuclei of most organisms, the homologous chromosomes do not pair. However, during the zygotene stage of meiotic prophase *synapsis* of homologous chromosomes takes place. This pairing begins at a number of points and extends "zipperlike" until complete. Pairing is accompanied by the formation of synaptonemal complexes (*q.v.*). When synapsis is finished, the apparent number of chromosome threads is half what it was before, and the visible bodies in the nucleus are now *bivalents* rather than single chromosomes.

During the pachytene stage, each paired chromosome separates into its two component sister chromatids (except at the region of the centromere). As a result of the longitudinal division of each homologous chromosome into two chromatids, there exist in the nucleus N groups of four chromatids lying parallel to each other called *tet-*

rads. A type of localized breakage followed by exchange between nonsister chromatids occurs. This process, called crossing over, is accompanied by the synthesis of an amount of DNA constituting less than 1% of the total in the nucleus. The exchange between homologous chromatids results in the production of crossover chromatids containing genetic material of both maternal and paternal origin.

During the diplotene stage one pair of sister chromatids in each of the tetrads begins to separate from the other pair. However, the chromatids are prevented from separating at places where exchanges have taken place. In such regions, the overlapping chromatids form a cross-shaped structure called a *chiasma* (pl. *chiasmata*). The chiasmata slip along laterally toward the ends of the chromatids, with the result that the position of a chiasma no longer coincides with that of the original crossover. This *terminalization* proceeds until during diakinesis all the chiasmata reach the ends of the tetrads and the homologues can separate during anaphase.

The chromosomes coil tightly during diakinesis and so shorten and thicken to produce a group of compact tetrads lying well spaced out in the nucleus, often near its membrane. Terminalization is completed, and the nucleolus disappears.

The nuclear envelope disappears during division I and the tetrads are arranged at the equator of the spindle. The chromatids of a tetrad disjoin in such a way that there is separation of maternal from paternal chromosomal material with the exceptions of regions distal to where crossing over has occurred. Division I produces two *secondary gametocytes,*

---

ends attached to the inner surface of the nuclear envelope. The chromosomes are uncoiled and maximally extended. In this state, the chromomeres and chromonemata can be seen readily. A centrosome encompassed by astral rays is present. It contains a mother and daughter centriole, oriented at right angles to each other.

(B) During zygonema, synapsis of homologous chromosomes takes place. This pairing begins at one or more points and extends "zipperlike" until complete. Subsequently, the cell contains two bivalents. The centrosome has divided into two daughter centrosomes, each containing a single centriole.

(C) During pachynema, each chromosome can be seen to be composed of two sister chromatids, except at the region of the centromere. As a result, the bivalents are converted into tetrads. A synaptonemal complex extends the length of the bivalent. The centrosomes move apart.

(D) During diplonema, the medial complex of the synaptonemal complex disappears, and in each tetrad one pair of sister chromatids begins to separate from the other pair. However, the chromatids are prevented from separating at places where interchanges have taken place. The points where the chromatids form cross-shaped configurations are called chiasmata.

(E) At diakinesis, the chromatids are shorter and thicker, and terminalization is occurring. The centrosomes have reached the poles, and the nuclear envelope is beginning to disappear.

(F) At metaphase I, the tetrads are arranged at the equator of the spindle.

(G) At late anaphase I, the homologous chromosomes have separated and have moved to each pole. However, the centromeres have not divided. As a consequence, maternal and paternal chromosomal material has been separated (except in regions distal to points of crossing over).

(H^1, H^2) Secondary spermatocytes containing dyads.

(I^1, I^2) During early anaphase II, centromeres divide and allow separation of sister chromatids.

(J^1, J^2, J^3, J^4) The four spermatids containing monads. Chromosomes once again uncoil and elongate. (J^1) and (J^4) each contain one single and one double crossover chromatid.

which contain *dyads* surrounded by a nuclear envelope.

Division II commences after a short interphase during which the chromosomes do not uncoil. The nuclear membrane disappears, and the dyads arrange themselves upon the metaphase plate. The chromatids of each dyad are equivalent (again with the exception of regions distal to points of crossing over); the centromere divides and thus allows each chromosome to pass to a separate cell. In animals, division II produces four *spermatids* (*q.v.*) or *ootids* (*q.v.*) which contain *monads* surrounded by a nuclear membrane. Meiosis therefore provides a mechanism whereby (1) an exchange of genetic material may take place between homologous chromosomes and (2) each gamete receives but one member of each chromosome pair.

The type of meiosis described above immediately precedes gametogenesis. This type *(gametic meiosis)* is characteristic of all animals. Fungi are characterized by *zygotic meiosis,* where meiosis immediately follows zygote formation. *Haplodiplomeiosis* refers to the situation seen in most plants where meiosis intervenes between a prolonged diploid phase and an abbreviated haploid phase. *See* **alteration of generations, interkinesis, oogenesis, sex, spermatogenesis.**

meiosporangium (*plural* **meiosporangia**) a sporangium in which meiosis occurs. *Compare with* **mitosporangium.**

meiospore a spore produced by meiosis.

meiotic cycle a first (reductional) division, followed by a second (equational) division. *See* **meiosis.**

meiotic drive any meiotic mechanism that results in the unequal recovery of the two types of gametes produced by a heterozygote.

Melandrium album a dioecious plant with a method of sex determination similar to man (that is, maleness is determined by the Y chromosome).

melanin a dark brown to black pigment responsible for the coloration of skin, hair, and the pigmented coat of the retina. The macromolecule consists of polymers of indole 5,6-quinone and 5,6-dihydroxyindole-2-carboxylic acid and is formed by the enzymatic oxidation of tyrosine or tryptophan. A segment of the structure is shown on page 193. *See* **eumelanin, phaeomelanin.**

melanism the hereditary production of increased melanin resulting in darker coloring. *See* **albinism, industrial melanism.**

melanocyte a pigment cell containing melanin granules.

melanocyte-stimulating hormone intermedin (*q.v.*).

melanoma a cancer composed of melanocytes (*q.v.*).

Melanophus femur-rubrum a species of grasshopper widely used for the cytological study of meiosis.

melanosome an intracellular organelle containing aggregations of tyrosinase, found in melanocytes.

melanotic tumor of *Drosophila* *See* **pseudotumor.**

Meleagris gallopavo the domesticated turkey.

melphalan a mutagenic, alkylating agent. See structure below.

melting in nucleic acid studies, the denaturation of double-stranded DNA to single strands.

melting-out temperature the temperature at which the bonds joining the constituent strands of a DNA/DNA or DNA/RNA duplex are broken and the molecules dissociate.

melting profile a curve describing the degree of dissociation of the strands in a DNA/DNA or DNA/RNA duplex in a given time as a function of temperature. The stabilities of duplexes are a function of their molecular weights, so that the melting profiles shift to the right as the chain lengths increase.

melting proteins *See* **helix-destabilizing proteins.**

melting temperature the temperature at which 50% of the double helices have denatured; the midpoint of the temperature range over which DNA is denatured. *See* **denaturation, Tm.**

memory in immunology, the ability to mount a specific secondary response to an antigen which had been encountered at some previous time. *See* **immune response.**

menarche the beginning of the first menstrual

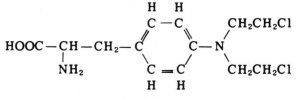

Melphalan.

202

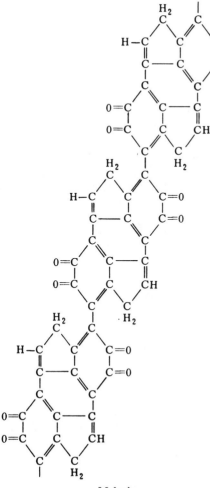

Melanin.

cycle during puberty in human females. *Compare with* **menopause.**

Mendeleev's table   the periodic table of chemical elements (page 236).

Mendelian character   a character that in inheritance follows Mendel's laws.

Mendelian genetics   referring to the inheritance of chromosomal genes following the laws governing the transmission of chromosomes to subsequent generations; also called *Mendelism.*

Mendelian population   an interbreeding group of organisms sharing a common gene pool.

mendelize   to segregate according to Mendel's laws.

Mendel's laws   1. *The Law of Segregation.* The factors of a pair of characters are segregated. In modern terms, this law refers to the separation into different gametes and thence into different offspring of the two members of each pair of alleles possessed by the diploid parental organism. **2.** *The Law of Independent Assortment.* The members of different pairs of factors assort independently. A restatement of the law in modern terms is that the members of different pairs of alleles are assorted independently into gametes during gametogenesis (provided they reside on different chromosomes), and that the subsequent pairing of male and female gametes is at random. *See* **Appendix C,** 1856, 1865, 1866, Mendel; 1900, de Vries, Correns, Tschermak; 1902, Sutton.

menopause   the cessation of menstrual cycles in human females, usually occurring between age 50 and 60.

meq   milliequivalent. *See* **gram equivalent weight.**

mer   *See* **mers.**

mercaptoethanol $SHCH_2CH_2OH$, a mitotic poison.

mercaptopurine   a synthetic purine analogue, one of the first inhibitors of DNA synthesis, shown to suppress the growth of cancer cells. Mercaptopurine also functions as an immunosuppressive agent, and it allowed the first successful organ transplants in humans. *See* **azathioprene.**

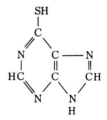

Mercenaria mercenaria   *See* **Pelecypoda.**

mericlinal chimera   an organism or organ composed of two genetically different tissues one of which partly surrounds the other.

meristems   the undifferentiated, mitotically active tissues of plants. The meristems at the tips of the roots and shoots are referred to as *apical meristems.*

meristic variation   variation in characters that can be counted, like the number of bristles, leaves, scales, etc.

meroblastic cleavage   cleavage producing cells, some of which are larger than others, because of a polarized distribution of yolk.

merogony   1. a form of asexual reproduction characterized by nuclear replication without plasmotomy (*q.v.*), resulting in the simultaneous production of two to many merozoites; a type of schizogony (*q.v.*) producing merozoites. **2.** the development of an experimentally produced egg fragment (containing a diploid nucleus or a haploid nucleus of male or female origin) into a small-sized embryo, called a merogone.

meroistic pertaining to an insect ovariole containing nutritive cells. If apical nurse cells are joined to the oocytes by nutritive cords, the ovariole is called acrotrophic or telotrophic. If nurse cells are included with each oocyte in a follicle, the ovariole is called polytrophic. *See* **insect ovary types.**

meromixis genetic exchange in bacteria involving a unidirectional transfer of a partial genome. *See* **F factor (fertility factor).**

merospermy the situation in which the nucleus of the fertilizing sperm does not fuse with the egg nucleus and later degenerates. Subsequent development is by gynogenesis.

merotomy the dissection of cells into several portions, with or without nuclei, as in experimental grafting with *Acetabularia* (*q.v.*).

merozoite the stage in the life cycle of the malaria parasite that invades the human erythrocyte. The parasite multiplies within the red cell and eventually causes it to burst, releasing 10 to 20 merozoites that invade more red cells. The merozoite produces immune decoy proteins. *See* **malaria, sporozoite.**

merozygote a partially diploid bacterial zygote containing an exogenotic chromosomal fragment donated by the F⁺ mate. The exogenote may also be introduced during transduction or sexduction. *See* **endogenote, exogenote.**

mers the unit that defines the number of bases in an oligonucleotide polymer. For example, oligonucleotides that contain 15 or 17 bases are referred to as 14 mers and 17 mers, respectively.

Mertensian mimicry *See* **mimicry.**

mesenchyme an embryonic type of connective tissue, consisting of amoeboid cells with many processes. Populations of such cells form a loose network. Most mesenchyme is derived from mesoderm. The mesenchyme produces the connective tissue and the circulatory system during the development of vertebrates.

Mesocricetus auratus the golden hamster, a rapidly breeding rodent used in the laboratory.

mesoderm the middle layer of embryonic cells between the ectoderm and endoderm in triploblastic animals. Mesoderm forms muscle, connective tis-

sue, blood, lymphoid tissue, the linings of all the body cavities, the serosa of the viscera, the mesenteries, and the epithelia of the blood vessels, lymphatics, kidney, ureter, gonads, genital ducts, and suprarenal cortex.

mesokaryotic referring to the dinoflagellate nucleus, in which the chromosomes are in a condensed, discrete condition at all times.

mesosome one of the invaginated segments of the plasma membranes in certain bacteria to which DNA molecules are attached.

mesothorax the middle of the three thoracic segments of an insect. It bears a pair of legs and (in winged insects) a pair of wings.

Mesozoic the 180 myr era during which dinosaurs arose, flourished, and became extinct. *See* **geologic time divisions.**

messenger RNA an RNA molecule that functions during translation (*q.v.*) to specify the sequence of amino acids in a nascent polypeptide. In eukaryotes, mRNA is formed in the nucleus from premessenger RNA molecules (*q.v.*). *See* **Appendix C,** 1961, Jacob and Monod, Brenner *et al.,* Gros *et al.;* 1964, Marbaix and Burny; 1967, Taylor *et al.;* 1969, Lockard and Lingrel; **exon, intron, polyadenylation, polysome, posttranscriptional modification.**

met methionine. *See* **amino acid.**

metabolic block a nonfunctional reaction in a metabolic pathway, as a consequence of a defective (mutant) enzyme whose normal counterpart catalyzes the reaction.

metabolic pathway a series of stepwise biochemical changes in the conversion of some precursor substance to an end product, each step usually catalyzed by a specific enzyme.

metabolic poison a compound poisoning a metabolic process. *See* **dinitrophenol.**

metabolism the sum of all the physical and chemical processes by which living cells produce and maintain themselves and by which energy is made available for the use of the organism.

metabolite a product of metabolism.

metacentric designating a chromosome with a centrally placed centromere.

metachromasy the phenomenon where one dye stains more than one color. A substance that stains metachromatically is called a chromatrope. Chromatropes are high-molecular-weight structures with serially arranged charged groups. Acidic mucopolysaccharides and nucleic acids are prime examples. Azure B (*q.v.*) is an example of a metachromatic dye. The color it produces depends on the way dye molecules stack on the chromatrope. As the amount of stacking increases, the color changes from green, to blue, to red. In an azure B-stained tissue section, chromosomes generally stain green, nucleoli and cytoplasmic ribosomes blue, and mucopolysaccharide-containing deposits red.

metachromatic dye a dye that stains tissues two or more colors. *See* **azure B, metachromasy.**

metafemale in *Drosophila*, a female phenotype of relatively low viability in which the ratio of X chromosomes to sets of autosomes exceeds 1.0; previously called a superfemale. *See* **intersex, metamale.**

metagenesis alternation of generations (*q.v.*) among animals. Metagenesis is commonly seen in invertebrates, especially coelenterates. Unlike the situation among plants, both generations are diploid.

metalloenzyme a protein combined with one or more metal atoms that functions as an enzyme.

metallothioneins small proteins that bind heavy metals and thereby protect cells against their toxic effects. Genes coding for metallothioneins are activated by the same metal ions that these proteins bind.

metamale in *Drosophila*, a poorly viable male characterized by cells containing one X and three sets of autosomes; previously called a supermale. *See* **intersex, metafemale.**

metamerism a phenomenon, common in annelids and arthropods, involving the repetition of a pattern of elements belonging to each of the main organ systems of the body. It is exhibited along the anteroposterior axis of the body. The term is also applied to a comparable repetition along the axis of an appendage.

metamorphosis the transformation from larval to adult form.

metaphase *See* **mitosis.**

metaphase arrest referring to the accumulation of metaphase figures in a population of cells poisoned with colchicine, Colcemid, or some other spindle poison.

metaphase plate the grouping of the chromosomes in a plane at the equator of the spindle during the metaphase stage of mitosis (*q.v.*).

metastable state an excited state of an atomic nu-

cleus, which returns to its ground state by the emission of radiation.

metastasis the spread of malignant neoplastic cells from the original site to another part of the body.

metatarsus the basal tarsal segment of the insect leg. In the male of *Drosophila melanogaster*, the metatarsus of each foreleg bears a sex comb.

metathorax the hindmost of the three thoracic segments of an insect. It bears a pair of legs and (in many winged insects) a pair of wings. In flies, it bears the halteres.

methanogens bacteria that live in oxygen-free environments and generate methane by the reduction of carbon dioxide. Examples are species belonging to the genera *Methanobacterium, Methanococcus,* and *Methanomicrobium*. They are placed in the archaebacteria (*q.v.*) on the basis of the nucleotide sequences of their 16S rRNAs.

methionine *See* **amino acid.**

method of least squares *See* **least squares.**

methotrexate a folic acid antagonist that kills cells by inhibiting the enzyme dihydrofolate reductase and thus blocking the synthesis of nucleic acids. Mammalian cells that develop resistance to methotrexate do so by amplifying the genes encoding dihydrofolate reductase. *See* **Appendix C,** 1978, Schimke *et al.*; **folic acid.**

methylated cap a modified guanine nucleotide terminating eukaryotic mRNA molecules. The cap is introduced after transcription by linking the 5′ end of a guanine nucleotide to the 5′ terminal base of the mRNA and adding a methyl group to position 7 of this terminal guanine. The addition of the terminal guanine is catalyzed by the enzyme *guanylyl transferase*. Another enzyme, *guanine-7-methyl transferase*, adds a methyl group to the 7 position of the terminal guanine. Unicellular eukaryotes have a cap with this single methyl group (cap 0). The predominant form of the cap in muticellular eukaryotes (cap 1) has another methyl group added to the next base at the 2′-*o* position by the enzyme *2′-o-methyl transferase*. More rarely, a methyl group is also added to the 2′-*o* position of the third base, creating cap 2 type. Capping occurs shortly after the initiation of transcription and precedes all excision and splicing events. The function of the cap is not known, but it may protect the mRNA from degradation by nucleases or provide a ribosome binding site. *See* **posttranscriptional modification.**

methylation the addition of a methyl group

(−CH₃) to DNA or RNA. *See* **restriction and modification model.**

methylcholanthrene   a carcinogenic hydrocarbon.

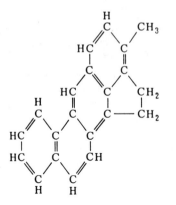

5-methyl cytosine (5-Me C)   a modified base found in the DNA of both prokaryotes and eukaryotes. 5-Me C is generated by a methylase enzyme that adds a methyl group to cytosine at certain sites in the DNA. In mammalian chromosomes, the bulk of 5-Me C is in simple sequence repetitive DNAs, but it is also present throughout all sequences. Structural genes appear to be hypomethylated in tissues where they are actively expressed and hypermethylated in tissues where they are inactive.

methyl green   a basic dye often used in cytochemistry to detect DNA. In 2 molar solution of magnesium chloride at pH 5.7 methyl green stains only undegraded DNA. *See* **pyronin Y.**

methyl guanosine   *See* **rare bases.**

methyl inosine   *See* **rare bases.**

2′-o-methyl transferase   *See* **methylated cap.**

metric traits   *See* quantitative character, continuous variation.

Mg   magnesium.

MHC   major histocompatibility complex (*q.v.*).

MIC   major immunogen complex (*q.v.*).

MIC2   a gene known to reside in the human pseudoautosomal region (*q.v.*)

micelle   a spherical array of amphipathic molecules in which the nonpolar tails form a hydrocarbon microdroplet enclosed in a shell composed of the polar heads.

Michaelis constant (K_m)   the substrate concentration at which the reaction rate of an enzyme is half maximal; the concentration at which half of the enzyme molecules in the solution have their active sites occupied by a substrate molecule.

Michurinism   the genetic theories expounded by the Russian horticulturist I.V. Michurin, particularly those dealing with the modification of the genetical constitution of a scion by grafting. Michurin's theories were subsequently incorporated into Lysenkoism (*q.v.*).

Micrasterias thomasiana   a disc-shaped unicellular desmid, whose cortical differentiation has been extensively studied.

micro-   **1.** a prefix meaning one millionth; used with units of measurements in the metric system. **2.** a prefix meaning microscopic or minute.

microbeam irradiation   the use of a beam of microscopic diameter to selectively irradiate portions of a cell with ionizing radiation or ultraviolet light.

microbial genetics   the genetic study of microorganisms.

microbiology   the scientific study of microorganisms.

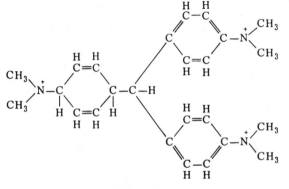

Methyl green.

microbody a cytoplasmic organelle bounded by a fragile membrane and containing a variety of oxidases (urate oxidase, catalase, etc.).

Microbracon hebetor a wasp (called *Habrobracon juglandis* or *Bracon hebetor* in the early genetic literature). It is the arrhenotokous species in which the genetic control of sex determination was elucidated.

microcarriers microscopic beads or spheres composed of dextran or other substances, that are used in tissue culture to attract and hold cells that must be anchored before they can proliferate. *See* **anchorage-dependent cells.**

microcinematography *See* **time-lapse microcinematography.**

micrococcal nuclease an endonuclease used to cleave eukaryotic chromatin preferentially between nucleosomes.

microconidia *See* **conidium.**

microevolution an evolutionary pattern usually viewed over a short period of time, such as changes in gene frequency within a population over a relatively few generations (industrial melanism, for example). *See* **macroevolution.**

microfilaments elongated intracellular fibers 5–7 nm in diameter, containing polymerized actin, and thought to function in maintenance of cell structure and movement.

Micrographia a book published in 1665 by R. Hooke in which the first description of cells is given.

micromanipulator an instrument allowing surgery on and injection into microscopic specimens. It is also useful in isolating single cells.

micron (μ) a unit of length (1×10^{-6} meters) convenient for describing cellular dimensions.

micronucleus the smaller reproductive nucleus as distinguished from the larger, vegetative macronucleus (*q.v.*) in the cells of ciliates. Micronuclei are diploid and are transcriptionally inactive. They participate in meiosis and autogamy.

microorganism an organism too small to be observed with the unaided eye.

micropyle 1. a canal through the coverings of the nucellus through which the pollen tube passes during fertilization. In a mature seed, the micropyle serves as a minute pore in the seed coat through which water enters when the seed begins to germinate. 2. the pore in the egg membranes of an insect oocyte that allows entry of the sperm.

microsome fraction a cytoplasmic component, obtained upon centrifugation of homogenized cells, consisting of ribosomes and torn portions of the endoplasmic reticulum. *See* **Appendix C,** 1943, Claude.

microspectrophotometer an optical system combining a microscope with a spectrophotometer. With this apparatus one can determine the amount of light of specified wavelength passing through a given cytoplasmic region relative to some standard area. From such measurements estimates can be made of the concentrations of dye-binding or ultraviolet-light-absorbing materials present in the cytoplasmic area in question. *See* **flying spot cytometer.**

microspore the first cell of the male gametophyte generation of seed plants. Each becomes a pollen grain.

microspore culture *See* **Appendix C,** 1973, Debergh and Nitsch; **anther culture.**

microsporidia parasitic species belonging to the phylum Cnidosporidia (*see* **Appendix A,** Protoctista). Microsporidian species of genus *Nosema* are parasites of silkworms, honeybees, and *Drosophila.* The feeding stage of a microsporidian consists of a minute amoeba with a nucleus, but no mitochondria. This is the simplest eukaryotic cell known. A unique characteristic of microsporidians is that the unicellular resting spore contains a coiled hollow tube, the polar filament, which is a specialized product of the Golgi apparatus. During infection, the polar filament turns inside out to form a hollow tube through which the infectious part of the spore is extended into the new host. Microsporidia are considered the most primitive eukaryotes because, like bacteria, they have 70S ribosomes containing 16S and 23S rRNA, rather than the larger ribosomes that characterize other eukaryotes. These 80S ribosomes contain 18S and 29S RNAs. *See* **ribosome.**

microsporocyte the pollen mother-cell, which undergoes two meiotic divisions to produce four microspores.

microsporogenesis the production of microspores.

microsporophyll the stamen. *See* **flower.**

microsurgery surgery done while viewing the object through a microscope and often with the aid of a micromanipulator. *See* **Appendix C,** 1952, Briggs and King; 1967, Goldstein and Prescott; 1980, Capecchi.

microtome a machine for cutting thin (1 to 10 mμ

thick) slices of paraffin-embedded tissue. These sections are later stained and examined with the light microscope. *See* **Appendix C,** 1870, His; **ultramicrotome.**

microtomy the technique of using the microtome in the preparation of sections for study under the microscope.

microtrabecular lattice a network of thin filaments interconnecting the three major cytoskeletal elements (microtubules, microfilaments, and intermediate filaments). This three-dimensional lattice can be visualized in freeze-etched preparations viewed with the electron microscope.

microtubule organizing centers (MCs) the structures or loci that give rise to microtubular arrays. Centrioles and kinetosomes function as MCs in some organisms. In others, the MC consists of a deposit of amorphous granulofibrillar material. MCs contain RNA, and RNA replication accompanies their multiplication.

microtubules long, nonbranching, thin cylinders with an outside diameter about 24 nm and a central lumen about 15 nm diameter. The lengths are at least several μm. The tubules are composed of strands called protofilaments, and there are usually 13 of these. Each protofilament in turn is composed of a linear array of subunits, and each subunit is a dimer containing an alpha and a beta tubulin molecule. Microtubules play key roles in cell division, secretion, intracellular transport, morphogenesis, and ciliary and flagellar motion. *See* **axoneme, tubulin.**

Microtus the genus of meadow mice and voles subjected to extensive cytogenetic study. *M. agrestis,* the field vole, has giant sex chromosomes.

microvillus a fingerlike projection of the plasmalemma of a cell.

mictic referring to an organism or species capable of biparental sexual reproduction.

middle lamella the outermost layer of a plant cell wall that connects it to its neighbor.

midparent value the mean of two parental values for a quantitative trait in a specific cross. *See* **quantitative inheritance.**

midpiece the portion of the sperm behind the head containing the nebenkern.

midspindle elongation the elongation during anaphase of the midregion of the mitotic spindle. This elongation serves to draw the chromosomes poleward once the movement of chromosomes along traction fibers has ceased.

migrant selection selection based on the different migratory abilities of individuals of different genotypes. If, for example, individuals carrying gene M found new colonies more often than those bearing gene m, then gene M is said to be favored by migrant selection.

migration in population genetics, the movement of individuals between different populations of a species, resulting in gene flow (*q.v.*).

migration coefficient the proportion of a gene pool represented by migrant genes per generation.

migration inhibition factor a lymphokine that inhibits the movement of macrophages under *in vitro* culture conditions.

Miller spreads chromosomal whole mounts spread for electron microscopy by a method developed by O. L. Miller. In this method, chromosomes from ruptured nuclei are centrifuged through a solution of 10% formalin in 0.1 M sucrose and onto a membrane-coated grid. The grid is then treated with an agent that reduces the surface tension as the grid dries. After being stained with phosphotungstic acid, the preparation is ready for viewing under the electron microscope.

Miller trees transcribing rRNA genes first seen in Miller spreads of the amplified rRNA of extrachromosomal nucleoli from salamander oocytes. Each rRNA transcription unit (rTU) was about 2.5 μm long, corresponding to a molecule of 8,000 bp. The rRNA molecules attached to the chromatin fiber give rTUs a Christmas tree morphology (hence the term Miller tree), and such rTUs are arranged in tandem, separated by fiber-free spacers. *See* **Appendix A,** 1969, Miller and Beatty; 1976, Chooi.

milliequivalent *See* **gram equivalent weight.**

milliliter (ml) one-thousandth of a liter; the volume contained in a cube of side 1 cm. One ml of water weighs one gram at 4°C.

Millipore filter a disc-shaped synthetic filter having holes of specified diameter through its surface. The available pore sizes range between 0.005 and 8 μ. The discs are used to filter microorganisms out of nutrient fluids that will not stand sterilization by autoclaving.

mimicry the similarity in appearance of one species of animal to another that affords one or both protection. In *Batesian mimicry,* one of two species is poisonous, distasteful, or otherwise protected from predators, and often conspicuously marked. The mimic is innocuous, gaining protection from predators by its similarity to the model. In *Mullerian mimicry,* both are distasteful to predators, and

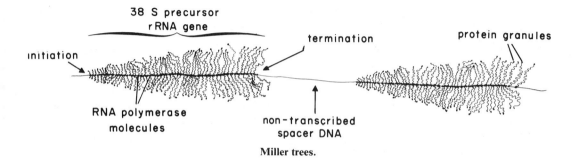

38 S precursor rRNA gene

initiation

termination

protein granules

RNA polymerase molecules

non-transcribed spacer DNA

Miller trees.

they gain mutually from having the same warning coloration, since predators learn to avoid both species after tasting one. *Peckhammian mimicry* is an aggressive mimicry in which the predator is the pretender (mimic); e.g., a female firefly of one species mimics the flashing sexual signals of another species, thus attracting a male of that other species which becomes a meal for the pretender. In *Mertensian mimicry,* one species is mildly poisonous (e.g., false coral snakes) and thus serves as a model for a fatally poisonous species (e.g., true coral snakes). Obviously, a predator can learn only if it survives the encounter. *See* **automimicry, frequency-dependent fitness.**

minicells small anucleate bodies produced by the aberrant division of bacteria. Certain mutant strains of *E. coli* produce large numbers of minicells. Although minicells lack genophores, they may contain one or more copies of the plasmids present in the parent cells. This permits the analysis of the plasmid DNA or of plasmid-encoded gene products in the absence of the host genome and its products. *See* **maxicells.**

minichromosomes beaded DNA structures of certain viruses, similar to the beaded nucleosomal structures characteristic of eukaryotic chromatin. The circular DNA duplexes of papovaviruses (*q.v.*) are bound throughout the replication cycle to host histones, except H1.

minigenes segments on chromosomes that code for the variable regions of immunoglobulin heavy and light chains. There are several hundred of these segments, but during the differentiation of B lymphocytes a single minigene is attached to the gene segment coding for the constant region. *See* **Appendix C,** 1965, Dreyer and Bennett; **immunoglobulin genes.**

minimal medium in microbiology a medium providing only those compounds essential for the growth and reproduction of wild-type organisms.

mink *See Mustela.*

minority advantage the phenomenon first observed in multiple choice mating experiments with *Drosophila melanogaster.* Males carrying certain genetic markers mate with relatively greater success when they are rare than when they are common. *See* **Appendix C,** 1951, Petit; **frequency-dependent selection.**

Minutes mutations in *Drosophila* that are dominant and lethal when homozygous. Heterozygotes have a prolonged developmental period, and the adults are small, semisterile, with short thin bristles, rough eyes, etched tergites, and abnormal wing venation. Many *Minutes* are small deficiencies. There are at least 60 different *Minute* loci scattered throughout the genome, and some represent genes encoding ribosomal proteins.

Miocene an epoch in the Tertiary period, after the Oligocene and before the Pliocene. Representatives of all mammalian families were present by this time. *See* **geologic time divisions.**

Mirabilis jalapa the four-o'clock, a variegated dicotyledonous plant extensively studied in terms of plastid inheritance. *See* **Appendix C,** 1909, Correns and Bauer.

mischarged tRNA a tRNA molecule to which an incorrect amino acid is attached.

mismatch *See* **mispairing.**

mismatch repair correction of nucleotide mispairings by "cut and patch repair" (*q.v.*).

mispairing the presence in one chain of a DNA double helix of a nucleotide not complementary to the nucleotide occupying the corresponding position in the other chain.

missense mutant a mutant in which a codon is mutated to one directing the incorporation of a different amino acid. This substitution may result in an inactive or unstable product. *Contrast with* **nonsense mutation.**

missing link an unknown or postulated interme-

199

diate in an evolutionary sequence of fossil forms. *See* **living fossil.**

Mississippian   *See* **Carboniferous.**

mistranslation   the insertion of an incorrect amino acid into a site on a growing polypeptide chain that is the result of environmental factors or mutations that effect either the tRNA, or the enzymes that attach specific amino acids to specific tRNAs, or the ribosome itself.

mitochondrial DNA   the mitochondrial genome consists of a circular DNA duplex, and there are generally 5 to 10 copies per organelle. The mammalian mtDNA has a molecular weight of about 11×10^6 d and therefore is less than 10^{-5} times that the size of the nuclear genome. The genetic code of mitochondria differs slightly from the "universal" genetic code. Since, of the two gametes, only the egg cell contributes significant numbers of mitochondria to the zygote, mtDNA is maternally inherited. *See* **Appendix C,** 1959, Chevremont *et al.;* 1966, Nass; 1968, Thomas and Wilkie; 1974, Dujon *et al.,* Hutchison, *et al.,* 1979, Avise *et al.,* Barrell *et al.;* 1981, Anderson *et al.;* **mtDNA lineages, universal code theory.**

mitochondrial sex locus (ω)   A gene on the mitochondrial DNA of yeast cells that can produce markedly different recombination frequencies when homozygous or heterozygous in mitochondrial "crosses."

mitochondrion   a semiautonomous, self-reproducing organelle that occurs in the cytoplasm of all cells of most, but not all, eukaryotes. For example, microsporidia (*q.v.*) lack mitochondria. Each mitochondrion is surrounded by a double limiting membrane. The inner membrane is highly invaginated, and its projections are called *cristae.* In most eukaryotes, the cristae are platelike, but certain protoctists (*viz.* ciliates, sporozoa, diatoms, chrysophytes) have tubular cristae. Mitochondria are the sites of the reactions of oxydative phosphorylation (*q.v.*), which result in the formation of ATP. Mitochondria contain distinctive ribosomes, tRNAs, aminoacyl-tRNA synthetases, and elongation and termination factors. Mitochondria depend upon genes within the nucleus of the cells in which they reside for many essential mRNAs. The proteins translated from these mRNAs in the cell cytosol are imported into the organelle. Mitochondria are believed to have arisen from aerobic bacteria that established a symbiotic relationship with primitive protoeukaryotes. *See* **Appendix C,** 1890, Altman; 1898, Benda; 1952, Palade; 1964, Luck and Reich; **chloramphenicol, citric acid cycle, electron transport chain, endosymbiont theory, glycolysis, leader**

sequence peptide, nebenkern, petites, ribosomes, serial symbiosis theory.

mitogen   a compound that stimulates cells to undergo mitosis.

mitomycin   a family of antibiotics produced by *Streptomyces caespitosus.* Mitomycin C prevents DNA replication by crosslinking the complementary strands of the DNA double helix.

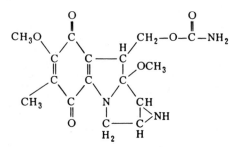

mitosis   mitosis or nuclear division is generally divided into four phases: *prophase, metaphase, anaphase,* and *telophase.*

Prophase. During mitotic prophase, the centriole divides and the two daughter centrioles move apart. The chromosomes become visible within the nucleus because they coil up to produce a series of compact gyres. Each chromosome is longitudinally double except in the region of the centromere, and each replicate strand of a chromosome is called a chromatid. The nucleolus and the nuclear envelope break down.

Metaphase. During this phase, the chromosomes move about within the spindle and eventually arrange themselves in the equatorial region of the spindle. The two chromatids are now ready to be separated and to move under the action of the traction fibers to the poles of the spindle.

Anaphase. During this phase, the centromere becomes functionally double, and the chromatids are converted to independent chromosomes that separate and move to opposite poles.

Telophase. During telophase, the spindle disappears and reconstruction of nuclear envelopes about the two groups of offspring chromosomes begins. As nuclear envelopes form around each group the chromosomes return to their extended state, and nucleoli reappear.

Next, cytokinesis occurs, and the cytoplasm is divided into two parts by a cleavage furrow in the case of animal cells or by a cell plate in the case of plant cells. The result of mitosis and cytokinesis is the production of two daughter cells with precisely the same nuclear content and approximately equal amounts of cytoplasm. *See* **Appendix C,** 1873, Schneider; 1879, 1882, Flemming.

mitosporangium a sporangium containing spores produced by mitotic divisions. *Compare with* **meiosporangium.**

mitospore a spore produced by mitosis and therefore having the same chromosome number as the mother cell.

mitotic apparatus an organelle consisting of three components: (1) the asters, which form about each centrosome, (2) the gelatinous spindle, and (3) the traction fibers, which connect the centromeres of the various chromosomes to either centrosome. *See* **Appendix C,** 1952, Mazia and Dan; **mitosis, multipolar spindle.**

mitotic center the agent that defines the poles toward which the chromosomes move during anaphase. The centrioles function as mitotic centers in most animal cells. In plants with anastral mitosis, the nature of the mitotic center is unclear.

mitotic chromosome *See* **human mitotic chromosomes.**

mitotic crossover somatic crossing over.

mitotic index the fraction of cells undergoing mitosis in a given sample.

mitotic poison a compound that prevents mitosis. *See* **colchicine, maleuric acid, mercaptoethanol, podophyllin, vincaleukoblastine.**

mitotic recombination somatic crossing over.

mixis biparental sexual reproduction.

mixoploidy the presence of more than one chromosome number in a cellular population.

MLD medial lethal dose (*q.v.*).

mM millimolar concentration.

Mn manganese.

MN blood group a human blood group system defined by red cell antigens specified by a gene on the short arm of chromosome 4.

Mo molybdenum.

mobile genetic element *See* **transposable genetic elements.**

Möbius strip a topological figure, named after the German astronomer, A. F. Möbius, made by putting a 180-degree twist in a long, rectangular strip, then pasting the ends together. The strip has only one surface and one edge. If a Möbius strip is cut longitudinally, it forms a double-sized ring. If a ring with a double twist is cut longitudinally, it forms interlocked rings. A Möbius strip demonstrates the behavior of a twisted ring chromosome when it attempts replication.

modal class the class that contains more individuals than any other in a statistical distribution.

mode modal class.

modification in nucleic acid metabolism, any changes made to DNA or RNA nucleotides after their original incorporation into a polynucleotide chain: e.g., methylation, deamination, formylation, etc. *See* **modification methylases.**

modification allele *See* **DNA restriction enzyme.**

modification methylases bacterial enzymes that bind to the DNA of the cell at specific sites determined by specific base sequences. Here they attach methyl groups to certain bases. This methylation pattern is unique to and protects the species from its own restriction endonucleases. Modification methylases are coded for by modification alleles. *See* **restriction and modification model.**

modified bases postsynthetically altered nucleotides of the four usual bases (A, T, G, and C) of DNA. *See* **modification methylases.**

modifier referring in the genetic literature to a gene that modifies the phenotypic expression of a nonallelic gene.

modular organisms organisms that consist of populations of units or modules that are attached together, but if separated are capable of independent growth and reproduction. An example would be plants like strawberries where segments of the branching root systems can generate new plants when severed from the parent. In such organisms, a mutation in a meristematic cell may be expressed during the subsequent growth of a module built from the mutant clone, and eventually the mutation may be contained in eggs or pollen. Thus, Weismann's distinction between soma and germ cells does not apply to modular organisms, and evolutionary changes in modular species may originate from both germinal and somatic mutations. *See* **Appendix C,** 1883, Weismann.

modulating codon specific triplets that code for rare tRNAs. The translation of a mRNA molecule is slowed down when a modulating codon is encountered.

moiety one of two roughly equal parts.

molecular biology a modern branch of biology concerned with explaining biological phenomena in molecular terms. Molecular biologists often use the biochemical and physical techniques to investigate genetic problems.

molecular clock *See* **DNA clock hypothesis, protein clock hypothesis.**

molecular cloning *See* **gene cloning, recombinant DNA technology.**

molecular genetics that subdivision of genetics which studies the structure and functioning of genes at the molecular level.

molecular hybridization base pairing between DNA strands derived from different sources or of a DNA strand with an RNA strand.

molecular mass *See* **relative molecular mass (Mr).**

molecular mimicry *See* **eclipsed antigens.**

molecular sieve a crystalline alumino-silicate pellet used to absorb water, carbon dioxide, hydrogen sulfide, and similar gases from gas mixtures and organic solvents. Molecular sieves are also sometimes used as ion exchange media.

molecular weight the sum of the atomic weights of all of the atoms in a given molecule. The term has largely been replaced by *relative molecular mass (q.v.)*.

molecule that ultimate unit quantity of a compound that can exist by itself and retain all the chemical properties of the compound.

molting hormone *See* **ecdysones.**

moltinism a polymorphism in terms of the number of larval molts undergone by a given species. For example, in *Bombyx mori* there are strains that are known to molt three, four, or five times.

molybdenum a biological trace element. Atomic number 24; atomic weight 95.94; most abundant isotopes ^{92}Mo, ^{94}Mo, ^{95}Mo, ^{96}Mo, ^{97}Mo, ^{98}Mo; radioisotope ^{99}Mo, half life 67 hours, radiation emitted—beta particles.

monad 1. a single organism, usually implying a free-living, unicellular, flagellate stage. 2. The haploid set of chromosomes found in the nucleus of an ootid or a spermatid. *See* **meiosis.**

Monarch butterfly *Danaus plexippus.* Monarch larvae generally feed upon plants, such as milkweeds, which contain molecules toxic to vertebrates. These molecules are sequestered by the insect, and they serve to render the animals unpalatable to avian predators. *See* **Basilarchia archippus.**

monestrous mammal *See* **estrous cycle.**

mongolism Down syndrome *(q.v.).*

Mongoloid pertaining to a race of mankind, char-acterized by a faintly yellowish skin, an epicanthic fold, sparse body hair, and black straight head hair.

monkey *See* **Cercopithecus aethiops, Macaca mulatta, woolly monkey.**

monoallelic referring to a polyploid in which all alleles at a given locus are the same. In a hexaploid, for example, $A_1A_1A_1A_1A_1A_1$.

monocentric designating a chromosome having a single centromere.

monochromatic light light of a single wavelength.

monochromatic radiation electromagnetic radiation of a single wavelength, or in which all of the photons have the same energy.

monocistronic mRNA messenger ribonucleic acid coding for a single polypeptide chain; the typical mRNA of eukaryotic cells.

monoclonal antibodies immunoglobulins derived from a single clone of plasma cells. Since all immunoglobulins produced by a given plasma cell (or clone thereof) are chemically and structurally identical, these antibodies constitute a pure population with highly specific antigen-binding properties. *See* **hybridoma.**

monocyte the largest leukocyte found in the blood; they are phagocytic, amoeboid cells. *See* **agranulocytes, macrophages.**

monoecy the condition where a plant bears both staminate and pistillate flowers (corn is an example).

monoenergetic radiation radiation of a given type (alpha, beta, neutron, gamma) in which all particles or photons have the same energy.

monogamy an animal reproductive strategy in which a specific male and female form an exclusive mating pair during a reproductive cycle, a season, or a lifetime. *Compare with* **polygamy.**

monogenic character a character determined by a single gene.

monohybrid an individual that is heterozygous (e.g., *Aa*) for a pair of alleles at the locus under study.

monohybrid cross a mating between two individuals, both of which are heterozygotes genetically identical at a given locus; e.g., *Aa* × *Aa*.

monolayer a single layer of cells growing on a surface.

monolepsis transmission to the offspring of characteristics of but one parent.

monomer a simple compound from which, by repetition of a single reaction, a polymer is made. For example, uridylic acid (U) can be polymerized to form polyuridylic acid (UUU . . . .).

monomorphic locus a genetic locus at which the most common allele exceeds a frequency of 0.95 in the gene pool of the population. *Compare with* **polymorphic locus.**

monomorphic population a population showing only one trait (of potentially variable expression) due to fixation of one allelic form of the gene responsible for that trait.

mononuclear leukocyte *See* **agranulocytes.**

monophasic lethal a mutation having one effective lethal phase (*q.v.*).

monophyletic group a natural taxon composed of two or more species. It includes the ancestral species (known or hypothesized) and all of its descendants. Members of a monophyletic group are sister taxa. *See* **polyphyletic group.**

monoploid **1.** the basic chromosome number in a polyploid series. **2.** a somatic cell or individual having only one set of chromosomes.

monosome **1.** a single mRNA-ribosome complex. *See* **polysome.** **2.** a chromosome lacking a homologue.

monosomy the condition in which one chromosome of one pair is missing. A monosomic diploid has one chromosome less than the normal diploid number (2N-1). An XO *Drosophila* is monosomic for the sex chromosome. *See* **Appendix C,** 1921, Bridges; 1926, Goodspeed and Clausen.

monospermy fertilization by only one spermatozoon.

monothetic group in systematics, organisms that are grouped together because they possess a unique set of features that is both sufficient and necessary for membership in the group thus defined. For example, all mammals and only mammals have hair and nurse their young with milk.

monotreme an egg-laying mammal such as the duckbilled platypus or the echidna.

monotrichous having a single flagellum.

monotypic a taxonomic group containing only one taxon of the next lower rank; for example, a monotypic genus contains only one species; a monotypic species has no recognized subgroups (races, subspecies); the opposite of polytypic.

monozygotic twins *See* **twins.**

Morgan unit a unit for expressing the relative distance between genes on a chromosome. One morgan (M) equals a crossover value of 100%. A crossover value of 10% is a decimorgan (dM); 1% is a centimorgan (cM); named in honor of Thomas Hunt Morgan.

Mormoniella vitripennis a calcidoid wasp also referred to in the scientific literature as *Nasonia brevicornis* and *Nasonia vitripennis.* The wasp is often used for classroom demonstrations of parthenogenesis (*q.v.*). In the laboratory, it is provided with pupae of the fly *Sarcophaga bullata* to parasitize.

morph any of the individuals of a polymorphic population; any phenotypic or genotypic variant.

morphogen any compound produced in a localized region of a developing organism, forming a concentration gradient as it diffuses, and which causes cells that receive it to enter a specific developmental pathway. *See* **retinoic acid.**

morphogenes genes involved directly or indirectly in the control of growth and morphogenesis: e.g., genes for hormones, inducers, mitogens, inhibitors, cell cycle control factors, etc.

morphogenesis the developmental processes leading to the characteristic mature form of a cell, an organism, or part of an organism. *See* **development, differentiation.**

morphogenetic movements the cell movements that change the shape of differentiating cells or tissues, *e.g.,* in an embryo (invagination, expansion, longitudinal extension, dorsal convergence, etc.).

morphogenetic stimulus a stimulus exerted by one part of the developing embryo upon another, leading to morphogenesis in the reacting part.

morphology the science dealing with the visible structures of organisms and the developmental and evolutionary history of these structures.

morphometric cytology the determination of quantitative parameters of cytological structures in tissue sections.

morula an embryo that consists of a cluster of cleaving blastomeres. A stage prior to the blastula.

mosaic an individual composed of two or more cell lines of different genetic or chromosomal constitution, both cell lines being derived from the same zygote; in contrast with a *chimera* (*q.v.*). *See* **Appendix C,** 1963, Davidson *et al.*

mosaic development in mosaic development, the fates of all the parts of the embryo are already fixed at or before fertilization, so that any localized ablation will later be manifested by the absence of the part in question. *See* **regulative development.**

mosaic evolution evolutionary change in one or more body parts without simultaneous changes in other parts. For example, in the evolution of birds from dinosaurs, the origin of feathers occurred long before specialized bones (e.g., keel) and powerful flight muscles developed.

motility symbiosis the situation where motility is conferred upon an organism by its symbiont. For example, many protists contain cortical populations of symbiotic spirochaetes whose coordinated movements propel the host through its aqueous medium.

mouse *See Mus musculus,* **oncomouse.**

mouse inbred lines laboratory strains of mice propagated by brother-sister matings for many generations and hence highly homozygous and genetically uniform. In some strains, the inbreeding program has been carried out for 40 years. The strains most commonly mentioned in the literature are: albino (A, Ak, BALB, R_{III}), black (C_{57} black, C_{58}), black agouti (CBA, C3H), brown (C_{57} brown), dilute brown (DBA/2), dilute brown piebald (I). *See* **Appendix C,** 1909, Little; 1942, Snell.

mouse L cells a strain of fibroblastlike cells carried in tissue culture. The cells originated from subcutaneous areolar and connective tissue derived from a male c3H mouse. *See* **Appendix C,** 1940, Earle.

mouse satellite DNA a DNA making up about 10% of the DNA isolated from a wide variety of mouse tissues. It forms a band slightly separated from the main peak when mouse DNA is spun to equilibrium in a CsCl density gradient. *See* **centrifugation separation.** Mouse satellite DNA consists of about one million copies per genome of a sequence some 400 nucleotide pairs in length. *In situ* hybridization experiments show that most of this DNA is located in the pericentric heterochromatin. *See* **Appendix C,** 1970, Pardue and Gall.

MPD maximum permissible dose (*q.v.*).

M phase *See* **cell cycle.**

Mr relative molecular mass (*q.v.*).

mRNA messenger RNA.

mRNA coding triplets *See* **amino acid, start codon, stop codon.**

MS2 an RNA phage from which an RNA-replicase has been isolated.

MSH melanocyte-stimulating hormone. *See* **intermedin.**

M strain the *M*aternally contributing strain of *Drosophila* in a P-M hybrid dysgenesis cross. M strains lack P factors. *See* **hybrid dysgenesis, P elements, P strain.**

MTA mammary tumor agent (*q.v.*).

mtDNA mitochondrial DNA (*q.v.*).

mtDNA lineages evolutionary trees derived from data on mitochondrial DNAs. Human mitochondria are maternally inherited, and therefore mtDNA is contributed by the female parent to the next generation. Furthermore, mitochondrial genes do not undergo recombination. For these reasons, it is much simpler to trace mutations in mt DNA than genomic DNA. In a study of mtDNA from 147 individuals from five geographically distinct human populations, it was possible to trace the mtDNA polymorphisms back to a common female ancestor that lived in Africa approximately 200,000 years ago. In the popular press, this woman is called the "African Eve." *See* **Appendix C,** 1987, Cann *et al.;* **mitochondrial DNA.**

M5 technique a technique used to detect induced sex-linked lethal and viable mutations in *Drosophila melanogaster.* The technique gets its name from the X chromosome used to balance the chromosomes bearing the induced mutations. The M5 or Muller 5 chromosome is the fifth of a series synthesized by H. J. Muller. It contains a complex inversion and the marker genes *B*ar, *a*pricot, and *sc*ute. For this reason, the M5 chromosome is sometimes abbreviated *Basc. See* **balanced stock.**

mt mRNA, mt rRNA, mt tRNA symbols for mitochondrial messenger, ribosomal, and transfer RNAs, respectively.

mu 1. map unit. 2. mate killers.

mucopolysaccharide a polysaccharide composed of sugars and sugar derivatives, such as amino sugars and uronic acids. *See* **chondroitin sulfuric acid.**

mucoprotein a protein containing more than 4% carbohydrate. *See* **glycoprotein.**

mulatto the hybrid from a Negro-white cross.

mule *See* **horse-donkey hybrids.**

Mullerian mimicry *See* **mimicry.**

multifactorial polygenic.

multiforked chromosome a bacterial chromosome containing more than one replication fork, due to a second initiation's having begun before completion of the first replication cycle.

multigene family a set of genes descended by duplication and variation from some ancestral gene.

Such genes may be clustered together on the same chromosome or dispersed on different chromosomes. Examples of multigene families include those that encode the histones, hemoglobins, immunoglobulins, histocompatibility antigens, actins, tubulins, keratins, collagens, heat shock proteins, salivary glue proteins, chorion proteins, cuticle proteins, yolk proteins, and phaseolins. *See* **reiterated genes.**

multimer   a protein molecule made up of two or more polypeptide chains, each referred to as a monomer. The terms dimer, trimer, tetramer, pentamer, etc., are used if the number of monomers per multimer is known.

multiparous   bearing or producing more than one offspring at a birth. *See* **parity.**

multiple allelism   *See* **allele.**

multiple choice mating   referring to an experimental design in studies of behavior genetics where a test organism is allowed to choose between two (or more) genetically different mates.

multiple codon recognition   *See* **wobble hypothesis.**

multiple-event curve   a curve (relating relative survival to radiation dose) that contains an initial flat portion. This finding indicates that there is little biological effect until a certain dose has accumulated, and suggests that the sensitive target must be hit more than once (or that there must be multiple targets, each of which must be destroyed) to produce a biologically measurable effect. *See* **single-event curve, target theory.**

multiple factor hypothesis   *See* **quantitative inheritance.**

multiple genes   *See* **polygene, multiple factor hypothesis, quantitative inheritance.**

multiple infection   simultaneous invasion of a bacterial cell by more than one phage, often of different genotypes in experiments designed to promote phage recombination; superinfection.

multiple myeloma   *See* **myeloma.**

multiple neurofibromatosis   *See* **neurofibromatosis.**

multiplicity of infection   the average number of phages that infect a bacterium in a specific experiment. The fraction of bacteria infected with 0, 1, 2, 3, . . . , *n* phage follows a Poisson distribution.

multiplicity reactivation   the production of recombinant virus progeny following the simultaneous infection of each host cell by two or more virus particles, all of which are incapable of multiplying because they carry lethal mutations induced by exposure to a mutagen.

multipolar spindle   a spindle with several poles found in cells with multiple centrioles. Such cells are seen infrequently, but they can be produced in large numbers by irradiation. *See* **mitotic apparatus.**

multitarget survival curve   *See* **extrapolation number.**

multivalent   designating an association of more than two chromosomes whose homologous regions are synapsed by pairs (as in autotetraploids and translocation heterozygotes).

multivoltine   producing more than one brood in a year, as in certain birds and moths.

mu **phage**   a phage "species" whose genetic material behaves like insertion sequences, being capable of transposition, insertion, inactivation of host genes, and causing rearrangements of host chromosomes.

murine   belonging to the family of rodents that contains the mice and rats.

murine mammary tumor virus   an oncogenic RNA virus. *See* **mammary tumor agent.**

Musa paradisaica sapientum the banana. *See* **parthenocarpy.**

Musca domestica   the housefly. DDT resistance in this species has been extensively studied by geneticists.

muscular dystrophy   a heterogeneous group of hereditary diseases affecting humans and other mammals that cause progressive muscle weakness due to defects in the biochemistry of muscle tissue. The most common type in humans is X-linked, recessive Duchenne muscular dystrophy, which affects about 1 in 3,500 boys. The normal gene, DMD, contains more than 2×10^6 base pairs. It transcribes an mRNA about 14 kb long that specifies a protein called dystrophin. This has a molecular weight of 400 kd and represents only 0.002% of the total protein in striated muscle. *See* **Appendix C,** 1987, Hoffman *et al.*

Mus musculus   the laboratory mouse. Its diploid chromosome number is 20, and extensive genetic maps are available for the 19 autosomes and the X chromosome. There are large collections of strains containing neurological mutants, loci associated with oncogenic viruses (especially retroviruses), loci that encode enzymes, and histocompatibility loci. *See* **Appendix C,** 1905, Cuénot; 1909, 1914, Little;

1936, Bittner; 1940, Earle; 1942, Snell; 1948, Gorer *et al.;* 1953, Snell; 1967, Mintz; 1975, Mintz and Illmensee; 1976, Hozumi and Tonegawa; 1980, Gordon *et al.;* 1987, Kuehn *et al.;* **mouse inbred lines, oncomouse.**

mustard gas   sulfur mustard (*q.v.*).

Mustela   the genus that includes *M. erminea,* the ermine; *M. lutreola,* the European mink; *M. vison,* the North American mink.

mutable gene   in multicellular organisms, a gene that spontaneously mutates at a sufficiently high rate to produce mosaicism.

mutable site   a site on a chromosome at which mutations can occur.

mutagen   a physical or chemical agent that raises the frequency of mutation above the spontaneous rate.

mutagenic   causing mutation.

mutagenize   to expose to a mutagenic agent.

mutant   an organism bearing a mutant gene that expresses itself in the phenotype of the organism.

mutant hunt   the isolation and accumulation of a large number of mutations affecting a given process, in preparation for mutational dissection of the gene(s) governing that process. For example, one might select for mutations that confer phage resistance in *E. coli.*

mutation   **1.** the process by which a gene undergoes a structural change. **2.** a modified gene resulting from mutation. **3.** by extension, the individual manifesting the mutation. *See* **Appendix C,** 1901, de Vries; **isocoding mutation, point mutation.**

mutational dissection   *See* **genetic dissection.**

mutational load   the genetic disability sustained by a population due to the accumulation of deleterious genes generated by recurrent mutation.

mutation breeding   induction of mutations by mutagens to develop new crop varieties that can increase agricultural productivity.

mutation distance   the smallest number of mutations required to derive one DNA sequence from another.

mutation event   the actual origin of a mutation in time and space, as opposed to the phenotypic manifestation of such an event, which may be generations later.

mutation frequency   the proportion of mutants in a population.

mutation pressure   the continued production of an allele by mutation.

mutation rate   the number of mutation events per gene per unit time (e.g., per cell generation).

mutator gene   a mutant gene that increases the spontaneous mutation rate of one or more other genes. The *Dotted* (*q.v.*) gene of maize was the first mutator gene to be reported. *See* **Appendix C,** 1938, Rhoades.

mutein   a mutant protein, such as a CRM (*q.v.*).

muton   the smallest unit of DNA in which a change can result in a mutation (a single nucleotide). *See* **Appendix C,** 1955, Benzer.

mutual exclusion   a phenomenon observed among ciliary antigens of certain protozoans in which only one genetic locus for a serotype is active at a given time. For example, in *Paramecium primaurelia* and *Tetrahymena pryriformis,* mutual exclusion of serotypes in heterozygotes occurs with allelic genes as well as with nonallelic genes.

mutualism   a symbiosis in which both species benefit.

mutually exclusive events   a series of alternative events in which only one can occur at a given time.

myc   a gene originally described in the avian MC29 myelocytomatosis virus, an oncovirus of the chicken. A homologous gene is located on the long arm of human chromosome 8. The viral gene is often symbolized *v-myc* and the cellular gene *c-myc* (pronounced "see-mick"). *See* **Burkitt lymphoma, oncogene.**

mycelium   the vegetative portion of a fungus composed of a network of filaments called hyphae.

Mycoplasma   a genus of bacteria that is characterized by the absence of a cell wall. *M. capricolum* is of interest because in this species UGA encodes tryptophan rather than serving as a termination codon. *See* **Appendix A,** Prokaryotae, Aphragmabacteria; **Appendix C,** 1985, Yamao; **pleuropneumonialike organisms, universal code theory.**

Mycostatin   a trade name for nystatin (*q.v.*).

myelin sheath   the insulating covering of an axon formed by the plasma membrane of a Schwann cell.

myeloblasts   cells that differentiate by aggregation to form multinucleated, striated muscle cells.

myeloma   cancer of plasma cells, presumably due to clonal proliferation of a single plasma cell that escapes the normal control of division. Such cells reproduce and secrete a specific homogeneous protein related to γ globulins. *See* **Bence-Jones proteins.**

myeloma protein a partial or complete immunoglobulin molecule secreted by a myeloma (*q.v.*).

Myleran a trade name for busulfan (*q.v.*).

myoglobin the monomeric heme protein of vertebrate muscle. Human myoglobin contains 152 amino acids. The myoglobin cistron is thought to be derived directly from the ancestral cistron that by duplication produced the gene that served as the ancestor of the alpha chain cistron. *See* **Appendix C,** 1960, Kendrew *et al.;* **hemoglobin.**

myoglobin gene the gene that encodes myoglobin. It is remarkable in that less than 5% of its structure codes for message. All the genes of the α- and β-hemoglobin families are made up of three coding regions interrupted by two introns. The myoglobin gene contains four exons and three introns, and each of these introns is much longer than any of those found in hemoglobin genes.

myosin the hexameric protein that interacts with actin (*q.v.*) to convert the energy from the hydrolysis of ATP into the force for muscle contraction. Actin functions both as a structural protein and an enzyme. A myosin molecule can catalyze the hydrolysis of 5 to 10 ATP molecules per second. Each myosin consists of a slender stem (about 135 nm long) and a globular head region (about 10 nm long). The molecule is formed from two identical heavy chains, each possessing about 2,000 amino acids. In the tail region, the heavy chains twist together to form an α helix, from which the two globular heads protrude. The C termini are distal to the heads. Two light-chain proteins, A_1 (190 amino acids) and A_2 (148 amino acids) attach to the globular heads of each heavy chain. The light chain proteins contain calcium-binding sites. The globular head regions contain the ATPase activity and can bind temporarily to actin to form a complex referred to as actomyosin. In avian and mammalian species, numerous isoforms of both myosin heavy and light chains have been isolated from muscle and nonmuscle tissues.

myosin genes the genes encoding the isoforms of the heavy and light myosin chains. In *Drosophila,* two myosin heavy chain genes have been identified: one encoding a muscle myosin (*Mhc*) and one encoding a cytoplasmic myosin (*Mhc-c*). The transcription unit of *Mhc* is 22 kb long and contains 19 different exons. Multiple transcripts are generated by alternative splicing (*q.v.*). Genes for the two light chains are also known. In mammals, the muscle myosin heavy-chain isoforms are encoded by a family containing at least 10 genes.

myria a rarely used prefix meaning ten thousand. Used with metric units of measurement.

myriapod a millipede or centipede.

Mytilus edulis *See* **Pelecypoda.**

myxomatosis a fatal virus disease affecting rabbits. The virus was introduced into wild populations of rabbits in Australia as a means of controlling them.

Myxomycota the phylum containing the plasmodial slime molds. These protoctists generate multinucleate plasmodia that feed by phagocytosis and subsequently form stalked, funguslike fruiting structures. From the standpoint of genetics, *Physarum polycephalum* is the best-known species.

n neutron.

N **1.** the haploid chromosome number. **2.** normal solution. **3.** nitrogen.

N-acetyl serine an acetylated serine thought to function in mammalian systems as N-formylmethionine does in bacterial translation.

$$CH_3 - \overset{\overset{\displaystyle O}{\|}}{C} - \overset{\overset{\displaystyle H}{|}}{N} - \underset{\underset{\displaystyle OH}{|}}{\underset{\underset{\displaystyle CH_2}{|}}{CH}} - COOH$$

NAD nicotinamide-adenine dinucleotide (*q.v.*).

NADP nicotinamide-adenine dinucleotide phosphate (*q.v.*).

Naegleria a genus of soil amoebae capable of transforming into flagellates. Species from this genus are often studied in terms of the morphogenesis of flagella.

nail patella syndrome a hereditary disease in man. Individuals afflicted with this disorder have misshapen fingernails and small kneecaps or lack them. The disease is due to a dominant gene residing on chromosome 9.

nalidixic acid an antibiotic that inhibits DNA replication in growing bacteria. It specifically inhibits the DNA gyrase of *E. coli*.

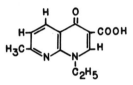

narrow heritability *See* **heritability.**

nascent polypeptide chain the forming polypeptide chain that is attached to the 50 S subunit of a ribosome through a molecule of tRNA. The free end of the nascent polypeptide contains the N-terminal amino acid. *See* **translation.**

nascent RNA an RNA molecule in the process of being synthesized (hence incomplete) or a complete, newly synthesized RNA molecule before any alterations have been made (e.g., prior to nuclear processing, *q.v.*).

Nasonia brevicornis another name for *Mormoniella vitripennis* (*q.v.*).

Nasonia vitripennis another name for *Mormoniella vitripennis* (*q.v.*).

native indigenous. A native species is not introduced into an area by man, either intentionally or accidentally.

natural immunity an outmoded concept that some immunities are inherited in the apparent absence of prior contact with an antigen. The prevailing paradigm is that all immunity ultimately requires contact with a sensitizing antigen and therefore is acquired.

natural killer (NK) cells large leukocytes found in the blood (where they make up about 10 percent of the total lymphocytes) and in spleen and lymph nodes. They are activated by interferon (q.v.), and they attack tumor cells without prior immunization. NK cells are distinct from B lymphocytes and T lymphocytes.

natural selection the differential fecundity (*q.v.*) in nature between members of a species possessing adaptive characters and those without such advantages. *See* **Appendix C,** 1818, Wells; 1858, Darwin and Wallace; 1859, Darwin; 1934, 1937, L'Héritier and Teissier; 1952, Bradshaw; 1954, Allison; **artificial selection, evolution, fundamental theorem of natural selection, industrial melanism.**

Nautilus See **living fossil.**

n_D refractive index.

Neanderthal man a subspecies of *Homo sapiens* that lived in Europe, North Africa, the Near East, Iraq, and Central Asia in the middle and upper Pleistocene. The fossils are named after the valley in western Germany where they were first discovered. *See* **Homo.**

Nearctic one of the six biogeographic realms

(*q.v.*) of the earth, comprising North America, Greenland, and extending to the Mexican plateau.

nebenkern   a two-stranded helical structure surrounding the proximal region of the tail filament of a spermatozoon. The nebenkern is derived from clumped mitochondria.

negative complementation   suppression of the wild-type activity of one subunit of a multimeric protein by a mutant allelic subunit.

negative contrast technique   *See* **negative staining.**

negative eugenics   *See* **eugenics.**

negative gene control   prevention of gene expression by the binding of a specific controlling factor to DNA. For example, in bacterial operons (either inducible or repressible), the binding of a repressor protein to the operator prevents transcription of structural genes in that operon. *See* **regulator gene.** *Compare with* **positive gene control.**

negative interference   a situation in which the coefficient of coincidence is greater than 1. In such cases, the occurrence of one exchange between homologous chromosomes appears to increase the likelihood of another in its vicinity.

negative regulation   *See* **negative gene control.**

negative staining   a staining technique for high-resolution electron microscopy of viruses. A virus suspension is mixed with a phosphotungstic acid solution and poured into an atomizer sprayer. The mixture is then sprayed upon electron microscope grids previously coated with a film of carbon. The phosphotungstic acid enters the contours of the specimen, which is viewed as a light object against a dark background. *See* **Appendix C,** 1959, Brenner and Horne.

negative supercoiling   *See* **supercoiling.**

neobiogenesis   the concept that life has been generated from inorganic material repeatedly in nature.

neo-Darwinism   the post-Darwinian concept that species evolve by the natural selection of adaptive phenotypes caused by mutant genes.

Neogene   a subdivision of the Tertiary period, incorporating the Pliocene and Miocene epochs. *See* **geologic time divisions.**

Neolithic   pertaining to the later Stone Age, during which agriculture and animal husbandry originated and flourished.

neomorph   a mutant gene producing a qualitatively new effect that is not produced by the normal allele.

neomycin   an antibiotic produced by *Streptomyces fradiae.*

neontology   the study of living (extant) species, as opposed to paleontology (the study of extinct species).

neoplasm   a localized population of proliferating cells in an animal that are not governed by the usual limitations of normal growth. The neoplasm is said to be *benign* if it does not undergo metastasis and *malignant* if it undergoes metastasis.

neotenin   synonym for allatum hormone (*q.v.*).

neoteny   the retention of larval characteristics throughout life with reproduction occurring during the larval period. In *Ambystoma mexicanum,* for example, the gill-breathing, water-dwelling larval salamander matures and reproduces sexually without undergoing metamorphosis to a lung-breathing, land-dwelling, adult form.

Neotropical   one of the six biogeographic realms (*q.v.*) of the earth, comprising Central and South America (south of the Mexican plateau) and the West Indies.

neuraminic acid   a nine-carbon amino sugar widely distributed in living organisms. One of the distinctions between eubacteria and archaebacteria is the presence of neuraminic acid in the cell walls of the former and its absence in the latter. In animals, neuraminic acid is found in mucolipids, mucopolysaccharides, and glycoproteins. Neuraminic acid-containing membrane components play a role in the attachment and penetration of virus particles into animal cells. *See* **ganglioside.**

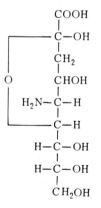

neurofibroma   a fibrous tumor of peripheral nerves.

neurofibromatosis   one of the most common single gene disorders affecting the human nervous system. The disease is characterized by the presence in the skin, or along the course of peripheral nerves, of multiple neurofibromas that gradually increase in number and size. Due to a dominant gene on chromosome 17. Also known as von Recklinghausen disease, its prevalence is about 1/30,000 births.

neurohormone a hormone synthesized and secreted by specialized nerve cells; e.g., gonadotropin-releasing hormone produced by neurosecretory cells located in the hypothalamus.

neurohypophysis the portion of the hypophysis that develops from the floor of the diencephalon.

neurological mutant a mutant producing malformations of the sense organs or the central nervous system or striking abnormalities in locomotion or behavior. Hundreds of neurological mutants have been collected in *Drosophila, Caenorhabditis,* and the mouse. *See* **Appendix C,** 1969, Hotta and Benzer; 1971, Suzuki *et al.;* 1981, Chalfie and Sulston; 1986, Tomlinson and Ready.

neuron a nerve cell.

neuropathy a collective term for a great variety of behavioral disorders that may have hereditary components.

neurosecretory spheres electron-dense spheres 0.1–0.2 μm in diameter, synthesized by and transported in the axoplasm of specialized neurons.

Neurospora crassa the ascomycete fungus upon which many of the classical studies of biochemical genetics were performed. In *Neurospora,* each set of meiotic products is arranged in a linear fashion, and therefore the particular meiotic division at which genetic exchange occurs can be determined by dissecting open the ascus and growing the individual ascospores (*see* **ordered tetrad**). The haploid chromosome number of this species is 7, and seven detailed linkage maps are available. Restriction maps also exist for the *Neurospora* mitochondrial DNA molecule. It contains about 60 kb and several of its genes have been mapped. *See* **Appendix C,** 1927, Dodge; 1941, Beadle and Tatum; 1944, Tatum *et al.;* 1948, Mitchell and Lein.

neurula the stage of development of a vertebrate embryo at which the neural axis is fully formed and histogenesis is proceeding rapidly.

neutral equilibrium *See* **passive equilibrium.**

neutral mutation **1.** a genetic alteration whose phenotypic expression results in no change in the organism's adaptive value or fitness for present environmental conditons. **2.** a mutation that has no measurable phenotypic effect as far as the study in question is concerned.

neutral mutation–random drift theory of molecular evolution a theory according to which the majority of the nucleotide substitutions in the course of evolution are the result of the random fixation of neutral or nearly neutral mutations, rather than the result of positive Darwinian selection. Many protein polymorphisms are selectively neutral and are maintained in a population by the balance between mutational input and random extinction. Neutral mutations are not functionless; they are simply equally effective to the ancestral alleles in promoting the survival and reproduction of the organisms that carry them. However, such neutral mutations can spread in a population purely by chance because only a relatively small number of gametes are "sampled" from the vast supply produced in each generation and therefore are represented in the individuals of the next generation. *See* **Appendix C,** 1968, Kimura.

neutron an elementary nuclear particle with a mass approximately the same as that of a hydrogen atom and electrically neutral; its mass is 1.0087 mass units.

neutron contrast matching technique a technique that involves determining the neutron scattering densities of particles irradiated in solutions containing various concentrations of light and heavy water. This technique was used on nucleosomes (*q.v.*), and it was found that under conditions where neutron scattering from DNA dominated the reaction, the radius of gyration was 50Å. When scattering from the histone proteins was dominant, the radius was 30Å. The larger radius for DNA proved that it was located on the surface of the nucleosome. *See* **Appendix C,** 1977, Pardon *et al.*

N-formylmethionine a modified methionine molecule that has a formyl group attached to its terminal amino group. Such an amino acid is "blocked" in the sense that the absence of a free amino group prevents the amino acid from being inserted into a growing polypeptide chain. N-formylmethionine is the starting amino acid in the synthesis of all bacterial polypeptides. *See* **Appendix C,** 1966, Adams and Cappecchi; **start codon.**

$$H-\overset{\overset{\displaystyle O}{\|}}{C}-\overset{\overset{\displaystyle H}{|}}{N}-CH-COOH$$
$$CH_2-CH_2-CH_2-S-CH_3$$

niacin an early name for nicotinic acid.

niche *See* **ecological niche.**

niche preclusion *See* **first-arriver principle.**

nick in nucleic acid chemistry, the absence of a phosphodiester bond between adjacent nucleotides in one strand of duplex DNA. *Compare with* **cut.**

nickase an enzyme that causes single-stranded breaks in duplex DNA, allowing it to unwind.

nick-closing enzyme *See* **topoisomerase.**

nick translation an *in vitro* procedure used to radioactively label a DNA of interest uniformly to a high specific activity. First, nicks are introduced into the unlabeled DNA by an endonuclease, generating 3′ hydroxyl termini. *E. coli* DNA polymerase I is then used to add radioactive residues to the 3′ hydroxy terminus of the nick, with concomitant removal of the nucleotides from the 5′ side. The result is an identical DNA molecule with the nick displaced further along the duplex. *See* **strand-specific hybridization probes.**

Nicotiana a genus containing about 60 species, many of which have been intensively studied genetically. Much interest has been generated from the finding that tumors arise spontaneously at high frequency in certain interspecific hybrids, such as those plants produced by the cross *N. langsdorffii* × *N. glauca.* The species of greatest commercial importance is *N. tabacum,* the source of tobacco. *See* **Appendix C,** 1761, Kölreuter; 1925, 1926, Goodspeed, Clausen.

nicotinamide-adenine dinucleotide a coenzyme (formerly called DPN or coenzyme 1) functioning as an electron carrier in many enzymatic oxidation-reduction reactions.

nicotinamide-adenine-dinucleotide phosphate an electron carrier formerly called TPN or coenzyme 2. The oxidized form is symbolized NAD^+, the reduced form by NADPH. *See* **citric acid cycle, cytochrome system.**

nicotine a poisonous, volatile alkaloid present in the leaves of *Nicotiana tabacum* and responsible for many of the effects of tobacco smoking. It functions in the plant as a potent insecticide.

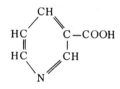

nicotinic acid one of the B vitamins. Also called niacin in the older literature.

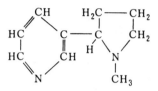

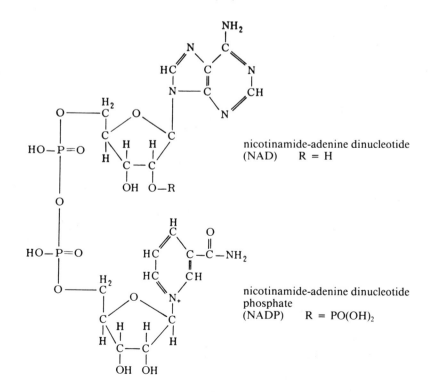

nicotinamide-adenine dinucleotide
(NAD)   R = H

nicotinamide-adenine dinucleotide
phosphate
(NADP)   R = PO(OH)₂

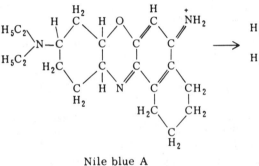

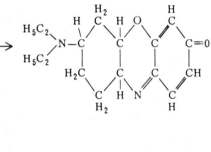

Nile blue A   Nile red

Niemann-Pick disease a group of human disorders characterized by enlargement of the spleen and liver and by the accumulation of sphingomyelin (*q.v.*) and other lipids throughout the body. The disease is due to a defective lysosomal sphingomyelinase coded for by an autosomal recessive gene. Amniocentesis and testing of fetal cells for sphingomyelinase activity permits monitoring of pregnancies at risk. Heterozygotes can be identified, since their leukocytes contain about 60% the normal activity of sphingomyelinase.

nif genes genes that enable the bacteria containing them to fix atmospheric nitrogen. Such genes are generally carried by the plasmids of nodulating bacteria, and they encode the enzyme nitrogenase. *See* **Rhizobium.**

nigericin *See* **ionophore.**

Nile blue a mixture of two dyes: Nile blue A, a water-soluble basic dye; and Nile red, a lysochrome formed by spontaneous oxidation of Nile blue A (an example of allochromacy, *q.v.*).

ninhydrin an organic reagent that reacts with and colors amino acids. Ninhydrin solutions are sprayed on chromatographs, and the separated amio acids and polypeptides are then rendered visible as ninhydrin-positive spots.

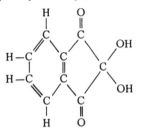

nitrocellulose filter a very thin filter composed of nitrocellulose fibers that selectively bind single-stranded DNA strongly, but not double-stranded DNA or RNA. The ssDNA binds along its sugar-phosphate backbone, leaving its bases free to pair with complementary bases contained in labeled ssDNA or RNA probes. *See* **DNA hybridization.**

nitrogen the fourth most abundant of the biologically important elements. Atomic number 7; atomic weight 14.0067; valence 3^-, 5^+; most abundant isotope ^{14}N; heavy isotope ^{15}N.

nitrogen fixation the enzymatic incorporation of nitrogen from the atmosphere into organic compounds.

nitrogen mustard di(2-chloroethyl) methylamine; an alkylating agent (*q.v.*) that is a potent mutagen and chromosome-breaking agent. *See* **sulfur mustard.**

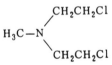

nitrogenous base a purine or pyrimidine; more generally an aromatic, nitrogen-containing molecule that has basic properties (is a proton acceptor).

nitrous acid HNO_2, a mutagen that converts the NH_2 groups of the purines and pyrimidines to OH groups.

NK cells *See* **natural killer (NK) cells.**

node 1. in vascular plants, a slightly enlarged portion of a stem where leaves and buds arise and where branches originate. 2. in a circular DNA superhelix, the point of contact in a figure-of-eight; if the left strand in the upper part of 8 is closest to the viewer at the node, it is called a positive node; if

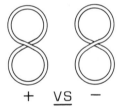

+ VS −

212

the left strand in the upper part of 8 is in back of the other strand at the node, it is called a negative node.

noise in colloquial usage, variation in an experiment attributed to uncontrolled effects; usually associated with a variance component called experimental error.

Nomarski differential interference microscope an optical system that, like the phase contrast microscope, permits the visualization of transparent structures in a living cell. However, in the Nomarski system the field is quite shallow so that there is freedom from phase disturbances from structures above and below the plane of focus. The observation method is comparable to that with extreme oblique illumination, and the specimen therefore appears in relief.

nomenclature the naming of taxa in accordance with certain international rules.

nonautogenous anautogenous (*q.v.*).

nonbasic chromosomal proteins acidic or neutral proteins (therefore not histones) associated with chromosomes: e.g., certain enzymes such as DNA polymerases.

non-Darwinian evolution genetic changes in populations produced by forces other than natural selection; a term usually associated with the neutralist view of evolution. *See* **neutral mutation–random drift theory of molecular evolution.**

nondisjunction the failure of homologous chromosomes (in meiosis I, primary nondisjunction) or sister chromatids (in meiosis II, secondary nondisjunction; or mitosis) to separate properly and to move to opposite poles. Nondisjunction results in one daughter cell receiving both and the other daughter cell none of the chromosomes in question. *See* **Appendix C**, 1914, Bridges.

nonessential amino acids *See* **essential amino acids.**

nonhomologous chromosomes chromosomes that do not synapse during meiosis.

noninducible enzyme constitutive enzyme (*q.v.*).

nonlinear tetrad a group of four meiotic products that are randomly arranged in the ascus. *See* **linear tetrad.**

non-Mendelian ratio in the progeny of a cross, unusual phenotypic ratios that fail to follow Mendel's laws, suggesting that gene conversion or another aberrant mechanism is responsible.

nonparametric statistics *See* **statistics.**

nonparental ditype *See* **tetrad segregation types.**

nonpermissive cells *See* **permissive cells.**

nonpermissive conditions environmental settings in which conditional lethal mutants fail to survive.

nonpolar referring to water-insoluble chemical groups, such as the hydrophobic side chains of amino acids.

nonrandom mating *See* **assortative mating, inbreeding, outbreeding.**

nonreciprocal recombination *See* **unequal crossing over.**

nonreciprocal translocation *See* **translocation.**

nonrecurrent parent the parent of a hybrid that is not again used as a parent in backcrossing.

nonrepetitive DNA segments of DNA exhibiting the reassociation kinetics expected of unique sequences; single sequence DNA.

nonselective medium a growth medium that allows growth of all genotypes present in a recombination or mutation experiment. *Compare with* **selective medium.**

nonsense codon synonymous with stop codon (*q.v.*).

nonsense mutation a mutation that converts a sense codon to a chain-terminating codon or vice versa. The results following translation are abnormally short or long polypeptides, generally with altered functional properties. *Contrast with* **missense mutation.**

nonsense suppressor a gene coding for a tRNA that is mutant in its anticodon and therefore able to recognize a nonsense (stop) codon; nonsense suppressors cause extension of polypeptide chain synthesis through stop codons. *See* **Appendix C**, 1969, Abelson *et al.*; **amber suppressor, ochre suppressor, readthrough.**

nonspherocytic hemolytic anemia *See* **glucose-6-phosphate dehydrogenase deficiency.**

nopaline *See* **opine.**

NOR nucleolar organizer region.

noradrenaline norepinephrine.

norepinephrine a hormone of the adrenal medulla

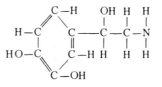

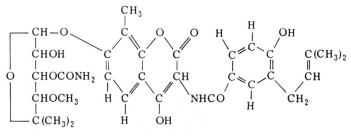

Novobiocin.

that causes vasoconstriction and raises the blood pressure.

***n* orientation**   one of two possible orientations for inserting a target DNA fragment into a vector; in the *n* orientation, the genetic map of both target and vector have the same orientation; in the *u* orientation, the target and the vector are in different orientations.

normal distribution   the most commonly used probability distribution in statistics. The formula for the normal curve is $Y = \dfrac{1}{\sqrt{2\pi}\,\sigma} e^{\frac{-(X-\mu)^2}{2\sigma^2}}$ where μ = the mean, σ = the standard deviation, e = the base of natural logarithms, π = 3.1416, and Y = the height of the ordinate for a given value of X. The graph of this formula, the normal curve, also called Laplacian or Gaussian, is bell shaped. The value of m locates the curve along the abscissa and that of σ determines its shape. The larger the standard deviation, the broader the curve. In nature, a vast number of continuous distributions are normally distributed.

normalizing selection   the removal of those alleles that produce deviations from the average population phenotype by selection against all deviant individuals. Such selection will reduce the variance in subsequent generations. Also called *stabilizing selection, centripetal selection.*

normal solution   one containing a gram equivalent weight of solute dissolved in sufficient water to make a liter of solution.

norm of reaction   the phenotypic variability produced by a given genotype under the range of environmental conditions common to the natural habitat of the species or under the standard culture or experimental conditons. *See* **adaptive norm.**

northern blotting   *See* **Southern blotting.**

Nosema   See **microsporidia.**

Notch   a series of overlapping deficiencies of the X chromosomes of *Drosophila melanogaster.* All deficiencies lack the 3C7 band, and females hetero-

zygous for the deficiency show distal notches of the wing. Hemizygous males die as embryos. *See* **Appendix C**, 1938, Slizynska.

Notophthalmus viridescens   the common spotted newt of the eastern United States. The amplification of rDNA has been extensively studied using the oocytes of this species. *See* **lampbrush chromosomes,** *Triturus.*

novobiocin   an antibiotic produced by *Streptomyces niveus.*

np   nucleotide pair.

nRNA   nuclear RNA.

nt   abbreviation for "nucleotide." *Compare with* **bp.**

N-terminal end   proteins are conventionally written with the amino (NH_2) end to the left. The assembly of amino acids into a polypeptide starts at the N-terminal end. *See* **translation.**

N-terminus   N-terminal end.

nu (*v*) body   particles arranged like beads on a string along interphase chromosomes. These are most clearly seen in electron micrographs of negatively stained Miller spreads (*q.v.*). Nu bodies correspond to the nucleosomes (*q.v.*) of the biochemist.

nuclear duplication   mitosis (*q.v.*).

nuclear emulsion   a photographic emulsion especially compounded to make visible the individual tracks of ionizing particles.

nuclear envelope   an envelope surrounding the nucleus, composed of two membranes enclosing a perinuclear cisterna. The outermost membrane is studded with ribosomes. The perinuclear cisterna is traversed by ring-shaped pores.

nuclear family   a pair of parents and their children.

nuclear fission   a transformation of atomic nuclei characterized by the splitting of a nucleus into at least two other nuclei and the release of amounts of energy far greater than those generated by conventional chemical reactions.

nuclear fusion the coalescence of two or more atomic nuclei with the release of relatively vast amounts of energy.

nuclear processing of RNA a primary transcript (*q.v.*) experiences excision of introns and splicing of exons. If it is to serve as an mRNA, a methylated cap (*q.v.*) is added to the 5′ end and a poly-A tail is usually added to the 3′ end prior to transport from the nucleus to the cytoplasm. *See* **posttranscriptional modification.**

nuclear reactor the apparatus in which nuclear fission may be sustained in a self-supporting chain reaction. A source of energy and radioisotopes.

nuclear RNA RNA molecules found in the nucleus either associated with chromosomes or in the nucleoplasm. *See* **chromosomal RNA, heterogeneous nuclear RNA.**

nuclear transfer the injection of a diploid somatic nucleus into an enucleated egg. The nature of the ensuing development reveals the developmental potentialities of the implanted nucleus. Various amphibian species are generally used (*Rana pipiens, Xenopus laevis*). *See* **Appendix C,** 1952, Briggs and King; 1962, 1967, Gurdon.

nuclease any enzyme that breaks down nucleic acids.

nucleic acid a nucleotide polymer. *See* **deoxyribonucleic acid, ribonucleic acid.**

nucleic acid bases *See* **bases of nucleic acids.**

nuclein the acidic, phosphorus-rich substance isolated from human white blood cells by Miescher. We now know that nuclein was a mixture of nucleic acids and proteins. *See* **Appendix C,** 1871, Miescher.

nucleocapsid a virus nucleic acid and its surrounding capsid. *See* **capsomere.**

nucleo-cytoplasmic ratio the ratio of the volume of nucleus to the volume of cytoplasm.

nucleoid **1.** a DNA-containing region within a prokaryote, mitochondrion, or chloroplast. **2.** in an RNA tumor virus, the core of genetic RNA surrounded by an icosahedral protein capsid.

nucleolus an RNA-rich, spherical body associated with a specific chromosomal segment, the nucleolus organizer. In maize, the nucleolus organizer resides on chromosome 6. The illustration shows this chromosome and its nucleolus as they appear in meiotic prophase. The nucleolus organizer contains the ribosomal RNA genes (*q.v.*), and the nucleolus is composed of the primary products of these genes, their associated proteins, and a variety of enzymes (e.g., RNA polymerase, RNA methylases, and RNA endonucleases). Under the electron microscope, the nucleolus is seen to consist of a fibrous central core (pars amorpha) and a granular cortex (pars granulosa). *See* **Appendix C,** 1838, Schleiden; 1934, McClintock; 1965, Ritossa and Spiegelman; **Miller trees, preribosomal RNA, rDNA amplification.**

nucleolus organizer *See* **nucleolus.**

nucleolus organizer region (NOR) *See* **nucleolus.**

nucleon a constituent particle of an atomic nucleus.

nucleoplasm the protoplasmic fluid contained in the nucleus.

nucleoprotein a compound of nucleic acid and protein. Either one of two main classes of basic proteins are found combined with DNA: one of low molecular weight (protamine) and one of high molecular weight (histone). The basic amino acids of these proteins neutralize the phosphoric acid residues of the DNA. *See* **Appendix C,** 1866, Miescher.

nucleosidase any enzyme that catalyzes the splitting of nucleosides into bases and pentoses.

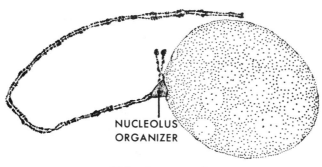

NUCLEOLUS
ORGANIZER

Maize chromosome 6.

nucleoside a purine or pyrimidine base attached to ribose or deoxyribose. The nucleosides commonly found in DNA or RNA are cytidine, cytosine deoxyriboside, thymidine, uridine, adenosine, adenine deoxyriboside, guanosine, and guanine deoxyriboside. Note that thymidine is a deoxyriboside and cytidine, uridine, adenosine, and guanosine are ribosides. *See* **rare bases, inosine.**

nucleosome a beadlike structure of eukaryotic chromosomes, consisting of a core of eight histone molecules (two each of proteins H2A, H2B, H3, and H4) wrapped by a DNA segment about 150 base pairs in length and separated from adjacent nucleosomes by a "linker" DNA sequence of about 50 bp). *See* **Appendix C,** 1974, Kornberg; 1977, Pardon *et al.,* Leffak *et al.;* **chromatosome, histones.**

nucleotide one of the monomeric units from which DNA or RNA polymers are constructed, consisting of a purine or pyrimidine base, a pentose, and a phosphoric acid group. The nucleotides of DNA are deoxyadenylic acid, thymidylic acid, deoxyguanilic acid, and deoxycytidylic acid. The corresponding nucleotides of RNA are adenylic acid, uridylic acid, guanylic acid, and cytidylic acid.

nucleotide pair a hydrogen-bonded pair of purine-pyrimidine nucleotide bases on opposite strands of a double-helical DNA molecule. Normally, adenine pairs with thymine and guanine pairs with cytosine; also called *complementary base pairs. See* **deoxyribonucleic acid.**

nucleotide pair substitution the replacement of a given nucleotide pair by a different pair, usually through a transition or a transversion *(both of which see).*

nucleus the spheroidal, membrane-bounded structure present in all eukaryotic cells which contains DNA, usually in the form of chromatin. *See* **Appendix C,** 1831, Brown; **nuclear envelope.**

nuclide a species of atom characterized by the constitution of its nucleus. This is specified by the number of protons and neutrons it contains.

nude mouse a laboratory mouse homozygous for the recessive mutation *nu,* which maps to chromosome 11. Such mice are characterized by the complete absence of hair and thymus glands. Nude mice lack T lymphocytes (*q.v.*), but have natural killer cells (*q.v.*) and B lymphocytes (*q.v.*), and they are unable to reject homografts. The nude mouse serves as a model system for the study of the immunological effects of thymus deprivation. *See* **rejection.**

null allele an allele that produces no functional product and therefore usually behaves as a genetic recessive. For example, in the human ABO blood group system, the recessive allele *(i)* produces no

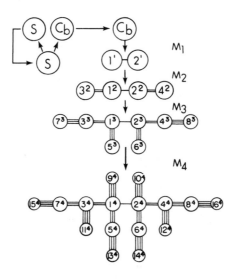

Nurse cells.

A diagram of the steps in the production of a clone of 16 cystocytes. The stem cell (S) divides into 2 daughters: one behaves like its parents; the other differentiates into a cystoblast (Cb). This cell, by a series of 4 mitoses ($M_1 - M_4$), each followed by incomplete cytokinesis, produces a branching chain of 16 interconnected cells. Cystocytes (represented by open circles) are given superscript designations (1,2,3,4) depending on whether they belong to the 1st, 2nd, 3rd, or 4th generation. The area in each circle is proportional to the volume of the cell. The number of lines connecting any two cells shows the division at which the ring canal joining them was formed.

detectable antigen, either in homozygous condition (blood group O) or in heterozygous condition with allele I^A (blood group A) or with allele I^B (blood group B).

null hypothesis method   the standard hypothesis used in testing the statistical significance of the difference between the means of samples drawn from two populations. The null hypothesis states that there is no difference between the populations from which the samples are drawn. One then determines the probability that one will find a difference equal to or greater than the one actually observed. If this probability is 0.05 or less, the null hypothesis is rejected, and the difference is said to be significant.

nulliplex   *See* **autotetraploidy.**

nullosomic   lacking both members of a pair of chromosomes.

number of degrees of freedom   *See* **degrees of freedom.**

numerical taxonomy   a system of classification using a multitude of characteristics to determine overall phenotypic similarity, each trait being given equal weight and without regard to phylogenetic relationships; also known as *phenetic taxonomy.*

nu particles   *See* **nucleosomes.**

nurse cells   cells in the insect ovary that function to nourish the oocyte. In insects with polytrophic meroistic ovaries, the oocyte and its accompanying nurse cells are descendants of a single cell, the germinal cystoblast. In *Drosophila melanogaster* (see p. 216), this cell divides four times to give a clone of 16 interconnected cells. One (either 1^4 or 2^4) becomes the oocyte, and the others, nurse cells. The nuclei of the 15 nurse cells undergo endomitotic cycles of DNA replication and are active in transcribing a variety of RNA molecules that enter the cytoplasm. Nurse cells eventually pump all their cytoplasm into the oocyte. *See* **insect ovary types.**

nutritional mutant   a mutation converting a prototroph into an auxotroph.

nutritive chord   *See* **insect ovary types.**

nystagmus   a jerky twitching of the eye. *See* **albinism.**

nystatin   a polyene antibiotic (*q.v.*).

O

O 1. Ordovician. 2. oxygen.

O antigens polysaccharide antigens of the cell walls of enterobacteria such as *Escherichia* or *Salmonella;* in contrast to the polysaccharide K antigens of bacterial capsules or the protein H antigens of bacterial flagella.

oat *See Avena.*

obligate restricted to a specified condition of life. For example, an obligate parasite cannot live in the absence of its host. *See* **facultative.**

Occam's razor a rule attributed to the medieval philosopher William of Occam. It states that when there are several possible explanations of a phenomenon, one selects as most probable the explanation that is the simplest and most consistent with the data at hand. Also called the **parsimony principle.**

oceanic island an island that has risen from the sea. *See* **continental island.**

ocellus one of the simple eyes located near the compound eyes of an insect; an eyespot in many invertebrates.

ochre codon a triplet of mRNA nucleotides (UAA) usually not recognized by any tRNA molecules; one of three stop codons that normally signal termination of translation. *See* **amber codon, opal codon.**

ochre mutation one of a group of mutations resulting in abnormally short polypeptide chains. Because of a base substitution, a codon specifying an amino acid is converted to UAA, which signals chain termination. UAA appears to be the codon primarily used for chain termination in *Escherichia coli. See* **amber mutation, nonsense mutation.**

ochre suppressor any mutant gene coding for a mutant tRNA whose anticodon can respond to the UAA stop codon by the insertion of an amino acid. *See* **amber suppressor.**

octad a fungal ascus containing eight linear ascospores; produced in some ascomycete species when the tetrad of meiospores undergoes a mitotic division following meiosis. *See* **ordered tetrad.**

octopine *See* **opine.**

ocular albinism a hereditary eye disease of humans occurring in two forms, one inherited as an autosomal recessive and the other as an X-linked recessive. In females heterozygous for X-linked ocular albinism, the retinas show a mosaic pattern of pigment distribution due to the random inactivation of X chromosomes at an early developmental stage. In males, the prevalence of X-linked ocular albinism is 1/50,000. *See* **dosage compensation.**

OD optical density. *See* **Beer-Lambert law.**

OD_{260} unit one absorbance (OD_{260}) unit is that amount of material per ml of solution that produces an absorbance of 1 in a 1.0-cm light path at a wavelength of 260 nm. *See* **absorbance.**

Oenothera lamarckiana the evening primrose. During meiosis, plants of this and related species, such as *O. grandiflora,* have their chromosomes arranged in rings rather than pairs. The evolution of this atypical cytogenetic behavior, the result of the accumulation of reciprocal translocations, has been extensively studied. *See* **Appendix C,** 1901, de Vries; 1930, Cleland and Blakeslee; **Renner complex.**

Ohno's hypothesis the proposal advanced by S. Ohno that the unique regulatory features of the X chromosomes dictate the evolutionary conservation of the primordial X-linkage group among mammals. Thus, if any gene locus is found to be X-linked in humans, it is likely to be X-linked in all other mammalian species.

oil-immersion objective the objective lens system used for highest resolution with the light microscope. The space between the coverslip over the object to be examined and the lens is filled with a drop of oil of the same refractive index as the glass.

Okazaki fragments *See* **replication of DNA.**

Oligocene an epoch of the early Tertiary period, after the Eocene and before the Miocene. The first anthropoids appeared during this epoch. *See* **geologic time divisions.**

oligo dA (oligo dT) a homopolymer chain of

deoxyriboadenylate (or deoxyribothymidylate) subunits of unspecified length, but generally 100–400 residues.

oligogene   a gene producing a pronounced phenotypic effect as opposed to a polygene (*q.v.*), which has an individually small effect.

oligomer   a molecule made up of a relatively few monomeric subunits.

oligomycin   a polyene antibiotic (*q.v.*).

oligonucleotide   a linear sequence of up to 20 nucleotides joined by phosphodiester bonds. *See* **polynucleotide.**

oligonucleotide-directed mutagenesis   a technique that allows a specific mutation to be inserted in a gene at a selected site. An olignucleotide sequence complementary to the segment of interest, but containing an alteration at a selected site, is chemically synthesized. Next this is hybridized to a complementary wild-type target gene contained in a single-stranded phage such as M13. The hybridized oligonucleotide fragment is then used as a primer by DNA polymerase I, which extends the molecule while taking instructions from the wild-type complementary strand. The result is a double helix containing a mutant and a wild-type strand. The heteroduplex is then used to transform bacterial cells. From these colonies, strains that contain the mutant homoduplexes can be recovered and propagated. This procedure is also called *site-specified mutagenesis*. *See* **Appendix C,** 1978, Hutchison *et al.*

oligopyrene sperm   *See* **sperm polymorphism.**

oligosaccharide   a polymer made up of a few (2–10) monosaccharide units.

oligospermia   an abnormally low concentration of sperm in the semen.

ommatidium   a facet of the insect compound eye. *See* **sevenless.**

ommochromes   *See* ***Drososphila* eye pigments.**

omnipotent suppressors   nonsense suppressors in yeast that are codon nonspecific, act only upon UAA and UAG mutations, and fall into two complementation groups. They are thought to be mutations of ribosomal components rather than suppressor mutations in tRNAs since these are codon specific.

oncogene   a gene that induces uncontrolled cell proliferation. Some oncogenes were originally of cellular origin but now reside in the genomes of retroviruses (*q.v.*). Here they have acquired the ability to transform cells to a neoplastic state. The *v-src*

gene of the Rous sarcoma virus (*q.v.*) and the *v-sis* gene of the simian sarcoma virus (*q.v.*) are examples. Oncogenes also have been isolated from tumors that have arisen spontaneously or have been induced by chemical carcinogens. Finally, there are oncogenes that reside in oncogenic viruses with DNA genomes. The polyoma virus (*q.v.*) and simian virus 40 (*q.v.*) are examples. Viral and cellular oncogenes arise from cellular proto-oncogenes (*q.v.*), which play a role in the control of normal cell proliferation. *See* **Appendix C,** 1981, Parker *et al.*; 1982, Reddy *et al.*; **myc, oncogenic virus, oncomouse, T24 oncogene.**

oncogene hypothesis   a proposal that carcinogens of many sorts act by inducing the expression of retrovirus genes already resident in the target cell. It is now known that while cells from different species harbor genes homologous to retrovirus oncogenes, the cellular genes were the progenitors of the viral oncogenes. The cellular genes are now called *proto-oncogenes* (*q.v.*), and they evidently function in the normal physiology of cells from evolutionarily diverse species. *See* **Appendix C,** 1969, Huebner and Todaro.

oncogenic virus   a virus that can transform the cells it infects so that they proliferate in an uncontrolled fashion. *See* **Appendix C,** 1910, Rous; 1981, Parker *et al.*; 1983, Doolittle *et al.*; **Maloney leukemia virus, mammary tumor agent, polyoma virus, Rauscher leukemia virus, Rous sarcoma virus, retrovirus, Shope papilloma virus, simian sarcoma virus, simian virus 40, transformation.**

oncolytic   capable of destroying cancer cells.

oncomouse   a laboratory mouse carrying activated human cancer genes. De Pont started selling oncomice late in 1988. They were the first transgenic animals to be patented. These mice carry the *ras* oncogene plus a mouse mammary tumor virus promoter. This ensures that the oncogene is activated in breast tissue, and the mice develop breast cancer a few months after birth.

oncornavirus   an acronym for *onco*genic *RNA virus*. *See* **retrovirus.**

one gene–one enzyme hypothesis   the hypothesis that a large class of genes exists in which each gene controls the synthesis or activity of but a single enzyme. *See* **Appendix C,** 1941, Beadle and Tatum; 1948, Mitchell and Lein.

one gene–one polypeptide hypothesis   the hypothesis that a large class of genes exists in which each gene controls the synthesis of a single polypeptide. The polypeptide may function independently or as a subunit of a more complex protein. This hypothesis replaced the earlier one gene–one enzyme hypothesis once heteropolymeric enzymes were discovered. For example, hexosaminidase (*q.v.*) is encoded by *two* genes. *See* **two genes–one polypeptide chain.**

one-step growth experiment an experiment in which phage were absorbed on *E. coli*, incubated, and the concentration of free phage particles determined at periodic intervals. The phage concentration remained constant for a time, then rose sharply, and then plateaued to give a "one step" growth curve. Therefore after infection there is a latent period, followed by the rupture of the host cell and release of the progeny phage. *See* **Appendix C,** 1939, Ellis and Delbrück.

ONPG *o*-nitrophenyl galactoside, an unnatural substrate for beta-galactosidase. It is cleaved by this enzyme (see diagram below) into galactose and *o*-nitrophenol (a yellow compound, easily assayed spectrophotometrically). ONPG has been extensively used to determine enzyme activity associated with mutants of the *lac* operon (*q.v.*) in *E. coli*. Unlike IPTG (*q.v.*), ONPG is not an inducer of the operon, so these two substances are often used in combination.

ontogeny the development of the individual from fertilization to maturity.

oocyte the cell that upon undergoing meiosis forms the ovum.

oogenesis the formation of eggs, including the meiotic behavior of oocytes, vitellogenesis, and the formation of egg membranes.

oogonium 1. the female gametangium of algae and fungi. *Contrast with* **antheridium. 2.** in animals, a mitotically active germ cell that serves as a source of oocytes. The stem cell shown on page 216 is an oogonium.

oolemma the plasma membrane of the ovum.

ooplasm the cytoplasm of an oocyte.

ootid nucleus one of the four haploid nuclei

formed by the meiotic divisions of a primary oocyte. Three of the nuclei are discarded as polar nuclei and the remaining functions as the female pronucleus. *See* **oriented meiotic division, polar body.**

opal codon the mRNA stop codon UGA. *See* **amber codon, ochre codon.**

opaque-2 a mutant strain of corn that produces proteins rich in lysine. Mutants of this kind are of potential use in combating kwashiorkor (*q.v.*). *See* **Appendix C,** 1964, Mertz *et al.*

open population a population that is freely exposed to gene flow (*q.v.*).

open reading frame *See* **reading frame.**

operational definition a definition in terms of properties significant to a given experimental situation, without consideration of the more fundamental characteristics of the defined subject.

operator a chromosomal region capable of interacting with a specific repressor, thereby controlling the functioning of adjacent cistrons. *See* **regulator gene.**

operon a unit consisting of one or more cistrons that function coordinately under the control of an operator gene. *See* **Appendix C,** 1961, Jacob and Monod; *lac* **operon.**

operon network a collection of operons and their associated regulator genes that interact in the sense that the products of structural genes in one operon serve to activate or suppress another operon by acting as repressors or effectors.

opine a compound, specifically synthesized by crown gall plant cells, that can be used by agrobacteria as specific growth substances. Examples are nopaline [N-α-(1,3-dicarboxylpropyl)-L-arginine] and octopine [N-α-(D-1-carboxyethyl)-L-arginine]. *See* **Agrobacterium tumefaciens.**

opisthe the posterior daughter organism produced in a transverse division of a protozoan.

opportunism a theory that (1) all potential modes

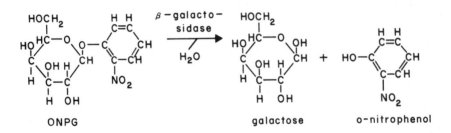

ONPG   galactose   o–nitrophenol

of existence will eventually be tried by some group and all potential niches will eventually become occupied, and (2) organisms evolve only as historical conditions permit and not according to what would theoretically be best.

opportunistic species   a species specialized to exploit newly opened habitats because of its ability to disperse for long distances and to reproduce rapidly.

opsin   a species-specific protein that combines with retinene to form rhodopsin (*q.v.*).

opsonin   any substance that promotes cellular phagocytosis. When antibodies bind to antigens by their Fab portions (*see* **immunoglobulin**), the shape of the molecule changes to expose the Fc region. Scavenger cells such as macrophages have Fc receptors on their surfaces. Thus, phagocytic cells can bind to and engulf antigen-antibody complexes. Neutrophils and macrophages have receptors for certain activated complement components. Thus, antigen-antibody-complement complexes also enhance phagocytosis through immune adherence. IgG antibodies are much more effective opsonins than IgM in the absence of complement, but IgM antibodies are more effective opsonins in the presence of complement.

optical antipodes   enantiomers (*q.v.*).

optical density   *See* **Beer-Lambert law.**

optical isomers   molecular isomers that in solution cause the rotation of the plane of a beam of plane-polarized light passed through the solution. The rotation is due to the asymmetry of the molecule. Molecules with this property are given the prefix *d* or *l* depending on whether the plane is rotated to the right (dextro) or to the left (levo).

orange G   an acidic dye often used in cytochemistry.

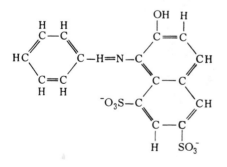

orangutan   *See* **Pongo pygmaeus.**

orcein   a dye used in cytology. *See* **aceto-orcein.**

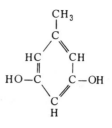

ordered octad   *See* **ordered tetrad.**

ordered tetrad   a linear sequence of four haploid meiotic cells (or pairs of each of four haploid cells produced by a postmeiotic division) within a fungal ascus. This physical arrangement allows identification of chromatids participating in crossover events. Drawing A (p. 222) illustrates that, in a tetrad heterozygous for alleles controlling ascospore pigmentation, single crossovers between these genes and the centromere will generate spores showing 2-2-2-2 and 2-4-2 segregation patterns. Drawing B illustrates that in *Neurospora* such patterns are observed, together with noncrossover asci showing 4-4 distributions.

Ordovician   a period in the Paleozoic era during which the marine invertebrates and especially the echinoderms diversified. Jawless fishes first appeared, and a mass extinction occurred at the end of the period. *See* **geologic time divisions.**

ORF   the symbol for *open reading frame. See* **reading frame.**

organ culture   the maintenance or growth of organ primordia or the whole or parts of an organ *in vitro* in a way that may allow further differentiation or the preservation of architecture or function or both. *See also* **in vivo** **culturing of imaginal discs.**

organelle   a complex intracellular structure of characteristic morphology and function, such as a mitochrondrion or plastid.

organic   **1.** pertaining to organisms (dead or alive) or to the chemicals made by them. **2.** chemical compounds based on carbon chains or rings. They may also contain oxygen, hydrogen, nitrogen, and various other elements.

organizer   a living part of an embryo that exerts a morphogenetic stimulus upon another part, bringing about its determination and morphological differentiation. *See* **Appendix C,** 1918, Spemann and Mangold.

organogenesis   the formation of organs.

A

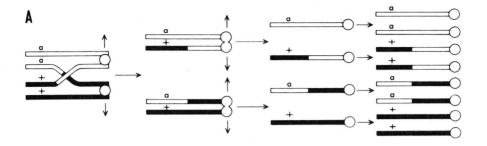

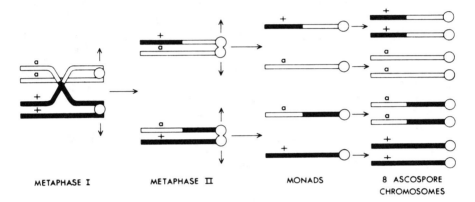

| METAPHASE I | METAPHASE II | MONADS | 8 ASCOSPORE CHROMOSOMES |

B

Ordered tetrad.

222

Oriental designating one of the six biogeographic realms (*q.v.*) of the globe, including the southern coast of Asia east of the Persian Gulf, the peninsula of India south of the Himalayas, eastern India, south China, Sumatra, Kalimantaro (Borneo), Java, Sulawesi (Celebes), and the Philippines.

oriented meiotic division an oocyte meiotic division, as in *Drosophila,* where the spindles are oriented in single file with their long axes perpendicular to the egg surface. The nucleus farthest from the surface functions as the oocyte pronucleus. Aberrant chromosomes that are differentially distributed to the other nuclei are eliminated.

Origin of Species an abbreviated name for the most famous book by Charles Darwin that documented the phenomenon of evolution and elaborated a theory to explain its mechanism. The full title of the book was *On the Origin of Species by means of Natural Selection, or the Preservation of Favored Races in the Struggle for Life.* The first edition was published in 1859, and no biological treatise written before or since has produced an impact upon society equal to it. The 1,250 copies of the first edition were sold out the first day. *The Origin* went through six editions, the last in 1872.

***ori* site** a 422 base-pair segment of the *E. coli* chromosome where replication is initiated.

ornithine cycle a cyclic series of reactions in which potentially toxic, nitrogenous products from protein catabolism are converted to urea that is innocuous. In the cycle diagrammed, ammonia is removed from the system and used in the conversion of ornithine to citrulline. Aspartic acid enters the cycle, and its amino group is incorporated into arginosuccinic acid before it can form ammonia. Arginosuccinic acid is converted to arginine, and the fumaric acid released enters the citric acid cycle

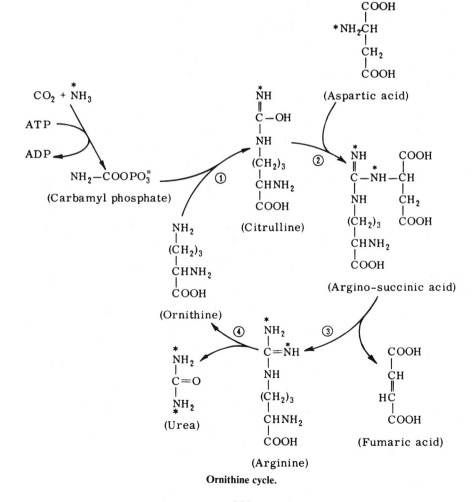

Ornithine cycle.

(*q.v.*). Urea splits off arginine and regenerates ornithine. In man, mutants are known that block the cycle at any one of its steps, as shown in the diagram on page 223. Blocking produces disorders that include: ornithine transcarbamylase deficiency, from blocking of step 1; citrullinuria (condensing enzyme deficiency), step 2; arginosuccinic aciduria (arginosuccinase deficiency), step 3; lysine intolerance (inhibition of arginase by excess lysine), step 4.

orphan drugs pharmaceuticals developed to treat diseases that afflict relatively few people.

orphan viruses viruses found in the digestive and respiratory tracts of healthy people; hence they are nonpathogenic (orphan = without an associated disease). *See* **reovirus.**

orphons dispersed, single pseudogenes (*q.v.*) derived from tandemly repeated families or gene clusters, such as those for histones or hemoglobins. Orphons may serve as a reservoir of sequences that can evolve new functions, and have probably been important factors in the evolution of higher organisms. *See* **hemoglobin genes.**

ortet the single ancestral organism that produced a clone of genetically identical organisms (ramets) by budding. *See* **modular organisms, ramets.**

orthochromatic dye a dye that stains tissues a single color in contrast to a metachromatic dye (*q.v.*).

orthogenesis a pattern of unidirectional change observed during the evolution of a related group of animals. For example, the fossil record shows a tendency toward an increase in the size of adults, when more recent species are compared with ancestral species in both the horse and elephant lineages. Trends of this sort were used in the past as evidence that evolution was driven toward a desired end by mystical forces. However, most students of evolution now believe that such patterns result from natural selection.

orthologous locus a gene that has evolved directly from an ancestral locus. *Contrast with* **paralogous locus.**

orthopteran an organism belonging to the Orthoptera, an order of the Hemimetabola containing cockroaches, locusts, grasshoppers, and similar insects.

orthoselection continuous selection on the members of a lineage over a long time, causing continued evolution in a given direction that may create an impression of "momentum" or "inertia" in evolutionary trends.

orthotopic transplantation the transplanation of grafts between identical sites in such a way that the graft maintains its normal orientation.

Oryctolagus cuniculus the rabbit, a mammal belonging to the Lagomorpha commonly reared in the laboratory and the subject of intensive genetic research. An extensive collection of mutations is available influencing a wide variety of morphological and physiological traits. The haploid chromosome number is 22, and about 60 genes have been distributed among 16 linkage groups.

Oryza sativa rice (*q.v.*).

Oryzias latipes the medaka, a freshwater fish common to Japan, Korea, and China. It is easily maintained in the laboratory and was the first fish in which Mendel's laws were shown to be valid. Y-linked inheritance was first demonstrated in the medaka and the guppy.

osmium tetroxide OsO_4, a compound often used as a fixative in electron microscopy.

osmosis diffusion of a solvent through a semipermeable membrane separating two solutions of unequal solute concentrations. The direction of solvent flow tends to equalize the solute concentrations.

***otu* mutation** a sex-linked female sterile gene in *Drosophila melanogaster* that is remarkable in that different mutant alleles can produce quite different ovarian pathologies. One class produces *ovarian tumors*; hence the *otu* symbol. Another class is characterized by germaria that lack oogonia. A third class generates oocytes that grow slowly and fail to complete development. The accompanying nurse cells contain polytene chromosomes, and the largest of these have 8,000 times the haploid amount of DNA.

Ouchterlony technique a gel-diffusion, antibody-antigen precipitation test that depends upon horizontal diffusion from two or more opposite sources. An agar slab is prepared and two or more wells are cut in it. One (A) is filled with an aqueous suspension of antibody molecules, while each of the other wells (B and C in the illustration on page 225) is filled with a different antigen preparation. The antigen and antibody molecules diffuse toward each other and eventually interact, forming curved precipitation lines. In example I, the antigens in wells B and C are different. In II, well B contains a single antigen and well C two antigens, one identical to that in B. *See* **Appendix C,** 1948, Ouchterlony.

Oudin technique a gel-diffusion, antibody-antigen precipitation test that depends on simple vertical

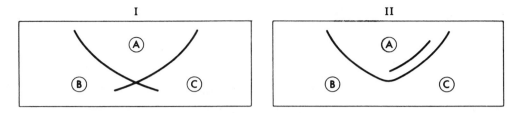

I   II

Ouchterlony technique.

diffusion in one dimension. A gel column is prepared containing a homogeneous distribution of antibody molecules. Above this is layered an aqueous suspension of antigen molecules. As these diffuse into the gel, a moving zone of antigen-antibody precipitate is formed. If several antigens and antibodies are present, separate zones of interaction will be seen. *See* **Appendix C,** 1946, Oudin.

outbreeding the crossing of genetically unrelated plants or animals; crossbreeding.

outcross *See* **outbreeding.**

outgroup a species or higher monophyletic taxon that is examined in the course of a phylogenetic study to determine which of two homologous character states may be inferred to be apomorphous. The most critical outgroup comparison involves the sister group of the taxon under study.

outlaw gene a gene favored by selection despite its disharmoneous effects on other genes in the same organism. *See* **meiotic drive.**

ovariectomy the surgical removal of one or both ovaries.

ovariole one of several egg tubes constituting the ovary of most insects.

ovary the female gonad in animals or the ovule-containing region of the pistil of a flower.

overdominance the phenomenon of heterozygotes having a more extreme phenotype than either homozygote; monohybrid heterosis. Overdominance generally refers to the situation in which AA' individuals are more fit than AA or $A'A'$ individuals.

overlapping code a hypothetical genetic code, first proposed by George Gamow, in which any given nucleotide is shared by two adjacent codons. The genetic code used in biological systems was later shown to be nonoverlapping. *See* **overlapping genes** for the few exceptions.

overlapping genes genes whose nucleotide sequences overlap to some degree. The overlap may involve control genes (e.g., tryptophan operator and promoter regions in *E. coli*) or structural genes

(e.g., in bacteriophage ϕX174, gene *E* lies entirely within gene *D*, but they are translated in different reading frames). *See* **Appendix C,** 1976, Burrell *et al.;* **bidirectional genes.**

overlapping inversion a compound chromosomal inversion caused by a second inversion that includes part of a previously inverted segment.

overwinding positive supercoiling of DNA, resulting in further tension in the direction of winding of the two strands of a duplex about each other.

ovicide any compound that destroys eggs, especially a compound that destroys insect eggs.

oviduct the tube carrying eggs from the ovary to the uterus.

ovine referring to members of the sheep family, especially the domestic sheep species *Ovis aries.*

oviparous laying eggs in which the embryo develops outside the mother's body and eventually hatches. *See* **ovoviviparous, viviparous.**

oviposition the laying of eggs by a female insect.

ovipositor an organ at the hind end of the abdomen in female insects, through which eggs are laid.

Ovis aries the domestic sheep. The blood group genetics of this species has been intensively investigated. *See* **sheep.**

ovisorption the resorption of oocytes.

ovogenesis oogenesis (*q.v.*).

ovotestis the organ of some hermaphroditic animals that functions both as ovary and testis; the gonad of an animal that undergoes consecutive sexuality (*q.v.*).

ovoviviparous bringing forth young that develop from eggs retained within the maternal body, but separated from it by the egg membranes. Many fish, reptiles, molluscs, and insects are ovoviviparous. *See* **oviparous, viviparous.**

ovulation the release of a ripe egg from the mammalian ovarian follicle, frequently at the stimulus of a pituitary hormone.

ovule the structure found in seed plants which develops into a seed after the fertilization of an egg cell within it.

ovum an unfertilized egg cell.

oxidation classically defined as the combination of a molecule with oxygen or the removal of hydrogen from it. Since electrons are transferred to the oxidizing reagent, which becomes reduced, oxidation and reduction (*q.v.*) are always coupled.

oxidation-reduction reactions chemical reactions in which electrons are transferred from a reductant to an oxidant; as a consequence of the transfer, the reductant is oxidized and the oxidant is reduced.

oxidative phosphorylation the enzymatic phosphorylation of ADP to ATP, which is coupled to the electron transport chain (*q.v.*). Thus, respiratory energy is transformed into phosphate-bond energy.

oxidoreductases enzymes that transfer electrons. Catalase (*q.v.*) is an oxidoreductase.

oxygen the second most abundant of the biologically important elements. Atomic number 8; atomic weight 15.9994; valence 2^-; most common isotope ^{16}O.

oxyhemoglobin oxygenated hemoglobin.

oxytocin a polypeptide secreted by the hypothalamus and stored in the neurohypophysis. Oxytocin causes smooth-muscle contraction and may help terminate pregnancy.

Oxytricha *See* **Stylonichia.**

P

p *See* **symbols used in human cytogenetics.**

P 1. probability. 2. phosphorus. 3. phosphate (when combined in an abbreviation such as ADP, ATP). 4. symbol for panmictic index (*q.v.*).

P₁ the symbol denoting the immediate parents of the F₁ generation. The symbols P₂ and P₃ are used to designate the grandparental and great-grandparental generations, respectively, if one starts from the F₁ and works backward.

³²P a radioactive isotope of phosphorus, widely used to label nucleic acids; it emits a strong beta particle and has a half-life of 14.3 days.

Pᵢ inorganic phosphate.

pachynema *See* **meiosis.**

pachytene stage *See* **meiosis.**

packing ratio the ratio of DNA length to the unit length of the fiber containing it.

paedogenesis a type of precocious sexual maturity occurring in the larval stage of some animals. The eggs of paedogenetic females develop parthenogenetically.

paedomorphosis an evolutionary phenomenon in which adult descendants resemble the youthful stages of their ancestors; the opposite of "recapitulation," in which the early stages of descendants resemble the adult stages of their ancestors. Paedomorphosis may be produced either by the acceleration of sexual development (progenesis) or by retarded somatic development. *See* **heterochrony.**

Paeonia californica the California peony, a species with naturally occurring translocation complexes.

PAGE polyacrylamide gel electrophoresis. *See* **electrophoresis, polyacrylamide gel.**

pair bonding an intimate and long-lasting association between male and female animals of the same species, generally facilitating the cooperative rearing of their offspring.

paired a recessive lethal mutation on the left arm

of chromosome 2 of *Drosophila melanogaster.* The *prd* gene, which belongs to the pair rule class of zygotic segmentation mutants (*q.v.*), has been cloned and sequenced. It contains multiple conserved domains, including a homeobox and a histidine-proline repeat, both of which have been identified in other genes that control segmentation. *See* **gene networking.**

pairing synapsis.

pairing segments the segments of the X and Y chromosomes which synapse and cross over. The remaining segments, which do not synapse, are called the differential segments.

pair rule genes *See* **zygotic segmentation mutants.**

Palearctic designating one of the six biogeographic realms (*q.v.*) of the globe, including Eurasia except Iran, Afghanistan, the Himalayas, and the Nan-ling Range in China, Africa north of the Sahara, Iceland, Spitzbergen, and the islands north of Siberia.

Paleocene the most ancient of the Tertiary epochs. Grasses and primates appeared at this time. Drifting of continents continued. *See* **geologic time divisions.**

Paleogene a subdivision of the Tertiary period, incorporating the Oligocene, Eocene, and Paleocene epochs. *See* **geologic time divisions.**

Paleolithic that phase of human history prior to the cultivation of plants during which tools were manufactured, food was obtained by hunting, fishing, or collecting wild nuts and fruits. The Paleolithic culture lasted from about 500,000 years ago up to the beginning of the Neolithic stage about 10,000 years ago.

paleontology the study of extinct forms of life through their fossils, as opposed to neontology (*q.v.*).

paleospecies the successive species in a phyletic lineage that are given ancestor and descendant status according to the geological strata in which they appear. *See* **anagenesis.**

Paleozoic the earliest era of the Phanerozoic eon.

$$HO-CH_2-\overset{\overset{\displaystyle CH_3}{|}}{\underset{\underset{\displaystyle CH_3}{|}}{C}}-\overset{\overset{\displaystyle OH}{|}}{\underset{\underset{\displaystyle H}{|}}{C}}-\overset{\overset{\displaystyle O}{\parallel}}{C}-\overset{\overset{\displaystyle H}{|}}{N}-CH_2-CH_2-COOH$$

Pantothenic acid.

Invertebrates flourished during this 320-million year interval. *See* **geologic time divisions.**

Paley's watch an argument developed by William Paley (1743–1805) for the existence of God based upon the commonsense notion that a watch is too complicated to have originated by accident; it presents its own evidence of having been purposely designed. This argument is commonly applied to living organisms by creationists.

palindrome a sequence of deoxyribonucleotide base pairs that reads the same (5′ to 3′) on complementary strands; tandem inverted repeats; example:

5′ AATGCGCATT 3′
3′ TTACGCGTAA 5′

Palindromes serve as recognition sites for restriction endonucleases, RNA polymerase, and other enzymes. *See* **cruciform structure.**

palynology the study of both living and fossil spores, pollen grains, and other microscopic propagules.

Pan the genus containing *P. troglodytes,* the common chimpanzee and *P. paniscus,* the pygmy chimpanzee. *Pan troglodytes* is the living primate genetically closest to man. *Pan* and *Homo* diverged from a common ancestor about 7 million years ago. The haploid chromosome number for *P. troglodytes* is 24, and about 40 genes have been distributed among 19 linkage groups.

Panagrellus redivivus a free-living nematode whose developmental genetics is under study, primarily for comparison with *Caenorhabditis elegans* (*q.v.*). It differs from *Caenorhabditis* in having XX females and XO males. Its cell lineage (*q.v.*) is known in part.

pancreozymin a hormone secreted by the duodenum that causes secretion of pancreatic enzymes.

pandemic designating a disease simultaneously epidemic in human populations in many parts of the world.

panethnic referring to a hereditary disease that is found in a variety of ethnic groups.

Pangaea *See* **continental drift.**

pangenesis a defunct theory of development, pop-ular in Darwin's time, proposing that small particles (pangenes) from various parts of the body distill into the gametes, resulting in a blending of parental characteristics in their offspring.

panhypopituitarism *See* **pituitary dwarfism.**

panicle *See* **raceme.**

panmictic index (P) a measure of the relative heterozygosity. $1 - P = F$, Wright's inbreeding coefficient (*q.v.*).

panmictic unit a local population in which mating is completely random.

panmixia panmixis.

panmixis random mating, as contrasted with assortative mating (*q.v.*).

panoistic ovary *See* **insect ovary types.**

pantothenic acid a water soluble vitamin that functions as a subunit of coenzyme A (*q.v.*).

papain a proteolytic enzyme isolated from the latex of the papaya plant.

paper chromatography *See* **chromatography.**

Papilio glaucus the swallowtail butterfly, a species extensively studied in terms of isolating mechanisms (*q.v.*).

papillary pattern the pattern of dermal ridges on the fingertips and palms.

papilloma a benign cutaneous neoplasm; a wart.

papovavirus a group of animal DNA viruses (including SV4O and polyoma) responsible for papillomas of the rabbit, dog, cattle, horse, and human.

parabiotic twins artificial "Siamese twins" produced by joining two animals surgically. Since their blood circulations will eventually anastomose, one can study the transmission of humoral agents from one "twin" to the other.

paracentric inversion an inversion (*q.v.*) that does not include the centromere.

Paracentrotus lividus a common sea urchin extensively used in studies of molecular developmental genetics. *See* **echinoderm.**

paracrystalline aggregate a regular linear arrangement of stacked molecules.

Paranemic spiral.

paradigm (pronounced *"paradime"*) a term with a variety of meanings in the scientific literature. In its weakest sense, it is used as a synonym for model, hypothesis, or theory. It is used most commonly to refer to a known example or incident that serves as a model or provides a pattern for a more general phenomenon. In a still more restricted sense, paradigm may refer to a ruling model that has replaced all others. Darwin's theory of evolution by natural selection is an example of such a paradigm. As time passes, the matching of new discoveries with the model may lead to a revision of the paradigm. An example of this would be the transformation of the one gene–one enzyme to the one gene–one polypeptide paradigm.

paraffin section a section of tissue cut by a microtome after embedding in a paraffin wax; the classical method of preparing tissues for microscopical study. *See* **Appendix C,** 1860, Klebs.

paragenetic referring to a chromosomal change that influences the expression of a gene rather than its structure. *See* **position effect, Lyonization.**

parallel evolution the occurrence of the same or a similar trend independently evolved in two or more lineages; the lineages are usually, although not necessarily, related to one another.

paralogous locus a gene that originated by duplication and then diverged from the parent copy by mutation and selection or drift. *Contrast with* **orthologous locus.**

Paramecium a common genus of ciliate protozoa found in stagnant ponds. Common species are *P. aurelia, P. caudatum,* and *P. bursaria.* These are favorite species for the study of nuclear-cytoplasmic interactions. *See* **Appendix C,** 1976, Dippell; **universal code theory.**

Paramecium aurelia a cigar-shaped, heavily ciliated protoctist, 100–150 μm long, living in still or running fresh waters. *Paramecium aurelia* consists of a group of 14 syngens. Each syngen is genetically isolated from every other syngen and is also biochemically unique. However, all syngens are so similar morphologically that they have not been given individual species names. Each syngen has two mating types. Mating types were first discovered in syngen 1. *See* **Appendix C,** 1937, Sonneborn; 1971, Kung; **conjugation, killer paramecia.**

parameter the value of some quantitative characteristic in an entire population. For example, the mean height of all males 20 years of age or older in a pygmy tribe. *Compare with* **statistic.**

parametric statistics. *See* **statistics.**

paramutation a phenomenon discovered in corn where one allele influences the expression of another allele at the same locus when the two are combined in a heterozygote. The first allele is referred to as "paramutagenic," the second as "paramutable." The paramutable allele behaves like an unstable hypomorph when the paramutagenic allele is present.

paranemic spiral a spiral made up of two parallel threads coiled in opposite directions. The two threads can be easily separated without uncoiling. *See* **plectonemic spiral, relational coiling.**

parapatric referring to populations or species that occupy adjacent areas with a narrow zone of overlap within which hybridization commonly occurs.

parapatric speciation a mode of gradual speciation in which new species arise from populations that maintain genetic contact during the entire process by a narrow zone of overlap. Also called *semigeographic speciation. Compare with* **allopatric speciation, alloparapatric speciation, peripatric speciation.**

paraphyletic 1. in classification, an incomplete clade, i.e., one from which one or more members of a halophyletic group have been omitted. A paraphyletic group is recognized by the absence of the homologs that define the excluded clades. For example, group AB in the illustration (page 230) is paraphyletic; it is defined by the absence of feature 3 (features 1 and 2 are present in group CD as well). 2. in evolution, a monophyletic group that does not include all the groups descended from a single common ancestor. For example, the AB group below is paraphyletic; but group CD is excluded, since it is not a direct descendant from common ancestor 1.

Parascaris equorum a nematode (commonly called the horse thread-worm) studied by early cytologists because it exhibited chromosome diminution (*q.v.*). *See* **Appendix C,** 1898, Boveri.

parasexuality any process that forms an offspring cell from more than a single parent, bypassing standard meiosis and fertilization. In fungi, for example, diploid nuclei arise from rare fusions of two genetically unlike nuclei in heterokaryons. Somatic

229

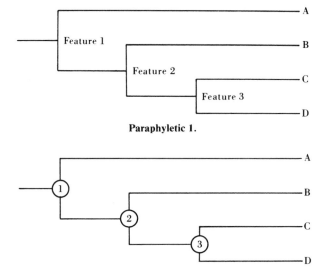

Paraphyletic 1.

Paraphyletic 2.

crossing over occurs, and eventually haploid nuclei showing new combinations of genes are formed from the diploids. *See* **Appendix C,** 1952, Pontecorvo and Roper. In viruses, parasexual recombination can occur if genetically different mutant strains multiply in a host cell after it has been infected by viruses of both types (*see* **Visconti-Delbrück hypothesis**). In bacteria, there are three phenomena that can lead to parasexual recombination: conjugation, transduction and transformation (*all of which see*). *Also see* **sexduction, transfection.**

parasitism a symbiotic association that benefits one member (the parasite) but is harmful to the other (the host).

parathyroid hormone a hormone controlling calcium and phosphorus balance, synthesized by the cells of the parathyroid gland.

paratope the site within an immunoglobulin Fab that specifically interacts with an antigenic determinant (epitope).

Parazoa a subdivision of the animal kingdom containing organisms, like sponges, that lack tissues organized into organs, and have an indeterminate shape. *See* **Appendix A.**

parenchyma **1.** plant tissues composed of thin-walled cells that fit loosely together, leaving intercellular spaces. **2.** a reticulum of cells between the organs of an animal. **3.** the cells performing the principal function of an organ.

parental ditype *See* **tetrad segregation types.**

parental imprinting the phenomenon whereby the degree to which a gene expresses itself depends upon the parent transmitting it. Huntington disease (*q.v.*) is an example. Individuals who receive this dominant gene show symptoms during adolescence if it is inherited from their father, but symptoms begin during middle age when the gene comes from their mother.

parity the fact of having borne children. A woman of parity 0 has borne no live children although she may have been pregnant one or more times. A woman of parity 1 has given birth only once. However, the number of children produced at this birth may have exceeded one.

paroral cone a protuberance in the oral region of a conjugating ciliate that juts into the body of the partner. The haploid nucleus residing in the paroral cone survives and all others degenerate. *See* **conjugation.**

pars amorpha *See* **nucleolus.**

parsimony principle the principle that the simplest sufficient hypothesis is to be preferred, even if others are possible. Also called *Occam's razor.*

pars intercerebralis that medial region in the insect forebrain containing neurosecretory cells.

Parthenium argentatum the guayule, a desert plant, reproducing sexually or asexually depending upon its genotype and studied by geneticists accordingly.

parthenocarpy the natural or artificially induced formation of fruit without seeds because of either (1) lack of pollination, (2) lack of fertilization, or (3) death of the embryo at an early stage of development. Edible bananas, for example, set fruit by parthenocarpy. Wild bananas are diploid (N = 11) and reproduced normally, whereas the commercially grown, edible bananas are triploid and both male and female are sterile.

parthenogenesis the development of an individual from an egg without fertilization. *See* **Appendix C, 1845, Dzierzon; artificial parthenogenesis, arrhenotoky, gynogenesis, heterogony, paedogenesis, thelytoky.**

partial denaturation an incomplete unwinding of the DNA double helix; GC-rich regions are more resistant to thermal disruption because three hydrogen bonds form between G and C, whereas only two form between A and T. *See* **deoxyribonucleic acid.**

partial diploid *See* **merozygote.**

partial dominance semidominance (*q.v.*).

particulate inheritance the Mendelian theory that genetic information is transmitted from one generation to another in the form of discrete units, so that the biological inheritance of offspring is not a solution in which the parental information is blended.

parturition the act of giving birth to young.

Pascal's pyramid *See* **binomial distribution.**

PAS procedure periodic acid Schiff procedure (*q.v.*).

passage number the number of times a culture has been subcultured.

passenger in recombinant DNA research, a DNA segment of interest that will be spliced into a DNA vehicle for subsequent cloning.

passive equilibrium an unstable equilibrium resulting from selective neutrality of alleles at a genetic locus, as occurs in Hardy-Weinberg equilibrium; also called *neutral equilibrium.*

passive immunity the immunity against a given disease produced by injection into a host of serum containing antibodies formed by a donor organism that possesses active immunity to the disease. *See* **active immunity.**

Patau syndrome trisomy 13. *See* **D₁ trisomy syndrome.**

patent period the interval during an infection when the causative agent can be detected. *Compare with* **latent period, prepatent period.**

paternal-X inactivation the method of dosage compensation found in marsupials where the paternal-X chromosome is inactivated in the somatic cells of females. *See* **random-X inactivation.**

path coefficient analysis a method invented by Sewall Wright for analyzing quantitatively the transmission of genes in regular and irregular breeding systems.

pathogenic producing disease or toxic symptoms.

pathovar a pathological variant of a bacterial species; symbolized *pv.* For example, the cause of citrus canker is *Xanthomonas campestris* pv. *citri.*

patroclinous designating an offspring that resembles the male more closely than the female parent. The sons of an attached-X female *Drosophila* are patroclinous in terms of their sex-linked genes. *See* **matrocliny.**

P blood group a system of human blood groups depending on an antigen on the red cells specified by the dominant gene *P. See* **blood group.**

pBR322 a plasmid cloning vector that grows under relaxed control in *E. coli.* It contains ampicillin- and tetracycline-resistance genes and several convenient restriction endonuclease recognition sites. *See* **Appendix C, 1979, Sutcliffe.**

PDGF platelet-derived growth factor (*q.v.*).

pea *See* **Pisum sativum.**

Peckhammian mimicry *See* **mimicry.**

Pecten irradians *See* **Pelecypoda.**

pectic acid a polymer made up of galacturonic acid subunits.

Pectic acid.

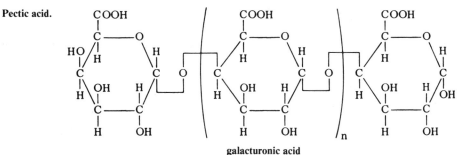

galacturonic acid

pectin a polysaccharide material within the cell wall and middle lamella. Pectin consists of pectic acid, many of whose COOH groups have been methylated.

pedigree a diagram setting forth the ancestral history or genealogical register. Symbols commonly used in such diagrams are illustrated below in the sample pedigree. Females are symbolized by circles and males by squares. Individuals showing the trait are drawn as solid figures. Offspring are presented beneath the parental symbols in order of birth from left to right. The arrow points to the propositus. The sex of individual II-3 is unknown. II-6 died at an early age and consequently her phenotype relative to the trait in question was unknown. II-7 and II-8 were dizygotic twins, whereas III-1 and III-2 were identical twins. Other symbols commonly encountered are shown below.

pedigree selection artificial selection of an individual to participate in mating based upon the merits of its parents or more distant ancestors.

pedogenesis *See* **paedogenesis.**

pelargonin a scarlet-colored anthocyan giving color to the petals of geraniums and asters.

Pelargonium zonale the geranium. Classical studies on the non-Mendelian inheritance of chloroplast mutants were performed on this species. *See* **Appendix C,** 1909, Correns and Bauer.

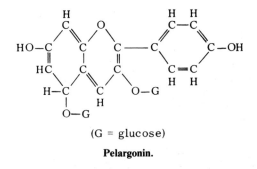

(G = glucose)

Pelargonin.

Pelecypoda the class containing the bivalve molluscs. The cytogenetics and quantitative genetics of certain species have been studied because of their economic importance. These include the American oyster (*Crassostrea virginica*), the clam (*Mercenaria mercenaria*), the mussel (*Mytilus edulis*), and the scallop (*Pecten jacobeus*).

P elements transposable elements in *Drosophila melanogaster* that are responsible for one type of hybrid dysgenesis (*q.v.*). P elements have been cloned in *E. coli* plasmids. When DNA molecules carrying P elements are microinjected into *Drosophila* embryos, some P elements integrate into the germ-line chromosomes and are transmitted to the progeny of the injected flies. *See* **Appendix C,** 1982, Bingham *et al.;* **transposable element.**

P element transformation the transfer of specific DNA segments into germ-line cells of *Drosophila*

Pedigree.

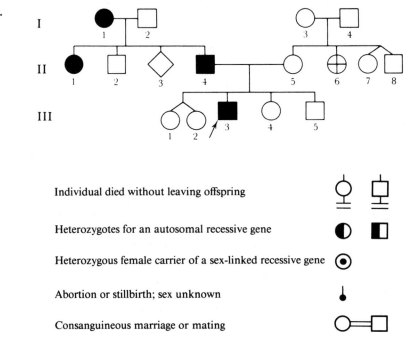

| Individual died without leaving offspring |
| Heterozygotes for an autosomal recessive gene |
| Heterozygous female carrier of a sex-linked recessive gene |
| Abortion or stillbirth; sex unknown |
| Consanguineous marriage or mating |

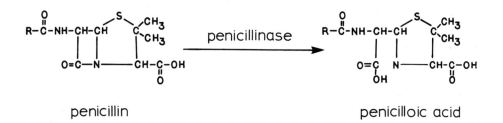

penicillin   penicilloic acid

using the transposable P element to carry exogenous DNA fragments. *See* **Appendix C,** 1982, Spradling and Rubin.

Pelger-Huet anomaly *See* **Pelger nuclear anomaly.**

Pelger nuclear anomaly a hereditary abnormality in man involving the nuclear morphology of the polymorphonuclear leukocytes. A similar syndrome occurs in the rabbit. Homozygotes show chondrodystrophy in the rabbit, but not in man.

pellagra a disease caused by deficiency of niacin.

Pelomyxa a genus of amoebae. *P. carolinensis* is a favorite species for nuclear transplantation studies. *P. palustris* is of interest, since it lacks mitochondria and instead harbors anaerobic bacteria in a permanent symbiotic relationship. *See* **serial symbiosis theory.**

penetrance the proportion of individuals of a specified genotype that show the expected phenotype under a defined set of environmental conditions. For example, if all individuals carrying a dominant mutant gene show the mutant phenotype, the gene is said to show complete penetrance.

penicillin any of a family of antibiotics derived from the mold *Penicillium notatum* and related species. Penicillins inhibit the synthesis of the wall of the bacterial cell. Bacteria poisoned with penicillin soon outgrow their cell walls and lyse. Different penicillins differ only in the side chain symbolized by R in the formula below. In the parent molecule, R = $C_6H_5CH_2CO-$. Ampicillin is a penicillin derivative that is effective against a larger variety of Gram negative bacteria than are most other penicillins. Here, R = $C_6H_5CH(NH_2)CO-$. Penicillin-resistant bacteria synthesize penicillinases. These enzymes attack the beta-lactam ring to produce penicilloic acid, which has no bacteriocidal activity. *See* **Appendix C,** 1929, Fleming; 1940, Florey *et al.*

penicillinases *See* **penicillin.**

penicillin enrichment technique *See* **penicillin selection technique.**

penicillin selection technique a method for isolat-

ing an auxotrophic mutant from a wild-type culture of bacteria by adding penicillin to minimal medium. Penicillin interferes with cell wall development, causing growing wild-type cells to rupture. However, nongrowing auxotrophic mutant cells are not killed. After one hour, about 99% of the wild-type cells have lysed, releasing their pool of metabolites into the medium. The culture must be filtered to remove these metabolites, because the auxotrophic mutants would use them for growth and be subject to penicillin-induced lysis; the filtering also removes the penicillin. Alternatively, the enzyme penicillinase can be used to destroy the penicillin. Surviving auxotrophs are then supplied with enriched medium. These cells produce the colonies that are harvested. *See* **Appendix C,** 1948, Lederberg and Zinder, Davis.

Penicillium notatum the ascomycote fungus that synthesizes penicillin (*q.v.*).

penicilloic acid *See* **penicillin.**

Pennsylvanian *See* **Carboniferous.**

pentabarbital Nembutal (*q.v.*).

pentosephosphate pathway a pathway of hexose oxidation that is an alternative to the glycolysis-citric acid cycle (*q.v.*). The pathway involves an interconversion of phosphates of 7-, 6-, 5-, 4-, and 3-carbon sugars, and of a sugar lactone and sugar acids in a cycle that effects the complete oxidation of glucose to CO_2 and H_2O with the formation of 36 molecules of ATP. *See* pages 234–235.

pepsin a proteolytic enzyme from the gastric mucosa that functions at low *p*H.

peptidase an enzyme catalyzing the hydrolytic cleavage of peptide bonds (*q.v.*).

peptide a compound formed of two or more amino acids.

peptide bond a covalent bond between two amino acids formed when the amino group of one is

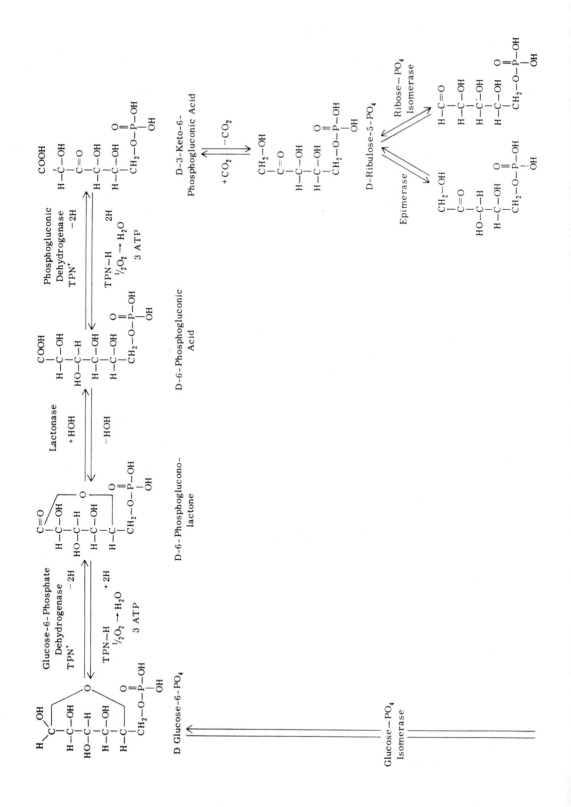

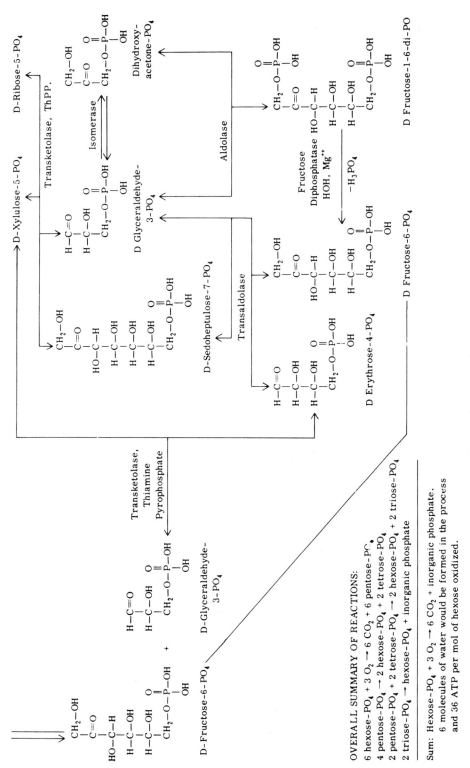

Pentosephosphate pathway.

OVERALL SUMMARY OF REACTIONS:

6 hexose-PO$_4$ + 3 O$_2$ → 6 CO$_2$ + 6 pentose-PO$_4$

4 pentose-PO$_4$ → 2 hexose-PO$_4$ + 2 tetrose-PO$_4$

2 pentose-PO$_4$ + 2 tetrose-PO$_4$ → 2 hexose-PO$_4$ + 2 triose-PO$_4$

2 triose-PO$_4$ → hexose-PO$_4$ + inorganic phosphate

Sum: Hexose-PO$_4$ + 3 O$_2$ → 6 CO$_2$ + inorganic phosphate.
6 molecules of water would be formed in the process
and 36 ATP per mol of hexose oxidized.

bonded to the carboxyl group of the other and water is eliminated. *See* **Appendix C,** 1902, Hofmeister and Fischer.

peptidyl transferase *See* **translation.**

peptidyl–RNA binding site *See* **translation.**

perennation the survival of plants from growing season to growing season with a period of reduced activity in between.

perennial a plant that continues to grow from year to year.

perfect flower *See* **flower.**

perfusion the introduction of fluids into organs by injection into their arteries.

pericarp the wall of the ovary after it has matured into a fruit; it may be dry and hard (as in the case of a nut) or fleshy (as in the case of a berry).

pericentric inversion an inversion that includes the centromere.

periclinal referring to a layer of cells running parallel to the surface of a plant part. *See* **anticlinal.**

periclinal chimera a plant made up of two genetically different tissues, one surrounding the other.

perikaryon that portion of the cell body of a neuron surrounding the nucleus as distinguished from the axon and dendrites.

perinuclear cisterna the fluid-filled reservoir enclosed by the inner and outer membranes of the nuclear envelope.

periodic acid Schiff procedure a staining procedure for demonstrating polysaccharides. Abbreviated PAS. *See* **Schiff's reagent.**

periodicity in molecular genetics, the number of base pairs per turn of the DNA double helix.

periodic table an arrangement of the chemical elements (*q.v.*) in order of increasing atomic number. Elements with similar properties are placed one under the other, yielding groups and families of elements. The earliest periodic table was drawn up by D. I. Mendeleev.

peripatric speciation a model proposing that speciation occurs in small populations isolated on the periphery of the distribution of the parental population, as opposed to parapatric speciation (*q.v.*). The isolated populations may undergo shifts in their gene frequencies under the influence of genetic drift. This is most likely to occur if new populations arise from a few founder individuals and no gene flow occurs between the isolates and the main population. *See* **Chronology C,** 1954, Mayr; **founder effect.**

Peripatus *See* **living fossil.**

peripatric living in a region peripheral to that of the main body of the species.

peristalsis waves of muscular contractions that

Periodic table.

	IA	IIA	IIIB	IVB	VB	VIB	VIIB	VIII			IB	IIB	IIIA	IVA	VA	VIA	VIIA	INERT GASES
1	1 H 1.00797																1 H 1.00797	2 He 4.0026
2	3 Li 6.939	4 Be 9.0122											5 B 10.811	6 C 12.01115	7 N 14.0067	8 O 15.9994	9 F 18.9984	10 Ne 20.183
3	11 Na 22.9898	12 Mg 24.312											13 Al 26.9815	14 Si 28.086	15 P 30.9738	16 S 32.064	17 Cl 35.453	18 Ar 39.948
4	19 K 39.102	20 Ca 40.08	21 Sc 44.956	22 Ti 47.90	23 V 50.942	24 Cr 51.996	25 Mn 54.9380	26 Fe 55.847	27 Co 58.9332	28 Ni 58.71	29 Cu 63.54	30 Zn 65.37	31 Ga 69.72	32 Ge 72.59	33 As 74.9216	34 Se 78.96	35 Br 79.909	36 Kr 83.80
5	37 Rb 85.47	38 Sr 87.62	39 Y 88.905	40 Zr 91.22	41 Nb 92.906	42 Mo 95.94	43 Tc (99)	44 Ru 101.07	45 Rh 102.905	46 Pd 106.4	47 Ag 107.870	48 Cd 112.40	49 In 114.82	50 Sn 118.69	51 Sb 121.75	52 Te 127.60	53 I 126.9044	54 Xe 131.30
6	55 Cs 132.905	56 Ba 137.34	57 †La 138.91	72 Hf 178.49	73 Ta 180.948	74 W 183.85	75 Re 186.2	76 Os 190.2	77 Ir 192.2	78 Pt 195.09	79 Au 196.967	80 Hg 200.59	81 Tl 204.37	82 Pb 207.19	83 Bi 208.980	84 Po (210)	85 At (210)	86 Rn (222)
7	87 Fr (223)	88 Ra (226)	89 ‡Ac (227)															

†Lanthanum series

58 Ce 140.12	59 Pr 140.907	60 Nd 144.24	61 Pm (147)	62 Sm 150.35	63 Eu 151.96	64 Gd 157.25	65 Tb 158.924	66 Dy 162.50	67 Ho 164.930	68 Er 167.26	69 Tm 168.934	70 Yb 173.04	71 Lu 174.97

‡Actinium series

90 Th 232.038	91 Pa (231)	92 U 238.03	93 Np (237)	94 Pu (242)	95 Am (243)	96 Cm (247)	97 Bk (247)	98 Cf (249)	99 Es (254)	100 Fm (253)	101 Md (256)	102 No (253)	103 Lw (257)

*( ) Numbers in parentheses are the mass numbers of the most stable or best-known isotope. Atomic weights are based on carbon 12, conforming to the conventions adopted in 1961 by the International Union of Pure and Applied Chemistry.

pass along tubular organs and serve to move the contents posteriorly.

perithecium the rounded or flask-shaped, fruiting body of certain ascomycete fungi and lichens. The mature fruiting body of *Neurospora,* for example, contains about 300 ascus sacs.

peritoneal sheath a network of anastomosing muscle fibers that holds together the ovarioles of an insect ovary.

peritrichous designating bacteria having flagella all over their surfaces.

permeability the extent to which molecules of a given kind can pass through a given membrane.

permease a membrane-bound protein in bacteria that is responsible for transport of a specific substance in or out of the cell; sometimes referred to as a *transport protein.* In *E. coli,* lactose permease actively transports lactose into the cell.

Permian the most recent of the Paleozoic periods. Reptiles flourished, including species with mammalian characteristics. Insects increased, while amphibians declined. The period ended with a mass extinction of many dominant life forms. All trilobites became extinct. *See* **geologic time divisions.**

permissible dose *See* **maximum permissible dose.**

permissive cells cells in which a particular virus may cause a productive infection (i.e., the production of progeny viruses). Cells in which infection is not productive are called *nonpermissive cells.* Some DNA tumor viruses may cause a productive infection in permissive cells of one species and a tumor in nonpermissive cells of another species.

permissive conditions environmental conditions under which a conditional lethal mutant (e.g., a temperature-sensitive mutant) can survive and produce wild-type phenotype.

permissive temperature *See* **temperature-sensitive mutation.**

Peromyscus a genus containing about 40 species of mice native to Central and North America. The genetic data accumulated so far have been primarily related to the deermouse, *P. maniculatus.* More limited data exist for *P. boylei, P. leucopus,* and *P. polionotus.* Studies of biochemical variation and cytogenetics are presently the areas of most genetical research within the genus.

peroxisomes membrane-bound intracellular organelles containing at least four enzymes involved in the metabolism of hydrogen peroxide. They are thought to be important in purine degradation, photorespiration, and the glyoxylate cycle. Sometimes called *microbodies.*

petites dwarf colonies of *Saccharomyces cerevisiae.* Such yeasts are slow growing because of mutations affecting mitochondria. *Segregational petites* contain mutated nuclear genes that result in mitochondrial defects. In the case of *cytoplasmic* or *vegetative petites,* it is the mitochondrial genome that contains mutations or deletions of various sizes. In situations where all mitochondrial-encoded gene products are missing, the yeast forms *promitochondria.* These have a normal outer membrane, but the inner membrane contains poorly developed cristae. In spite of the fact that the organelle cannot function in oxidative phosphorylation, it contains nuclear-encoded proteins such as DNA and RNA polymerases, all the enzymes of the citric acid cycle, and many inner membrane proteins. *See* **Appendix C,** 1949 Ephrussi *et al.*

Petri dish a round, shallow, covered glass container in which microorganisms or dividing eukaryotic cells are cultured on a nutrient gel (usually agar). *See* **plaque.**

pH *See* **hydrogen ion concentration.**

phaeomelanin one of the pigments found in the coats of mammals. It is derived from the metabolism of tyrosine and is normally yellow in color. The amount of this pigment inserted into the hairs is quantitatively and qualitatively controlled by the *agouti* locus. *See* **melanin.**

phage bacteriophage, bacterial virus.

phage conversion *See* **prophage-mediated conversion.**

phage cross a procedure requiring the multiple infection of a single bacterium with phages that differ at one or more genetic sites. Upon lysis of the host, recombinant progeny phage are recovered that carry genes derived from both parental phage types. *See* **Visconti-Delbrück hypothesis.**

phage induction the stimulation of prophage to enter the vegetative state, accomplished by exposing lysogenic cells to ultraviolet light. Hydrogen peroxide, x-rays, and nitrogen mustard (*q.v.*) also act as inducing agents. *See* **zygotic induction.**

phagocyte a cell that incorporates particles from its surroundings by phagocytosis.

phagocytosis the engulfment of solid particles by cells; the ingestion of microorganisms by leukocytes.

phagolysosome an organelle formed by the fusion of a phagosome (*q.v.*) and a lysosome (*q.v.*).

phagosome a membrane-bounded cytoplasmic particle produced by the budding off of localized invaginations of the plasmalemma. Recently phago-

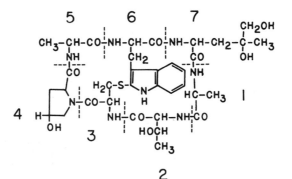

5   6   7

Phalloidin.

cytosed particles are segregated within the cell in phagosomes.

phalloidin the most common phallotoxin produced by *Amanita phalloides*. It is a cyclic heptapeptide made up of (1) alanine, (2) d-threonine, (3) cysteine, (4) hydroxyproline, (5) alanine, (6) tryptophan, and (7) γ-δ-dihydroxyleucine.

phallotoxins together with amatoxins (*q.v.*), the main toxic components produced by *Amanita phalloides* (*q.v.*). Phalloidin, one of the chief phallotoxins, forms tight complexes with filamentous actin. *See* **rhodaminylphalloidin.**

phanerogam an outmoded term referring to a plant belonging to the Spermatophyta (*q.v.*). *See* **cryptogam.**

Phanerozoic the geologic eon encompassing the Paleozoic, Mesozoic, and Cenozoic eras. During the 570-million-year interval an abundant fossil record was left in the rocks. *See* **geologic time divisions.**

phantom a volume of material approximating the density and effective atomic number of tissue. Radiation dose measurements are made within or on a phantom as a means of determining the radiation dose within or on a body under similar exposure conditions.

pharmacogenetics the area of biochemical genetics dealing with genetically controlled variations in responses to drugs.

phase contrast microscope light rays passing through an object of high refractive index will be retarded in comparison with light rays passing through a surrounding medium with a lower refractive index. The retardation or *phase change* for a given light ray is a function of the thickness and the index of refraction of the material through which it passes. Thus, in a given unstained specimen, transparent regions of different refractive indices retard the light rays passing through them to differing degrees. Such phase variations in the light focused on the image plane of the light microscope are not visible to the observer. The phase contrast microscope is an optical system that converts such phase variations into visible variations in light intensity or contrast. The phase microscope therefore allows cytologists to observe the behavior of living, dividing cells. *See* **Appendix C,** 1935, Zernicke.

phaseolin a glycoprotein that constitutes up to 50% of the storage protein in the cotyledons of the bean *Phaseolus vulgaris*. It is encoded by a family of about 10 genes. *See* **Appendix C,** 1981, Kemp and Hall.

Phaseolus the genus that includes *P. aureus,* the Mung bean; *P. limensis,* the lima bean; *P. vulgaris,* the red kidney bean.

phene a phenotypic character controlled by genes.

phenetic taxonomy a system of classification based upon phenotypic characteristics without regard to phylogenetic relationships; also known as *numerical taxonomy.*

phenocopy the alteration of the phenotype, by nutritional factors or the exposure to environmental stress during development, to a form imitating that characteristically produced by a specific gene. Thus, rickets due to a lack of vitamin D would be a phenocopy of vitamin D-resistant rickets.

phenocritical period the period in the development of an organism during which an effect produced by a gene can most readily be influenced by externally applied factors.

phenogenetics developmental genetics.

phenogram a branching diagram linking taxons by estimates of overall similarity based on evidence from a sample of characters. Characters are not evaluated as to whether they are primitive or derived.

phenogroup any combination of antigenically active factors of a blood group system that is inherited as a unit. The B and C blood group systems in cattle are examples. *See* **haplotype** and **Appendix C,** 1951, Stormont *et al.*

phenomic lag phenotypic lag.

phenon a set of organisms grouped together by methods of numerical taxonomy (*q.v.*).

phenotype the observable properties of an organism, produced by the genotype in conjunction with the environment. *See* **Appendix C,** 1909, Johannsen.

phenotypic lag the delay of the expression of a newly acquired character. Mutations may appear in a bacterial population several generations after the administration of a mutagen. Phenotypic lag may be due to any of the following reasons: (1) The mutagen may inactivate the gene at once, but this inactivity may not become apparent until the products previously made by the gene are diluted to a sufficient degree. A number of cell divisions occur before the concentration of these products falls below some critical level. (2) The "mutagen" may itself be inactive. It may undergo a series of reactions to yield a compound that is the true mutagen. The latent period would then be the time required for those reactions to take place. (3) The mutagen may cause the gene to become unstable. At a later time, it will return to a stable wild type or mutant state. (4) The microbe may be multinucleate, and a mutation may occur in but one nucleus. The latent period then would be the time required for nuclear segregation.

phenotypic mixing the production of a virus with a phenotype that does not match its genotype. During the assembly of a virus, nucleic acid and protein components are drawn randomly from two pools. In a host infected simultaneously by mutant and wild-type viruses, progeny phages are assembled without regard to matching the coat components to the genes in the nucleic acid core. Thus, discrepancies sometimes arise, and a virus is produced with coat proteins that are not specified by its genome. *See* **pseudovirion, reassortant virus.**

phenotypic plasticity a phenomenon in which a given genotype may develop different states for a character or group of characters in different environments; genotype-environment interaction (*q.v.*). *See* **norm of reaction.**

phenotypic sex determination control of gonad development by nongenetic stimuli. For example, the incubation temperature of fertilized eggs determines the type of sexual development in some turtle species.

phenotypic variance the total variance observed in a character. *See* **genetic variance, variance.**

phenylalanine *See* **amino acid.**

phenylketonuria a hereditary disorder of amino acid metabolism in humans, inherited as an autosomal recessive. Homozygotes cannot convert phenylalanine to tyrosine due to a lack of the liver enzyme phenylalanine hydroxylase. Brain dysfunction characteristic of the disease can be avoided by early dietary restriction of phenylalanine. Prevalence is 1/11,000. Abbreviated PKU. Phenylketonuria is sometimes called Følling disease after A. Følling, who discovered the underlying metabolic disorder in 1934.

phenylthiocarbamide a compound used to test for human taste sensitivity. *See* **taste blindness.**

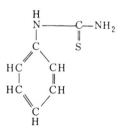

phenylthiourea phenylthiocarbamide (*q.v.*).

pheoplasts a brown plastid of brown algae, diatoms, and dinoflagellates. Also spelled *phaeoplast.*

pheromone a chemical exchanged between members of the same species that affects behavior. Examples of such pheromones are the sex attractants, alarm substances, aggregation-promotion substances, territorial markers, and trail substances of insects.

Philadelphia chromosome an aberrant chromosome observed in metaphase figures from bone marrow cells of patients suffering from chronic myelogenous leukemia. The chromosome aberration was at first thought to involve a terminal deletion of chromosome 22. It is now clear that the Philadelphia chromosome is a translocation in which approximately one half the long arm of chromosome 22 is moved to a terminal position on another chromosome. About 90% of the time, the long arm of chromosome 9 is the recipient. *See* **Appendix C,** 1971, O'Riordan *et al.*

philopatry a mode of dispersal in which propagules remain at or very near their point of origin.

phi X174 (ϕX174) an icosahedral bacteriophage of *E. coli*. Its circular, single-stranded DNA is made up of 5,386 nucleotides and contains 10 genes (5 of which overlap). *See* **Appendix C**, 1959, Sinsheimer; 1967, Goulian *et al.*; 1977, Sanger *et al.*; **overlapping genes.**

phloem the vascular tissue that conducts nutrient fluids in vascular plants.

phocomelia absence of the proximal portion of a limb or limbs, the hands or feet being attached to the trunk by a single bone. A genetic form occurs in humans that is inherited as an autosomal recessive. Phocomelia may also be caused by exposure of the developing embryo to the drug thalidomide (*q.v.*).

phosphate bond energy the energy liberated as one mole of a phosphorylated compound undergoes hydrolysis to form free phosphoric acid. *See* **ATP.**

phosphodiester any molecule containing the linkage

$$
\begin{array}{c}
\quad\;\; O \\
\quad\;\; \| \\
R-O-P-O-R' \\
\quad\;\; | \\
\quad\;\; O\,-
\end{array}
$$

where R and R' are carbon-containing groups, O is oxygen, and P is phosphorus. This type of covalent chemical bond involves the 5′ carbon of one pentose sugar (ribose or deoxyribose) and the 3′ carbon of an adjacent pentose sugar in RNA or DNA chains. *See* **deoxyribonucleic acid.**

phosphodiesterase I a 5′ exonuclease that removes, by hydrolysis, 5′ nucleotides from the 3′ hydroxy-terminus of oligonucleotides.

phospholipid a lipid containing phosphate esters of glycerol or sphingosine.

phosphorescence *See* **luminescence.**

phosphorus an element universally found in small amounts in tissues, a component of nucleic acids. Atomic number 15; atomic weight 30.4735; valence 5^+; most abundant isotope ^{31}P; radioisotope ^{32}P (*q.v.*).

phosphorylation the combination of phosphoric acid with a compound.

phosphovitin a phosphoprotein of molecular weight 40,000. Two molecules of phosphovitin and one of lipovitellin comprise the basic subunit of the amphibian yolk platelet. *See* **vitellogenin.**

photoactivated crosslinking a technique for crosslinking a nucleic acid (e.g., a tRNA chain) to

a polypeptide chain with which it is functionally associated (e.g., its cognate synthetase) by irradiating the synthetase-tRNA complex with ultraviolet light. The technique is used to locate the points of intimate contact between the two molecules. *See* **RNase protection.**

photoautotroph an organism that can produce all of its nutritonal and energy requirements using only inorganic compounds and light.

photoelectric effect a process in which a photon ejects an electron from an atom so that all the energy of the photon is absorbed in separating the electron and in imparting kinetic energy to it. *Contrast with* **Compton effect.**

photographic rotation technique a technique used to establish the symmetry of a structure (such as a virus) observed in an electron micrograph. The micrograph is printed n times, the enlarging paper being rotated $360°/n$ between successive exposures. Structures with n-fold radial symmetry show reinforcement of detail, whereas the micrograph will show no reinforcement when tested for $n - 1$ or $n + 1$ symmetry.

photolysis decomposition of compounds by radiant energy, especially light.

photomicrography the technique of making photographs through a light microscope.

photon a quantum of electromagnetic energy.

photoperiodism the response of organisms to varying periods of light. In plants, for example, the photoperiod controls flowering. *See* **phytochrome.**

photophosphorylation the addition of phosphate to AMP and ADP through the energy provided by light during photosynthesis.

photoreactivating enzyme an exonuclease catalyzing a photochemical reaction that removes UV-induced thymine dimers from DNA.

photoreactivation reversal of injury to cells caused by ultraviolet light, accomplished by postirradiation exposure to visible light waves. *See* **Appendix C**, 1949, Kelner; **thymine dimer.**

photoreceptor a biological light receptor. *See* **cone, ommatidium, rod.**

photosynthesis the enzymatic conversion of light energy into chemical energy in green plant cells resulting in the formation of carbohydrates and oxygen from carbon dioxide and water.

phototrophic photoautotrophic (*q.v.*).

phototropism a tropism in which light is the stimulus.

phragmoplast a differentiated region of the plant cell that forms during late anaphase or early telophase between the separating groups of chromosomes. The phragmoplast contains numerous microtubules that function to transport material used by the developing cell plate. Once the cell plate forms it divides the phragmoplast into two parts, and subsequently the cell plate is transformed into the middle lamella of the mature cell wall.

Phycomyces a genus belonging to the same family as the common bread molds *Mucor* and *Rhizopus.* Two species, *P. nitens* and *P. blakesleeanus,* are favorites for genetic studies.

phyletic evolution the gradual transformation of one species into another without branching; anagenesis; vertical evolution.

phyletic speciation *See* **phyletic evolution.**

phylogenetic tree a diagram that portrays the hypothesized genealogical ties and sequence of historical ancestor/descendant relationships linking individual organisms, populations, or taxa. When species are considered they are represented as line segments, and points of branching correspond to subsequent speciation events. When possible, the lineage is presented in relation to a geological time scale. *See* **Appendix C,** 1936, Sturtevant and Dobzhansky; 1963, Margoliash; **cladogram.**

phylogeny the relationships of groups of organisms as reflected by their evolutionary history.

phylum *See* **classification.**

Physarum polycephalum *See* **Myxomycota.**

physical map a map of the linear order of genes on a chromosome with units indicating their distances determined by methods other than genetic recombination (e.g., nucleotide sequencing, overlapping deletions in polytene chromosomes, electron micrographs of heteroduplex DNAs, etc.).

physiological saline an isotonic, aqueous solution of salts used for temporarily maintaining living cells.

physiology the study of the dynamic processes of living organisms.

phytochrome the molecule responsible for the photoperiodic control of flowering. During the day, a form of phytochrome that absorbs light at the far-red end of the spectrum accumulates in plants. This form of the pigment inhibits flowering in short-day plants (*q.v.*) and stimulates flowering in long-day plants (*q.v.*). During darkness this compound reverts to a red-absorbing form that is stimulatory to the flowering of short-day plants and inhibitory to the flowering of short-day plants and inhibitory to

long-day plants. Whether a plant belongs to the short- or long-day class is genetically controlled.

phytohemagglutinin a lectin (*q.v.*) extracted from the red kidney bean, *Phaseolus vulgaris,* that agglutinates human erythrocytes and stimulates lymphocytes to undergo mitosis. *See* **Appendix C,** 1960, Nowell.

phytohormone a plant hormone. *See* **auxin, cytokinin, gibberellin.**

phytotron a group of rooms used for growing plants under controlled, reproducible, environmental conditions.

pI isoelectric point (*q.v.*).

picornavirus a group of extremely small RNA viruses. The name is derived from the prefix *pico* (meaning small) + *RNA* + *virus.* The polio virus belongs to this group.

piebald designating an animal, especially a horse, having patches of black and white or of different colors. A pinto horse is piebald.

Pieris a genus of small butterflies extensively studied by ecological geneticists.

pig *See Sus scrofa.*

pigeon, domestic breeds the homing pigeon, carneaux, dragoon, white Maltese, white king, fantail, pouter, tumbler, roller, Jacobin, barb, carrier pigeon, and the ptarmigan. The species is *Columbia livia domestica.*

pilin the protein from which a pilus is constructed.

pillotinas large symbiotic spirochaetes living in the hind gut of termites. These spirochaetes are of interest because they contain microtubules, and the presence of microtubules in spirochaetes supports the symbiotic theory of the origin of undulipodia. *See* **motility symbiosis.**

pilus (*plural* **pili**) a filamentous, hollow appendage extending from the surface of a bacterial cell. Some pili serve as conjugation tubes: e.g., on "male" (F$^+$ and Hfr) cells, the F pilus (or sex pilus) functions as a conjugation tube through which donor DNA is transmitted to a recipient, "female" (F$^-$) cell. *See* **F factor.**

pin the type of flower characterized by long styles and low anthers found among distylic species such as seen in the genus *Primula.* See **thrum.**

pinocytosis the engulfment of liquid droplets by a cell through the production of pinosomes (*q.v.*).

pinosome a membrane-bounded cytoplasmic vesicle produced by the budding off of localized invaginations of the plasma membrane. Recently pinocytosed fluids are segregated within the cell in pinosomes.

pioneer the first plant or animal species to become established in a previously uninhabited area.

pistil the female reproductive organ of the flower (*q.v.*), consisting of ovary, style, and stigma.

pistillate designating a flower having one or more pistils and no stamens.

Pisum sativum the garden pea, Mendel's experimental organism. *See* **Appendix C,** 1822–24, Knight, Gross, Seton; 1856, 1865, Mendel.

pitch the number of base pairs in one complete revolution of a DNA double helix.

pith parenchymatous tissue in the center of roots and stems.

Pithecanthropus erectus currently called *Homo erectus erectus* (*q.v.*).

Pitressin a trademark for vasopressin.

pituitary dwarfism in humans, a form of ateliosis inherited as an autosomal recessive. There seems to be three types. In *primordial dwarfism,* growth hormone (*q.v.*) alone is deficient, whereas in *panhypopituitarism* there is a deficiency of all anterior pituitary hormones. In *Laron dwarfism,* excessive amounts of growth hormone are present, suggesting that the defect involves the growth hormone receptors.

pituitary gland the master endocrine gland, which lies beneath the floor of the brain, within the skull, of vertebrates. *See* **adenohypophysis, neurohypophysis.**

pK dissociation constant (*q.v.*).

PKU phenylketonuria (*q.v.*).

placebo an inactive substance given to certain patients randomly chosen (without their knowledge) from a group. A new medicinal compound is administered to the other patients. The effectiveness of the compound is determined by comparing the progress of the treated patients with those receiving the placebo.

placenta an organ consisting of embryonic and maternal tissues in close union through which the embryo of a viviparous animal is nourished.

Plantae the plant kingdom. Its members are all eukaryotes made up of cells containing green plastids. *See* **Appendix A; Eukaryotes.**

plaque a clear, round area on an otherwise opaque layer of bacteria or tissue-cultured cells where the cells have been lysed by a virulent virus. In the example shown below, the Petri dish contains a gel in which a nutrient broth is suspended. Covering the surface of the medium is a lawn of bacteria. These arose from a layer of 1×10^8 bacteria that was deposited upon the agar surface. Subsequently, a small number of viruses was spread over these cells. The holes represent points at which a phage particle was present. The initial phage particle was adsorbed to a bacterium and after a short period the bacterium lysed, releasing new phage. These new phage in turn attacked neighboring bacteria and produced more phage. The process continued until holes visible to the naked eye appeared. Different phages can sometimes be recognized by the morphology of the plaques they produce. The numerous small plaques in the figure are from T_6 bacteriophage, while the four large plaques are from phage T_7. Animal viruses will also attack monolayer cultures of animal cells on Petri plates, so it is possible to assay virus titer in the same way by plaque counts. *See* **Appendix C,** 1932, Ellis and Delbrück; 1952, Dulbecco.

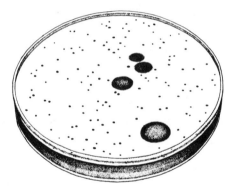

plaque assay a technique for counting the number of complete, infective phage in a culture by an appropriate dilution to ensure that no more than one phage can infect a given host cell, followed by counting the number of plaques that develop on a bacterial lawn.

plaque-forming cells in immunology, antibody-secreting cells that can cause a hemolytic plaque with the aid of complement on a lawn of erythrocytes. The term is also applied to cells in certain assay systems where the cell killing that creates the plaque is cell mediated, rather than antibody dependent.

plasma *See* **blood plasma.**

plasmablasts highly proliferative cells that are developmental intermediates between small B lymphocytes and immunoglobulin-secreting mature plasma cells.

plasma cell a terminally differentiated immunoglobulin-secreting cell of the B lymphocyte lineage.

plasmacytoma synonymous with myeloma (*q.v.*).

plasmagene a self-replicating cytoplasmic gene within an organelle or a symbiont of a eukaryotic cell or in a plasmid of a bacterial cell.

plasmalemma plasma membrane (*q.v.*).

plasma lipoproteins a multicomponent complex of proteins and lipids circulating in the plasma. Plasma lipoproteins are grouped by their densities into four classes: high-density lipoproteins (HDL), low-density lipoproteins (LDL), intermediate-density lipoproteins (IDL), and very-low-density lipoproteins (VLDL). LDLs are about 25% protein and 75% lipid, and about half of the lipid is cholesterol. LDLs serve as the major cholesterol transport system in human plasma. *See* **Appendix C,** 1975, Goldstein and Brown; **familial hypercholesterolemia.**

plasma membrane the membrane surrounding a cell. *See* **unit membrane.**

plasma protein any of the dissolved proteins of vertebrate blood plasma that are responsible for holding fluid in blood vessels by osmosis. *See* **Appendix A,** 1955, Smithies.

plasma thromboplastin component *See* **hemophilia.**

plasma transferrins beta globins that bind and transport iron to the bone marrow and tissue-storage areas. Numerous heritable transferrin variants are known.

plasmid an extrachromosomal genetic element found in a variety of bacterial species that generally confers some evolutionary advantage to the host cell (i.e., resistance to antibiotics, production of colicins, etc.). Plasmids are double-stranded, closed DNA molecules ranging in size from 1 to 200 kb. Plasmids whose replication is coupled to that of the host so that only a few would be present per bacterium are said to be under *stringent control.* Under *relaxed control,* the number of plasmids per host cell may be from 10 to 100.

plasmid cloning vector a plasmid used in recombinant DNA experiments as an acceptor of foreign DNA. Plasmid cloning vectors are generally small and replicate in a relaxed fashion. They are marked with antibiotic resistance genes and contain recognition sites for restriction endonucleases in regions of the plasmid that are not essential for its replication. One widely used plasmid cloning vector is pBR322 (*q.v.*). *See* **Appendix C,** 1973, Cohen *et al.*; **Ti plasmid.**

plasmid conduction the process whereby a conjugative plasmid can help a nonmobilizable plasmid to be transferred from a donor to a recipient cell. Nonmobilizable plasmids cannot prepare their DNA for transfer, but can become mobilized by recombination with a conjugative plasmid to form a single transferable DNA molecule. *See* **plasmid donation, relaxation complex.**

plasmid donation the process by which a nonconjugative plasmid is transferred from a donor to a recipient cell via the effective contact function provided by a conjugative plasmid. In *E. coli,* for example, the ColE1 plasmid does not have genes for the establishment of effective contact, but an F plasmid in the same cell is conjugative and can provide this function.

plasmid engineering *See* **recombinant DNA research.**

plasmid fusion *See* **replicon fusion.**

plasmin an enzyme, cleaved from plasminogen, that hydrolyzes fibrin. *See* **blood clotting.**

plasminogen a blood proenzyme that is activated by cleavage of a single arg-val peptide bond to form the functional enzyme plasmin (*q.v.*).

plasmodesmata cytoplasmic threads that form delicate protoplasmic connections between adjacent plant cells.

plasmodium a multinucleate mass of cytoplasm lacking internal cell boundaries. The term is used to refer to the amoeboid stage in the life cycle of Sporozoa (*q.v.*) and to the vegetative stage of Myxomycota (*q.v.*).

Plasmodium *See* **malaria.**

plasmogamy the fusion of protoplasts of two haploid cells without the fusion of their nuclei, as in certain fungi.

plasmon all extrachromosomal hereditary agents considered collectively.

plasmosome a term in the older literature referring to the nucleolus (*q.v.*).

plasmotomy fission, unrelated to nuclear division, of a multinucleated protist into two or more multinucleated sibling cells.

Plectonemic spiral.

plastid a self-replicating cytoplasmic organelle of algal and plant cells, such as a chloroplast, chromoplast, elaioplast, or leukoplast.

plastoquinone a group of quinones (*q.v.*) involved in the transport of electrons during photosynthesis in chloroplasts.

plate 1. a flat, round dish (Petri plate) containing agar and nutrients for the culture of bacteria. 2. to spread or inoculate cells on the surface of semisolid medium in such a culture dish. 3. a geological plate. *See* **plate tectonics.**

platelet-derived growth factor a protein synthesized by platelets that is released into the serum during blood clotting. PDGF represents the major growth factor in human serum, and it is a potent mitogen for connective tissue and glial cells. There are extensive similarities between the amino acid sequences of PDGF and the product of the *v-sis* oncogene of the simian sarcoma virus (*q.v.*). This suggests that *v-sis* resulted from viral recombination with a host gene encoding PDGF. *See* **Appendix C,** 1983, Doolittle *et al.;* **proto-oncogenes.**

platelets anucleate, oval, colorless corpuscles present in blood. Platelets, which are 1/3–1/2 the size of erythrocytes, originate from projections pinched off the surface of megakaryocytes (*q.v.*), and function in blood clotting.

plate tectonics movement of large segments of the earth's crust (plates) as a consequence of convection currents in the hot interior. Together with the phenomenon of sea-floor spreading, these geological events provide mechanisms that help explain biogeographical distributions of plants and animals. *See* **Appendix C,** 1968, Morgan *et al.*

plating efficiency *See* **absolute plating efficiency, relative plating efficiency.**

platyrrhine referring to members of the broadnosed primate infraorder Platyrrhini that includes the New World monkeys. *Compare with* **catarrhine.**

playback experiment an experiment designed to recover a DNA strand that has been saturated with RNA (*see* **RNA-driven hybridization**), and then using it in a further reassociation reaction to show that its $C_0t_{1/2}$ (*q.v.*) corresponds to that expected of nonrepetitive DNA.

plectonemic spiral a spiral in which two parallel threads coil in the same direction about one an-

other and cannot be separated unless uncoiled. The component strands of a DNA duplex are plectonemically coiled. *See* **paranemic spiral, relational coiling.**

pleiomorphism the occurrence of variable phenotypes in a genetically uniform group of organisms. *See* **phenotypic plasticity.**

pleiotropy the phenomenon where a single gene is responsible for a number of distinct and seemingly unrelated phenotypic effects.

Pleistocene the ice age, lasting from 10,000 B.C. to the beginning of the Pliocene. One of the two epochs of the Quaternary period. *Homo erectus* appeared, then *Homo sapiens. See* **geologic time divisions.**

pleomorphic having more than one form or shape.

plesiomorphic 1. in classification, referring to a character state that occurs in the group of organisms being considered, but also outside the group. Traits of this type cannot be used to define the group or to indicate that its members were derived from a common ancestor. 2. in evolution, an original primitive feature thought to have arisen in an ancestor of all the taxa being considered. *See* **apomorphic, cladogram.**

Pleurodeles salamanders of two species belonging to this genus, *P. waltlii* and *P. poireti,* have been studied both genetically and cytologically. Working maps of the oocyte lampbrush chromosomes of both species are available.

pleuropneumonia-like organisms a group of bacteria that do not form cell walls. PPLOs are included in the phylum Aphragmabacteria (*see* **Appendix A**). One PPLO, *Mycoplasma pneumoniae,* is the cause in humans of an atypical pneumonia.

Pliocene the most recent epoch in the Tertiary period during which many of the earlier mammals became extinct and the first hominids appeared. *See* **Australopithecine, geologic time divisions.**

-ploid a combining form used in cytology and genetics to designate a particular multiple of the chromosome set of the nucleus of an organism, as 16-ploid, 32-ploid, etc.

ploidy *See* **polyploidy.**

plumage pigmentation genes a group of genes controlling pigmentation of chicken feathers. Pigment will not be produced unless the gene *C* is present. A second gene *I*, which inhibits pigment formation, is located on a different chromosome. The White Leghorn breed of chickens has the genotype *IICC*, whereas the White Plymouth Rock is *iicc*.

pluripotent   pertaining to an embryonic tissue that has a large number of possible developmental fates because determination has not yet taken place. *See* **totipotency.**

plus and minus techniques   *See* **DNA sequencing techniques.**

plus (+) and minus (−) viral strands   **1.** in a single-stranded RNA virus, a plus strand is one having the same polarity as viral mRNA and containing codon sequences that can be translated into viral protein. A minus strand is a noncoding strand that must be copied by an RNA-dependent RNA polymerase to produce a translatable mRNA. **2.** in a single-stranded DNA virus, a plus strand is one contained in the virus particle or any strand having the same base sequence. A minus strand has a base sequence complementary to the plus strand; mRNA can be transcribed from the minus strand.

P-M hybrid dysgeneses   *See* **hybrid dysgenesis, M strain, P strain.**

pneumococcal transformation   *See* **transformation.**

pod corn   *Zea mays tunicata,* a primitive variety of corn characterized by kernels, each of which is enclosed in a husk.

podophyllin   a mitotic poison isolated from the plant *Podophyllum peltatum.*

Podospora anserina   a coprophilous fungus with a haploid chromosome number of 7 and seven well-defined linkage groups. The genetic control of sexual and vegetative reproduction has been extensively studied in this species.

point mutation   **1.** in classical genetics, any mutation that is not associated with a cytologically detectable chromosomal aberration or one that has no effect on crossing over (and therefore is not an inversion) and complements nearby lethals (and therefore is not a deficiency). **2.** in molecular genetics, a mutation caused by the substitution of one nucleotide for another. *See* **rotational base substitution.**

Poisson distribution   a function that assigns probabilities to the sequence of outcomes of observing no events of a specified type, one event, two events, and so on without limit. Events following a Poisson distribution are completely randomized. Suppose that within a defined spatial or temporal region one looks at the events occurring in nonoverlapping subregions. A Poisson distribution for the number of events observed in the total region implies that the events occurring in the subregions do not affect each other. A Poisson distribution will not be found if the events are correlated positively (in the case of clumping) or negatively (in the case of mutual repulsion). The Poisson is specified by the average number of events per observation, and its mean and variance are equal. The formula of the function is $P_i = (m^i e^{-m})/i!$ where m is the mean number of events; $i!$ is the factorial $i(i-1)(i-2) \ldots (2)(1)$; e is the base of natural logarithms, and i is the number for which the probability P_i is given. If m = 1.2, for example, the distribution is as shown in the table that appears at the bottom of this page.

Many natural distributions follow a Poisson, including the number of radioactive disintegrations of a radioisotope in a fixed period of time, or the number of larvae of a particular invertebrate species captured by towing a plankton net through a specified volume of sea water.

pokeweed mitogen   a lectin (*q.v.*) extracted from rhizomes of the pokeweed, *Phytolocca americana,* that stimulates the proliferation of lymphocytes, particularly mouse T and B lymphocytes.

pol I, II, III   *See* **DNA polymerase.**

polar   referring to water-soluble chemical groups such as a hydrophilic side chain of an amino acid.

polar body   the minute cell produced and discarded during the development of an oocyte. A polar body contains one of the nuclei derived from the first or second division of meiosis, but has practically no cytoplasm. *See* **ootid nucleus.**

polar fusion nucleus   in plants, the product of the fusion of the two polar nuclei. This, after fusing with a male nucleus, gives rise to the tripoid endosperm nucleus. *See* **double fertilization.**

polar gene conversion   a phenomenon in which a gradient of conversion frequencies exists from one end of a gene to the other; sites closer to one end of a gene usually have higher conversion frequencies than do those farther from that end.

polar granules   RNA-rich granules in the posterior ooplasm of the insect egg. In *Drosophila,* only pole

Number observed	0	1	2	3	4	5	6	7	Above >7
Probability	0.3012	0.3614	0.2169	0.0867	0.0260	0.0062	0.0012	0.0002	.0002

cells engulfing such granules can subsequently form germ cells.

polarity gradient　the quantitative effect of a polarity mutation in one gene on the expression of later genes in the operon. The effect is a function of the distance between the nonsense codon and the next chain-initiation signal.

polarity mutant　a mutant gene that is able to reduce the rate of synthesis of the proteins that normally would be produced by wild-type alleles of the genes lying beyond it on the chromosome. Such genes exert their effect during the translation of a polycistronic message (*q.v.*). *See* **regulator gene, translation.**

polarization microscope　a compound light microscope used for studying the anisotropic properties of objects and for rendering objects visible because of their optical anisotropy.

polar nuclei　*See* **ootid nucleus, pollen grain, polocyte.**

polaron　a chromosomal segment within which polarized genetic recombination takes place by gene conversion.

polar tubules　microtubules of the spindle apparatus that originate at the centriolar or polar regions of the cell. *See* **chromosomal tubules.**

pole cell　one of the cells that are precociously segregated into the posterior pole of the insect embryo before blastoderm formation. Among these cells are the progenitors of the germ cells.

Polish wheat　*Triticum polonicum* (N = 14). *See* **wheat.**

pollen grain　a microspore in flowering plants that germinates to form the male gametophyte (pollen grain plus pollen tube), which contains three haploid nuclei. One of these fertilizes the ovum, a second fuses with the two polar nuclei to form the 3N endosperm, and the third (the vegetative nucleus) degenerates once double fertilization has been accomplished.

pollen mother cell　microsporocyte.

pollen-restoring gene　a gene that permits normal microsporogenesis to occur in the presence of a cytoplasmic male sterility factor.

pollen tube　the tube formed from a germinating pollen grain that carries male gametes to the ovum. *See* **Appendix C,** 1830, Amici.

pollination　the transfer of pollen from anther to stigma. *See* **Appendix C;** 1694, Camerarius; **pollen grain.**

polocyte　the small degenerate sister cell of the secondary oocyte. This cell generally divides into two polar bodies, which disintegrate. *See* **polar body.**

polyacrylamide gel　a gel prepared by mixing a monomer (acrylamide) with a cross-linking agent (N,N′-methylenebisacrylamide) in the presence of a polymerizing agent. An insoluble three-dimensional network of monomer chains is formed. In water, the network becomes hydrated. Depending upon the relative proportions of the ingredients, it is possible to prepare gels with different pore sizes. The gels can then be used to separate biological molecules like proteins of a given range of sizes.

polyacrylamide gel electrophoresis　*See* **electrophoresis.**

polyadenylation　enzymatic addition of several adenine nucleotides to the 3′ end of mRNA molecules as part of the processing that primary RNA transcripts undergo prior to transport from the nucleus to the cytoplasm. The added segment is referred to as a "poly-A tail." Histone mRNAs lack poly-A tails. *See* **Appendix C,** 1971, Darnell *et al.;* **posttranscriptional modification.**

polyandry　the state of having more than one male mate at one time.

poly-A tail　*See* **polyadenylation.**

polycentric chromosome　polycentromeric chromosome. *See* **centromere.**

polycentromeric chromosome　*See* **centromere.**

polycistronic message　a giant messenger RNA molecule specifying the amino acid sequence of two or more proteins produced by adjacent cistrons in the same operon. *See* **transcription unit.**

polyclonal　an adjective applied to cells or molecules arising from more than one clone; e.g., an antigenic preparation (even a highly purified one) elicits the synthesis of various immunoglobulin molecules. These antibodies would react specifically with different components of the complex antigen molecule. Thus, the antibody preparation generated by such an antigen would be polyclonal in the sense that it would contain immunoglobulins synthesized by different clones of B lymphocytes.

polyclone　*See* **compartmentalization.**

Polycomb　a gene in *Drosophila* located in the proximal region of 3L. *Pc* encodes a protein that acts as a repressor of the genes at the *Antennapedia* complex (*q.v.*) and the *bithorax* complex (*q.v.*). This repressor binding can be visualized on the giant polytene chromosomes by immunochemical staining. *See* **Appendix C,** 1989, Zink and Paro.

polycomplex structures, observed in certain insects, within oocyte nuclei, formed by the fusion of components from synaptonemal complexes (*q.v.*) that have detached from the diplotene chromosomes.

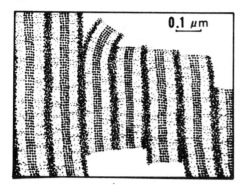

0.1 μm

polycythemia a disease in man characterized by overproduction of red blood cells.

polydactyly the occurrence of more than the usual number of fingers or toes.

polyembryony the formation of multiple embryos from a zygote by its fission at an early developmental stage. Monozygotic twins constitute the simplest example of polyembryony. Monozygotic quadruplets are commonly formed by armadillos. In certain parasitic wasps, as many as 2,000 embryos can be formed by polyembryony from a single zygote.

polyene antibiotic any antibiotic active against fungi, but not bacteria (nystatin and oligomycin are examples).

polyestrous mammal *See* **estrous cycle.**

polyethylene glycol a chemical used to promote the fusion of tissue-cultured cells, as in the production of a hybridoma (*q.v.*).

polygamy polandry and/or polygyny. *Compare with* **monogamy.**

polygene one of a group of genes that together control a quantative character. *See* **Appendix C,** 1941, Mather; **oligogene, quantitative inheritance.**

polygenic character a quantitatively variable phenotype dependent on the interaction of numerous genes.

polyglucosan a polymer such as glycogen made up of a chain of glucose units.

polygyny the mating of a male with more than one female. *Compare with* **monogamy, polyandry.**

polyhedrin *See* **baculoviruses.**

polylinker site a stretch of DNA engineered to have mutliple sites for cleavage by specific restriction endonucleases (*q.v.*).

polymer a macromolecule composed of a covalently bonded collection of repeating subunits or monomers linked together during a repetitive series of similar chemical reactions. Each strand of DNA is a linear polymer of nucleotide monomers. A linear polypeptide chain is a polymer of amino acid monomers.

polymerase any enzyme that catalyzes the formation of DNA or RNA molecules from deoxyribonucleotides and ribonucleotides, respectively (e.g., DNA polymerase, RNA polymerase).

polymerase chain reaction a technique for copying the complementary strands of a target DNA molecule simultaneously for a series of cycles until the desired amount is obtained. First, primers are synthesized that have nucleotide sequences complementary to the DNA that flanks the target region. The DNA is heated to separate the complementary strands and then cooled to let the primers bind to the flanking sequences. A heat-stable DNA polymerase is added, and the reaction is allowed to proceed for a series of replication cycles. Twenty will yield a millionfold amplification; thirty cycles will yield an amplification factor of one billion. *See* **Appendix C,** 1985, Saiki, Mullis *et al.;* **Taq DNA polymerase.**

polymerization the formation of a polymer from a population of monomeric molecules.

polymerization start site the nucleotide in a DNA promoter sequence from which the first nucleotide of an RNA transcript is synthesized.

polymorphic locus a genetic locus, in a population, at which the most common allele has a frequency less than 0.95. *Compare with* **monomorphic locus.**

polymorphism the existence of two or more genetically different classes in the same interbreeding population (Rh-positive and Rh-negative humans, for example). The polymorphism may be transient, or the proportions of the different classes may remain the same for many generations. In the latter case, the phenomenon is referred to as *balanced polymorphism*. If the classes are located in different regions, *geographic polymorphism* exists. *See* **Appendix C,** 1954, Allison; 1966, Lewontin and Hubby.

polymorphonuclear leukocyte *See* **granulocyte.**

polyneme hypothesis the concept that a newly formed chromatid contains more than one DNA duplex. *Contrast with* **unineme hypothesis.**

polynucleotide a linear sequence of 20 or more joined nucleotides. *See* **oligonucleotide.**

polynucleotide kinase an enzyme that phosphorylates the 5′ hydroxyl termini produced by endonucleases (*q.v.*).

polynucleotide phosphorylase the first enzyme involved in the synthesis of nucleic acids to be isolated. It links ribonucleotides together in a random fashion and is used to produce artificial mRNA molecules. *See* **Appendix C,** 1955, Grunberg-Manago and Ochoa.

polyoma virus a virus that induces tumors in newborn mice, rats, and hamsters and can also transform cultured mouse or rat cells. The genome of the virus is a double-stranded, supercoiled, circular DNA molecule containing about 5,300 base pairs. *See* **Appendix C,** 1983, Rassoulzadegan *et al.;* **oncogenic virus, transformation.**

polyp 1. the sedentary form of a coelenterate. 2. a small stalked neoplasm projecting from a mucous surface (for example, an intestinal polyp).

polypeptide a polymer made up of amino acids linked together by peptide bonds (*q.v.*).

polyphasic lethal a mutation characterized by two or more lethal phases separated by developmental periods in which it produces no deaths.

polyphenism the occurrence of several phenotypes in a population that are not due to genetic differences between the individuals in question.

polypheny pleiotropy (*q.v.*).

polyphyletic group a group of species classified together, some members of which are descended from different ancestral populations. *Contrast with* **monophyletic group.**

polyploid designating an individual having more than two sets of chromosomes.

polyploidy the situation where the number of chromosome sets is greater than two. *See* **Appendix C,** 1917, Winge; 1937, Blakeslee and Avery; **allopolyploid, autopolyploid, colchicine, wheat.**

polyprotein a cistronic product that is posttranslationally cleaved into several independent proteins. For example, an enkephalin precursor protein contains six copies of met-enkephalin and one copy of leu-enkephalin. *See* **enkephalins.**

polyribonucleotide phosphorylase *See* **polynucleotide phosphorylase.**

polyribosome polysome.

polysaccharide a carbohydrate formed by the po-

lymerization of many monosaccharide units. Starch, cellulose, and glycogen (*q.v.*) are examples of polysaccharides.

polysomaticism the phenomenon where an individual contains diploid and polyploid cells in the same tissue.

polysome a multiple structure containing a group of ribosomes held together by a molecule of messenger RNA. A contraction of *polyribosome. See* **Appendix C,** 1962, Warner *et al.*

polysomy the reduplication of some but not all of the chromosomes of a set beyond the normal diploid number. A metafemale *Drosophila* is polysomic (trisomic for the X).

polyspermy the penetration of more than one sperm into one ovum at the time of fertilization.

Polysphondylium pallidum *See* **Acrasiomycota.**

polytene chromosome a giant cablelike chromosome consisting of many identical chromatids lying in parallel. The chromatin is hypercoiled in localized regions, and since the chromatids are in register, a pattern of bands is produced vertical to the long axis of the chromosome. Polytene chromosomes are found within a limited number of organisms. They are present in the macronucleus anlage of some ciliates, in the synergids and antipodal cells of the ovules of certain angiosperms, and in various tissues of dipterans. The *Drosophila* salivary gland chromosomes (*q.v.*) have been studied most extensively. *See* **Appendix C,** 1881, Balbiani; 1912, Rambousek; 1934, Bauer; 1952, Beermann; 1959, Pelling; 1969, Ammermann; 1980, Gronemeyer and Pongs; *Anopheles,* **Balbiani ring,** *Calliphora erythrocephala, Chironomus, Culex pipiens, Glyptotendipes barbipes, otu* **mutation,** *Rhynchosciara, Sciara, Smittia.*

polyteny the interphase replication of each chromosome to produce giant chromosomes that are polytene cables.

polythetic group a group of organisms that share a large number of features, no single one of which is either essential for group membership or is sufficient to make an organism a member of the group.

polytopic pertaining to the distribution of subspecies in two or more geographically discontinuous areas.

polytrophic meroistic ovary *See* **insect ovary types.**

polytypic species a species subdivided into a number of specialized races.

pome a fleshy, many-seeded fruit such as the

apple or pear, in which the enlarged end of the flower stalk forms much of the flesh.

Pompe disease   a hereditary glycogen storage disease in humans arising from a deficiency of the lysosomal enzyme α-1,4-glucosidase, due to a recessive gene on chromosome 17. Prevalence of the disease is 1/100,000.

Pongidae   the family of primates containing all anthropoid apes.

Pongo pygmaeus   the orangutan, a primate with a haploid chromosome number of 24. About 30 biochemical marker genes have been distributed among 20 linkage groups. *See* **Hominoidea.**

popcorn   *See* **corn.**

population   a local (geographically defined) group of conspecific organisms sharing a common gene pool; also called a *deme*.

population biology   the study of the patterns in which organisms are related in space and time. Such disciplines as ecology, taxonomy, ethology, population genetics, and others that deal primarily with the interactions of organisms or groups of organisms (demes, species, etc.) are included under this term.

population cage   a special cage in which *Drosophila* populations can be reared for many generations. The cage is designed so that samples of the population can be conveniently withdrawn and food supplies can be replenished. *See* **Appendix C,** 1934, L'Héritier and Teissier.

population density   **1.** in ecology, the number of individuals of a population per unit of living space (e.g., per acre of land, per cubic meter of water, etc.). **2.** in cell or tissue culture, the number of cells per unit area or volume of a culture vessel. *See* **saturation density.**

population doubling level   in cell or tissue culture, the total number of population doublings of a cell line or strain since its initiation *in vitro.*

population doubling time   *See* **doubling time.**

population genetics   the study of the genetic composition of populations. Population geneticists try to estimate gene frequencies and detect the selective influences that determine them in natural populations. They also build mathematic models to elucidate the interaction of factors such as selection, population size, mutation, and migration upon the fixation and loss of linked and unlinked genes. *See* **Appendix C,** 1908, Hardy, Weinberg; 1930–32, Wright, Fisher, Haldane.

population structure   the manner in which a population is subdivided into local breeding groups or demes, the sizes of such demes in terms of the number of breeding individuals, and the amount of migration or gene flow between demes.

porcine   referring to members of the pig family, especially the domestic pig *Sus scrofa.*

porphyrin   any of a class of organic compounds in which four pyrrole nuclei are connected in a ring structure usually associated with metals (like iron or magnesium). Porphyrins form parts of the hemoglobin, cytochrome, and chlorophyll molecules. *See* **heme.**

position effects   the change in the expression of a gene accompanying a change in the position of the gene with respect to neighboring genes. The change in position may result from crossing over or from a chromosomal aberration. Position effects are of two types: the *stable* (S) type and the *variegated* (V) type. S-type position effects are also called *cis-trans* position effects. S-type position effects involve cistrons that possess at least two mutated sites separable by intragenic recombination. In the cis configuration (m^1 m^2/+ +) a normal phenotype is observed, whereas in the trans configuration (m^1 +/+ m^2) a mutant phenotype is produced. A reasonable explanation for such an observation would be that the mRNA transcribed from a (+ +) chromatid would function normally, whereas the mRNAs transcribed from (m^1 m^2), (m^1 +), or (+ m^2) chromatids would not. V-type position effects generally involve the suppression of activity of a wild-type gene when it is placed in contact with heterochromatin because of a chromosome aberration. Under some conditions, the gene may escape suppression, and consequently the final phenotype may be variegated, with patches of normal and mutant tissues. *See* **Appendix C,** 1925, Sturtevant; 1936, Schultz; 1945, Lewis.

positive assortative mating   *See* **assortative mating.**

positive control   control by a regulatory protein that must bind to an operator before translation can take place.

positive eugenics   *See* **eugenics.**

positive gene control   enhancement of DNA transcription through binding of specific expressor molecules to promoter sites. For example, the binding of CAP-cAMP complexes to promoters of bacterial genes involved in catabolism of sugars other than glucose facilitates binding of RNA polymerase to these operons when glucose is absent. *See* **glucose-sensitive operons.** *Compare with* **negative gene control.**

positive interference the interaction between crossovers such that the occurrence of one exchange between homologous chromosomes reduces the likelihood of another in its vicinity. *Compare with* **negative interference.** *See* **Appendix C,** 1916, Müller.

positive supercoiling *See* **supercoiling.**

positron a particle of the atomic nucleus equal in mass to the electron and having an equal but opposite (positive) charge.

postcoitum after mating.

postmating isolation mechanism *See* **postzygotic isolation mechanism.**

postmeiotic fusion a method for restoring diploidy in eggs produced by parthenogenesis, involving union of two identical haploid nuclei formed by a mitotic division of the egg nucleus.

postmeiotic segregation in ascomycete fungi such as *Neurospora,* the formation of heteroduplex regions (by meiotic crossing over) that results in aberrant 4:4 pattern of asci in which adjacent pairs of ascospores produced by mitotic division after meiosis have different genetic compositions. *See* **tetrad segregation types.**

postreductional disjunction referring to the separation of alleles at particular heterozygous loci during the first meiotic division. If the loci are represented by A and A', in the case of postreductional disjunction the two chromatids that enter one sister nucleus have one A and one A' allele, whereas in the case of prereductional disjunction both have A alleles or both A' alleles.

postreplication repair repair to a DNA region after a replication fork has passed that region or in nonreplicating DNA.

posttranscriptional modification those modifications made to pre-mRNA molecules before they leave the nucleus; also called *nuclear processing.* A gene containing three exons (E_1, E_2, and E_3) and two introns (I_1 and I_2) is diagrammed below. RNA polymerase II transcribes the 3'-5' strand of the gene to form a 5'-3' pre-mRNA molecule. Next, a methylated cap (MC) is added to the 5' end of the primary transcript, and a poly-A tail is added to the 3' end. Finally, the introns are removed and the exons spliced together during reactions that occur within a spliceosome, and the mature mRNA leaves the nucleus. *See* **alternative splicing, exon, intron, methylated cap, polyadenylation, RNA polymerase, small nuclear RNAs, spliceosome.**

posttranslational modifications alterations to polypeptide chains after they have been synthesized: e.g., removal of the formyl group from methionine in bacteria, acetylation, hydroxylation, phosphorylation, attachment of sugars or prosthetic groups, oxidation of cysteines to form disulfide bonds, cleavage of specific regions that convert proenzymes to enzymes, etc. *See* **cystine, N-formyl methionine.**

postzygotic isolation mechanism any factor that

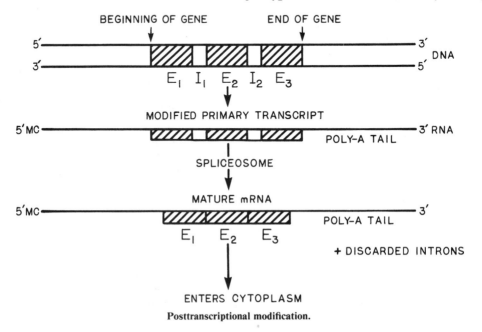

Posttranscriptional modification.

250

tends to reduce or prevent interbreeding between genetically divergent populations or species, but functioning after fertilization has occurred; includes hybrid inviability, hybrid sterility, and hybrid breakdown.

potassium an element universally found in small amounts in tissues. Atomic number 19; atomic weight 39.102; valence 1$^+$; most abundant isotope ^{39}K; radioisotope ^{42}K, half-life 12.4 hours, radiations emitted—beta particles and gamma rays.

potato *Solanum tuberosum.* Together with corn, wheat, and rice it is one of the four most valuable of the world's crops.

Potorous tridactylus the rat kangaroo, a marsupial favorable for chromosomal studies because of the small number and individuality of its chromosomes.

poultry breeds Plymouth Rock, New Hampshire, White Leghorn, Blue Andalusian, Rhode Island Red, Rhode Island White, Australorp, and Orpington. *See **Gallus domesticus.***

PP inorganic pyrophosphate.

P particle *See **kappa.***

pp60c-src the protein kinase encoded by the *c-src* gene (*q.v.*).

P1 phage a temperate bacteriophage that is widely used in transduction experiments with *E. coli.* Its genome consists of a linear double-stranded DNA molecule of about 90 kb. The molecule is terminally redundant and cyclically permuted. *See **cyclically permuted sequences.***

P22 phage a temperate phage of *Salmonella typhimurium.* The first example of transduction involved this bacteriophage. *See **Appendix C**, 1952, Zinder and Lederberg.

PPLO *pleuropneumonia-like organism* (*q.v.*).

ppm parts per million.

ppt precipitate.

pp60v-src the protein encoded by the oncogene of the Rous sarcoma virus. It is a phosphoprotein of molecular weight 60 kd, hence the *pp60* in the name; the *v-src* indicates that it is encoded by viral gene *src.* The molecule is a protein kinase that phosphorylates tyrosine subunits in cellular proteins, particularly those that form the adhesion portions of the plasmalemma.

preadaptation a new character or function arising from a mutation in an organism that becomes a useful adaptation after an environmental change, often occurring many generations later.

pre-adoptive parents *See **germinal choice.***

Precambrian the eon between the Phanerozoic and the Hadean eons. *See **geologic time divisions.*** The protists arose and evolved during this 3.2-billion-year interval.

precursor ribosomal RNA *See **pre-ribosomal RNA.***

preferential association an immunological theory that specific viral antigens interact more strongly with certain allelic products of the major immunogene complex than with others. This preferential association may make the virus more immunogenic, and hosts with the strongly interacting allelic product would tend to be more immune to viral infection than would those with weakly interacting allelic products.

preformation *See **epigenesis.***

prehensile adapted for grasping.

premating isolation mechanism *See **prezygotic isolation mechanism.***

premature initiation a second initiation of replication occurring before the first is completed; a phenomenon observed in bacteria grown in a complex nutrient broth or in some phage species that make replicas very rapidly.

prematurely condensed chromosomes interphase chromosomes that are experimentally forced to undergo rapid condensation to metaphase dimensions. This is done by fusing an interphase cell with a cell in mitosis. The interphase cell is induced to enter mitosis, and its chromosomes contract accordingly. *See **Appendix C**, 1970, Johnson and Rao.

premessenger RNA the giant RNA molecule transcribed from a structural gene. It will undergo posttranscriptional modification (*q.v.*) before it leaves the nucleus.

prepatent period the interval between infection with a pathogen or a parasite and the time when the causative agent of the ensuing disease can be detected by conventional diagnostic techniques. *See* **latent period, patent period.**

prepattern a morphogenetic pattern superimposed upon a population of cells arranged in a two-dimensional array. Specific types of differentiation are stimulated in certain cells located in defined areas. *See **Appendix C**, 1961, Tokunaga.

preprimosome *See **primosome.***

prepupal period the period between puparium formation and the eversion of the imaginal discs of the insect.

prereductional disjunction *See* **postreductional disjunction.**

preribosomal RNA the giant RNA molecule transcribed from a ribosomal RNA gene (*q.v.*). In *Drosophila,* it is 38S, in *Xenopus* it is 40S, and in HeLa cells it is 45S. After transcription, preribosomal-RNA is cleaved one or more times to generate the 5.8S, 18S, and 28S rRNAs that become components of ribosomes.

presumptive in embryology, referring to the presumed fate of an embryonic tissue in a normal development. For example, if a tissue is "presumptive neural tube," this means that in the course of normal development it will become neural tube tissue.

prezygotic isolation mechanism any factor that tends to reduce or prevent interbreeding between members of genetically divergent populations or species and functioning before fertilization occurs; includes ecological, temporal, ethological, and other isolating factors.

Pribnow box a segment upstream from the startpoint of prokaryotic structural genes to which the sigma subunit of the RNA polymerase binds. The segment is 6 bp long, and the nucleotides most commonly found are TATAAT. *See* **Appendix C,** 1975, Pribnow; **canonical sequence, Hogness box, promoter.**

primaquine-sensitivity *See* **glucose-6-phosphate dehydrogenase deficiency, malaria.**

primary culture a culture started from cells, tissues, or organs taken directly from the organism.

primary immune response *See* **immune response.**

primary ionization the ionization produced by the primary particles passing through matter as contrasted to the "total ionization," which includes the "secondary ionizations" of delta rays (*q.v.*).

primary nondisjunction sex chromosomal nondisjunction in diploid organisms with the XX, XY system of sex determination. In the homogametic sex, gametes are produced with two X chromosomes or none. In the heterogametic sex, primary nondisjunction during the first meiotic division produces gametes with no sex chromosome (O) or with an X and a Y. Primary nondisjunction during the second meiotic division produces XX and O or YY and O gametes.

primary sex ratio the ratio of male to female zygotes at conception.

primary sexual character an organ that functions in producing gametes; the ovaries and the testes.

primary speciation the splitting of one species into two, usually resulting from natural selection favoring different gene complexes in geographically isolated populations.

primary structure the specific sequence of monomeric subunits (amino acids or nucleotides) in a macromolecule (protein or nucleic acid, respectively). *See* **protein structure.**

primary transcript an RNA molecule as it was initially transcribed from DNA. In eukaryotic cells, a primary transcript usually contains introns (*q.v.*) that will be absent in the mature form of the RNA. *See* **post-transcriptional modification.**

primase in *E. coli,* the product of *dnaG* gene, responsible for initiation of precursor fragment synthesis in the lagging strand during discontinuous replication. Primase makes the RNA primer that is subsequently elongated by DNA polymerase III. The primase in *E. coli* consists of a single polypeptide of 60,000 d. Unlike RNA polymerase, primase is not inhibited by rifampicin (*q.v.*) and can polymerize deoxyribonucleotides as well as ribonucleotides *in vitro. See* **dna mutations, replication of DNA.**

primate a mammal belonging to the order Primates, which includes man, the apes, and monkeys. *See* **Appendix A.**

primed in immunology, sensitization by contact of competent lymphocytes with antigens to which they are programmed to respond.

primed synthesis technique a method for nucleotide sequencing involving enzymatically controlled extension of a primer DNA strand. *See* **DNA sequencing techniques.**

primer DNA single-stranded DNA required for replication by DNA polymerase III in addition to primer RNA (*q.v.*).

primer RNA a short RNA sequence synthesized by a primase from a template strand of DNA and serving as a required primer onto which DNA polymerase III adds deoxyribonucleotides during DNA replication. Primers are later enzymatically removed and the gaps closed by DNA polymerase I, and the remaining nicks are sealed by ligase. *See* **primase, replication of DNA.**

primordial dwarfism *See* **pituitary dwarfism.**

primordium the early cells that serve as the mitotic progenitors of an organ during development.

primosome a complex of proteins (including primase) required for the priming action that initiates synthesis of each Okazaki fragment in eukaryotic DNA replication. The complex minus primase is called a *preprimosome. See* **replication of DNA.**

Primula a genus of cowslips and primrose whose population genetics has been extensively studied.

prion an infectious protein known to be the cause of scrapie, a transmissible, degenerative disease of the nervous system of sheep and goats. The word prion is a contraction of "protein" and "infection." Prions were formerly called "slow viruses," but now are known to be devoid of nucleic acid and therefore are neither viruses nor viroids.

pro proline. *See* **amino acid.**

probability of an event the long-term frequency of an event relative to all alternative events, and usually expressed as a decimal fraction. Probabilities range between zero (if the event never occurs) and 1 (if the event always occurs and no alternative event ever occurs). In some cases we know a probability *a priori,* as in the case of a coin toss. In the long run, the coin will come up tails with a frequency of 0.5. More often, a probability must be estimated by averaging the results of many trials. *See* **conditional probability, independent probabilities.**

proband propositus (*q.v.*).

proband method a method in human genetics for comparing the proportion in families of children in which a proband shows a specific trait with the proportion expected if the trait were inherited as a single gene. For example, if one considers a group of families, each with both parents heterozygous for a recessive gene and each with two children, the proportion of affected children is 57%, not 25%. This is because the families are chosen in the first place through an affected child, and all sibships in which just by chance no affected individuals occurred have been left out. Thus, there is an *ascertainment bias* that loads the results in favor of the trait. *See* **Appendix C,** 1910, Weinberg.

probe in molecular biology, any biochemical labeled with radioactive isotopes or tagged in other ways for ease in identification. A probe is used to identify or isolate a gene, a gene product, or a protein. For example, a radioactive mRNA hybridizing with a single strand of its DNA gene, a cDNA hybridizing with its complementary region in a chromosome, or a monoclonal antibody combining with a specific protein. *See* **cDNA library, hybridoma, strand-specific hybridization probes.**

procaryote *See* **prokaryote.**

Procaryotes *See* **Prokaryotes.**

processed gene an eukaryotic pseudogene (*q.v.*) lacking introns and containing a poly-A segment near the downstream end, suggesting that it arose by some kind of reverse copying from processed

nuclear RNA into double-stranded DNA; also called *retrogene.*

processing **1.** posttranscriptional modifications of primary transcripts. **2.** antigen processing involves partial degradation by macrophages (and, in some cases, coupling with RNA) before the immunogenic units appear on the macrophage membrane in a condition that is stimulatory to cognate lymphocytes.

processive enzyme an enzyme that remains bound to a particular substrate during repetitions of the catalytic event.

proctodone a hormone, thought to be secreted by cells of the anterior intestine of insects, that terminates diapause (*q.v.*).

procumbent designating a plant stem that lies on the ground for all or most of its length (as in the case of vines). *See* **runner.**

productive infection viral infection of a cell that produces progeny via the vegetative or lytic cycle.

productivity fertility. In *Drosophila* the term is used specifically to refer to the number of progeny surviving to the adult stage among those produced per mated parental female in a specified time interval.

proenzyme a zymogen (*q.v.*).

proflavin an acridine dye that can function as a mutagen, causing reading frame shifts (*q.v.*).

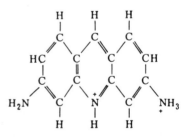

progenote the common ancestor of archaebacteria, eubacteria, and eukaryotes.

progeny the offspring from a given mating; members of the same biological family with the same mother and father; siblings.

progeny test the evaluation of the genotype of a parent by a study of its progeny under controlled conditions.

progeria a hereditary, premature aging disease of humans. Death usually occurs before age 14. Tissue-cultured cells from patients with progeria have a markedly reduced replicative life span and have a reduced ability to repair damaged DNA. Also called *Hutchinson-Gilford syndrome.*

progesterone a steroid hormone secreted by the corpus luteum (*q.v.*) to prepare the uterine lining for implantation of an ovum; also later secreted by the placenta (*q.v.*); essential for the maintenance of pregnancy.

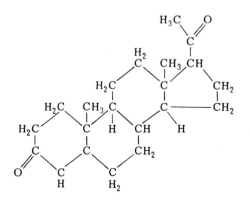

progestin *See* **progestogens.**

progestogens a group name for substances having progesteronelike activity; also termed *progestins. See* **progesterone.**

prognosis a forecast of the course and termination of a disease.

proinsulin a protein synthesized and processed by the beta cells of the pancreas. The molecule contains both the A and B peptides of insulin (*q.v.*) and an intervening C peptide containing 30 amino acids. Specific proteases cleave the precursor at two points, releasing the connecting peptide and the intact insulin molecule.

prokaryon synonymous with nucleoid (*q.v.*).

prokaryote member of the superkingdom Prokaryotes (*q.v.*).

Prokaryotes (*also* **Procaryotes**) the superkingdom containing all microorganisms that lack a membrane-bound nucleus containing chromosomes. Cell division involves binary fission. Centrioles, mitotic spindles, and mitochondria are absent. Aside from pillotinas (*q.v.*) prokaryotes also lack microtubules. This superkingdom contains one kingdom, the Monera (*q.v.*). *See* **Appendix A,** Prokaryotes; **Appendix C,** 1937, Chatton; *contrast with* **Eukaryotes.**

prolactin lactogenic hormone (*q.v.*).

proline *See* **amino acid.**

promiscuous DNA DNA segments that have been

transferred between organelles, such as mitochondria and chloroplasts, or from a mitochondrial genome to the nuclear genome of the host as a result of transpositional events happening millions of years ago. An example is a section of mitochondrial DNA present in the nuclear genome of *Strongylocentrotus purpuratus* (*q.v.*). *See* **Appendix C,** 1983, Jacobs *et al.*

promitochondria aberrant mitochondria characteristically found in yeasts grown under anaerobic conditions. Promitochondria have incomplete inner membranes and lack certain cytochromes. *See* **petities.**

promoter 1. a region on a DNA molecule to which an RNA polymerase binds and initiates transcription. In an operon, the promoter is usually located at the operator end, adjacent but external to the operator. The nucleotide sequence of the promoter determines both the nature of the enzyme that attaches to it and the rate of RNA synthesis. *See* **Appendix C,** 1975, Pribnow; **down promoter mutations, regulator gene, up promoter mutations.** 2. a chemical that, while not carcinogenic itself, enhances the production of malignant tumors in cells that have been exposed to a carcinogen.

pronase an enzyme from *Streptomyces* that digests mucoproteins.

pronucleus the haploid nucleus of an egg, sperm, or pollen grain. *See* **Appendix A,** 1877, Fol.

proofreading in molecular biology, any mechanism for correcting errors in replication, transcription, or translation that involves monitoring of individual units after they have been added to the chain; also called *editing.*

pro-oocyte one of the two cystocytes containing four ring canals that form synaptonemal complexes in *Drosophila melanogaster.* Upon entering the vitellarium, the anterior pro-oocyte switches to the nurse cell developmental pathway, leaving the posterior cell as the oocyte. *See* **nurse cells.**

propagule any kind of reproductive particle.

properdin pathway *See* **complement.**

prophage in lysogenic bacteria, the structure that carries genetic information necessary for the production of a given type of phage and confers specific hereditary properties on the host. *See* **Appendix C,** 1950, Lwoff and Gutman.

prophage attachment site either of the two attachment sites flanking an integrated prophage or the nucleotide sequences in a bacterial chromosome at which phage DNA can integrate to form a prophage.

prophage induction *See* **induction.**

prophage-mediated conversion the acquisition of new properties by a bacterium once it becomes lysogenized. A prophage, for example, confers upon its bacterial host an immunity to infection by related phages. Lysogenized bacteria also often show changes in their antigenic properties or in the toxins they produce.

prophase *See* **mitosis.**

propositus (*female,* **proposita**) the clinically affected family member through whom attention is first drawn to a pedigree of particular interest to human genetics; also called *proband.*

prosimian a member of the most primitive primate suborder, the Prosimii, containing tree shrews and tarsiers.

Prosobranchiata one of the three subdivisions of the mollusc class Gastropoda. *See* **Appendix A.**

prospective significance the normal fate of any portion of an embryo at the beginning of development.

prostaglandin a group of naturally occurring, chemically related, long-chain fatty acids that exhibit a wide variety of physiological effects (contraction of smooth muscles, lower blood pressure, antagonism of certain hormones, etc.). The first prostaglandin was originally isolated from the prostate gland (hence the name), but they are now known to be produced by many tissues of the body.

prosthetic group that portion of a complex protein that is not a polypeptide. Usually the prosthetic group is the active site of such a protein. The heme groups of hemoglobin are examples of prosthetic groups.

protamines basic proteins that replace histones in sperm cells of many animal species. *See* **nucleoprotein.**

protan *See* **color blindness.**

protandry 1. the maturation of the pollen-bearing organs before the female organs on a monoecious plant. 2. sequential hermaphroditism in animals, with the male stage preceding the female stage (*compare with* **protogyny**). 3. the appearance of male animals earlier in the breeding season than females.

protanomaly *See* **color blindness.**

protanopia *See* **color blindness.**

protease an enzyme that digests proteins.

protein a molecule composed of one or more polypeptide chains, each composed of a linear chain of amino acids covalently linked by peptide bonds. *See* **Appendix C,** 1838, Mulder, Berzelius; 1902, Hofmeister and Fisher; **amino acid, peptide bond, protein structure, translation.**

protein clock hypothesis the postulation that amino acid substitutions occur at a constant rate for a given family of proteins (e.g., cytochromes, hemoglobins) and hence that the degree of divergence between two species in the amino acid sequences of the protein in question can be used to estimate the length of time that has elapsed since their divergence from a common ancestor.

protein engineering any biochemical technique by which novel protein molecules are produced. These techniques fall into three categories: (1) the *de novo* synthesis of a protein, (2) the assembly of functional units from different natural proteins, and (3) the introduction of small changes, such as the replacement of individual amino acids, into a natural protein. *See* **Appendix C,** 1965, Merrifield and Stewart.

protein kinase an enzyme that attaches phosphate groups to serine, threonine, or tyrosine molecules built into proteins.

proteinoid an amino acid polymer with a molecular weight as high as 10,000 d formed under "pseudoprimeval conditions" by heating to 70°C a dry mixture containing phosphoric acid and 18 amino acids. Such proteinoids are acted upon by proteolytic enzymes and have nutritive value for bacteria, but are nonantigenic.

protein structure The *primary* structure of a protein refers to the number of polypeptide chains in it, the amino acid sequence of each, and the position of inter- and intra-chain disulfide bridges. The *secondary* structure refers to the type of helical configuration possessed by each polypeptide chain resulting from the formation of intramolecular hydrogen bonds along its length. The *tertiary* structure refers to the manner in which each chain folds upon itself. The *quaternary* structure refers to the way two or more of the component chains may interact.

protein synthesis *See* **translation.**

proteolytic causing the digestion of proteins into simpler units.

proter the anterior daughter organism produced by the transverse division of a protozoan.

Proterozoic the more recent of the two eras making up the Precambrian eon. Stromatolites (*q.v.*) occur in early Proterozoic strata, and by the end of the era animals as advanced as coelenterates and annelids were present. The origin of eukaryotes presumably occurred midway through the era. *See* **Appendix C,** 1954, Barghoorn and Tyler; **geologic time divisions.**

prothallus (prothalium) the independent gametophyte of a horsetail or fern. *See* **Appendix A,** (Plantae, Tracheophyta).

prothetely an experimentally induced abnormality in which an organ appears in advance of the normal time because of a partially inhibited metamorphosis; for example, the formation of pupal antennae on a caterpillar.

prothoracic gland a gland located in the prothorax of insects that secretes ecdysone (*q.v.*). *See* **ring gland.**

prothoracicotropic hormone a peptide hormone produced by neurosecretory cells in the dorsum of the insect brain that stimulates the prothoracic gland (*q.v.*) to synthesize and secrete ecdysones. Symbolized PTTC.

prothrombin an inactive form of thrombin. *See* **blood clotting.**

protist an informal term used to refer to any single-celled (usually eukaryotic) organism.

protocooperation population or species interaction favorable to both, but not obligatory for either one.

Protoctista one of the five kingdoms of living organisms. It contains the eukaryotic microorganisms and their immediate descendants, i.e., the nucleated algae, flagellated water molds, slime molds, and protozoa. *See* **Appendix A,** Subkingdom Eukaryotes.

protogyny sequential hermaphroditism with the ovary functioning before the testis. *Compare with* **protandry.**

protomers single polypeptide chains (either identical or nonidentical) of a multimeric protein.

protomitochondria *See* **petites.**

proton an elementary particle of the atomic nucleus with a positive electric charge (equal numerically to the negative charge of the electron) and a mass of 1.0073 mass units.

proto-oncogene a cellular gene that functions in controlling the normal proliferation of cells and either (1) shares nucleotide sequences with any of the known viral *onc* genes, or (2) is thought to represent a potential cancer gene that may become carcinogenic by mutation, or by overactivity when coupled

to a highly efficient promoter. Some proto-oncogenes (e.g., *c-src*) encode protein kinases that phosphorylate tyrosines in specific cellular proteins. Others (e.g., *c-ras*) encode proteins that bind to guanine nucleotides and possess GTPase activity. Still other oncogenes encode growth factors or growth factor receptors. *See* **platelet-derived growth factor.**

protoplasm the substance within the plasma membrane of a cell; the nucleus and surrounding cytoplasm.

protoplast the organized living unit of a plant or bacterial cell consisting of the nucleus (or nucleoid), cytoplasm, and surrounding plasma membrane, but with the cell wall left out of consideration. Protoplasts can be generated experimentally; e.g., the walls of *E. coli* cells can be removed by lysozyme treatment. Aphragmabacteria (*see Mycoplasma*) lack cell walls and in this sense are protoplasts.

protoplast fusion a mechanism for achieving genetic transformation by joining two protoplasts or joining a protoplast with any of the components of another cell.

Protostomia a subdivision of Bilateria containing animals in which the mouth arises from the blastopore. *See* **Appendix A; Deuterostomia.**

prototroph 1. an organism that is able to subsist on a carbon source and inorganic compounds. For most bacteria, the carbon source could be a sugar; green plants use carbon dioxide. 2. a microbial strain that is capable of growing on a defined minimal medium; wild-type strains are usually regarded as prototrophs.

Protozoa a kingdom erected in Cavalier-Smith's classification to contain the majority of unicellular heterotrophic eukaryotes. Protozoa contain 80 S ribosomes, they lack chloroplasts, and their undulipodia lack mastigonemes (*q.v.*). *See* **classification, Archaeozoa, Chromista.**

provirus 1. a virus that is integrated into a host cell chromosome and is transmitted from one cell generation to another without causing lysis of the host. 2. more specifically, a duplex DNA sequence in an eukaryotic chromosome (corresponding to the genome of an RNA retrovirus) that is transmitted from one cell generation to another without causing lysis of the host. Such proviruses are often associated with transformation of cells to the cancerous state.

proximal toward or nearer to the place of attachment (of an organ or appendage). In the case of a chromosome, the part closest to the centromere.

Prunus   the genus that includes *P. amygdalus,* the almond; *P. armeniaca,* the apricot; *P. avium,* the cherry; *P. domestica,* the plum; *P. persica,* the peach.

pseudoalleles   genes that behave as alleles in the cis-trans test (*q.v.*) but can be separated by crossing over. *See* **Appendix A,** 1949, Green and Green.

pseudoautosomal genes   *See* **human pseudoautosomal region.**

Pseudocoelomata   a subdivision of the Protostomia containing animals having a body cavity that is not lined with peritoneum. The space is formed by dispersion of mesenchyme. *See* **Appendix A.**

pseudocopulation   the mode of pollination in certain orchids in which structures of the flower closely resemble a female insect, and the male insects attempting copulation serve to transfer pollen from one flower to another.

pseudodiploid   a condition in which the chromosome number of a cell is the diploid number characteristic of the organism but, as a consequence of chromosomal rearrangements, the karyotype is abnormal and linkage relationships may be disrupted.

pseudodominance   the phenotypic expression of a recessive allele on one chromosome as a consequence of deletion of the dominant allele from the homologue.

pseudoextinction   disappearance of a taxon by virtue of its being evolved by anagenesis into another taxon.

pseudogamy   the parthenogenetic development of an ovum following stimulation (but not fertilization) by a male gamete or gametophyte; synonymous with *gynogenesis.*

pseudogene   a gene bearing close resemblance to a known gene at a different locus, but rendered nonfunctional by additions or deletions in its structure that prevent normal transcription and/or translation. Pseudogenes are usually flanked by direct repeats of 10 to 20 nucleotides; such direct repeats are considered to be a hallmark of DNA insertion. Two classes of pseudogenes exist: (1) *Traditional pseudogenes* (as exemplified in the globin gene families) appear to have originated by gene duplication and been subsequently silenced by point mutations, small insertions, and deletions; they are usually adjacent to functional copies and show evidence of being under some form of selective constraint for several millions of years after their formation. (2) *Processed pseudogenes* lack introns, possess a remnant of a poly-A tail, are often flanked by short direct repeats, and are usually unassociated with functional copies; all of which suggests their formation by the integration into germ-line DNA of a reverse-transcribed processed RNA. *See* **Appendix C,** 1977, Jacq *et al.;* **hemoglobin genes, orphons, processed gene..**

pseudohermaphroditism   a condition in which an individual has gonads of one sex and secondary sexual characters of the other sex or of both sexes. Pseudohermaphrodites are designated as male or female with reference to their sex chromosome constitution or the type of gonadal tissue present.

Pseudomonas   a genus of Gram-negative, motile bacteria, each with a single polar flagellum. Two species have been the subjects of genetic study, *P. aeruginosa* and *P. putida.* Lysogeny (*q.v.*) is common in *P. aeruginosa,* rare in *P. putida.*

pseudotumor   an aggregation of blackened cells in *Drosophila* larvae, pupae, and adults of certain genotypes. Such "tumors" result from encapsulation during the larval stage of certain tissues by hemocytes and subsequent melanization of these masses.

pseudouridine   *See* **rare bases.**

pseudovirion   a synthetic virus consisting of the protein coat from one virus and the DNA from a foreign source. *See* **phenotypic mixing, reassortment virus.**

pseudo-wild type   the wild phenotype of a mutant, produced by a second (suppressor) mutation.

P site   *See* **translation.**

psoralens   photosensitive cross-linking reagents that act on specific base-paired regions of nucleic acids. *See* **trimethylpsoralen.**

P strain   the paternally contributing strain of *Drosophila* in a P-M hybrid dysgenesis cross. P strains differ genetically from M strains in that they contain multiple P factors in their genomes. *See* **hybrid dysgenesis, M strain, P elements.**

[32]P suicide   inactivation of phages due to the decay of radiophosphorus molecules incorporated into their DNA.

psychosis   a generic term covering any behavioral disorder of a far-reaching and prolonged nature. *See* **manic-depressive psychosis.**

PTC   abbreviation for *phenylt*hio*c*arbamide or *p*lasma*t*hromboplastin *c*omponent (*both of which* see).

pteridine   *See* **Drosophila eye pigments.**

pteridophytes   the ferns, horsetails, club mosses, and other vascular spore-bearing plants.

pteroylglutamic acid   folic acid (*q.v.*).

pterygote   an insect belonging to a division that includes all winged species. Some pterygotes (e.g., fleas) are wingless, but they are believed to have been derived from winged ancestors. *See* **apterygotes, Appendix A.**

PTTH   *prothoracicotropic hormone* (*q.v.*).

Pu   abbreviation for any purine (e.g., adenine or guanine). *See* **R3.**

puff   *See* **chromosomal puff.**

pulse-chase experiment   an experimental technique in which cells are given a very brief exposure (the pulse) to a radioactively labeled precursor of some macromolecule, and then the metabolic fate of the label is followed during subsequent incubation in a medium containing only the nonlabeled precursor (the chase).

pulsed-field gradient gel electrophoresis   a technique for separating DNA molecules by subjecting them to alternately pulsed, perpendicularly oriented electrical fields. The technique has allowed separation of the yeast genome into a series of molecules that ranged in weight between 40 and 1800 kb and represent intact chromosomes. *See* **Appendix C,** 1984, Schwartz and Cantor.

pulvillus   the last segment of the foot in an insect. It has a pad with a claw on either side.

punctuated equilibrium   a term describing a pattern seen in the fossil record of relatively brief episodes of speciation followed by long periods of species stability. Although this pattern conflicts with the pattern of gradualism (*q.v.*), no special developmental, genetic, or ecological mechanisms are required to explain it. Both the gradual and the punctuated shifting equilibrium pattern can be simulated from mathematical equations that include only terms from random mutation, natural selection, and population size. *See* **Appendix C,** 1985, Newman *et al.*

Punnett square   the checkerboard method commonly used to determine the types of zygotes produced by a fusion of gametes from the parents. The results allow the computation of genotypic and phenotypic ratios. Named after R. C. Punnett, its inventor.

pupal period   the developmental period between pupation and eclosion.

puparium formation   the formation of a pupal case by the tanning of the skin molted from the last-instar larval insect.

pupation   that stage in the metamorphosis of the insect signaled by the eversion of the imaginal discs.

purebred   derived from a line subjected to inbreeding (*q.v.*).

pure culture   a culture that contains only one species of microorganism. *See* **Appendix C,** 1881, Koch.

pure line   a strain of an organism that is homozygous because of continued inbreeding.

purine   *See* **bases of nucleic acids.**

puromycin   an antibiotic that, because of its structural resemblance to the terminal aminoacylated adenosine group of aminoacyl tRNA, becomes incorporated into the growing polypeptide chain and causes the release of incompleted polypeptide chains (which are terminated with a puromycin residue) from the ribosome.

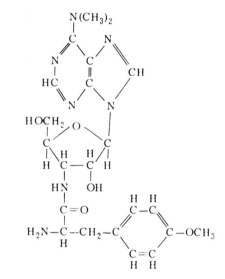

P value   probability value. A decimal fraction showing the number of times an event will occur in a given number of trials. *See* **probability of an event.**

Py   abbreviation for any pyrimidine (e.g., thymine, cytosine, uracil). *See* **Y.**

pycnosis   the contraction of the nucleus into a compact, strongly staining mass, taking place as the cell dies.

pycocin   *See* **bacteriocin.**

pyloric stenosis   the constriction of the valve between the stomach and intestine, a congenital disorder of high heritability.

pyrenoid   a small, round protein granule surrounded by a starch sheath found embedded in the chloroplasts of certain algae and liverworts.

pyrethrins diterpene insecticides found in plant tissues. These compounds were first extracted from pyrethrum (chrysanthemum) flowers.

pyridoxal phosphate the coenzyme of both amino acid decarboxylating enzymes and transaminating enzymes.

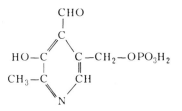

pyridoxine vitamin B$_6$.

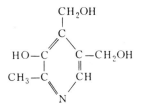

pyrimidine *See* **bases of nucleic acids.**

pyrimidine dimer the compound formed by UV irradiation of DNA whereby two thymine residues, or two cytosine residues, or one thymine and one cytosine residue occupying adjacent positions in the polynucleotide strand become covalently joined. *See* **thymine dimer.**

pyronin Y a basic dye often used in cytochemistry. In 2 M magnesium chloride at *p*H 5.7, pyronin Y stains only undegraded RNA. *See* **methyl green.**

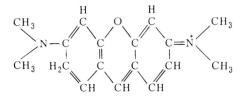

pyrrole molecules ring-shaped compounds containing one nitrogen and four carbon atoms that are components of porphyrin (*q.v.*) molecules.

Q

q *See* **symbols used in human cytogenetics.**

Q₁₀ temperature coefficient; the increase in a reaction or other process (expressed as a multiple of the initial rate) produced by raising the temperature 10° C.

Qa in the mouse, a series of loci located very close to the major histocompatibility complex (H-2) whose products are expressed on the surfaces of some lymphocyte classes and subclasses.

Q-bands *See* **chromosome banding techniques.**

Q beta an RNA phage infecting *E. coli.* Its genome consists of a circular, single-stranded RNA molecule. This strand acts both as a template for the replication of a complementary strand and as an mRNA molecule that directs the translation of viral poteins. *See* **Appendix C,** 1965, 1967, Spiegelman *et al.;* 1973, Mills *et al.;* 1983, Miele *et al.;* **Q beta replicase.**

Q beta replicase the RNA-directed RNA polymerase that catalyzes the replication of the Q beta virus genome.

quadrivalent a meiotic association of four homologous chromosomes; synonymous with *tetravalent.*

quadruplex *See* **autotetraploidy.**

quail *See Coturnix coturnix japonica.*

quantasome a photosynthetically active particle found in the grana of chloroplasts. Each quantasome is an oblate ellipsoid with axes of about 100 Å and 200 Å. Chlorophyll (*q.v.*) is localized within quantasomes.

quantitative character a character showing quantitative inheritance (beef and milk production in cattle, egg production in hens, DDT resistance in *Drosophila,* stature, weight, and skin pigmentation in man).

quantitative inheritance phenotypes that are quantitative in nature and continuous in distribution are referred to as quantitative characters (*q.v.*). During their genetic transmission, there is an absence of clear-cut segregation into readily recogniz-able classes showing typical Mendelian ratios. An often-used example is ear length in maize, as illustrated by the histograms on page 261. When crosses are made between individuals from lines showing large quantitative differences in ear length, the offspring are intermediate. When F₁ individuals are crossed, the F₂ population has a mean that is very similar to the F₁ mean, but some individuals produce ears as long or as short as the grandparents. Such results are explained by the "multiple factor hypothesis," which assumes that the quantitative character depends upon the cumulative action of multiple genes (or polygenes), each on a separate chromosome, and each producing a unitary effect. In the corn example, a simple model would employ three genes, each existing in two allelic forms. Each capital letter gene might be responsible for three units of "growth potential," and each small letter gene, for one unit. Thus the capital letter genes are all interchangeable in the sense that each produces the same phenotypic effect, and the same is true for the small-letter genes. The long- and short-eared parental individuals would be AABBCC and aabbcc, respectively, and their offspring would be AaBbCc. These would show little variability, because all plants would be genetically identical. The segregation of the alleles in the F₂ population would produce 27 different genotypic classes, and the cumulative action of the genes would generate 7 phenotypic classes. The most common genotype (making up 1/8th of the total population) would be AaBbCc, genetically identical to the F₁ plants. But there would also be plants of genotype AABBCC and aabbcc (each making up 1/64th of the population) and these would be phenotypically and genetically identical to the grandparents. There would also be individuals with various intermediate ear lengths of genotypes (AABBCc, aabbcC, AAbbCC, etc.) and the result would be an F₂ population with a mean equivalent to the F₁, but with a distribution whose width depended on the number of segregating alleles. By comparing the variances of the F₁ and F₂ populations, one can estimate the number of segregating gene pairs responsible for the trait. *See* **Appendix C,** 1889, Galton; 1909, Nilsson Ehle; **Wright's polygenic estimate.**

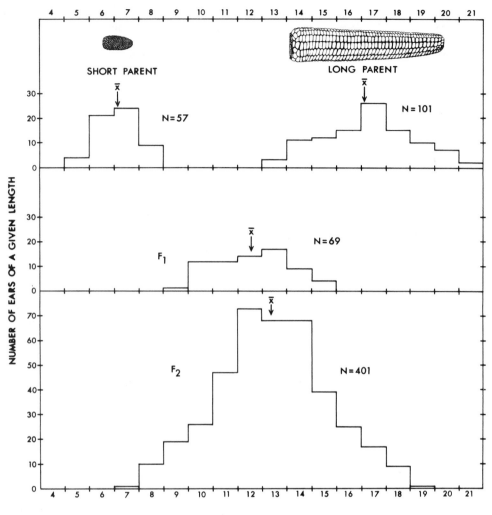

Quantitive inheritance in maize. Distribution of ear lengths in parents, F₁ and F₂ generations of a cross between Tom Thumb popcorn and Black Mexican sweet corn. (After Emerson and East, 1913)

quantum according to the quantum theory, energy is radiated in discrete quantities of definite magnitude called quanta and absorbed in a like manner.

quantum speciation the rapid evolution of new species, usually within small, peripheral isolates, with founder effects and genetic drift playing important roles. *See* **evolution.**

quartet a group of four nuclei or of four cells arising from the two meiotic divisions.

Quaternary the most recent of the two geologic periods making up the Cenozoic era. *See* **geologic time divisions.**

quaternary protein structure *See* **protein structure.**

Quercus the genus of oaks including: *Q. alba,* the white oak; *Q. coccinea,* the scarlet oak; *Q. palustris,* the pin oak; *Q. suber,* the cork oak.

quick-stop mutants of *E. coli* that immediately cease replication when the temperature is increased to 42°C.

quinacrine an acridine derivative used in the treatment of certain types of cancer and malaria. It is also used as a fluorochrome in chromosome cytology. *See* **Appendix C,** 1970, Caspersson *et al.;* 1971, O'Riordan *et al.*

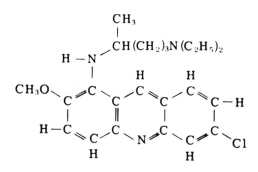

quinone a compound belonging to a class of molecules that function in biological oxidation-reduction systems.

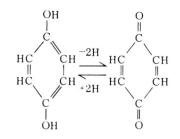

q.v. which see. An abbreviation for Latin, *quod vide.*

R

r 1. reproductive potential. 2. ring chromosome; *see* **symbols used in human genetics. 3.** roentgen. **4.** correlation coefficient; *see* **correlation.**

R 1. a chemical radical. Used to show the position of an unspecified radical in a generalized structural formula of a group of organic compounds. 2. a drug-resistant plasmid conferring resistance to one or more antibiotics on bacteria in which it resides. 3. the single-letter symbol for purine. *Compare with* **Y.**

rabbit *see Oryctolagus cuniculus.*

race a phenotypically and/or geographically distinctive subspecific group, composed of individuals inhabiting a defined geographical and/or ecological region, and possessing characteristic phenotypic and gene frequencies that distinguish it from other such groups. The number of racial groups that one wishes to recognize within a species is usually arbitrary but suitable for the purposes under investigation. *See* subspecies.

raceme an inflorescence as in the hyacinth, in which the flowers are borne on pedicles arising from the rachis. A branched raceme, such as may be seen in oats, rice, wheat, and rye, is called a panicle. *See* **spike.**

rad abbreviation of *r*adiation *a*bsorbed *d*ose. A unit defining that energy absorbed from a dose of ionizing radiation equal to 0.01 joule per kilogram. 1 rad = 0.01 Gy.

Radiata a subdivision of the Eumetazoa containing animals, such as jelly fish and coral polyps, characterized by radial symmetry. *See* **Appendix A.**

radiation the emission and propagation of energy through space or a medium in the form of waves. When unqualified, radiation usually refers to electromagnetic radiations (radio waves, infrared, visible light, ultraviolet, x rays, and gamma rays), and, by extension, ionizing particles.

radiation absorbed dose *See* **rad.**

radiation chimera an experimentally produced animal containing hemopoietic cells of a genotype different from that of the rest of the organism. Recipients receive a single dose of radiation that kills the stem cells of the bone marrow and much of the differentiated hemopoietic tissue. Very shortly thereafter, they receive an intravenous inoculation of bone marrow or fetal liver cells from nonirradiated donors. The injected stem cells home to the recipient's bone marrow sites and begin repopulating them, and ultimately they replace the recipient's hemopoietic tissues.

radiation dosage *See* **dose.**

radiation genetics the scientific study of the effects of radiation on genes and chromosomes. The science began with the demonstration for *Drosophila* and corn that X-rays produced deleterious mutations. *See* **Appendix C,** 1927, Muller; 1928, Stadler.

radiation-induced chromosomal aberration a chromosomal aberration (*q.v.*) induced through breakage caused by ionizing radiation. In the table on page 264 are shown the origin and mitotic behavior of a variety of radiation-induced aberrations. Original break positions are indicated by short diagonal lines.

radiation sickness a syndrome characterized by nausea, vomiting, diarrhea, psychic depression, and death following exposure to lethal doses of ionizing radiation. The median lethal radiation dose for man is between 400 and 500 r. Such a dose leaves only about 0.5% of the body's reproducing cells still able to undergo continued mitosis. Since each cell continues to function normally in the physiological sense, death is not immediate. Damage shows up first in tissues with a high mitotic rate (the blood-cell-forming tissues of the bone marrow, for example). Death occurs when the surviving cells are unable to restore by mitosis the needed numbers in time to maintain the physiological functioning of the various vital tissues.

radical scavenger a molecule with a high affinity for free radicals. If a radical scavenger is added to a biological system prior to irradiation, it may act as a protective agent.

radioactive decay the disintegration of the nucleus of an unstable nuclide accompanied by the spontaneous emission of charged particles and/or photons.

CHROMOSOME ABERRATIONS				CHROMATID ABERRATIONS			
TYPE	INTERPHASE	PRE-METAPHASE	ANAPHASE	TYPE	PROPHASE	PRE-METAPHASE	ANAPHASE
CHROMOSOME BREAK				CHROMATID BREAK			
INTERSTITIAL DELETION				ISO-CHROMATID BREAK			
EXCHANGES — INTRACHANGE (GROSS)				EXCHANGES — INTRACHANGE			
EXCHANGES — INTERCHANGES — ASYMMETRICAL				EXCHANGES — INTERCHANGES — ASYMMETRICAL			
EXCHANGES — INTERCHANGES — SYMMETRICAL				EXCHANGES — INTERCHANGES — SYMMETRICAL			

After K. Sax, 1940. *Genetics* 25:41–68. As modified by N. H. Giles, 1954, in *Radiation Biology.* Vol 1: p.716. A. Hollaender, editor, McGraw-Hill, New York.

Radiation-induced chromosomal aberrations.

radioactive isotope an isotope with an unstable nucleus that stabilizes itself by emitting ionizing radiations. The use of radioisotopes in biology dates back to 1943 when the X-10 reactor at the Oak Ridge Laboratory in Tennessee started their commercial production. *See* **autoradiography, labeled compound, radioimmunoassay, tritium.**

radioactive series a succession of nuclides, each of which transforms by radioactive disintegration into the next until a stable nuclide results.

radioactivity the spontaneous disintegration of certain nuclides accompanied by the emission of one or more types of radiation, such as alpha particles, beta particles, and gamma photons.

radioautograph autoradiograph (*q.v.*).

radioautographic efficiency autoradiographic efficiency (*q.v.*).

radioautography autoradiography (*q.v.*).

radiobiology a branch of biology that deals with the effects of radiation on biological systems. It includes radiation genetics (*q.v.*).

radiogenic element an element derived from another element by atomic disintegration.

radiograph a shadow image made on photographic emulsion by the action of ionizing radiation. The image is the result of the differential attenuation of the radiation during its passage through the object being radiographed. A chest X-ray negative is a radiograph.

radioimmunoassay a highly sensitive technique for the quantitative determination of antigenically active substances that are present in very small amounts, such as hormones. The concentration of an unknown, unlabeled antigen is determined by comparing its inhibitory effect on the binding of radioactively labeled antigen to specific antibody with the inhibitory effect of known standards. Symbolized RIA. *See* **Appendix C,** 1957, Berson and Yalow.

radiological survey the evaluation of the radiation hazards incident to the production, use, or existence of radioactive materials or other sources of radiation under a specified set of conditions.

radiomimetic chemical a chemical that mimics ionizing radiations in terms of damage to nucleic acids. Radiomimetic compounds include **sulfur mustards, nitrogen mustards,** and **epoxides,** (*all of which see*).

radioresistance the relative resistance of cells, tissues, organs, or organisms to the injurious action of radiation. Ultraviolet-resistant bacteria, for example, can excise ultraviolet-induced thymine dimers from their DNA.

radiotracer *See* **radioactive isotope, labeled compound.**

radon the name used to refer to the many isotopes of element 86. Radon is an inert gas that is readily soluble in water. All its isotopes are radioactive with short half-lives, and all decay with the emission of densely ionizing alpha particles (*q.v.*). While such particles are too weak to penetrate the skin, they are very dangerous when radon is ingested or inhaled. Radon is found in nature because it is continuously formed by the radioactive decay of the longer-lived, precursor elements uranium and thorium, which occur in certain minerals. The most common radon isotope in man's environment is ^{222}Rn, which has a half-life of 3.8 days. Radiation from radon is responsible for over half of the average exposure to humans from ionizing radiation.

ramets buds that can detach from a plant or animal and result in the asexual production of offspring genetically identical to each other and the parent. Ramets can also refer to the specific offspring produced by asexual budding from a single ancestral organism (the ortet). *See* **modular organisms.**

Rana frogs of this genus have been used widely in research. The leopard frog, *R. pipiens,* is the most common species bred in the laboratory, and many of its mutations have been recovered and analyzed. Mutant strains are also available for *R. sylvatica, R. esculenta,* and *R. temporaria. R. esculenta* is the only anuran for which working maps of the lampbrush chromosomes are available.

r and K selection theory a theory in population ecology that attempts to establish whether environmental conditions favor the maximization of r (the intrinsic rate of natural increase) or of K (the car-

rying capacity of the environment). When populations can expand without food reserves limiting their growth, then r selection is in control. When food reserves limit population size, K selection takes over, and increase in one genotype must be at the expense of another. Whereas r selection operates in ecological situations where food reserves fluctuate drastically, and species are favored that reproduce rapidly and produce large numbers of offspring. K selection operates in populations that are close to the environmental carrying capacity, and species are favored that reproduce slowly and generate a few offspring that are well adapted to a relatively stable environment.

random assortment *See* **assortment.**

random genetic drift genetic drift (*q.v.*).

random mating a population mating system in which every male gamete has an equal opportunity to join in fertilization with every female gamete, including those gametes derived from the same individuals (if the species is monoecious or hermaphroditic); panmixis.

random primers randomly generated oligodeoxyribonucleotides, some of which anneal to complementary sequences in the template nucleic acid and serve as primers in reactions involving reverse transcriptase.

random sample a sample of a population selected so that all items in the population are equally likely to be included in the sample.

random sampling error *See* **experimental error, sampling error.**

random-X inactivation the method of dosage compensation (*q.v.*) found in eutherian mammals. *See* **paternal-X inactivation.**

Ranunculus a genus of flowering plants including the buttercup. The cytology of meiosis has been extensively studied in species from this genus.

Raphanobrassica a fertile allotetraploid developed by Karpechenko from a cross between the radish *(Raphanus sativus)* and the cabbage *(Brassica oleracea).*

rapidly reannealing DNA, rapidly reassociating DNA repetitious DNA (*q.v.*).

rapid-lysing (*r*) mutants mutants of T-even phage that enhance the rate at which *E. coli* host cells are lysed; on a bacterial lawn, *r*-plaques are larger than wild-type plaques (*r*$^+$). *See* **plaque.**

rare bases purines (other than adenine and guanine) and pyrimidines (other than cytosine and ura-

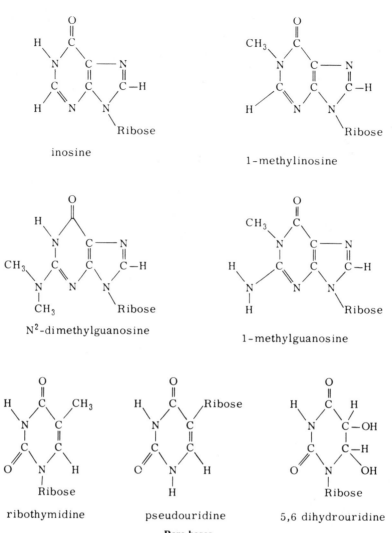

inosine

1-methylinosine

N^2-dimethylguanosine

1-methylguanosine

ribothymidine

pseudouridine

5,6 dihydrouridine

Rare bases.

cil) found in transfer RNA (*q.v.*). *See* **bases of nucleic acids.**

rare earth any of the series of very similar metals ranging in atomic number from 57 to 71. *See* **periodic table.**

Rassenkreis *See* **circular overlap, polytypic species.**

rat *See Rattus.*

rat kangaroo *See Potorous tridactylus.*

Rattus the genus of rats including *R. norvegicus,* the brown rat, and *R. rattus,* the black rat. The white laboratory rat is an albino form of *R. norvegicus.* The laboratory rat has 22 chromosome pairs including the sex chromosomes X and Y. There-

fore, the 9 linkage groups described so far represent only a small picture of the net genome.

Rauscher leukemia virus a retrovirus isolated from the plasma of leukemic mice. *See* **retrovirus.**

R bands *See* **chromosome banding techniques.**

RBC red blood cell.

RBE relative biological effectiveness (*q.v.*).

rDNA **1.** in general, any DNA regions that code for ribosomal RNA components. **2.** specifically, a tandem cluster of eukaryotic rRNA genes with a sufficiently atypical base composition to allow its isolation directly from sheared genomic DNA. In recent literature, rDNA is also used to refer to hybrid molecules formed by uniting two or more heterologous DNA molecules. To avoid confusion, the cymbol rtDNA should be used for such recombinant DNA molecules and rDNA should be reserved for ribosomal DNA. *See* **Appendix C,** 1967, Birnstiel.

rDNA amplification The genes for rRNA are preferentially replicated during oogenesis in amphibia. In *Xenopus laevis,* for example, there are 2,000 rDNA repeats integrated into the chromosomes of the oocyte. However, there are 2,000,000 DNA repeats distributed among about 1,000 extrachromosomal nucleoli that lie near the periphery of the nucleus of each diplotene oocyte. These amplified genes arose from single copies of the chromosomal rDNA, and during pachynema they replicated extrachromosomally by a rolling circle (*q.v.*) mechanism. These extrachromosomal nucleoli function to transcribe the rRNAs stored in the growing oocyte. Amplification of rDNA also occurs commonly in insects with panoistic ovaries and in the macronuclei of protozoa, such as *Tetrahymena. See* **Appendix C,** 1968, Gall, Brown and Dawid; **insect ovary types, Miller trees.**

reading the unidirectional process by which mRNA sequences are decoded (translated) into amino acid sequences (polypeptide chains).

reading frame a nucleotide sequence that starts with an initiation codon, partitions the subsequent nucleotides into amino acid-encoding triplets, and ends with a termination codon. The interval between the start and stop codons is called the *open reading frame* (ORF). If a stop codon occurs soon after the initiation codon, the reading frame is said to be *blocked.*

reading frame shift Certain mutagens (acridine dyes, for example) intercalate themselves between the strands of a DNA double helix. During subsequent replication, the newly formed complementary strands may have a nucleotide added or subtracted. A cistron containing an additional base or missing a base will transcribe a messenger RNA with a *reading frame shift.* That is, during translation the message will be read properly up to the point of loss or addition. Thereafter, since the message will continue to be read in triplets, all subsequent codons will specify the wrong amino acids (and some may signal chain termination). *See* **translation, nonsense mutation, amino acid, acridine orange, acriflavin.**

reading mistake the incorrect placement of an amino acid in a polypeptide chain during protein synthesis.

readthrough 1. transcription beyond a normal terminator sequence in DNA, due to occasional failure of RNA polymerase to recognize the termination signal or due to the temporary dissociation of a termination factor (such as rho in bacteria) from the terminator sequence. 2. translation beyond the chain-terminator (stop) codon of a mRNA, as occurs by a nonsense suppressor (*q.v.*) tRNA.

reannealing in molecular genetics, the pairing of single-stranded DNA molecules that have complementary base sequences to form duplex molecules. Reannealing and annealing (*q.v.*) differ in that the DNA molecules in the first case are from the same source and in the second case from different sources. *See* **Appendix C,** 1960, Doty *et al;* **Alu family, mouse satellite DNA, reassociation kinetics, repetitious genes.**

reassociation reannealing (*q.v.*).

reassociation kinetics a technique that measures the rate of reassociation of complementary strands of DNA derived from a single source. The DNA under study is fragmented into pieces several hundred base pairs in length and then disassociated into single strands by heating. Subsequently, the temperature is lowered and the rate of reannealing (*q.v.*) is monitored. Reassociation of DNA is followed in the form of a cot curve, which plots the fraction of molecules that have reannealed against the log of cot. Cot values are defined as $C_0 \times t$, where C_0 is the initial concentration of single-stranded DNA in moles of nucleotides per liter and t is the reannealing time in seconds. Typical cot curves are shown on page 268. DNAs reannealing at low cot values (10^{-4}–10^{-1}) are composed of highly repetitive sequences, DNAs reannealing at cot values between 10^0 and 10^2 are moderately repetitive, and DNAs reannealing at higher cot values are nonrepetitive. *See* **Appendix C,** 1968, Britten and Kohne; **Alu family, delta T50H, mouse satellite DNA, repetitious DNA.**

reassortant virus a virion consisting of DNA from one virus and protein from another viral species; e.g., through genetic engineering, a hybrid virus has been made containing genes from the human influenza virus and capsid proteins that provoke immunity, but also containing avian influenza virus genes that slow the rate of viral replication. *See* **phenotypic mixing, pseudovirion.**

recapitulation the theory first put forth by Ernst Haeckel that an individual during its development passes through stages resembling the adult forms of its successive ancestors. The concept is often stated "ontogeny recapitulates phylogeny" and is sometimes referred to as the *biogenic law.*

RecA protein the product of the *RecA* locus of *E.*

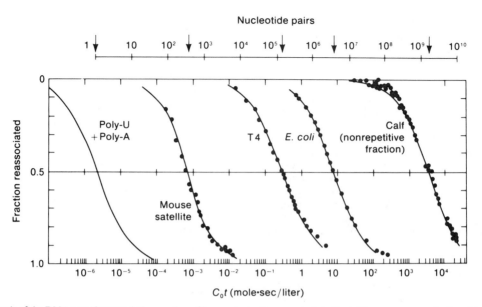

Nucleotide pairs

For each of the DNA samples tested, the number of base pairs in the genome is indicated by an arrow on the logarithmic scale at the top of the graph. The poly-U + poly-A sample is a double helix of RNA, with one strand containing only A and the other strand only U. The mouse satellite DNA is a fraction of nuclear DNA in mouse cells that differs in its physical properties from the bulk of the DNA. The calf DNA represents only those sequences that are present in single copies per haploid genome. The denatured DNA samples were fragmented by mechanical shear to chain lengths of about 400 nucleotides and incubated at a temperature near 60°C. The fraction reassociated was measured by the decrease in UV absorption as double strands formed. (After R. J. Britten and D. E. Kohne, "Repeated Sequences in DNA," Science 161:529, 1968. Copyright 1968 by the American Association for the Advancement of Science.)

Reassociation kinetics.

coli, which acts as a protease in the SOS response (*q.v.*) and also functions in recombination repair.

receptor element *See* **controlling elements.**

receptor-mediated endocytosis endocytosis that involves the binding of a ligand, such as vitellogenin (*q.v.*) to a plasma membrane receptor followed by the lateral movement of the ligand-receptor complex through the membrane toward a coated pit. The cytoskeleton of each coated pit is a basketlike network of hexagons and pentagons formed by the assembly of three-legged protein complexes called *triskelions.* Each triskelion is composed of three molecules of *clathrin,* a 185-kd protein, and three smaller polypeptides. Once a clathrin-coated pit contains a large number of ligand-receptor complexes, it invaginates further into the cytoplasm and eventually a small vesicle is pinched off the pit. This endocytotic vesicle is called a *receptosome.* Once the ligands have been internalized in a receptosome, the receptor molecules are returned intact to the plasma membrane.

receptor-mediated translocation a hypothesis concerning the translocation of nascent polypeptides across the endoplasmic reticulum membrane. As

shown in the diagram (page 269), soon after the signal sequence peptide of the nascent chain emerges from the ribosome it is recognized by a specific receptor called the signal recognition particle (SRP). The second component of the translocation process is the docking protein. It is bound to the surface of the ER membrane, and it serves as a receptor for the SRP. Since the SRP binds to both the docking protein and the signal sequence of the protein being translated, it serves to bring the ribosome into the vicinity of the ER membrane. Subsequently, the ribosome binds to a ribosome receptor on the ER and the nascent polypeptide is threaded through a pore in the membrane and into the ER lumen. A peptidase then removes the signal peptide from the newly synthesized protein molecule. *See* **leader sequence peptide, signal hypothesis, signal recognition particle, translation.**

receptosome *See* **receptor-mediated endocytosis.**

recessive complementarity *See* **complementary genes.**

recessive gene in diploid organisms, a gene that is phenotypically manifest in the homozygous state but is masked in the presence of its dominant allele.

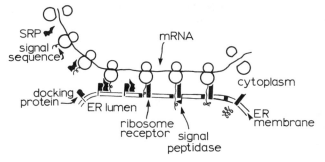

Receptor-mediated translocation.

Usually the dominant gene produces a functional product, while its recessive allele does not. Therefore, the normal phenotype is produced if the dominant allele is present (in one or two doses per nucleus), and the mutant phenotype appears only in the absence of the normal allele (i.e., when the recessive gene is homozygous). By extension, the terms dominant and recessive are used in the same sense for heterokaryons and merozygotes.

recessive lethal   an allele that kills the cell or organism that is homozygous or hemizygous for it. *See* **lethal mutation.**

reciprocal crosses   crosses of the forms A ♀ X B ♂ and B ♀ X A ♂, where the individuals symbolized by A and B differ in genotype or phenotype or both. Reciprocal crosses are employed to detect sex linkage, maternal inheritance, or cytoplasmic inheritance (*all of which see*).

reciprocal genes   complementary genes (*q.v.*).

reciprocal hybrids   hybrid offspring derived from reciprocal crosses of parents from different species.

reciprocal recombination   in the gametes of dihybrids, the production of new linkage arrangements that are different from those of the maternal and paternal homologues. For example, if the nonallelic mutants *a* and *b* were present in the coupling configuration *AB/ab,* crossovers would generate the reciprocal recombinant gametes *Ab* and *aB* in equal numbers.

reciprocal translocation   *See* **translocation.**

rec⁻ **mutant**   a class of mutations characterized by defective recombination. Such mutants are also radiation-sensitive, which suggests that enzymes functioning during the naturally occurring breakage and rejoining characterizing meiotic crossing over may also repair damage caused by mutagens.

recoil energy   the energy imparted to the positively charged ion formed during the radioactive transmutation of an atom. A high-energy beta particle is emitted concurrently.

recombinant   1. the new individuals or cells arising as the result of recombination. 2. recombinant DNA or a clone containing recombinant DNA.

recombinant DNA   a composite DNA molecule created *in vitro* by joining a foreign DNA with a vector molecule.

recombinant DNA technology   techniques for joining DNA molecules *in vitro* and introducing them into living cells where they replicate. These techniques make possible (1) the isolation of specific DNA segments from almost any organism and their amplification in order to obtain large quantities for molecular analysis, (2) the synthesis in a host organism of large amounts of specific gene products that may be useful for medicine or industry, and (3) the study of gene structure-function relationships by *in vitro* mutagenesis of cloned DNAs. *See* **Appendix C,** 1972, Jackson *et al.;* 1973, Cohen *et al.;* 1974, Murray and Murray; 1975, Asilomar Conference; Benton and Davis; 1976, Efstratiadis *et al., Kan et al.;* 1977, Gilbert; 1979, Goeddel *et al.;* 1980, Chakrabarty, Berg *et al.;* 1981, Wagner, Kemp and Hall; 1982, Eli Lilly; 1985, Smithies *et al.;* **expression vectors, gene cloning.**

recombinant inbred (RI) lines   inbred lines, each derived independently from an F₂ generation produced from crossing two unrelated, inbred, progenitor lines. Each RI line has a characteristic combination of genes with a different pattern of alternative alleles at multiple loci. This technique has been used in mice to fix chance recombinants in a homozygous state in a group of strains derived from two unrelated but highly inbred progenitor strains.

recombinant joint   the edge of a heteroduplex region where two recombining DNA molecules are connected.

recombinant RNA technology   techniques that unite foreign RNA molecules or splice different RNAs from the same species. For example, a heterologous RNA sequence can be constructed by ligation of two or more different RNA molecules with T4 RNA ligase.

recombination the occurrence of progeny with combinations of genes other than those that occurred in the parents, due to independent assortment or crossing over.

recombination frequency the number of recombinants divided by the total number of progeny. This frequency is used as a guide in assessing the relative distances between loci on a genetic map.

recombination nodules electron-dense organelles seen attached to synaptonemal complexes. These presumably play a role in crossing over.

recombination repair formation of a normal DNA molecule by exchanging correct for incorrect segments between two damaged molecules.

recombination suppression *See* **crossover suppressor.**

recombinators any sequences of nucleotides that promote genetic recombination in their neighborhood. An example would be the chi sequence (*q.v.*) in the *E. coli* chromosome.

recon the smallest unit of DNA capable of recombination. *See* **Appendix C,** 1955, Benzer.

record of performance a record of an animal with respect to certain economically important characteristics. Such data are used by livestock breeders in the artificial selection and development of improved breeds.

recurrence risk the risk that a genetic defect that has appeared once in a family will appear in a child born subsequently.

recurrent parent backcross parent.

red blood cell erythrocyte. *See* **hemoglobin, sickle-cell anemia.**

Red Queen hypothesis one of two major mathematical models concerning the likely evolutionary state of communities under conditions of constancy in the physical environment. *Stationary models* predict that evolution would grind to a halt. The Red Queen hypothesis predicts that evolution would continue because (1) the most important component of the species environment is other species in the community, and (2) not all species will be at their local adaptive peaks, and hence are capable of further evolution even though the physical environment has stabilized. Any evolutionary advance made by one species will, through a close network of interactions, represent a deterioration in the biotic environment of all other species in that same community. Consequently, these other species become subject to selective pressures to achieve evolutionary advances of their own, simply to catch up. The name for this hypothesis is derived from the Red Queen in *Through the Looking Glass,* who said: "Now here, you see, it takes all the running you can do to keep in the same place." *See* **lag load, zero sum assumption.**

reductase an enzyme responsible for reduction in an oxidation-reduction reaction.

reduction classically defined as the addition of hydrogen or electrons. Most biological reductions involve hydrogenations, and hydrogen transfer reactions are usually mediated by NADPH. In cases involving electron transfer, cytochromes (*q.v.*) are reduced. *See* **nicotinamide-adenine-dinucleotide phosphate, oxidation.**

reduction divisions the division that halves the zygotic chromosome number. *See* **Appendix C,** 1883, van Beneden; 1887, Weismann; **meiosis.**

reductionism a philosophy that each phenomenon in the natural world can be understood from a knowledge of its component parts. *See* **mechanistic philosophy.**

redundant cistrons cistrons frequently repeated on a chromosome. Examples are the cistrons in the nucleolus organizer coding for the ribosomal RNA molecules.

redundant code *See* **degenerate code.**

redundant DNA *See* **repetitious DNA.**

refractive index the ratio of the velocity of light in a vacuum to its velocity in a given substance. *See* **phase contrast microscope.**

regression coefficient the rate change of the dependent variable with respect to the independent variable. The change in mutation frequency per unit change in radiation dose, for example, would be determined by the regression coefficient of the regression line (*q.v.*).

regression line a line that defines how much an increase or decrease in one factor may be expected from a unit increase in another. *See* **line of best fit, scatter diagram.**

regressive evolution changes in the genetic structure of a population that result in loss of complexity, as exemplified by the loss of pigment and eyes in some cave-dwelling animals.

regulation the power of an embryo to continue normal or approximately normal development or regeneration in spite of experimental interference by ablation, implantation, transplantation, etc.

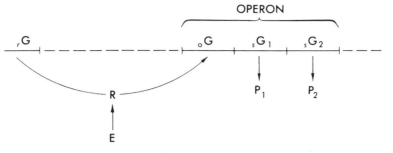

Regulator gene.

regulative development embryonic development in which the fates of all parts of the embryo are not fixed before fertilization. In such development an ablated part can be repaired, or even separated blastomeres can form identical twins. *See* **mosaic development.**

regulator element *See* **controlling elements.**

regulator gene a gene whose primary function is to control the rate of synthesis of the products of other distant genes. The regulator gene ($_rG$) controls the synthesis of a protein repressor (R), which inhibits the action of an operator gene ($_oG$) and thus turns off the operon it controls. In the above diagram, the horizontal line represents a chromosome upon which four genes reside. The left gene can be distant from the other three closely linked genes (in fact, $_rG$ can be on a different chromosome). Genes $_sG_1$ and $_sG_2$ are structural genes or cistrons of the conventional sort that produce specific proteins P_1 and P_2, respectively, through the formation of specific messenger RNA molecules. The repressor is present in exceedingly small amounts. It possesses two sites, one of which can attach to the operator and one of which can bind an effector (E) molecule. Once bound to E, however, the repressor changes shape and cannot attach to the operator. The effector molecule is generally a substrate of an enzyme produced by $_sG_1$ or $_sG_2$. Thus the system is an inducible one, since synthesis of P_1 and P_2 proceeds only in the presence of E. *See* **allosteric effect, constitutive mutations,** *cro* **repressor, derepression, inducible system, operon, repressible system.**

regulatory sequence a DNA sequence involved in regulating the expression of the structural gene(s) in the common operon. Examples include attenuators, operators, and promoters.

regulon a noncontiguous group of genes under control of the same regulator gene.

reiterated genes genes that are present in multiple copies that are clustered together on specific chromosomes. Ribosomal RNA genes, transfer RNA

genes, and histone genes are examples of such tandem multigene families.

rejection in immunology, destruction of a cell or tissue graft by the immune system of the recipient, directed against antigens on the graft that are foreign to the recipient. *See* **Appendix C,** 1914, Little; 1927, Bauer; 1948, Snell; **histocompatibility molecules.**

relational coiling the loose plectonemic coiling of two chromatids about one another. The two chromatids cannot separate until uncoiling is completed. *See* **plectonemic spiral.**

relative biological effectiveness the ratio of the doses of different ionizing radiations required to produce the same biological effect.

relative molecular mass (Mr) the mass of a molecule relative to the dalton (*q.v.*). Mr has no units and has replaced the term *molecular weight* in the recent chemical literature.

relative plating efficiency the percentage of inoculated cells that give rise to colonies, relative to a control where the absolute plating efficiency is arbitrarily set as 100. *See* **absolute plating efficiency.**

relaxation complex a group of three proteins tightly bound to some *E. coli* supercoiled plasmids that convert supercoiled DNA to a nicked open circle. When heated or treated with alkali, proteolytic enzymes, or detergents, one of these proteins nicks one strand at a specific site, thereby relaxing the supercoil to a nicked open circular form. During relaxation, the two smaller proteins are released, but the largest protein becomes covalently attached to the 5'-P end of the nick. Nicking plays a role in transfer of the plasmid during conjugation. The site of the nick establishes the *transfer origin.*

relaxation of selection cessation of selection in an experimental situation.

relaxation proteins *See* **helix-destabilizing proteins.**

relaxed control *See* **plasmid.**

release factors specific proteins that read termination codons and cause the release of the finished polypeptide.

releaser in ethology (*q.v.*), the particular physical attributes or behavior of one animal that stimulates another animal to perform a specific response.

relic coil the relaxed spirals often seen in prophase chromosomes. These coils are believed to be left over from the tightly coiled condition that the chromosome maintained at the previous metaphase.

relict surviving beyond others of a kind, as a species that persists in one region after becoming extinct elsewhere or a surviving species of a group of which others are extinct.

relief in photography, etc., to produce the effect of a third dimension in a two-dimensional image, as by the use of shadows. *See* **shadow casting.**

rem *r*oentgen *e*quivalent *m*an, the dose of ionizing radiation that has the same biological effect as one *rad* of x rays. It is equal to *rads* × *rbe* (relative biological effectiveness).

renaturation the return of a protein or nucleic acid from a denatured state to its native three-dimensional configuration.

Renner complex a group of chromosomes (and the genes within them) that are distributed as a unit generation after generation. Such complexes are found in species belonging to the genera *Oenothera* and *Rhoeo,* for example, where whole sets of chromosomes are involved in a series of interchanges. At first meiotic metaphase, one sees rings of bivalents, rather than independent tetrads.

reovirus a virus whose name is derived from *r*espiratory and *e*nteric *o*rphan virus. Reoviruses have been found in humans, but their relation to any disease is uncertain. The term "orphan" is used for viruses "without a disease." They are one of the few viruses that contain double-stranded RNA.

rep *r*oentgen *e*quivalent *p*hysical (*q.v.*).

repair *See* **DNA repair.**

repair synthesis enzymatic excision and replacement of regions of damaged DNA as when UV-induced thymine dimers are removed. *See* **Appendix C,** 1964, Setlow and Carrier, Boyce and Howard-Flanders; 1965, Clark and Margulies; **cut-and-patch repair.**

rep DNA repetitious DNA (*q.v.*).

repeated epitope *See* **sporozoite.**

repeated gene families synonymous with multigene families (*q.v.*).

repeating unit the length of a nucleotide sequence that is repeated in a tandem cluster.

repeats small tandem duplications (*q.v.*).

repetition frequency the number of copies of a given DNA sequence present in the haploid genome.

repetitious DNA nucleotide sequences occurring repeatedly in chromosomal DNA. Analysis of reassociation kinetics (*q.v.*) reveals that repetitious DNA can belong to the highly repetitive or middle repetitive categories. The highly repetitive fraction contains sequences of several nucleotides repeated millions of times. It is a component of constitutive heterochromatin (*q.v.*). Middle repetitive DNA consists of segments 100–500 bp in length repeated 100 to 10,000 times each. This class of rep DNA contains the genes transcribed into rRNAs and tRNAs. *See* **Appendix C,** 1966, Waring and Britten; 1968, Britten and Kohne; 1970, Pardue and Gall; 1978, Finnegan *et al.*

repetitive DNA repetitious DNA (*q.v.*).

repetitive genes *See* **multigene families.**

replacement sites positions within a gene at which point mutations alter the amino acid specification.

replacement vector *See* **lambda cloning vehicle.**

replica plating a technique used to produce identical patterns of bacterial colonies on a series of Petri plates. A Petri plate containing bacterial colonies is inverted, and its surface is pressed against a cylindrical block covered with velveteen. In this way, 10 to 20% of the bacteria are transferred to the fabric. Subsequently, bacteria-free plates are inverted and pressed against the velveteen disc to pick up samples of the colonies. About eight replicas may be printed from a single pad in this way. If the medium contained in each of the secondary plates differs in its selective properties, then the hundreds of different bacterial clones transferred may be scored simultaneously for their responses to a given agent. *See* **Appendix C,** 1952, Lederberg and Lederberg.

replicase *See* **RNA replicase.**

replication a duplication process requiring copying from a template.

replication-defective virus a virus defective in one or more genes essential for completing the infective cycle.

replication eye (bubble) an eye-shaped replicated DNA region within a longer, unreplicated region. *See* **D loop.**

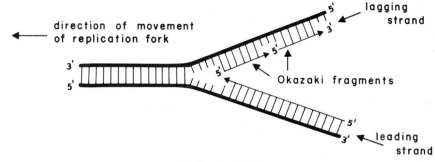

direction of movement of replication fork

lagging strand

Okazaki fragments

leading strand

Replication of DNA.

replication fork   *See* **replication of DNA.**

replication of DNA   during DNA replication the two strands of the duplex molecule separate to form a *replication fork.* DNA polymerase (*q.v.*) then adds complementary nucleotides starting at the 3' end. The strand that is continuously replicated in this way is referred to as the *leading strand.* The other strand is replicated discontinuously in short pieces. These Okazaki fragments are later connected by DNA ligase (*q.v.*) to form the lagging strand. *See* **Appendix C,** 1968, Okazaki; 1970, Smos and Inman; **primase.**

replication origin   a nucleotide sequence at which DNA synthesis begins; termed an *ori* site. Circular bacterial genophores have a single *ori* site, whereas there are many *ori* sites on each eukaryotic chromosome. *See* **replicon.**

replicative forms   double-stranded nucleic acid molecules seen at the time of the replication of single-stranded DNA and RNA viruses.

replicator   *See* **replicon.**

replicon   a genetic element that behaves as an autonomous unit during DNA replication. In bacteria, the chromosome functions as a single replicon, whereas eukaryotic chromosomes contain hundreds of replicons in series. Each replicon contains a segment to which a specific RNA polymerase binds and a replicator locus at which DNA replication commences. The polymerase makes an RNA primer called an initiator. *See* **Appendix C,** 1963, Monod and Brenner; 1968, Huberman and Riggs; **DNA fiber autoradiography, primase.**

replicon fusion   the joining of two complete replicating systems, mediated by a transposon. If two plasmids are joined (one of which carries a transposon), the process is called *plasmid fusion.*

replisome   a structure containing DNA polymerase and other enzymes that assembles at the replication fork of a bacterial chromosome to initiate DNA synthesis.

rep protein   a helicase (*q.v.*). identified in the *rep* mutant strain of *E. coli* that hydrolyzes ATP while forcing the strands of the DNA helix apart.

representation   *See* **abundance.**

repressible enzyme   an enzyme whose rate of production is decreased when the intracellular concentration of certain metabolities is increased.

repressible system   a regulatory system in which the product of a regulator gene (the repressor) blocks transcription of the operon only if it first reacts with an effector molecule (called the repressing metabolite). Thus, mRNA synthesis occurs only in the absence of the effector. *See* **regulator gene.**

repressing metabolite   *See* **repressible system.**

repression   1. the inhibition of transcription or translation when a repressor protein binds to an operator locus on DNA or to a specific site on a mRNA. 2. the cessation of synthesis of one or more enzymes when the products of the reactions they catalyze reach a critical concentration.

repressor   a protein (synthesized by a regulator gene) that binds to an operator locus and blocks transcription of that operon. *See* ***cro* repressor, regulator gene.**

reproduction probability   the average number of children of patients with a specific hereditary disease in relation to the average number of children of comparable individuals who do not have the hereditary disease. The reproduction probability is a measure of the selective disadvantage of a hereditary disease.

reproductive death   the suppression of the proliferative ability of a cell that otherwise would divide indefinitely.

reproductive isolation   the absence of interbreeding between members of different species. *See* **isolating mechanisms.**

reproductive potential   the theoretical logarithmic

rate of population growth when unimpeded by environmental limitations; also known as the biotic potential; symbolized r.

reproductive success   for a given individual, the number of offspring that survive to reproduce.

repulsion   *See* **coupling, repulsion configurations.**

RER   *rough-surfaced endoplasmic reticulum.* (*q.v.*).

RES   *reticuloendothelial system.*

residual genotype   background genotype (*q.v.*).

residual homology   in species hybrids, the homology remaining between those chromosomes that have been derived from a common ancestral chromosome and have since diverged through the accumulation of mutations. *See* **homoeologous chromosomes.**

resistance factor   a class of episomes that confer antibiotic resistance to the recipient bacterium. *See* **R plasmid.**

resistance transfer factor (RTF)   *See* **R plasmid.**

resolvase   an enzyme catalyzing the site-specific recombination between two transposons present as direct repeats in a *cointegrate structure* (*q.v.*).

resolving power   the ability of any magnifying system to reveal fine detail. This ability is often measured as the minimum distance between two lines or points at which they are resolved as two rather than as a single blurred object. The maximum resolving power of the light microscope is about 0.2 μm; that of the electron microscope, about 0.5 nm.

resource tracking   a hypothesis involving host-parasite coevolution according to which ectoparasites track a particular resource, such as a type of skin, hair, or feathers. If in addition there is opportunity for a given species of parasite to disperse to unrelated host species, then there will be no direct parallel relationship between the taxonomy of the hosts and that of their parasites (*contrast with* **Fahrenholz's rule**). In birds, for example, the same species of mite may be found on birds from different orders, and a single species of bird may be parasitized by different lice species. Thus, birds and their parasites show little phylogenetic parallelism.

respiration   the aerobic, oxidative breakdown and release of energy from fuel molecules.

respiratory pigment   a substance that combines reversibly with oxygen, thus acting as a carrier of it (hemoglobin, for example).

responder   in immunology, an animal capable of mounting an immunological response to a particular antigen.

resting cell (nucleus)   any cell (nucleus) not undergoing division. The cell (nucleus) is nevertheless very active metabolically.

restitution   the spontaneous rejoining of experimentally induced, broken chromsomes to produce the original configuration.

restitution nucleus   1. a nucleus containing double the expected number of chromosomes owing to a failure of the mitotic apparatus to function properly. 2. An unreduced product of meiosis. A diploid nucleus, resulting from the failure of the first or second meiotic division. *See* **meiosis.**

restricted transduction   *See* **transduction.**

restriction   ability of a bacteriophage to infect bacteria belonging to certain strains but not others. *See* **DNA restriction enzyme.**

restriction allele   *See* **DNA restriction enzyme.**

restriction and modification model   a theory proposed by W. Arber to explain host-controlled restriction of bacteriophage growth. According to this model, the DNA of the bacterium contains specific nucleotide sequences that are recognized and cleaved by the restriction endonucleases carried by that cell. The bacterium also contains methylases that methylate these sequences. This chemical modification thus protects the DNA of the bacterium from its own endonucleases. However, these serve to degrade foreign DNA introduced by phages. *See* **Appendix C,** 1972, Kuhnlein and Arber.

restriction endonuclease   any one of many enzymes that cleave foreign DNA molecules at specific recognition sites. Restriction endonucleases are coded for by genes called restriction alleles. The enzymes are named by a symbol that indicates the bacterial species from which they were isolated, followed by a Roman numeral that gives the chronological order of discovery when more than one enzyme came from the same source. Some restriction endonucleases, the organisms from which they were isolated, and their target nucleotide sequences are shown on page 275. The arrows indicate the cleavage sites.

$$\overset{*}{C}\ A$$

$\overset{*}{A}$ is N^6-methyladenine; (T)(G) signifies that either base can occupy that position. Note that *Bam* HI and *Eco* RI cleave the strands of DNA at specific sites four nucleotides apart. Such staggered cleavage yields DNA fragments with protruding 5′ termini.

Bain HI *Bacillus amyloliquefaciens*

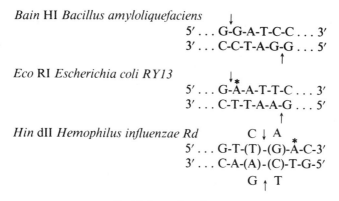

5′ . . . G-G-A-T-C-C . . . 3′
3′ . . . C-C-T-A-G-G . . . 5′

Eco RI *Escherichia coli RY13*

5′ . . . G-A-A-T-T-C . . . 3′
3′ . . . C-T-T-A-A-G . . . 5′

Hin dII *Hemophilus influenzae Rd*

C ↓ A

5′ . . . G-T-(T)-(G)-A-C-3′
3′ . . . C-A-(A)-(C)-T-G-5′

G ↑ T

Restriction endonucleases.

Such ends are said to be "sticky" or "cohesive" because they will hydrogen bond to complementary 3′ ends. As a result, the end of any DNA fragment produced by an enzyme, such as *Eco* RI, can anneal with any other fragment produced by that enzyme. This property allows splicing of foreign genes into *E. coli* plasmids. Enzymes like *Hin* dII produce flush or blunt-ended fragments. Restriction endonucleases are used extensively to map DNA regions of interest. *See* **Appendix C,** 1968, Smith *et al.;* 1970, Smith and Wilcox; 1971, Dana and Nathans; 1972, Mertz and Davis, Hedgpeth *et al.;* **polylinker site.**

restriction fragment a fragment of a longer DNA molecule digested by a restriction endonuclease.

restriction fragment length polymorphisms variations occurring within a species in the length of DNA fragments generated by a specific endonuclease. Such variations are generated by mutations that create or abolish recognition sites for these enzymes. For example, restriction endonuclease mapping of human structural genes for beta hemoglobin chains has shown that patients with the sickle cell mutation produce abnormal restriction fragments. Since restriction enzyme analyses can be performed on DNA from amniotic fluid cells, RFLPs provide a new approach to prenatal diagnoses. *See* **Appendix C,** 1978, Kan and Dozy; **alphoid sequences, DNA fingerprint technique, variable numbers of tandem repeats locus.**

restriction map a diagram portraying a linear array of sites on a DNA segment at which one or more restriction endonuclease enzymes cleave the molecule.

restriction site a deoxyribonucleotide sequence at which a specific restriction endonuclease cleaves the molecule.

restrictive conditions any environmental condition (e.g., temperature or type of host) under which a conditional mutation either cannot grow or expresses the mutant phenotype. *See* **temperature sensitive mutation.**

restrictive transduction *See* **transduction.**

reticulate evolution the netlike lineage relation seen for a series of related allopolyploid species. The cross-links represent places where hybridization has occurred and allotetraploid species have arisen. Reticulate evolution is common in plants. *See* **dendritic evolution.**

reticulocyte an immature erythrocyte at an active stage in the synthesis of hemoglobin.

reticuloendothelial system a network of phagocytic cells residing in the bone marrow, spleen, and liver of vertebrates, where they free the blood or lymph of foreign particles.

retinene a light-absorbing, carotenoid pigment that is derived from vitamin A (*q.v.*).

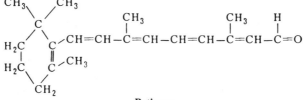

Retinene.

retinoblastoma a malignant neoplasm composed of primitive retinal cells usually occurring in children less than three years old. Retinoblastoma occurs in hereditary and non-hereditary forms. In the case of hereditary retinoblastoma, patients are generally affected bilaterally. The gene involved (*Rb*) is located on the long arm of chromosome 13. In contrast to oncogene activation or alteration, *Rb* functions to suppress tumor formation, since the loss or inactivation of both alleles results in the disease. *See* **anti-oncogene.**

retinoic acid a morphologically active compound that exists in a concentration gradient along the limb bud of tetrapods and stimulates the development of limbs and digits. Retinoic acid is vitamin A (*q.v.*) with the terminal CH_2OH replaced by a COOH group.

retinol synonymous with vitamin A (*q.v.*).

retrodiction the act of predicting the yet undiscovered results of past events. Useful theories of evolution should allow retrodictions that can be validated by looking in the right places within the fossil record.

retrogene *See* **processed gene.**

retrogression evolution toward a less complex state; characteristic of some parasitic groups. For example, tapeworms, having no digestive system, are thought to have evolved from free-living flatworms with a digestive system.

retroposon a mobile genetic element that transposes via an RNA intermediate. The best understood retroposons are the retroviruses (*q.v.*).

retroregulation the ability of downstream DNA sequences to regulate translation of an mRNA.

retroviruses RNA viruses that utilize reverse transcriptase (*q.v.*) during their life cycle. This enzyme allows the viral genome to be transcribed into DNA. The name *retro*virus alludes to this "backwards" transcription. The transcribed viral DNA is integrated into the genome of the host cell where it replicates in unison with the genes of the host chromosome. The cell suffers no damage from this relationship unless the virus carries an oncogene (*q.v.*). If it does, the cell will be transformed into a cancer cell. Among the oncogenic retroviruses are those that attack birds (such as the Rous sarcoma virus), rodents (the Maloney and Rauscher leukemia viruses and the mammary tumor agent), carnivores (the feline leukemia and sarcoma viruses), and primates (the simian sarcoma virus). The virus responsible for the current AIDS epidemic is the retrovirus HIV (*q.v.*). Retroviruses violate the central dogma (*q.v.*) during their replication.

reversals situations in phylogenetic analysis in which a derived character state changes (reverses) to a preexisting character state. For example, among the vertebrates, limbs are considered to be derived from a limbless state; but snakes are thought to have evolved from tetrapod (four-legged) ancestors and in the process have lost their limbs.

reverse loop pairing *See* **inversion.**

reverse mutation a change in a mutant gene which restores its ability to produce a functional protein.

reverse selection selection in an experimental situation for a trait opposite to the one selected earlier (for example, selection first for increased numbers of thoracic bristles in *Drosophila,* then for decreased numbers).

reverse transcriptase RNA-dependent DNA polymerase (*q.v.*).

reverse transcription DNA synthesis from an RNA template, mediated by reverse transcriptase. *See* **RNA-dependent DNA polymerase.**

reversion reverse mutation (*q.v.*).

revertant 1. an allele that undergoes reverse mutation. 2. an organism bearing such an allele.

reverted Bar *See Bar.*

R$_f$ in paper chromatography, a ratio given by the distance traveled by the solute divided by the distance traveled by the solvent. For a given solute molecule the R$_f$ varies with the solvent, and therefore the solvent must be specified for any R$_f$ value.

RF 1. replicative form; 2. recombination frequency.

R factor resistance factor (*q.v.*).

RFLP pronounced "rif lip." *See* **restriction fragment length polymorphisms.**

rhesus factor *See* **Rh factor.**

rhesus monkey *See Macaca mulatta.*

rheumatoid factor a distinctive gamma globulin commonly present in the serum of patients with rheumatoid arthritis.

Rh factor an antigen occurring on the erythrocytes of certain human beings. The Rh system actually contains several antigens. The most important one was first found in the Rhesus monkey—hence the name. Persons of genotype *r/r* produce no antigen and are classified as Rh-negative. *R/R* and *R/r* individuals produce the antigen. A pregnant mother who is Rh-negative but is carrying an Rh-positive child may produce antibodies against the child *in utero,* causing the child to develop a hemolytic disease called erythroblastosis fetalis. The Rh genetic locus is on the distal end of the long arm of chromosome 1. *See* **Appendix C,** 1939, Levine and Stetson; **RhoGAM.**

Rhizobium a genus of nitrogen-fixing bacteria that live as symbionts in the root nodules of leguminous plants. The most studied species are *R. meliloti* and *R. trifoli.* The genes involved in host specificity, nodulation, and nitrogen fixation are carried on large plasmids, and restriction maps have been constructed of those regions carrying symbiotically important genes. *See* **Appendix C,** 1981, Hombrecher *et al.;* **bacteroids,** *nif* **genes, plasmid, restriction map.**

Rhizopoda the phylum of protoctists containing singled celled amoebas. *See Amoeba proteus,* **Appendix A.**

rhodoplast the red plastid of red algae.

rhodopsin the light-sensitive chromoprotein found in the rod cells of the retina. Rhodopsin consists of the protein opsin in combination with retinene (*q.v.*). In the primary event of visual excitation, rhodopsin (also called *visual purple*) is bleached to a yellow compound. In humans, the gene encoding rhodopsin (RHO) resides on the long arm of chromosome 3. The cone pigment genes (*q.v.*) show considerable sequence similarities to RHO. *See* **color blindness.**

rhodamine B a fluorochrome commonly used to tag compounds that bind to specific cell components. *See* **rhodaminylphalloidin.**

rhodaminylphalloidin a fluorescent derivative of phalloidin used to specifically stain actin filaments in whole mounts of cells. *See* **fluorescence microscopy, phallotoxins.**

rho factor an oligomeric protein in *E. coli* that attaches to certain sites on its DNA to assist in termination of transcription.

RhoGAM the trade name for anti-Rh gamma globulin used in the prevention of Rh hemolytic disease. *See* **Appendix C,** 1964, Gorman *et al.;* **Rh factor.**

rho(ρ)-independent terminators *E. coli* DNA sequences recognized by RNA polymerase as transcription termination signals in the absence of rho (ρ) factors.

Rhynchosciara a genus of primitive flies belonging to the Sciaridae. Two species, *R. angelae* and *R. hollaenderi,* can be reared in the laboratory and have been extensively studied because of the gigantic polytene chromosomes found in various larval tissues (such as the salivary glands, Malpighian tubules, and anterior mid gut).

RIA radioimmunoassay (*q.v.*).

riboflavin vitamin B$_2$, a subunit of flavin adenine dinucleotide and flavin mononucleotide.

riboflavin phosphate flavin mononucleotide (*q.v.*).

ribonuclease the enzyme that hydrolyzes ribonucleic acid.

ribonuclease A an endoribonuclease that specifically attacks pyrimidines at the 3′ phosphate group and cleaves the 5′ phosphate linkage to the adjacent nucleotide. The end products of digestion are pyrimidine 3′ phosphates and oligonucleotides with pyrimidine 3′ phosphate termini.

ribonuclease P a bacterial enzyme whose catalytic activity depends upon an RNA molecule. Historically, ribonuclease P was the first RNA found to exhibit the characteristics of a true enzyme (i.e., it acts

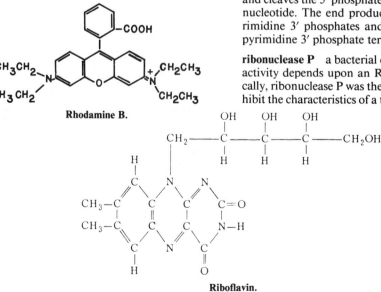

Rhodamine B.

Riboflavin.

catalytically but is not changed or consumed during the reaction). The enzyme cleaves precursor transfer RNA molecules to produce mature tRNAs. Ribonuclease P is composed of about five times more RNA than protein by weight. The RNA component alone can perform the catalytic function, whereas the protein component alone cannot. This enzyme has been found in *E. coli* and in *Bacillus subtilis*. *See* **Appendix C,** 1983, Guerrier-Takada *et al.;* 1989, Cech and Altmann; **ribozyme.**

ribonuclease T1 an endoribonuclease that specifically attacks the 3′ phosphate groups of guanosine nucleotides and cleaves the 5′ phosphate linkage to the adjacent nucleotide. The end products of digestion are guanosine 3′ phosphates and oligonucleotides with guanosine 3′ phosphate termini.

ribonucleic acid (RNA) any of a family of polynucleotides characterized by their component sugar (ribose) and one of their pyrimidines (uracil). RNA molecules are single stranded and have lower molecular weights than DNAs. There are three classes of RNAs: (1) messenger RNA (*q.v.*), (2) ribosomal RNA (*see* **ribosome**), and (3) transfer RNA (*q.v.*). *See* **Appendix C,** 1941, Brachet and Caspersson.

ribonucleoprotein a complex macromolecule containing both RNA and protein and symbolized by RNP.

ribonucleotide an organic compound that consists of a purine or pyrimidine base bounded to ribose, which in turn is esterified with a phosphate group.

ribose a five-carbon sugar.

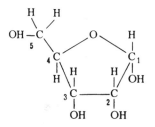

ribosomal binding site *See* **Shine-Dalgarno sequence.**

ribosomal DNA *See* **rDNA.**

ribosomal DNA amplification *See* **rDNA amplification.**

ribosomal precursor RNA *See* **preribosomal RNA.**

ribosomal protein *See* **ribosome.**

ribosomal RNA (rRNA) *See* **ribosome.**

ribosomal RNA genes rRNA genes reside as tandem repeating units in the nucleolus organizer regions of eukaryotic chromosomes. Each unit is separated from the next by a nontranscribed spacer. Each unit contains three cistrons coding for the 28S, 18S, and 5.8S rRNAs. The transcriptional polarity of the unit is 5′-18S-5.8S-28S-3′. Ribosomal RNA genes are often symbolized by rDNA. *See* **preribosomal RNA, rDNA amplification.**

ribosomal stalling *See* **attenuation.**

ribosome one of the ribonucleoprotein particles, 10–20 mμ in diameter, that are the sites of translation. Ribosomes consist of two unequal subunits bound together by magnesium ions. Each subunit is made up of roughly equal parts of RNA and protein. Each ribosomal subunit is assembled from one molecule of ribosomal RNA that is noncovalently bonded to 20 to 30 smaller protein molecules to form a compact, tightly coiled particle. In eukaryotes, the rRNAs of cytoplasmic ribosomes are formed by cistrons localized in the nucleolus organizer region of chromosomes. At least four classes of ribosomes exist that can be characterized by the sedimentation constants of their component rRNAs. Animal ribosomes also contain a 5.8S rRNA, which is hydrogen bonded to the 28S rRNA and is derived from the same intermediate precursor as the 28S rRNA. *See* **Appendix C,** 1956, Palade and Siekevitz; 1959, McQuillen *et al.;* 1961, Littauer, Waller and Harris; 1964, Brown and Gurdon; 1965, Sabatini *et al.;* **translation.**

5-ribosyluracil pseudouridine. *See* **rare bases.**

ribothymidine *See* **rare bases.**

SEDIMENTATION COEFFICIENT (S)

Ribosomal Source	Ribosome	Large Subunit	Small Subunit	Large rRNA	Small rRNA
Bacteria, chloroplasts	70	50	30	23	16
Plant cytoplasm	80	~60	~40	25	16
Animal cytoplasm	80	~60	~40	29	18
Mitochondria	81	61	47	25	19

Characteristics of four classes of ribosomes.

278

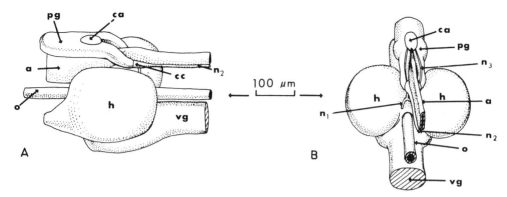

Ring gland

ribozyme an RNA molecule with catalytic activity. An example is the self-splicing rRNA of *Tetrahymena thermophila*. The gene for the 26S rRNA contains an intron 413 base pairs long. The precursor rRNA molecule transcribed from this gene includes a copy of this intervening sequence, which must be deleted by RNA splicing. If the pre-rRNAs are incubated *in vitro* with certain cations and guanosine, the intervening sequences are excised and the mature sequences are ligated. Ribonuclease P (*q.v.*) is another example of a ribozyme. Perhaps in the primordial biosphere ribozymes functioned as the first enzymes, and as life evolved proteins began to fine-tune the catalysis and eventually replaced ribozymes as enzymes. *See* **Appendix C,** 1981, Cech *et al.;* 1989, Cech and Altman.

rice *Oryza sativa.* Together with wheat, corn, and potatoes it is one of the world's four most important crops.

Ricinus communis the castor bean. The genetics of sex determination has been extensively studied in this species.

rickets a deficiency disease of growing bone due to insufficient vitamin D in the diet. *See* **vitamin D-resistant rickets.**

rifampicin the most commonly used of the rifamycins (*q.v.*).

rifamycins a group of antibiotic molecules produced by *Streptomyces mediterranei* that interfere with the beta subunit of prokaryotic RNA polymerases and thereby inhibit initiation of transcription. *See* **RNA polymerase.**

right splicing junction the boundary between the right (3′) end of an intron and the left (5′) end of an adjacent exon in mRNA; also called the *acceptor splicing site.*

ring canals canals connecting sister cystocytes in the *Drosophila* egg chamber. Such canals are surrounded by a ring of protein material. They function to allow a stream of cytoplasm to flow from nurse cells to oocyte. *See* **nurse cells.**

ring chromosome **1.** an aberrant chromosome with no ends. **2.** a ring-shaped chromosomal association seen during diakinesis in normal metacentric tetrads with two terminal chiasmata.

Ringer solution a physiological saline containing sodium, potassium, and calcium chlorides used in physiological experiments for temporarily maintaining cells or organs alive *in vitro.* Ringer solution is sometimes simply designated as "ringer."

ring gland a gland lying above the hemispheres (h) of the brain of the larval *Drosophila*. Its lateral extremities encircle the aorta (a) like a ring, hence its name. The gland contains three endocrine tissues: the corpus allatum (ca), the prothoracic gland (pg), and the corpus cardiacum (cc). The diagrams above show the ring gland viewed from the side (A) and from above (B). Other symbols are as follows: n_1 = afferent nerve to the corpus cardiacum from the brain; n_2 = efferent nerve from the corpus cardiacum; n_3 = nerve from corpus cardiacum to the corpus allatum; o = oesophagus; vg = ventral ganglion. *See* **allatum hormones, ecdysone.**

RK a symbol used in *Drosophila* studies to indicate *rank* or valuation of a given mutant. For example, RK1 indicates the best and most used mutants, with sharp classification, excellent viability, and accurate genetic localization. RK5 mutants show poor penetrance, low viability, and their chromosomal loci are not accurately determined.

rII locus the segment of the chromosome of T_4 bacteriophage that was the first to be subjected to fine structure genetic mapping. *See* **Appendix C,** 1955, Benzer.

R loop during molecular hybridization, the single-stranded sense strand of DNA that is prevented

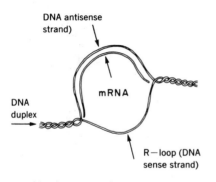

DNA antisense strand)

mRNA

DNA duplex

R—loop (DNA sense strand)

No intervening sequence

One intervening sequence

R-loop mapping.

from reannealing because its complementary antisense strand is base-paired with an mRNA exon as a heteroduplex. *See* **coding strand.**

R-loop mapping a technique for visualizing under the electron microscope the complementary regions shared by a specific eukaryotic RNA and a segment of one strand of a DNA duplex. The RNA-DNA hybrid segment displaces one of the DNA strands, causing it to form a *loop;* hence the name of the technique. Double-stranded regions appear thicker than single-stranded regions in electron micrographs. Introns cannot hybridize with mature mRNA (from which introns have been removed); thus, one intron results in two R loops; two introns yield three R loops, etc.

RNA ribonucleic acid.

RNA coding triplets *See* **amino acid, start codon, stop codon.**

RNA-dependent DNA polymerase an enzyme that synthesizes a single strand of DNA from deoxyribonucleoside triphosphates, using RNA molecules as templates. Such enzymes occur in oncogenic RNA viruses. This class of enzymes, also known as *reverse transcriptases,* can be used experimentally to make complementary DNA (cDNA) from purified RNA. The functioning of these polymerases contradicts the central dogma (*q.v.*) in the sense that the direction of information exchange between DNA and RNA is reversed. *See* **Appendix C,** 1970, Baltimore and Temin.

RNA-driven hybridization an *in vitro* technique that uses an excess of RNA molecules to ensure that all complementary sequences in single-stranded DNA undergo molecular hybridization. *See* **DNA-driven hybridization.**

RNA gene a DNA segment coding for one of the various types of non-messenger RNA (rRNA, 5S RNA, or tRNA).

RNA ligase an enzyme, such as T_4 RNA ligase (*q.v.*), that can join RNA molecules together. *See* **Appendix C,** 1972, Silber *et al.*

RNA-P I, II, III *See* **RNA polymerase.**

RNA phage an RNA bacteriophage such as MS2 and $Q\beta$.

RNA polymerase an enzyme that transcribes an RNA molecule from the antisense strand of a DNA molecule. Two kinds of RNA polymerases are known in prokaryotes: one produces the RNA primer required for DNA replication; the other transcribes all three types of RNA (mRNA, tRNA, rRNA). In eukaryotes, each type of RNA is transcribed by a different RNA polymerase. RNA polymerase I (RNA-PI) resides in the nucleolus and catalyzes the synthesis of rRNA. RNA-PII is localized in the nucleoplasm, where it catalyzes the synthesis of mRNA. RNA-PII is specifically inhibited by alpha amanitin (*q.v.*). RNA-PIII makes tRNA, 5S RNA, and other small RNA molecules. The structure of the RNA polymerase of *E. coli* is known in great detail. It has the general formula $\omega\alpha\alpha\sigma\beta'\beta$. The component proteins, their molecular weights, the encoding genes, and their locations are shown below.

Protein	MW (daltons)	Gene	Map positions (minutes)
omega	11,000	?	?
alpha	36,500	rpo A	72
sigma	70,000	rpo D	66.5
beta'	151,000	rpo C	89.5
beta	155,000	rpo B	89.5

The catalytic site for RNA polymerization is thought to reside in the beta protein, and this is also the place where rifampicin (*q.v.*) binds. The sigma protein functions in promoter recognition and the initiation of RNA synthesis. Eukaryotic RNA polymerases are still more complex, since they contain about ten, rather than five subunits. *See* **Appendix C, 1961, Weiss and Nakamoto; 1969, Burgess** *et al.;* **coding strand, Miller trees, terminator, transcription unit.**

RNA primer *See* **primer RNA.**

RNA processing *See* **nuclear processing of RNA, posttranscriptional modification.**

RNA puff *See* **chromosomal puff.**

RNA replicase an RNA-dependent RNA polymerase. *See* **MS2** and **Qβ.**

RNase any enzyme hydrolyzing RNA.

RNase protection a technique for locating the points of effective contact between a nucleic acid chain and a cognate polypeptide chain; the complex (e.g., tRNA and its cognate aminoacyl-tRNA synthetase) is treated with a group of RNases that digest all of the RNA except those regions in contact with the synthetase. *See* **photoactivated cross-linking.**

RNA splicing the removal of non-coding regions from a large precursor RNA molecule, and the nucleotide sequences transcribed from non-adjacent DNA segments are then joined together to produce a smaller mature RNA. *See* **nuclear processing of RNA.**

RNP ribonucleoprotein.

rNTP ribonucleoside 5′-triphosphates.

Robertsonian translocation *See* **centric fusion.**

rod one of the elongate, unicellular photoreceptors in the vertebrate eye, involved with vision in dim light. Rods do not discriminate color differences. *See* **rhodopsin.**

roentgen the quantity of ionizing radiation that liberates 2.083×10^9 ion pairs in a cubic centimeter of air (at 0°C and at a pressure of 760 mm of mercury) or approximately two ion pairs per cubic micron of a substance such as protein (which has a density of 1.35). A gram of tissue exposed to 1 roentgen of gamma rays absorbs about 93 ergs.

roentgen equivalent physical (rep) the amount of ionizing radiation that will result in the absorption in tissue of 93 ergs per gram.

rogue a variation from the standard variety, usually inferior.

rolling circle a model mechanism for the replication of DNA molecules, so named because the growing point can be imagined as rolling around a circular template strand. The circular DNA is shown below in A. In B, a nick opens one strand, and the free 3′-OH end is extended by DNA polymerase. The newly synthesized strand displaces the original parental strand as it grows (C, D). By E, the polymerase has completed one revolution, and by F, two revolutions. The result is a molecule containing three unit genomes, one old and two new. The displaced strand can then serve as a template for a complementary strand. This mechanism is used to generate concatemeric duplex molecules (e.g., phage lambda, amplified rDNA in amphibian oocytes, etc.). This type of DNA replication is sometimes called *sigma replication* because the structure produced by the rolling circle resembles the Greek lower case letter *σ*. *Compare with* **theta replication.** *See* **Appendix C, 1968, Gilbert and Dressler.**

Romalea microptera the lubber grasshopper. Meiosis has been extensively studied in this species.

root cap a cap of cells covering the apex of the growing point of a root and protecting it as it is forced through soil.

root hair a tubular outgrowth of an epidermal cell of a root which functions to absorb water and nutrients from the soil.

root nodules a small swelling on roots of legumes

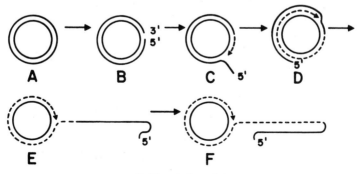

Rolling circle model.

281

(*q.v.*) produced as a result of infection by symbiotic nitrogen-fixing bacteria. *See Rhizobium.*

Rosa the genus that includes the rose species, which have been extensively hybridized. Commercially grown species include *R. centifolia, R. damascena, and R. multiflora.*

Rose chamber a closed culture vessel permitting long-term observation of explanted cells under phase microscopy. The fluid culture medium may be periodically renewed without disturbing the growing cells.

R_0t the product of RNA concentration and the time of incubation in an RNA-driven hybridization; the analog of C_0t values used to describe DNA-driven hybridization reactions.

rotational base substitution a break is induced by radiation at a corresponding point in both complementary strands of a DNA molecule. The bond broken is between the base and the sugar molecule to which it is attached. Thus two complementary bases (held together by hydrogen bonds) are detached from their backbones. If the pair rotates before it is reinserted in the molecule, the resulting transversional mutation would be termed a rotational base substitution. Many radiation-induced point mutations may result in this way.

rotation technique *See* **photographic rotation technique.**

rough surfaced endoplasmic reticulum *See* **endoplasmic reticulum.**

Rous sarcoma virus an avian retrovirus discovered by P. Rous. The genome of the RSV contains four genes: *gag,* which codes for the protein of the capsid; *pol,* which codes for a reverse transcriptase (*q.v.*); *env,* which codes for the protein that forms spikes on the envelope of the viral membrane, and *src.* This is an oncogene, so named because it induces sarcomas. The *src* gene codes for a protein kinase, $pp60^{v-src}$ (*q.v.*), which is localized in the plasmalemma. Vertebrate cells contain a gene homologous to the *src* gene. To distinguish the two, the viral gene is abbreviated *v-src* and the cellular gene *c-src.* The two genes differ in that *v-src* has an uninterrupted coding sequence, whereas *c-src* contains seven exons separated by six introns. The *c-src* gene is a proto-oncogene (*q.v.*). *See* **Appendix C,** 1910, Rous; 1975, Wang *et al.;* 1981, Parker *et al.;* 1989, Bishop and Varmus; **oncogene hypothesis, retrovirus.**

royal hemophilia classical hemophilia (*q.v.*) transmitted by a defective X chromosome first carried by Queen Victoria of Great Britain and passed on to plague three generations of European royalty.

R plasmid an extrachromosomal DNA molecule that confers on bacteria resistance to one or more antibiotics. It consists of two components: the resistance transfer factor (RTF) required for transfer of the plasmid between bacteria, and the r-determinants (genes conferring antibiotic resistance) that are part of transposons (*q.v.*).

rpo *See* **RNA polymerase.**

rRNA ribosomal RNA. *See* **ribosome.**

rRNA transcription unit *See* **Miller trees.**

r strategy a type of life cycle exploiting high reproductive rate to achieve survival. *See* **r and K selection theory.**

RSV Rous sarcoma virus (*q.v.*).

rtDNA *See* **rDNA.**

RTF resistance transfer factor. *See* **R plasmid.**

rTU rRNA transcription unit. *See* **Miller trees.**

ruffled edges *See* **lamellipodia.**

runner a procumbent shoot that takes root, forming a new plant that eventually is freed from connection with the parent by decay of the runner. The runner serves as a vegetative propagule. *See* **modular organisms.**

runting disease a pathological condition seen in young experimental animals inoculated with allogeneic immunocompetent cells that produce a graft-versus-host reaction (*q.v.*).

rut the period of sexual activity; estrus.

S

s **1.** selection coefficient (*q.v.*) **2.** standard deviation (*q.v.*) **3.** sedimentation coefficient (*q.v.*) **4.** second.

S **1.** Svedberg unit (*q.v.*) **2.** Silurian. **3.** sulfur. **4.** DNA synthesis phase of the cell cycle (*q.v.*).

^{35}S a beta-emitting radioactive isotope of sulfur with a half-life of 87.1 days; commonly used to label proteins via their sulfur-containing amino acids cysteine and methionine.

S$_1$, S$_2$, S$_3$, etc. the representation for continued selfing (self-fertilization) of plants. S$_1$ designates the generation obtained by selfing the parent plant; S$_2$, the generation obtained by selfing the S$_1$ plant, etc.

Saccharomyces cerevisiae baker's yeast, a favorite ascomycete for genetic study. Approximately 400 genes have been located on its 17 chromosomes. A circular gene map is also available for the yeast mitochondrion. The circular mtDNAs contain about 80 kb. In yeast mitochondria, CUA codes for threonine instead of leucine. *See* **Appendix A**, Fungi, Ascomycota; **Appendix C**, 1949, Ephrussi *et al.;* 1974, Dujon *et al.;* 1979, Cameron *et al.;* **artificial chromosomes, cassettes, genetic code, omnipotent suppressors, petites, universal code theory.**

salivary gland chromosomes polytene chromosomes found in the interphase nuclei of the salivary gland cells in larval diptera. These chromosomes undergo complete somatic pairing; consequently, the mature salivary gland chromosome consists of two homologous polytene chromosomes fused side by side. *See* **Appendix C**, 1881, Balbiani; 1912, Rambousek; 1933, Painter; 1934, Bauer; 1935, Bridges.

salivary gland squash preparation **1.** a rapid method of preparing insect polytene chromosomes for microscopic investigation without sectioning them. The organ is simply squashed in a drop of stain placed between slide and coverslip. *See* **aceto-orcein.** **2.** a method of preparing giant polytene chromosomes for localization of specific DNA sequences via *in situ* hybridization. Larval salivary glands are quickly squashed in an appropriate fixative between slide and coverslip, frozen, and the coverslip removed. The squashed specimens are then dehydrated and prepared for hybridization with labeled nucleic acid probes. *See in situ* **hybridization.**

S allele *See* **self-sterility genes.**

Salmo the genus containing various fish species of economic importance; especially *S. salar*, the Atlantic salmon and *S. gairdneri,* the rainbow trout.

Salmonella together with *Escherichia* and *Shigella,* one of the genera of enteric bacteria containing favorite species for genetic study. *Salmonella typhimurium,* the mouse typhoid bacillus, has been the most studied, but considerable work has also been done with *S. abony, S. pullorum, S. montevideo, S. minnesota,* and *S. typhosa.* Generalized transduction was first shown in *S. typhimurium* using phage P22. *See* **Appendix C**, 1952, Zinder and Lederberg.

saltation **1.** a theory that new species originate suddenly from one or more mutations with large phenotypic effects ("macromutations"); referred to by R. Goldschmidt as "hopeful monsters." **2.** quantum speciation (*q.v.*). *See* **evolution.**

saltatory replication lateral amplification of a chromosome segment to produce a large number of copies of a specific DNA sequence. *See* **gene amplification, rDNA amplification.**

salvage pathways metabolic pathways for the synthesis of nucleosides and nucleotides utilizing preformed purines and pyrimidines or nucleosides. Examples include the conversion of a base to a nucleoside, interconversion of bases and nucleosides, conversion of nucleosides to nucleotides, and interconversions by base alterations. *Compare with* ***de novo*** **pathway.**

samesense mutation a point mutation (usually in the third position of a codon) that does not change the amino acid specificity of the codon so altered; a "silent" mutation. *See* **degenerate code.**

sampling error variability due to the limited size of the samples.

Sandhoff disease a lethal hereditary disease in humans due to a recessive gene (HEXB) on autosome

5. The gene codes for the beta subunit of hexosaminidase A. Neurologic deterioration results from storage of gangliosides due to the enzyme deficiency. Symptoms are similar to those of Tay-Sachs disease (*q.v.*).

Sanger-Coulson method *See* **DNA sequencing techniques.**

saprophyte a plant living on and deriving its nutrition from dead organic matter.

sarcoma a cancer of mesodermal origin (e.g., connective tissue). *See* **Rous sarcoma virus, simian sarcoma virus.**

sarcomere the repeating unit, about 2.5 μm long, within striated muscle fibers, containing a set of interacting actin and myosin filaments.

sarcosomes the mitochondria of the flight muscles of insects.

sat DNA satellite DNA (*q.v.*).

satellite a distal chromosomal segment separated from the rest of the chromosome by a thin chromatic filament or stalk called the *secondary constriction.*

satellite DNA any fraction of the DNA of an eukaryotic species that differs sufficiently in its base composition from that of the majority of the DNA fragments to separate as one or more bands distinct from the bands containing the majority of the DNA during isopycnic CsCl gradient centrifugation. Satellite DNAs obtained from chromosomes are either lighter (A + T rich) or heavier (G + C rich) than the majority of the DNA. Satellite DNAs are usually highly repetitious. *See* **Alu family, mouse satellite DNA, reassociation kinetics, repetitious DNA.**

satellite RNAs *See* **small satellite RNAs.**

saturation density the maximum cell number attainable under specified culture conditions in a culture vessel. This term is usually expressed as the number of cells per square centimeter in an anchorage-dependent culture, or per cubic centimeters in a suspension culture.

saturation hybridization an *in vitro* reaction in which one polynucleotide component is in great excess, causing all complementary sequences in the other polynucleotide component to enter a duplex form. *See* **DNA-driven hybridization reaction, RNA-driven hybridization.**

scaffold *See* **chromosome scaffold.**

scanning electron microscope *See* **electron microscope.**

scanning hypothesis a theory to explain initiation of translation in eukaryotes according to which a 40S ribosomal subunit attaches at or near the mRNA cap and then drifts in the 3′ direction until it encounters an AUG start codon; at this point, an initiation complex is formed and the reading frame becomes established.

scape a leafless flower-bearing stem arising from ground level (as in the dandelion and daffodil).

scarce mRNA *See* **complex mRNA.**

scatter diagram a diagram in which observations are plotted as points on a grid of x and y coordinates to see if there is any correlation (*q.v.*). For example, one might plot for a number of species the LD 50 (*q.v.*) against the average DNA content per somatic cell nucleus. Finding a correlation suggests that the variables may be interrelated. No correlation suggests that the variables chosen have no bearing on one another.

scattering the change of direction of particles or waves as a result of a collision or interaction. Scattering of electrons by the specimen, for example, is responsible for the electron microscopic image.

SCE sister chromatid exchange (*q.v.*).

Schiff reagent a regent that attaches to and colors aldehyde-containing compounds. Used in the PAS and Feulgen procedures (*q.v.*).

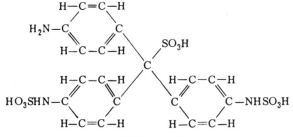

Schiff reagent.

Schistocerca gregaria a locust whose population dynamics has been extensively studied.

schistosomiasis also called *bilharziasis* or *snail fever;* a helminthic infection of humans involving 200 million persons in Africa, the Middle East, the Orient, and South America. The causative agent is the blood fluke *Schistosoma mansoni,* and its snail vector is *Biomphalaria glabrata.* The genetic control of susceptibility of *B. glabrata* to *S. mansoni* is a subject of active investigation.

schizogony a series of rapid mitoses without increase in cell size which gives rise to schizonts.

schizont a sporozoan spore arising from schizogony.

schizophrenia the name given to a constellation of symptoms including hallucinations and delusions, disorders of thinking and concentration, and erratic behavior. Schizophrenia appears to be a family of diseases of high heritability that together afflict about 1 percent of all humans.

Schizophyllum commune a basidiomycete that has been the subject of considerable genetic research. Two extensive linkage groups have been elucidated.

Schizosaccharomyces pombe a fission yeast, a species suitable for tetrad analysis (*q.v.*).

Schwann cell the cell that enfolds a myelinated nerve fiber. *See* **myelin sheath.**

Sciara a genus of fungus gnats extensively studied in terms of the cytogenetics of chromosome diminution (*q.v.*). The giant polytene chromosomes of the larval salivary glands of *S. coprophila* have been mapped and certain DNA puffs have been extensively studied.

scintillation counter *See* **liquid scintillation counter.**

scission a severance of both strands of a DNA molecule; a cut (*q.v.*). *Compare with* **nick.**

scleroproteins the very stable fibrous proteins, present mainly as surface coverings of animals. Keratin and collagen are examples of scleroproteins.

scrapie *See* **prion.**

scurvy *See* **ascorbic acid.**

scutellum 1. the single cotyledon of a grass embryo. 2. a shield-shaped metathoracic tergite of *Drosophila.*

SD 1. standard deviation. 2. segregation distortion (*q.v.*).

SD sequence *See* **Shine-Dalgarno sequence.**

SDS sodium dodecyl sulfate (*q.v.*).

SDS-PAGE technique *See* **electrophoresis.**

SE, S.E. standard error (*q.v.*).

Se *Secretor* gene (*q.v.*).

Searle translocation a reciprocal X-autosome translocation in the mouse which exhibits paternal X inactivation in the somatic cells of female heterozygotes. All other X-autosome translocations studied in the mouse show random inactivation of either normal or rearranged chromosome. *See* **Lyon hypothesis.**

seasonal isolation a type of ecological reproductive isolation in which different species become reproductively active at different times; temporal isolation.

secondary constriction a thin chromatic filament connecting a chromosomal satellite to the rest of the chromosome.

secondary gametocytes *See* **meiosis.**

secondary immune response *See* **immune response.**

secondary nondisjunction sex chromosomal nondisjunction in an XXY individual resulting in gametes containing either two X chromosomes, one X, one Y, or an X and a Y.

secondary protein structure *See* **protein structure.**

secondary sex ratio the ratio of males to females at birth; in contrast to the primary sex ratio at conception.

secondary sexual character a characteristic of animals other than the organs producing gametes that differs between the two sexes (mammary glands, antlers, external genitalia, etc.). *See* **primary sexual character.**

secondary speciation the fusion through hybridization of two species that were formerly geographically isolated, followed by establishment of a new adaptive norm through natural selection.

second cousin *See* **cousin.**

second division segregation ascus pattern in ascomycetes, a 2-2-2-2 or 2-4-2 linear order of spore phenotypes within an ascus. These patterns indicate that a pair of alleles (e.g., those controlling spore pigmentation) separated in the second meiotic division because crossing over occurred between that locus and the centromere. *See* **ordered tetrad.**

second law of thermodynamics   *See* **thermodynamics.**

second messenger   small molecules or ions generated in the cytoplasm in response to binding of a signal molecule to its receptor on the outer surface of the cell membrane. Two major classes of second messengers are known: one involves cyclic adenosine monophosphate and the other employs a combination of calcium ions and either inositol triphosphate or diacylglycerol. *See* **G proteins.**

second site mutation   *See* **suppressor mutation.**

secretin   a hormone that stimulates the secretion of pancreatic juice. The epithelial cells of the duodenum release secretin when activated by the acidic contents of the stomach.

secretion   the passage out of a cell or gland of compounds synthesized within it.

Secretor **gene**   a dominant autosomal gene in man that permits the secretion of the water-soluble forms of the A and B blood-group antigens into saliva and other body fluids. The *Se* gene is not linked to the *I* locus. *See* **A, B antigens.**

sedimentation coefficient (s)   the rate at which a given solute molecule suspended in a less dense solvent sediments in a field of centrifugal force. The sedimentation coefficient is a rate per unit centrifugal field. The s values for most proteins range between 1×10^{-13} sec and 2×10^{-11} sec. A sedimentation coefficient of 1×10^{-13} sec is defined as one Svedberg unit (S). Thus a value of 2×10^{-11} sec would be denoted by 200S. For a given solvent and temperature, s is determined by the weight, shape, and degree of hydration of the molecule. *See* **Appendix C,** 1923, Svedberg.

seed   a mature ovule containing an embryo in an arrested state of development, generally with a food reserve.

seeding efficiency   in cell culture, the percentage of the cells in an inoculum that attach to the culture vessel within a specific length of time; synonymous with *attachment efficiency.*

segmental alloploid   *See* **allosyndesis.**

segmental interchange   a translocation.

segmented genome   a viral genome fragmented into two or more nucleic acid molecules. For example, the alfalfa mosaic virus has four different RNA segments, each packaged in a different virion. Successful infection requires that at least one RNA of each type enters the cell. Such a virus is said to be *heterocapsidic.* If all fragments of a segmented genome are present in the same virion (e.g., influenza virus), the virus is said to be *isocapsidic.*

segment identity genes   genes that determine the type of differentiation the cells in a specific *Drosophila* segment will undergo. These genes express themselves later in development than the zygotic segmentation genes. While mutations in these genes cause the deletion of certain body parts and are generally lethal, mutations of segment identity genes allow the mutants to survive but with inappropriate structures developing in specific segments. For example, a bizarre four-winged fly results if the mutation converts halteres (*q.v.*) to wings. The segment identity genes are located in two clusters on the right arm of chromosome 3 (the *Antp* complex and the *bx* complex). *See* ***Antennapedia, bithorax,* homeotic mutations, zygotic segmentation mutations.**

segment polarity genes   *See* **zygotic segmentation mutants.**

segregational lag   delayed phenotypic expression of an induced mutation in one nucleoid of a multinucleoid bacterium. The lag period is the time required for the fission of the parent to produce a cell containing only the mutant chromosome.

segregational load   the genetic disability sustained by a population due to genes segregating from advantageous heterozygotes to less fit homozygotes.

segregational petites   *See* **petites.**

segregation distortion   a distortion of expected Mendelian ratios in a cross due to dysfunction or lethality in gametes bearing certain alleles. This form of meiotic drive (*q.v.*) is represented in *Drosophila melanogaster* by the segregation distorter (SD) mutation on chromosome 2. SD/sd$^+$ heterozygous males produce both SD and sd$^+$ spermatozoa, but only those carrying SD are functional. SD cannot achieve fixation, however, because it is lethal when homozygous.

segregation of chromosomes   *See* **disjunction.**

segregation of genes   *See* **Mendel's laws.**

segregation ratio distortion   the distortion of the 1:1 segregation ratio produced by a heterozygote. Such distortions may arise because of abnormalities of meiosis that result in an *Aa* individual's producing unequal numbers of *A*- and *a*-bearing gametes, or it may arise from *A*- and *a*-bearing gametes, being unequally effective in producing zygotes.

selection   the process determining the relative share allotted individuals of different genotypes in the propagation of a population. The selective effect of a gene can be defined by the probability that carriers of the gene will reproduce. *See* **alloprocoptic selection, artificial selection, balanced selection, directional selection, disruptive selection, frequency-**

dependent selection, group selection, indirect selection, kin selection, normalizing selection, r and K selection, sexual selection, stabilizing selection theory.

selection coefficient (s) the proportionate reduction in the average gametic contribution to the next generation made by individuals of one genotype relative to those of another genotype (usually the most fit). For example, if the best adapted genotypes are *AA* and *Aa,* and they are not being selected against, $s = 0$, and their fitness = $(1 - s)$ = 1. If individuals of genotype *aa* leave only on the average 80% as many progeny (proportionate to their numbers in the population) as the other genotypes, then the selection coefficient against *aa* individuals is 0.2 or 20%, and their fitness is $(1 - 0.2)$ = 80%.

selection differential the difference between the average value of a quantitative character in the whole population and the average value of those selected to be parents of the next generation. *See* **record of performance.**

selection pressure the effectiveness of natural selection in altering the genetic composition of a population over a series of generations.

selective advance the increment in the average value (measured for a quantitative character being selected in a population) from one generation to the next, usually a fraction of the selection differential (*q.v.*).

selective medium a medium designed to allow growth of only those cells of a specific genotype. *Compare with* **nonselective medium.**

selective neutrality a situation in which the phenotypic manifestations of certain mutant alleles are equivalent to that of the wild-type allele in terms of their fitness values. *See* **neutral gene theory, silent mutations.**

selective plating a method for selectively isolating recombinants. Two different auxotropic mutants are plated upon a minimal medium. Only the recombinant class receiving the normal allele of each mutant can multiply under these conditions.

selective silencing any mechanism that consistently eliminates plasmagenes of one parent from the zygote, such as the destruction of chloroplasts or chloroplast DNA in some algae and plants or of sperm mitochondria in some animals.

selective system any experimental technique that aids in the detection and isolation of a specific (usually rare) genotype. *See* **penicillin selection technique.**

selective variant in microbial genetics, a mutation that confers upon the organism the ability to exist under conditions that kill off all organisms not possessing the mutation. Examples of selective variants are mutations conferring resistance to antibacterial agents or the ability to synthesize some essential metabolite lacking in the medium.

selector genes *See* **compartmentalization.**

self to undergo self-pollination (*q.v.*) or self-fertilization (*q.v.*).

self-assembly the spontaneous aggregation of multimeric biological structures involving formation of weak chemical bonds between surfaces with complementary shapes. For example, most of the components of the phage T_4 capsid (head, tail, base plate, and tail fibers) are self-assembled.

self-compatible said of a plant that can be self-fertilized.

self-fertilization the fusion of male and female gametes from the same individual.

self-incompatibility self-sterility (*q.v.*).

selfish DNA functionless DNA regions that exist merely to perpetuate themselves. According to the selfish DNA theory, an eukaryotic organism is a "throwaway survival machine" used by selfish DNA to replicate itself. Spacer DNA, satellite DNA, and some other kinds of repetitive DNA may be examples; also called *junk DNA.*

self-pollination the transfer of pollen to the stigmas of the same plant.

self-splicing rRNA *See* **ribozyme.**

self-sterility the inability of some hermaphrodites to form viable offspring by self-fertilization.

self-sterility genes genes that prevent the deleterious effects of inbreeding in monoecious plants by controlling the rate of growth of the pollen tube down the style. During this process the sporophytic tissue discriminates against gametophytic tissue containing a common sterility gene. *See* **Appendix C,** 1925, East and Mangelsdorf.

SEM scanning electron microscope. *See* **electron microscope.**

semelparity reproduction that occurs only once in the life of an individual (e.g., annual plants, Pacific salmon). *Compare with* **iteroparity.**

semen a biochemically complex nutrient fluid containing spermatozoa which is transferred to the female during copulation.

semiconservative replication the method of replication of DNA in which the molecule divides longitudinally, each half being conserved and acting as a template for the formation of a new strand. *See* **Appendix C,** 1957, Taylor *et al.;* 1958, Meselson and Stahl; 1963, Cairns; 1964, Luck and Reich. *Compare with* **conservative replication.**

semidiscontinuous replication a mode of DNA replication in which one new strand is synthesized continuously, while the other is synthesized discontinuously as Okazaki fragments. *See* **replication of DNA.**

semidominance the production of an intermediate phenotype in individuals heterozygous for the gene concerned; also known as *partial dominance. See* **incomplete dominance.**

semidwarf a term used to distinguish mutant strains of wheat that are of agricultural importance from the extremely short dwarfs of purely genetic interest. Semidwarfs grow from half to two-thirds the height of, and have greater yields than those of standard varieties.

semigeographic speciation *See* **parapatric speciation.**

semilethal mutation a mutation causing death of more than 50%, but not of all individuals of mutant genotype.

seminiferous tubule dysgenesis Klinefelter syndrome (*q.v.*).

semiochemistry the study of the chemical signals that mediate interactions between members of different species. *See* **pheromone.**

semipermeable membrane any membrane that permits passage of molecules selectively.

semispecies incipient species.

semisterility a situation in which half or more of all zygotes are inviable (as in *Oenothera* crosses that maintain only heterozygotes). *See* **balanced lethal system.**

semisynthetic antibiotic a natural antibiotic that has been chemically modified in the laboratory to enhance its stability.

Sendai virus a virus, first isolated in Japan, that causes an important and widespread infection of laboratory mice; it belongs to the parainfluenza group of mixoviruses. The virus is widely used in cell fusion studies. The viruses so modify the surfaces of infected cells that they tend to fuse. Even UV-killed viruses adsorb on host cells and promote their fusion. *See* **Appendix C,** 1965, Harris and Watkins.

senescent aging.

sense codon any of the 61 triplet codons in mRNA that specify an amino acid.

sense strand *See* **coding strand.**

sensitive developmental period a period during development when there is an enhanced chance that genetic malfunction will bring development to a standstill. In *Drosophila* these sensitive periods correspond to the onset of embryonic, larval, pupal, or adult development, and it is during such periods that many new systems are differentiated and put to immediate test. Gastrulation (*q.v.*) is a sensitive period for amphibians.

sensitive volume **1.** that portion of an ionization chamber that responds to radiation passing through it. **2.** that biological volume in which an ionization must occur to produce a given effect (such as a mutation).

sensitizing agent an agent which, when added to a biological system, increases the amount of damage done by a subsequent dose of radiation.

sepiapterin *See* ***Drosophila* eye pigments.**

ser serine. *See* **amino acid.**

SER *s*mooth-surfaced *e*ndoplasmic *r*eticulum.

serial homology the resemblance between different members of a single, linearly arranged series of structures within an organism (the vertebrae are an example).

serial symbiosis theory the theory that eukaryotic cells evolved from bacterial ancestors by a series of symbiotic associations. In its most modern form, it suggests that the mitochondria and microtubule organizing systems of present-day eukaryotes evolved from bacteria and spirochaetes that lived as symbionts in a line of single-celled eukaryotes that were the ancestors of both fungi and animals. A subline of these protoctists subsequently entered an endosymbiosis with cyanobacteria. These evolved into chloroplasts, and the algae and plant lineages developed from this group. *See* **Appendix C,** 1978, Schwartz and Dayhoff; 1981, Margulis; 1986, Shih *et al.;* **endosymbiont theory,** *Pelomyxa.*

sericins a group of proteins found in silk (*q.v.*).

sericulture the culture of *Bombyx mori* (*q.v.*) for the purpose of silk production.

serine *See* **amino acid.**

serine proteases a family of homologous enzymes which require the amino acid serine in their active site and appear to use the same mechanism for catalysis. Members include enzymes involved in digestion (trypsin, chymotrypsin, elastase), blood coagulation (thrombin), clot dissolution (plasmin), complement fixation (C1 protease), pain sensing (kallikrein), and fertilization (acrosomal enzymes).

serology the study of the nature, production, and interactions of antibodies and antigens.

serotonin a cyclic organic compound (5-hydroxy-tryptamine) that causes certain smooth muscles to contract rapidly and increases capillary permeability. The symptoms of anaphylaxis are due in large part to serotonin released from platelets that accumulate in the capillary bed of the lung. Serotonin also plays an important role in the metabolism of the central nervous system.

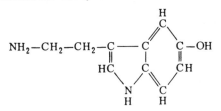

serotype an antigenic property of a cell (bacterial cell, red blood cell, etc.) identified by serological methods.

serotype transformation *See* **antigenic conversion.**

serum the fluid remaining after the coagulation of blood.

seta *See* **chaeta.**

sevenless a gene in *Drosophila melanogaster* that controls the development of R7, the seventh photoreceptor cell within an ommatidium (*q.v.*). In the absence of the wild-type allele of *sevenless,* R7 develops as a cone cell. It appears that the protein encoded by this cell fate gene is a membrane-bound receptor that transmits positional information that controls the type of differention the R7 cell undergoes. *See* **Appendix C,** 1986, Tomlinson and Ready.

Sewall Wright effect genetic drift (*q.v.*).

sex 1. in its broadest sense, sex is any process that recombines in a single organism genes derived from more than a single source. In prokaryotes, sex may involve genetic recombination between two autopoietic cells, or between an autopoietic cell (like *E. coli*) and a nonautopoietic episome (like phage lambda). Eukaryotic sex always involves two autopoietic organisms and leads to the alternating generation of haploid and diploid cells. Meiosis results in the formation of haploid gametes that unite in the process of fertilization to restore the diploid condition. Prokaryotic sex probably evolved more than 3 billion years ago, in the Archean era, while

meiotic sex evolved among the protoctists late in the Proterozoic era, about 1 billion years ago. 2. a classification or organisms or parts of organisms according to the kind of gamete produced; larger, nutrient-rich gametes are female; smaller, nutrient-poor gametes are male. Meiosis in some organisms produces morphologically indistinguishable *isogametes,* in which case the sexes are arbitrarily designated "plus" and "minus." An individual that produces both male and female gametes is *monoecious* (plant) or *hermaphroditic* (animal).

sex cells gametes.

sex chromatin a condensed mass of chromatin representing an inactivated X chromosome. Each X chromosome in excess of one forms a sex chromatin body in the mammalian nucleus. *See* **Barr body, late-replicating X chromosome.**

sex chromosomes the homologous chromosomes that are dissimilar in the heterogametic sex. *See* **X chromosome, W, Z chromosomes, Y chromosomes.**

sex comb a row of bristles arranged like the teeth of a comb on the forelegs of *Drosophila melanogaster* males. *See* **gynandromorph, prepattern.**

sex-conditioned character a phenotype that is conditioned by the sex of the individual. For example, a sex-conditioned, autosomal gene may behave as a dominant in males and as a recessive in females. Furthermore, in the homozygous female the condition may be expressed to a minor degree. Human baldness is an example of a sex-conditioned character. Also called a sex-influenced character.

sex determination the mechanism in a given species by which sex is determined. In many species sex is determined at fertilization by the nature of the sperm that fertilizes the egg. Y-bearing sperm produce male zygotes; X-bearing sperm, female zygotes. In man the Y chromosome is the masculinizing agent, and feminization will occur only when it is absent. *See* **Appendix C,** 1902, McClung; 1925, Bridges; **genic balance, Klinefelter syndrome, Turner syndrome.**

sexduction the process whereby a fragment of genetic material from one bacterium is carried with the sex factor *F* to a second bacterium.

sex factor *See* **fertility factor.**

sex hormone any hormone produced by or influencing the activity of gonads: e.g., gonadotropins, estrogens, androgens. Sex hormones are responsible for development of certain secondary sexual characteristics (e.g., growth of facial hair and muscular development in men).

sex index in *Drosophila,* the number of X chromosomes per set of autosomes (e.g., a male has a sex index of 0.5, a female is 1.0, a metamale 0.33, and a metafemale 1.5).

sex-influenced character sex-conditioned character.

sex-limited character a phenotype expressed in only one sex, although it may be due to a sex-linked or autosomal gene. Examples: the recessive, female sterile genes of *Drosophila;* the genes influencing milk and egg production in farm animals.

sex linkage a special case of linkage occurring when a gene that produces a certain phenotypic trait (often unrelated to primary or secondary sexual characters) is located on the X chromosome. The result of this situation is that in certain crosses the phenotypic trait in question may be observed only in individuals of the heterogametic sex, differences between reciprocal crosses (*q.v.*) may also be observed, and the trait will be observed much less frequently among members of the homogametic sex. Genes residing on the Y chromosome will influence only the heterogametic sex. *See* **Appendix C,** 1820, Nasse; 1910, 1911, Morgan; **sex chromosomes.**

sex pilus *See* **F pilus.**

sex ratio the relative proportion of males and females of a specified age distribution in a population.

sex ratio organisms spiroplasmas (*q.v.*) responsible for male-specific lethality in certain *Drosophila* species. SROs are transmitted in the ooplasm, and they kill male embryos. The Y chromosome is not associated with SRO-induced male lethality, and therefore only embryos with two X chromosomes can withstand the infection. Sex ratio spiroplasmas are referred to as sex ratio spirochaetes in the earlier literature.

sex reversal the change from functioning as one sex to functioning as the other sex. The change may be a normal occurrence (*see* **consecutive sexuality**) or it may be experimentally or environmentally induced.

sexual differentiation the process by which sexual determination is phenotypically expressed through the proper development of sexual organs and characteristics.

sexual dimorphism *See* **dimorphism.**

sexual isolation ethological isolation (*q.v.*).

sexual reproduction reproduction involving fusion of haploid gamete nuclei, which result from meiosis.

sexual selection a theory first proposed by Charles Darwin that in certain species there occurs a struggle between males for mates and that characteristics enhancing the success of those bearing them would have value and be perpetuated irrespective of their general value in the struggle for existence. In the current literature, sexual selection is generally subdivided into *intrasexual* and *epigamic* selection. Intrasexual selection is what Darwin had in mind and involves the power to conquer other males in battle. In epigamic selection, the female is the active selective agent, making a choice from among a field of genetically variable males of her own species. The results of epigamic selection may be seen in the elaborate sexual displays of certain male birds. *See* **Appendix C,** 1981, Lande.

sexuparous producing offspring by sexual reproduction. The term is used to describe the sexual phase in a species that alternates sexual reproduction with parthenogenesis. This phenomenon occurs in aphids, for example.

shadow casting the use of a vacuum evaporator (*q.v.*) to deposit a coating of a heavy metal on a submicroscopic particle. The coating is cast at an angle so that metal will build up on one side of the particle and cover the specimen support except in the shadow of the particle. The length and shape of the shadow allows calculation of the dimensions of the particle.

shearing in molecular biology, the process whereby a DNA sample is broken into pieces of fairly uniform size (e.g., by subjecting them to shearing forces in a Waring blender).

sheep common breeds are Merino, Rambouillet, Dorset, Debouillet, Lincoln, Leicester, Cotswold, Romney, Corridale, Columbia, Romeldale, Panama, Montadale, Polwarth, Targee, Hampshire, Shropshire, Southdown, Suffolk, Cheviot, Oxford, Tunis, Ryeland, and Blackface Highland. The species is *Ovis aries.*

shift *See* **translocation.**

shifting balance theory a theory, proposed by Sewell Wright, that maintains that biological evolution proceeds most rapidly when subpopulations of a species remain isolated for a time sufficient to acquire distinctive adaptations, followed by reestablishment of gene flow, the broadening of genetic diversity, and the enhancement of evolutionary flexibility.

Shigella dysenteriae the dysentery bacillus. Many *E. coli* phages also attack this species.

Shine-Dalgarno (SD) sequence part or all of the bacterial mRNA leader sequence AGGAGG preceding the AUG initiation codon; it is complementary to a sequence at the 3′ end of 16S rRNA and thereby serves as a binding site for ribosomes. *See* **Appendix C**, 1974, Shine and Dalgarno.

Shope papilloma virus an icosahedral particle about 53 nm in diameter containing a double-stranded DNA molecule weighing 5×10^6 d. The virus produces papillomas in rabbits.

short-day plant a plant with a flowering period that is accelerated by daily exposures to light of less than twelve hours.

short-period interspersion a genomic pattern in which moderately repetitive DNA sequences (each about 300 bp in length) alternate with nonrepetitive sequences of about 1,000 bp.

shotgun experiment the random collection of a sufficiently large sample of cloned fragments of the DNA of an organism to create a cloned "gene library" for that species from which cloned molecules of interest can later be selected.

shuttle vector a cloning vector able to replicate in two different organisms—e.g., in *E. coli* and yeast. These DNA molecules can therefore shuttle between the different hosts. Also called *bifunctional vectors.*

sib a shortened form for sibling.

siblings brothers and/or sisters; the offspring of the same parents.

sibling species species that are almost identical morphologically but are reproductively isolated; also called *cryptic species.*

sibmating a brother-sister mating.

sibship all the siblings in a family.

sickle-cell anemia a generally fatal form of hemolytic anemia seen in individuals homozygous for an autosomal, codominant gene H^S. The erythrocytes of such individuals contain an abnormal hemoglobin, Hb^S. These cells undergo a reversible alteration in shape when the oxygen tension of the plasma falls slightly, and they assume elongate, filamentous, and sicklelike forms. Such red cells show a greatly shortened life span, since they tend to clump together and are rapidly destroyed. About 0.2% of the Black babies born in the United States suffer from sickle-cell anemia. *See* **Appendix C,** 1949, Pauling *et al.;* 1957, Ingram; 1978, Kan and Dozy.

sickle-cell trait the benign condition shown by individuals carrying both the normal gene, H^A, and H^S. The erythrocytes of such individuals produce both Hb^A and Hb^S. Such heterozygotes are healthy,

and their erythrocytes can be caused to sickle only under conditions where the oxygen concentration is drastically reduced. H^A/H^S individuals suffer far less severely from *Plasmodium falciparum* infections than do H^A/H^A individuals. This malaria plasmodium enters the erythrocyte and lives by engulfing cytoplasm into its food vacuole. Parasites are unable to feed efficiently upon H^A/H^S cells because Hb^S molecules are insoluble compared to Hb^A and cause a great increase in the viscosity of the cytoplasm. About 9% of the Black population of the United States shows the sickle-cell trait. *See* **Appendix C,** 1954, Allison; **hemoglobin, hemoglobin S, malaria.**

siderophilin plasma transferrin (*q.v.*).

sievert the amount of ionizing radiation that liberates one joule of energy per kilogram of tissue. Symbolized Sr. The millesievert is often used in estimates of annual dose rates to human populations. The average U.S. citizen receives approximately 4 mSr per year, with radon (*q.v.*) accounting for 55% and dental and medical x rays about 15%.

Sigma (Σ) the summation of all quantities following the symbol.

sigma (σ) factor a polypeptide subunit of the RNA polymerase of *Escherichia coli.* This molecule by itself has no catalytic function, but it serves to recognize specific binding sites on DNA molecules for the initiation of RNA transcription. *See* **Appendix C,** 1969, Burgess *et al.;* **Pribnow box, RNA polymerase.**

sigma replication *See* **rolling circle.**

sigma virus a virus that confers CO_2 sensitivity upon *Drosophila melanogaster.*

signal hypothesis the notion that the N-terminal amino acid sequence of a secreted polypeptide is critical for attaching the nascent polypeptide to membranes. *See* **Appendix C,** 1975, Blobel and Dobberstein; **receptor-mediated translocation.**

signal recognition particle a nucleoprotein particle that functions during receptor-mediated translocation (*q.v.*). The particle contains a 7S RNA molecule and six different proteins. More than 75% of the total 7S RNA in animal cells is present in SRPs. There is a high degree of homology between the 7S RNAs of different animal species.

signal sequence synonym for leader sequence (*q.v.*).

significance of results if the probability values are equal to or less than 0.05 but greater than 0.01, the results are said to be *significantly different;* probability values $\leq 0.01 > 0.001$ are *highly significant;*

and those ≤ 0.001 are called *very highly significant* by convention. *See* **P value.**

silent allele an allele that has no detectable product. *See* **null allele.**

silent mutation a gene mutation that has no consequence at the phenotypic level; i.e., the protein product of the mutant gene functions just as well as that of the wild-type gene. Functionally equivalent amino acids may sometimes substitute for one another (e.g., leucine might be replaced by another nonpolar amino acid such as isoleucine). *See* **samesense mutation.**

silk the cocoon filament spun by the fifth-instar larva of *Bombyx mori.* Each cocoon filament contains two cylinders of fibroin, each surrounded by three layers of sericin. Fibroin is secreted by the cells of the posterior portion of the silk gland. These cells undergo 18 to 19 cycles of endomitotic DNA replication before they begin transcribing fibroin mRNAs. The fibroin gene is present in only one copy per haploid genome. It resides on chromosome 23 and is about 18 kb long. The fibroin gene is fundamentally an extensive array of 18 bp repeats coding for Gly-Ala-Gly-Ala-Gly-Ser. The similarity of the fibroin gene to a satellite DNA suggests that the gene grew to its current size and continues to evolve by unequal crossing over. The sericin proteins receive their name because of the abundance of serine, which makes up over 30% of the total amino acids. There are at least three sericins and one of these is encoded by a gene on chromosome 11. All sericins are secreted by the cells from the middle region of the silk gland. Several mutations are known that influence silk production. Some (*Fib* and *Src-2*) represent mutations in the cistrons coding for fibroin and sericin molecules. Others (*Nd, Nd-s,* and *flc*) seem to have defects in the intracellular transport and secretion of fibroin. *See* **Appendix C,** 1972, Suzuki and Brown.

silkworm the larva of *Bombyx mori (q.v.).*

Silurian the Paleozoic period during which vascular plants and arthropods invaded the land. In the oceans, agnathans diversified and placoderms arose. *See* **geologic time divisions.**

simian sarcoma virus retrovirus first found in the woolly monkey (*q.v.*). This oncogenic virus carries the *v-sis* oncogene, which encodes a transforming protein p28^{sis}. Nucleotide sequences related to *v-sis* have been located on human chromosome 22 at a site subsequently called *c-sis. See* **Appendix C,** 1983, Doolittle *et al.;* **platelet-derived growth factor, proto-oncogene, Rous sarcoma virus.**

simian virus any of a group of viruses that attack nonhuman primates.

simian virus 40 a DNA virus that readily infects cultured primate cells. SV40 is lytic in monkey cells, but temperate in mouse cells, causing occasional neoplastic transformations. The virus replicates in the nuclei of host cells and may become stably integrated into the host genome. The virus genome consists of a circular DNA molecule containing 5227 base pairs. The complete nucleotide sequence for this virus has been determined, and it contains both conventional genes, overlapping genes, and split genes. *See* **Appendix C,** 1971, Dana and Nathans; 1978, Reddy *et al.;* **enhancer, oncogenic virus, transformation.**

simple-sequence DNA satellite DNA (*q.v.*).

simplex *See* **autotetraploidy.**

Sinanthropus pekinensis a group of extinct hominids, originally found near Peking, China, but no longer considered a distinct species; now included in the species *Homo erectus. See* **Peking man.**

single-copy plasmids plasmids maintained in bacterial cells in a ratio of one plasmid for each host chromosome.

single-event curve a dose-response curve in radiobiology which gives a linear relation when the log of survival is plotted against radiation dose. *See* **multiple event curve, target theory.**

single-strand assimilation the process whereby a single strand of DNA displaces its homologous strand in a duplex, forming a D loop (*q.v.*). The reaction is mediated by RecA protein (*q.v.*) and is involved in recombination and heteroduplex formation.

single-stranded DNA binding protein in *E. coli,* a tetrameric protein of 74,000 d molecular weight that binds to the single-stranded DNA generated when a helicase (*q.v.*) opens the double helix. This stabilizes the single-stranded molecule and prevents reannealing or the formation of intrastrand hydrogen bonds. *Compare with* **helix-destabilizing proteins.**

single-strand exchange pairing of one strand of duplex DNA with a complementary strand in another DNA molecule, displacing its homologue in the other duplex. *See* **5-bromodeoxyuridine.**

sire the male parent in animal breeding.

sister chromatid exchange crossing over between the sister chromatids of a meiotic tetrad or between the sister chromatids of a duplicated somatic chromosome.

sister chromatids identical nucleoprotein molecules joined by a centromere. *See* **chromatid.**

sister group a species or higher monophyletic taxon that is hypothesized to be the closest genealogical relative of a given taxon. Sister taxa are derived from an ancestral species not shared by any other taxon.

site the position occupied by a mutation within a cistron.

site-specified mutagenesis a technique that introduces nucleotide alterations of known composition and location into a gene under study. *See* **oligonucleotide directed mutagenesis.**

site-specific recombination a process in which two specific (not necessarily homologous) DNA sequences are joined: e.g., during phage integration/ excision or resolution of cointegrate structures during transposition.

sliding filament model a theoretical mechanism that explains muscle contraction by the making and breaking of cross bridges between adjacent thick (myosin) and thin (actin) filaments. A similar model is postulated to explain the lengthening of the microtubules of the spindle apparatus, whereby separation of chromosomes or chromatids is affected during cell division.

slime molds *See* **Acrasiomycota, Myxomycota.**

slow component in a reassociation reaction, the last component to reassociate, usually consisting of nonrepetitive (unique) DNA sequences.

slow stop mutants temperature-sensitive *dna* mutants of *E. coli* that complete the current round of DNA replication when placed at the restrictive temperature, but cannot initiate another round of replication.

Slp sex-limited protein; a serum protein that is normally found only in male mice and encoded by a gene in the major histocompatibility complex (H-2).

small angle x-ray diffraction the technique used in the analysis of widely spaced repetitions, such as the groups of atoms that form monomeric subunits of a polymer. *See* **x-ray crystallography, large angle x-ray diffraction.**

small cytoplasmic RNAs the cytoplasmic counterparts of small nuclear RNAs (*q.v.*); found in small ribonucleoprotein particles (*q.v.*) in their native state.

small nuclear RNAs a family of small RNA molecules that bind specifically with a small number of proteins to form small nuclear ribonucleoprotein particles. These snRNPs (pronounced "snurps") play a role in the posttranscriptional modification of RNA molecules. *See* **Appendix C,** 1979, Lerner and Steitz; **spliceosome, Usn RNAs.**

small satellite RNAs small RNA molecules, associated with some RNA plant viruses, that depend on the supporting RNA plant virus to provide a protective coat protein and presumably at least some of the proteins necessary for replication of the satellite RNA. All known satellite RNAs have 400 or fewer nucleotides in their simplest (monomeric) form.

Smittia a genus of chironomid possessing giant polytene chromosomes (*q.v.*) and hence subjected to cytological study. *See* **Chironomus.**

smooth endoplasmic reticulum *See* **endoplasmic reticulum.**

smut 1. a fungus disease of cereals characterized by black masses of spores. 2. any basidiomycete fungus of the order Ustilagnales that causes smut disease.

snapdragon *See* ***Antirrhinum majus.***

sneak synthesis *See* background constitutive synthesis.

snRNA a small nuclear RNA (*q.v.*).

snRNPs small nuclear ribonucleoproteins. *See* **small nuclear RNAs.**

S1 nuclease an endonuclease from *Aspergillus oryzae* that selectively degrades single-stranded DNA to yield 5′ phosphoryl mono- or oligonucleotides.

snurps *See* **small nuclear RNAs.**

social Darwinism a theory originated by the British philosopher Herbert Spencer, proposing that most of the "progress" in human societies has been brought about by competition (economic, military) and the "survival of the fittest." Spencer believed that human progress required a struggle and competition, not only between individuals, but also between social classes, nations, states, and races, and he ranked human races and cultures according to their assumed levels of evolutionary attainment.

social evolution a continued increase in the complexity of human society resulting from the selection, transmission, and utilization of the useful information gained in each generation.

sociobiology the study of animal behavior from a genetic perspective.

sodium an element universally found in small amounts in tissues. Atomic number 11; atomic weight 22.9898; valence 1+; most abundant isotope ^{23}Na; radioisotopes: ^{24}Na, half-life 15 hours, radiations emitted—beta particles and gamma rays; ^{22}Na, half-life 2.6 years, radiations emitted—positrons and gamma rays.

sodium dodecyl sulfate $CH_3(CH_2)_{11}OSO_3Na$ (sodium lauryl sulfate), an anionic detergent used in the SDS-PAGE method of protein fractionation. *See* **electrophoresis.**

Solanum tuberosum *See* **potato.**

solenoid structure a supercoiled DNA structure produced during nuclear chromosomal condensation in eukaryotes. *See* **Appendix C,** 1976, Finch and Klug.

solution hybridization liquid hybridization (*q.v.*).

soma the somatic cells of a multicellular organism in contrast to the germ cells.

somatic cell any cell of the eukaryotic body other than those destined to become sex cells. In diploid organisms, most somatic cells contain the 2N number of chromosomes; in tetraploid organisms, somatic cells contain the 4N number, etc.

somatic cell genetic engineering correction of genetic defects in somatic cells by genetic engineering: e.g., insertion of genes for insulin production into defective pancreatic cells. Such correction would not be hereditary.

somatic cell genetics the genetic study of asexually reproducing body cells, utilizing cell fusion techniques, somatic assortment, and somatic crossing over. *See* **Appendix C,** 1964, Littlefield; 1965, Harris and Watkins; 1967, Weiss and Green; 1969, Boon and Ruddle; 1985, Smithies *et al.*

somatic cell hybrid a hybrid cell resulting from cell fusion (*q.v.*).

somatic crossing over crossing over during mitosis of somatic cells that leads to the segregation of heterozygous alleles. *See* **Appendix C,** 1936, Stern.

somatic doubling the doubling of the diploid chromosome set. Such doubling may be induced experimentally by applying the alkaloid colchicine in a lanolin paste to somatic tissues that are undergoing mitosis.

somatic mutation a mutation occurring in any cell that is not destined to become a germ cell. If the mutant cell continues to divide, the individual will come to contain a patch of tissue of genotype different from the cells of the rest of the body.

somatic pairing the conjoining of the homologous chromosomes in somatic cells, a phenomenon seen in dipterans. The fact that the polytene chromosomes of *Drosophila* undergo somatic pairing makes possible the identification of chromosomal rearrangements, the mapping of deficiencies, and, as a result, the cytological localization of genes. *See* **Diptera.**

somatoclonal variation the appearance of new traits in plants that regenerate from a callus in tissue culture. Some of the variations represent single nucleotide changes; others involve chromosomal translocations, losses, or duplications. Much of the variation occurs during tissue culture, rather than as a result of unmasking the variation present in the parent plant. *See* **gametoclonal variation.**

somatostatin a peptide hormone, present in hypothalamic extracts, that inhibits the release of growth hormone (somatotropin) by the anterior lobe of the pituitary gland, and of insulin and glucagon by the pancreas. The gene for this 14-residue peptide has been chemically synthesized, spliced into a plasmid, and cloned in *E. coli.* Biologically active somatostatin has been recovered from these clones.

somatotropin growth hormone.

Sordaria fimicola an ascomycete fungus often used in studies of gene conversion (*q.v.*).

SOS boxes the operator sequences in *E. coli* DNA that are recognized by a repressor called the LexA protein. This protein represses several loci involved in DNA repair functions. *See* **SOS response.**

SOS response an error-prone mechanism of repairing damaged DNA in *E. coli* by the coordinated induction of several enzymes. Damaged DNA somehow activates an enzyme called RecA protease, and this protease cleaves a protein called LexA repressor. Many genes involved in repair functions become activated when this repressor is cleaved. *See* **SOS boxes.**

Southern blotting a technique, developed by E.M. Southern, for transferring electrophoretically resolved DNA segments from an agarose gel to a nitrocellulose filter paper sheet via capillary action. Subsequently, the DNA segment of interest is probed with a radioactive, complementary nucleic acid, and its position is determined by autoradiography. A similar technique, referred to as *northern blotting,* is used to identify RNAs. For example, an electropherogram containing a multitude of different mRNAs could be probed with a radioactive cloned gene. In cases where proteins have been separated electrophoretically, a specific protein on an electropherogram can be identified by the *western blotting* procedure. In this case, the probe is a radioactively labeled antibody raised against the protein in question. *See* **Appendix C**, 1975, Southern; 1977, Alwine *et al.*

spacer DNA untranscribed segments of eukaryotic and some viral genomes flanking functional genetic regions (cistrons). Spacer segments usually contain repetitive DNA. The function of spacer DNA is not presently known, but it may be important for synapsis. *See* **transcribed spacer.**

special creation a nonscientific philosophy asserting that each species has originated through a separate act of divine creation by processes that are not now in operation in the natural world.

specialized **1.** an organism having a narrow range of tolerance for one or more ecological conditions. **2.** a species having a relatively low potential for further evolutionary change; the opposite of generalized.

specialized transduction *See* **transduction.**

speciation **1.** the splitting of an ancestral species into daughter species that coexist in time; horizontal evolution or speciation; cladogenesis. **2.** the gradual transformation of one species into another without an increase in species number at any time within the lineage; vertical evolution or speciation; phyletic evolution or speciation. *See* **alloparapatric speciation, allopatric speciation, evolution, parapatric speciation, peripatric speciation, sympatric speciation.**

species **1.** biological (genetic) species: reproductively isolated systems of breeding populations. **2.** paleospecies (successional species): distinctly different appearing assemblages of organisms as a consequence of species transformation (*q.v.*). **3.** taxonomic (morphological; phenetic) species: phenotypically distinctive groups of coexisting organisms. **4.** microspecies (agamospecies): asexually reproducing organisms (mainly bacteria) sharing a common morphology and physiology (biochemistry). **5.** biosystematic species (ecospecies) populations that are isolated by ecological factors rather than ethological isolation (*q.v.*).

species group superspecies (*q.v.*).

species selection a form of group selection (*q.v.*) in which certain species (produced by cladogenesis) continue the cladogenic process and others become extinct.

species transformation the transformation of a species (A) into another (species B) during the passage of time. Species transformation does not increase the number of species, since species A and B do not coexist in time. *See* **anagenesis, speciation, vertical evolution.**

specific activity the ratio of radioactive to non-radioactive atoms or molecules of the same kind. Sometimes given as the number of atoms of radioisotope per million atoms of stable element. Also expressed in curies per mole.

specific immune suppression an immune response in which the initial exposure to a particular antigen results in the loss of the ability of the organism to respond to subsequent exposures of that antigen, but not to different antigens. *See* **immunological tolerance.**

specific ionization the number of ion pairs per unit length of path of the ionizing radiation in a given medium (per micron of tissue, for example).

specificity selective reactivity between substances: e.g., between an enzyme and its substrate, between a hormone and its cell-surface receptor, or between an antigen and its corresponding antibody.

specificity factors proteins that temporarily associate with the core component of RNA polymerase and determine to which promoters the enzyme will bind (e.g., the sigma factor, *q.v.*). *See* **antispecificity factor.**

specimen screen the support for sections to be viewed under the electron microscope consisting of a disc made of copper or gold mesh.

spectrin a protein that is a major component of the plasma membranes of animal cells. It is composed of two different polypeptide chains, alpha and beta, which form heterodimers. Each polypeptide contains tandemly repeated sequences that can fold upon themselves and so give the spectrin filament great flexibility. In the cell membrane, spectrin filaments form a pentagonal network in which their ends attach to junctions made of actin and other protein.

spectrophotometer an optical system used in biology to compare the intensity of a beam of light of specified wave length before and after it passes through a light-absorbing medium. *See* **microspectrophotometer.**

spelt *Triticum spelta* (N = 21), the oldest of the cultivated hexaploid wheats, grown since the latter days of Roman Empire. *See* **wheat.**

S period *See* **cell cycle.**

sperm a spermatozoan or spermatozoa.

spermateleosis spermiogenesis (*q.v.*).

spermatheca the organ in a female or a hermaphrodite which receives and stores the spermatozoa donated by the mate.

spermatid one of four haploid cells formed during meiosis in the male. Spermatids without further division transform into spermatozoa, a process known as spermiogenesis (*q.v.*).

spermatocyte a diploid cell that undergoes meiosis and forms four spermatids. A primary spermatocyte undergoes the first of the two meiotic divisions and gives rise to two secondary spermatocytes. Each of these divides to produce two haploid spermatids.

spermatogenesis an inclusive term covering both male meiosis and spermiogenesis.

spermatogonia mitotically active cells in the gonads of male animals that are the progenitors of primary spermatocytes.

Spermatophyta in older taxonomies the division of the plant kingdom containing the contemporary dominant flora. Spermatophytes are characterized by the production of pollen tubes and seeds. All angiosperms and gymnosperms are included in the Spermatophyta. *See* **Appendix A.**

spermatozoan a haploid male gamete produced by meiosis. *See* **Appendix C,** 1677, van Leeuwenhoek; 1841, Kölliker; 1877, Fol.

sperm bank a depository where samples of human semen are stored in liquid nitrogen at −196°C; when needed, perhaps years later, a sample can be thawed and used in artificial insemination.

spermiogenesis the formation of sperm from the spermatids produced during the meiotic divisions of spermatocytes.

sperm polymorphism the production of normal and aberrant sperm during spermatogenesis. The normal sperm are called *eupyrene,* those containing subnormal numbers of chromosomes are *oligopyrene,* and those lacking a nucleus altogether are *apyrene.* Apyrene and oligopyrene sperm are formed by certain snails (*Viviparus malleatus* is an example) and moths *(Bombyx mori),* but the function of these abnormal gametes is unknown.

sperm sharing a phenomenon occurring in Brazilian freshwater snails of the genus *Bioaphalaria* in which a simultaneous hermaphrodite (acting mechanically as a male) transfers sperm to its partner that was collected when it functioned as a female in a previous mating. Sperm sharing may occur both within and between species. The term *sperm commerce* refers to the transfer of a sperm donor's own sperm along with exogenous sperm from a previous mating. *See* **hermaphrodite.**

Sphaerocarpus a genus of liverworts. Classic tetrad analyses have been made on species from this genus. *See* **Appendix A,** Plantae, Bryophyta; **Appendix C,** 1939, Knapp and Schreiber.

S phase *See* **cell cycle.**

spheroplast a protoplast (*q.v.*) to which some cell wall remnants are attached. For example, a rod-shaped bacterium treated with lysozyme becomes spherical because the enzyme removes peptidoglycan components that give rigidity to the cell wall.

sphingomyelin a molecule belonging to a family of compounds that occur in the myelin sheath of nerves. All sphingomyelins contain sphingosine, phosphorylcholine, and a fatty acid.

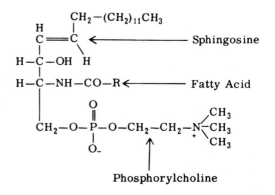

sphingosine an amino dialcohol component of the sphingolipids, which are abundant in the brain.

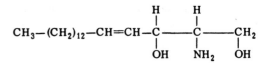

spike an inflorescence, such as the catkin of the pussy willow, in which the flowers arise directly from a central axis, the rachis. *See* **raceme.**

spikelet in grasses, a secondary spike bearing few flowers.

spindle a collection of microtubules responsible for the movement of eukaryotic chromosomes during mitosis (*q.v.*).

spindle attachment region (*also* **spindle fiber attachment, spindle fiber locus**) centromere (*q.v.*).

spindle fiber one of the microtubular filaments of a spindle.

spindle poison *See* **colchicine.**

spiral cleavage a type of embryonic development seen in invertebrates such as annelids and molluscs. The first and second divisions of the zygote are in vertical planes but at right angles to one another, producing a quartet of blastomeres. The next divisions are horizontal, cutting off successive quartets. However, each quartet is slightly displaced from the one above, giving a spiral appearance to the embryo. The direction of the spiral is genetically determined.

spirochete (*also* **spirochaete**) bacteria that are nonflagellated, spiral, and move by flexions of the body. *See* **Appendix A,** Eubacteria, Spirochaetae; *Treponema.*

spiroplasmas helical, motile bacteria that resemble spirochaetes. Unlike spirochaetes, spiroplasmas lack a cell wall, and they are therefore included in the Aphragmabacteria (*see* **Appendix A**). Spiroplasmas are responsible for certain plant diseases and cause male-specific lethality among the progeny of female *Drosophila* carrying them. *See* **sex ratio organisms.**

splice junctions segments containing a few nucleotides that reside at the ends of introns and function in excision and splicing reactions during the processing of transcripts from split genes. The sequence at the 5′ end of any intron transcript is called the *donor junction* and the sequence at the 3′ end the *acceptor junction.* U1 RNA (*q.v.*) contains a segment adjacent to its 5′ cap that exhibits complementarity to the sequences at the donor and acceptor splice junctions of introns. U1 binds to such segments causing introns to loop into a structure that allows intron excision and exon splicing. *See* **Usn RNAs.**

spliceosome the organelle in which the excision and splicing reactions that remove introns from premessenger RNAs occur. *See* **intron, exon, posttranslational modification, small nuclear RNAs, splice junctions, Usn RNAs.**

splicing 1. RNA splicing: the removal of introns and the joining of exons from eukaryotic primary RNA transcripts to create mature RNA molecules of the cytoplasm. 2. DNA splicing. *See* **recombinant DNA research.**

splicing homeostasis a phenomenon in which a maturase (*q.v.*) helps to catalyze the excision of an intron from its own primary RNA transcript. In so doing, the maturase destroys its own mRNA and thereby limits its own level of activity.

split genes genes containing coding regions (exons) that are interrupted by non-coding regions (introns). This type of genetic organization is typical of most eukaryotic genes and some animal viral genomes, but introns are not found in prokaryotic organisms. *See* **Appendix C,** 1977, Hogness *et al.*

sp. n. new species.

spontaneous generation the origin of a living system from nonliving material. *See* **Appendix C,** 1668, Redi; 1769, Spallanzani; 1861, Pasteur.

spontaneous mutation a naturally occurring mutation.

spontaneous reaction exergonic reaction (*q.v.*).

sporangium a structure housing asexual spores.

spore 1. sexual spores of plants and fungi are haploid cells produced by meiosis. 2. asexual spores of fungi are somatic cells that become detached from the parent and can either germinate into new haploid individuals or can act as gametes. 3. certain bacteria respond to adverse growth conditions by entering a spore stage until more favorable growth conditions return. Such spores are metabolically inert and exhibit a marked resistance to the lethal effects of heat, drying, freezing, deleterious chemicals, and radiation.

spore mother cell a diploid cell that by meiosis gives rise to four haploid spores.

sporogenesis the production of spores.

sporophyte the spore-producing, 2N individual. In the higher plants the sporophyte is the conspicuous plant. In lower plants like mosses, the gametophyte is the dominant and conspicuous generation. *See* **alternation of generations.**

Sporozoa a class of parasitic protoctists in the phylum Apicomplexa that reproduce sexually with an alternation of generations (*q.v.*). Both haploids and diploids undergo schizogony (*q.v.*) to produce small infective spores. All species of *Plasmodium* belong to the Sporozoa. *See* **Appendix A, malaria.**

sporozoite the stage in the life cycle of the malaria

parasite that infects man. Lance-shaped sporozoites reside in the salivary gland of the *Anopheles* mosquito and are delivered to the bloodstream of the victim when the mosquito takes a meal. The major surface antigen of the sporozoite is the circumsporozoite (CS) protein. In *Plasmodium knowlesi,* the CS protein contains a 12-amino-acid epitope that is repeated 12 times. When host antibodies bind to the CS protein, it sloughs off and is renewed. Thus, the CS protein serves as an immune decoy. The nucleotide sequence of the gene encoding the entire CS protein has been determined. Unlike most eukaryotic genes, it is not interrupted by introns. *See* **Appendix C,** 1983, Godson *et al.;* **malaria, merozoite.**

sporulation **1.** the generation of a bacterial spore. **2.** production of meiospores by fungi and many other eukaryotic organisms.

spreading position effect the situation in which a number of genes in the vicinity of a translocation or inversion seem to be simultaneously inactivated. *See* **Appendix C,** 1963, Russell.

38, 40, 45S preribosomal RNAs *See* **preribosomal RNA.**

src the oncogene of the Rous sarcoma virus (*q.v.*). *See* **c-src.**

30, 40, 47, 50, 60, 61S ribosomal subunits *See* **ribosome.**

70, 80, 81S ribosomes *See* **ribosome.**

4S RNA transfer RNA (*q.v.*).

5S RNA an RNA molecule composed of 120 nucleotides that is a component of the large ribosomal subunit. It is transcribed independently of the ribosomal RNA gene (*q.v.*). In man, the 5S RNA locus is near the telomere of the short arm of chromosome 1. In *Drosophila melanogaster,* it is on 2R at 56 E-F. *See* **Appendix C,** 1963, Rosset and Monier, 1970, Wimber and Steffensen; 1973, Ford and Southern; *Drosophilia* **salivary gland chromosomes.**

5.8S RNA an RNA molecule composed of 120 nucleotides that is a component of the large ribosomal RNA genes (*q.v.*) of the nucleolus organizer and functioning as a component of the large ribosomal subunit. *See* **ribosomal RNA gene.**

7S RNA *See* **signal recognition particle.**

5S RNA genes genes that are transcribed into 5S RNAs. Such genes occur in tandemly linked clusters in all eukaryotes. In *Xenopus laevis,* 5S RNA genes account for 0.5% of the entire genome. There are three separate 5S RNA multigene families. Two of these, the *major oocyte* and *trace oocyte* families,

are expressed only in oocytes, while a third, *somatic* 5S DNA, is expressed in all types of somatic cells. The major oocyte, trace oocyte, and somatic 5S DNAs are present in 20,000, 1,300, and 400 copies, respectively per haploid genome.

16, 18, 19, 23, 25, 28S rRNAs *See* **preribosomal RNA, ribosomal RNA genes, ribosome.**

SSC sister-strand crossover. *See* **sister-chromatid exchange.**

ssDNA single-stranded DNA.

stabilizing selection normalizing selection (*q.v.*).

stable equilibrium an equilibrium state of alleles at a genetic locus to which the population returns following temporary disturbances of the equilibrium frequencies. For example, a locus with overdominance should form a stable equilibrium as long as selection favoring heterozygotes remains constant.

stable isotope a nonradioactive isotope of an element.

stacking **1.** the planar alignment of adjacent flattish nitrogen bases in a DNA double helix. **2.** stacking of dye molecules on RNA to yield metachromasy (*q.v.*).

staggered cuts the result of breaking two strands of duplex DNA at different positions near one another, as occurs by action of many restriction endonucleases (*q.v.*).

stamen the pollen-bearing organ of the angiosperm flower. It consists of a filament bearing a terminal anther. *See* **flower.**

standard deviation (s) a measure of the variability in a population of items. The standard deviation of a sample is given by the equation $s = \sqrt{\Sigma(x - \bar{x})^2/N - 1}$, where N is the number of items in the sample and $\Sigma(x - \bar{x})^2$ is the sum of the squared deviations from the mean.

standard error (SE) a measure of variation of a population of means. $SE = s/\sqrt{N - 1}$, where N = the number of items in the population and s = standard deviation.

standard type the most common form of an organism.

Stanford-Binet test used to gauge intelligence, it consists of a series of questions and problems grouped for applicability to ages up to sixteen years. Some questions require verbal recognition and others recognition of form and manual skills. The subject's performance is expressed in terms of his mental age. *See* **intelligence quotient.**

298

Starch.

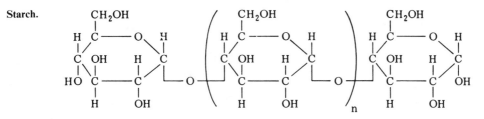

glucose

Staphylococcus aureus the pathogenic bacterium causing "staph" infections.

starch the storage polysaccharide of most plants. It is a polymer made up of α-D-glucose molecules.

start codon a group of three adjacent ribonucleotides (AUG) in an mRNA coding for the methionine in eukaryotes (formylated methionine in bacteria) that initiates polypeptide formation; also called an *initiation codon. See* **genetic code.**

startpoint in molecular genetics, the base pair on DNA that corresponds to the first nucleotide incorporated into the primary RNA transcript by RNA polymerase.

startsite synonym for startpoint (*q.v.*).

stasigenesis referring to a period during the paleontological history of a lineage during which little or no significant evolutionary change occurred.

stasipatric speciation speciation resulting from the dispersion of a favorable chromosomal rearrangement that yields homozygotes that are adaptively superior in a particular part of the geographical range of the ancestral species.

stasis in evolutionary studies, the persistence of a species over a span of geological time without significant change.

stationary phase a period of little or no growth that follows the exponential growth phase (*q.v.*) in a culture of microorganisms or in a tissue culture.

statistic the value of some quantitative characteristic in a sample from a population. *Compare with* **parameter.**

statistics the scientific discipline concerned with the collection, analysis, and presentation of data. The analysis of such data depends upon the application of probability theory. Statistical inference involves the selection of one conclusion from a number of alternatives according to the result of a calculation based on observations. *Parametric* methods in statistical analysis assume that the data follow a defined probability distribution (for example, a **normal, binomial,** or **Poisson distribution,**

all of which see), and the results of the calculations are valid only if the data are so distributed. The Student's *t* test (*q.v.*) is an example of a parametric procedure. *Nonparametric* methods in statistical inference are free from assumptions as to the shape of the underlying probability distribution. The Mann-Whitney rank sum test and the sign test are examples of nonparametric procedures. *See* **analysis of variance, chi-square test, Gaussian curve, null hypothesis, Student's *t* test.**

status quo hormone synonym for allatum hormone (*q.v.*).

steady-state system a system whose components seem unchanging because material is entering and leaving the system at identical rates.

stem cell **1.** one of the mitotically active somatic cells that serve to replenish those that die during the life of the metazoan organism. **2.** one of the mitotically active germinal cells that produce a continuing supply of gametes.

stem structure in molecular biology, the base-paired (unlooped) segment of a single-stranded RNA or DNA hairpin (*q.v.*).

stereochemical structure the three-dimensional arrangement of the atoms in molecules.

stereoisomers molecules that have the same structural formula, but that differ in the spatial arrangement of dissimilar groups bonded to a common atom.

steric relating to stereochemical structure (*q.v.*).

sterile **1.** unable to reproduce. **2.** free from living microorganisms; axenic.

sterile male technique a technique used in controlling noxious insects. Large numbers of artificially reared males are given nonlethal but sterilizing doses of ionizing radiation and then released in nature. The natural populations are so overwhelmed by these males that females are almost always fertilized by them. As a result, the fertilized eggs produced are rendered inviable, and a new generation cannot be produced.

sterilization **1.** elimination of the ability to reproduce. **2.** the process of killing or removing all living microorganisms from a sample.

steroid   a lipid belonging to a family of saturated hydrocarbons containing seventeen carbon atoms arranged in a system of four fused rings. The hormones of the gonads and adrenal cortex, the bile acids, vitamin D, digitalis, and certain carcinogens are steroids.

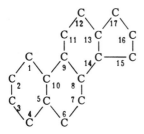

steroid receptor   a cytoplasmic receptor protein that can bind to a specific steroid hormone. The receptor-hormone complex then moves into the nucleus and binds to a specific DNA site to regulate gene activity.

steroid sulfatase (STS) gene   a pseudoautosomal gene in the mouse. *See* **human pseudoautosomal region.**

sterol   a compound with the general chemical ring structure of a steroid, but with a long side chain and an alcohol group. Cholesterol (*q.v.*) is an example of a sterol.

sticky ends   complementary single-stranded projections from opposite ends of a DNA duplex or from different duplex molecules that are terminally redundant. Sticky ends allow the splicing of hybrid molecules in recombinant DNA experiments. Many restriction endonucleases (*q.v.*) create sticky ends by making staggered cuts (*q.v.*) in a palindromic restriction site. Also called *cohesive ends. See* **Appendix C,** 1970, Smith and Wilcox.

stigma   the receptive surface usually at the apex of the style of a flower on which compatible pollen grains germinate.

stillbirth   the birth of a dead fetus.

stochastic process   a process that can be visualized as consisting of a series of steps at each of which the movement made is random in direction.

stock   1. that part of a plant, usually consisting of the root system together with part of the stem, onto which is grafted a scion. **2.** an artificial mating group, as, for example, a laboratory stock of mutant *Drosophila.*

stop codon   a ribonucleotide triplet signaling the termination of the translation of a protein chain (UGA, UAG, UAA). *See* **Appendix C,** 1965, Brenner *et al.*

strain   an intraspecific group of organisms possessing only one or a few distinctive traits, usually genetically homozygous (pure-breeding) for those traits, and maintained as an artificial breeding group by humans for domestication (e.g., agriculture) or for genetic experimentation. There is no clear distinction between the terms *strain* and *variety,* but the latter is generally applied when the differences between such intraspecific groups is substantial.

strand displacement   a replication mechanism, used by certain viruses, in which one DNA strand is displaced as a new strand is being synthesized.

strand-specific hybridization probes   specifically designed RNA transcripts used for blot or *in situ* hybridization experiments. A special plasmid vector is synthesized that contains a promoter for a phage RNA polymerase and an adjacent polylinker site (*q.v.*) which allows insertion of a DNA fragment in a specific direction. The vector is then cleaved with an appropriate restriction enzyme, and the gene fragment to be analyzed is ligated into the vector and propagated in *E. coli.* After purification, the plasmid DNA is used as a template for transcription by the specific phage RNA polymerase. By using appropriately labeled ribonucleoside triphosphates, radioactive transcripts of high specific activity are produced. These have two advantages over DNA probes obtained by nick translation (*q.v.*). (1) Since the RNA is strand specific, one strand of DNA can be analyzed at a time. (2) The sensitivity of hybridization is increased, since the RNA will not self-anneal. DNA probes, on the other hand, compete with their own complementary strands.

stratigraphic time divisions   geologic time divisions (*q.v.*).

streak plating   a technique of spreading microorganisms over the surface of a solidified medium for the purpose of isolating pure cultures.

streptavidin   a biotin-binding protein synthesized by *Streptomyces avidinii. See* **biotinylated DNA.**

Streptocarpus   the genus containing the Cape primroses. The inheritance of flower pigmentation has been thoroughly studied in various species in this genus. *See* **anthocyanins.**

Streptococcus pyogenes   the pathogenic bacterium causing "strep" infections.

streptolydigins   a group of antibiotics that, when

300

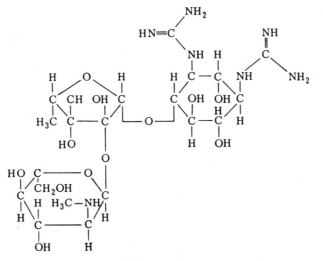

Streptomycin.

bound to the beta subunit of bacterial RNA polymerase, prevent transcriptional elongation.

Streptomyces a genus of saprophytic bacteria inhabiting the soil. Many species produce useful antibiotics, such as streptomycin and the tetracyclines. One hundred or so genetic markers have been identified in the most studied species, *S. coelicolor. See* **actinomycete, streptavidin, streptonigrin.**

streptomycin an antibiotic produced by *Streptomyces griseus,* that binds to the 30S subunit of the bacterial ribosome and leads to faulty translation of the advancing messenger tape. *See* **ribosome, translation.**

streptomycin suppression seen in bacterial mutants with an altered ribosomal protein (S12). This enables them to initiate polypeptide synthesis in the presence of streptomycin, and it also reduces the extent of misreading induced by that antibiotic. Such cells are converted from streptomycin-sensitive to streptomycin-resistant.

streptonigrin an antibiotic produced by *Streptomyces flocculus* that causes extensive chromosomal breakage.

stress fibers bundles of parallel-aligned, actin-containing microfilaments underlying the plasma membrane of cultured eukaryotic cells. Stress fibers permit cells to attach to the substratum and generate the stress or tension that causes them to assume a flattened shape. *See* **fibronectin.**

stringency the condition with regard to temperature, ionic strength, and the presence of certain organic solvents such as formamide (*q.v.*), under

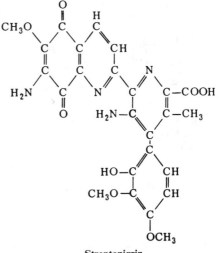

Streptonigrin.

which nucleic acid hybridizations are carried out. With conditions of high stringency, pairing will occur only between nucleic acid fragments that have a high frequency of complementary base sequences. Conditions of weaker stringency must be used if the nucleic acids come from organisms that are genetically diverse. Thus, if one were trying to isolate an alcohol dehydrogenase gene from a silkworm genomic library using a cloned gene from *Drosophila melanogaster* as a probe, less stringent conditions would be used than if the library came from *D. virilis.*

stringent control *See* **lambda cloning vehicle.**

stringent response the cessation of tRNA and ribosome synthesis by bacteria under poor growth conditions.

stroma the protein background matrix of a chloroplast or mitochondrion.

stromatolites living or fossil microbial mats dominated by cyanobacteria and fine sediment (usually calcium carbonate) trapped by these photosynthetic microbial communities. The oldest stromatolites are more than 3 billion years old and are among the oldest known fossils.

Strongylocentrotus purpuratus a common sea urchin used in studies of molecular developmental genetics. *See* **Lyctechinus pictus, echinoderm.**

strontium^{90} a radioisotope of strontium with a half-life of 28 years generated during the explosion of nuclear weapons. ^{90}Sr is one of the major sources of radiation due to fallout.

structural change chromosomal aberration (*q.v.*).

structural gene a DNA sequence coding for RNA or a protein; regulatory genes are structural genes whose products control the expression of other genes.

structural heterozygote a cell or an individual multicellular organism characterized by a pair of homologous chromosomes, one normal and the other containing an aberration, such as an inversion or a deficiency.

structural protein any protein that substantially contributes to shape and structure of cells and tissues: e.g., the actin and myosin components of muscle filaments, the proteins of the cytoskeleton, collagen, etc.

struggle for existence the phrase used by Darwin to describe the competition between animals for environmental resources such as food or a place to live, hide, or breed. Darwin wrote in *On the Origin of Species,* "I use the term *struggle for existence* in a large and metaphorical sense . . . including (which is more important) not only the life of the individual, but success in leaving progeny."

Student's *t* test a statistical method used to determine the significance of the difference between two sample means. The method was developed by the British statistician W.S. Gosset, who used the pseudonym "Student" in his publications. *See* page 56 for *t* distribution.

style a slender column of tissue arising from the top of the ovary and through which the pollen tube grows.

Stylonychia a genus of ciliates in which the macronuclear anlage undergo endomitotic DNA replication to form giant, banded, polytene chromosomes. Subsequently, the macronucleus undergoes a major reorganization of its DNA. The polytene chromosomes are destroyed and over 90% of the DNA is eliminated. The remaining DNA molecules are present as gene-sized pieces, and these undergo a series of replications as the macronucleus matures. Therefore, the macronucleus comes to contain multiple copies of a subset of the genes found in the micronucleus. A similar sort of chromatin elimination occurs in ciliates of the related genus *Oxytricha.* In *Stylonychia lemnae,* UAA and UAG encode the amino acid glutamine rather than serving as termination codons. *See* **Appendix C,** 1969, Ammermann; **genetic code.**

subculture a culture made from a sample of a stock culture of an organism transferred into a fresh medium.

subdioecy a sexual state of certain plants in which some unisexual individuals show imperfect sexual differentiation.

sublethal gene *See* **subvital mutation.**

submetacentric a chromosome that appears J-shaped at anaphase because the centromere is nearer one end than the other.

subpopulations breeding groups within a larger population or species, between which migration is restricted to a significant degree.

subspecies 1. a taxonomically recognized subdivision of a species. 2. geographically and/or ecologically defined subdivisions of a species with distinctive characteristics. *See* **race.**

substitutional load the cost to a population in genetic deaths of replacing an allele by another in the course of evolutionary change.

substitution vector *See* **lambda cloning vector.**

substrain a population of cells derived from a cell strain by isolating a single cell or groups of cells having properties or markers not shared by all cells of the strain.

substrate 1. the specific compound acted upon by an enzyme. 2. substratum.

substrate-dependent cells *See* **anchorage-dependent cells.**

substrate race a local race of organisms selected by nature to agree in coloration with that of the substratum.

substratum the ground or other surface upon which organisms walk, crawl, or are attached.

subtertian malaria *See* **malaria.**

subvital mutation a gene that significantly lowers

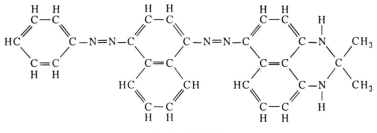

Sudan black B.

viability, but causes the death before maturity of less than 50% of those individuals carrying it. *Contrast with* **semilethal mutation.**

sucrose the sugar of commerce, a disaccharide composed of glucose and fructose.

sucrose gradient centrifugation *See* **centrifugation separation.**

Sudan black B a commonly used lysochrome.

sue **mutations** *See* **suppressor-enhancing mutations.**

sugar *See* **carbohydrate.**

sulfanilamide a compound that bears a close resemblance to a p-amino-benzoic acid. Sulfanilamide competes with p-amino-benzoic acid during the enzymatic synthesis of folic acid (*q.v.*). By blocking folic acid synthesis in bacteria, sulfanilamide serves as an effective chemotherapeutic agent.

sulfatide lipidosis an autosomal-recessive disease in man due to a defect in the production of the lysosomal enzyme arylsulfatase A. The clinical symptoms are paralysis, blindness, and dementia, leading to death during childhood.

sulfur an element universally found in small amounts in tissues. Atomic number 16; atomic weight 32.064; valence 2^-, 4^+, 6^+; most abundant isotope ^{32}S; radioisotope ^{35}S (*q.v.*).

sulfur-containing amino acids cysteine, cystine, methionine. *See* **amino acids.**

sulfur-dependent thermophiles a group of archaebacteria (*q.v.*) that generally live in sulfur-rich hot springs and generate energy by metabolizing sulfur. Species belonging to the genus *Sulfolobus* oxidize sulfur, whereas those of the genera *Thermoproteus* and *Thermococcus* reduce it.

sulfur mustard the first chemical mutagen to be discovered. *See* **Appendix C,** 1946, Auerbach and Robson; **nitrogen mustard.**

supercoiling the coiling of a covalently closed cir-

cular duplex DNA molecule upon itself so that it crosses its own axis. A supercoil is also referred to as a *superhelix.* The B form of DNA is a right-handed double helix. Winding of the DNA duplex in the same direction as that of the turns of the double helix is called *positive supercoiling.* Twisting of a duplex DNA molecule in a direction opposite to the turns of the strands of the double helix is called *negative supercoiling.*

superdominant overdominant (*q.v.*).

superfemale metafemale (*q.v.*).

supergene a chromosomal segment protected from crossing over and so transmitted from generation to generation as if it were a single recon.

supergene family sets of genes that share many of their nucleotide sequences, implying a common ancestry.

superhelix *See* **supercoiling.**

superinfection the introduction of such a large number of viruses into a bacterial culture that each bacterium is attacked by several phages.

supermale metamale (*q.v.*).

supernatant the fluid lying above a precipitate in a centrifuge, following the centrifugation of a suspension.

supernumerary chromosome a chromosome present, often in varying numbers, in addition to the characteristic invariable complement of chromosomes. *See* **Appendix C,** 1928, Randolph; **B chromosomes.**

superovulation the simultaneous release of more than the normal number of eggs from an ovary. This can be induced artificially by hormone treat-

$$CH_2-CH_2-Cl$$
$$|$$
$$S$$
$$|$$
$$CH_2-CH_2-Cl$$

Sulfur mustard.

303

ment in cattle and other livestock for embryo transfer (*q.v.*) to surrogate mothers.

superrepression an uninducible state for a gene usually attributed to (1) a defective operator locus to which a functional repressor protein cannot bind, or (2) a mutant regulatory gene whose repressor protein product is insensitive to the inducer substance; a phenomenon that causes a gene to be permanently "turned off."

superspecies a complex of related allopatric species (often called a species group). Such species are grouped together because of their morphological similarities. That the superspecies constitutes a natural grouping is demonstrated by finding in the genus *Drosophila* that whenever species hybrids are produced under laboratory conditions the parents are from the same species group.

supersuppressor a mutation that can suppress the expression of specific alleles of mutations at a number of different chromosomal sites; usually a nonsense suppressor.

supervital mutation a mutation that increases the viability of individuals bearing it above the wild-type level.

suppression 1. the restoration of a lost or aberrant genetic function (*see* **suppressor mutation**). 2. in immunology, a specific or nonspecific state of induced nonresponsiveness in the immune system. *See* **immunological suppression, suppressor T cell.**

suppressor-enhancing mutations genetic alterations that boost the activity of both temperature-sensitive as well as temperature-insensitive suppressors.

suppressor mutation a mutation that compensates for some other mutation, resulting in a normal or near-normal phenotype in the double mutant. Two main types of suppressor mutations occur: intergenic and intragenic. An *intergenic suppressor mutation* suppresses the effect of a mutation in another gene. Some intergenic suppressors change the physiological conditions so that the proteins encoded by the suppressed mutants can function. Other intergenic suppressors actually change the amino acid sequence of the mutant protein. For example, the intergenic suppressor may cause a base substitution in a tRNA gene. This results in an anticodon that reads a mutated codon of the mRNA of the suppressed mutant so as to insert a functionally acceptable amino acid in the protein responsible for the phenotype. An *intragenic suppressor mutation* suppresses the effect of a mutation in the same gene in which it is located. Some intragenic suppressors restore the original reading frame

after a frameshift. Other suppressor mutations produce new amino acid substitutions at different sites from those produced by the first mutation. However, the second changed amino acid compensates functionally for the first. Intragenic suppressor mutations are also called *second site mutations.*

suppressor T cell a subpopulation of T lymphocytes (designated Ts) whose function is to suppress the response of other lymphocytes to a particular antigen.

surface-dependent cells *See* **anchorage-dependent cells.**

surrogate mother a woman (or other female mammal) that receives an embryo transplant from another donor.

survival of the fittest the corollary of Darwin's theory of natural selection; namely, that as a result of the elimination by natural selection of those individuals least adapted to the environment, those that ultimately remain are the fittest.

survival value the degree of effectiveness of a given phenotype in promoting the ability of that organism to contribute offspring to the future populations.

suspension culture a type of *in vitro* culture in which the cells multiply while suspended in liquid medium. *See* **anchorage-dependent cells.**

Sus scrofa the pig. Domesticated pigs are generally given the subspecies name *domestica*. The haploid chromosome number is 19, and about 40 genes have been distributed among 12 linkage groups. *See* **swine** for a listing of breeds. *See* **Appendix A, Mammalia, Artiodactyla.**

SV 40 simian virus 40 (*q.v.*).

Svedberg *See* **sedimentation coefficient.**

sweepstakes route a potential migration pathway along which species disperse with difficulty. Chance events play a large role in colonization of new areas along this route. For example, birds blown far out to sea by a storm may accidentally land on an oceanic island and colonize it, but it is unlikely that this will happen a second time for that species.

sweet corn *See* **corn.**

swine any of a number of domesticated breeds of the species *Sus scrofa*. Popular breeds include Berkshire, Chester White, Duroc, Hampshire, Hereford, Ohio Improved Chester, Poland China, Spotted Poland China, Tamworth, Yorkshire.

switch gene a gene that causes the epigenotype to switch to a different developmental pathway.

switching sites break points at which gene segments combine in gene rearrangements.

swivelase *See* **gyrase, topoisomerase.**

symbiont an organism living in a mutually beneficial relationship with another organism from a different species: e.g., the coexistence of algae and fungi in lichens.

symbiosis any interactive association between two or more species living together. *See* **commensalism, mutualism, parasitism, serial symbiosis theory.**

symbiotic theory of the origin of undulipodia the theory proposed by L. Margulis that the ancestral eukaryote acquired undulipodia (*q.v.*) as the result of a motility symbiosis with spirochaetes.

symbols used in human cytogenetics A–G, the chromosome groups; 1–22, the autosome numbers; X, Y, the sex chromosomes; p, the short arm of a chromosome; q, the long arm of a chromosome; ace, acentric; cen, centromere; dic, dicentric; inv, inversion; r, ring chromosome; t, translocation; a plus (+) or minus (−) when placed before the autosome number or group letter designation indicates that the particular chromosome is extra or missing; when placed after a chromosome arm, a plus or minus designation indicates that the arm is longer or shorter than usual; a diagonal (/) separates cell lines when describing mosaicism. Examples: 45,XX, −C = 45 chromosomes, XX sex chromosomes, a missing chromosome from the C group; 46, XY, t (Bp−;Dq+) = a reciprocal translocation in a male between the short arm of a B and the long arm of a D group chromosome; inv (Dp+, q−) = a pericentric inversion involving a D chromosome; 2p+ = an increase in the length of the short arm of a chromosome 2; 46,XX, r = a female with one ring X chromosome; 45,X/46, XY = a mosaic of two cell types, one with 45 chromosomes and a single X, one with 46 chromosomes and XY sex chromosomes. *See* **human mitotic chromosomes.**

symmetrical replication bidirectional replication (*q.v.*).

sympatric speciation an uncommon process by which populations inhabiting (at least in part) the same geographic range become reproductively isolated.

sympatric species species whose areas of distribution coincide or overlap.

sympatry living in the same geographic location. *Compare with* **allopatry.**

symplesiomorphic character an ancestral or plesiomorphic character shared by two or more different organisms.

synapomorphic character a derived or apomorphic character shared by two or more different organisms.

synapsis the pairing of homologous chromosomes during the zygotene state of meiosis that results from the construction of a synaptonemal complex (*q.v.*). *See* **Appendix C**, 1901, Montgomery.

synaptonemal complex a tripartite ribbon consisting of parallel, dense, lateral elements surrounding a medial complex. The lateral elements lie in the central axes of the paired homologous chromosomes of a pachytene bivalent. The medial complex contains a system of interdigitating protein filaments that are oriented perpendicularly to the lateral elements and serve to maintain their parallel configuration during meiotic synapsis. *See* **Appendix C**, 1956, Moses and Fawcett; *Gowen's crossover suppressor,* **meiosis.**

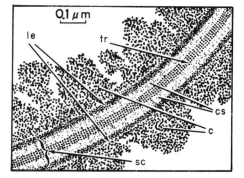

A drawing of a segment of a bivalent as seen under the electron microscope. (c) chromatin; (cs) central space; (le) lateral element; (sc) synaptonemal complex; (tr) transverse rods of the medial complex.

Synaptonemal complex.

syncaryon   synkaryon (*q.v.*).

syncytium   a mass of protoplasm containing many nuclei not separated by cell membranes.

syndactyl   having webbed digits either as a normal aspect of the species or, in man, pathologically.

syndesis   meiotic chromosomal synapsis.

syndrome   a group of symptoms that occur together, characterizing a disease.

synergid   one of two haploid cells that lie beside the ovum in the embryo sac (*q.v.*). Synergids of angiosperms function to nourish the ovum.

synergism   the phenomenon in which the action of two agents used in combination is more effective than the sum of their individual actions.

synezis   the clumping of chromosomes into a dense knot that adheres to one side of the nucleus. Synezis is a common occurrence during leptonema in microsporocytes.

syngamy   the union of the nuclei of two gametes following fertilization to produce a zygote nucleus; karyogamy.

syngen   *See Paramecium aurelia.*

syngeneic   pertaining to genetically identical organisms such as identical twins or the members of a highly inbred strain. Because syngeneic animals have the same antigens on their tissues, they can exchange skin or organ grafts successfully. *Compare with* **allogeneic, congenic strain.**

syngraft   a graft wherein the recipient receives a graft of tissue from a genetically identical donor (e.g., from an identical twin or from a member of the same highly inbred line). *Contrast with* **allograft, autograft, xenoplastic transplantation.**

synkaryon   **1.** the zygote nucleus resulting from the fusion of two gametic nuclei. **2.** the product of nuclear fusion in somatic cell genetic experiments.

synomone   *See* **allomone.**

synonym   in taxonomy, a different name for the same species or variety.

syntenic genes   genes thought to reside on the same chromosome because of their behavior in somatic cell hybridization experiments. A gene that is concurrently lost or retained along with a marker gene during the mitoses of hybrid cells is assumed to reside on the same chromosome as the marker gene.

synthetase   an enzyme catalyzing the synthesis of a molecule from two components, with the coupled breakdown of ATP or some other nucleoside triphosphate.

synthetic lethal   a lethal chromosome derived from normally viable chromosomes by crossing over.

synthetic linkers   short, chemically synthesized DNA duplexes containing sites for one or more restriction endonucleases. Synthetic linkers are used most commonly in the cloning of blunt-ended DNA molecules.

synthetic polyribonucleotides   RNA molecules made without a nucleic acid template, either by enzyme action or chemical synthesis. *See* **Appendix C**, 1961, Nirenberg and Matthaei; 1967, Khorana; **polynucleotide phosphorylase.**

systematics   the study of classification; taxonomy based on evolutionary relationships.

T

t   *See* **symbols used in human cytogenetics.**

t   the Student's *t* statistic that is used for testing the difference between the means of two samples. *See* **Student's *t* test,** and page 56.

T   thymine or thymidine.

tachyauxesis   *See* **heterauxesis.**

tachytelic evolution   *See* **evolutionary rate.**

tandem duplication   an aberration in which two identical chromosomal segments lie one behind the other. The order of the genes in each segment is the same.

tandem repeat   tandem duplication.

T antigen   a "tumor" antigen found in the nuclei of cells infected or transformed by certain oncogenic viruses such as polyoma. The antigen is thought to be a protein coded for by a virus cistron.

Taq DNA polymerase   a DNA polymerase synthesized by the thermophilic bacterium *Thermus aquaticus.* This enzyme, which is stable up to 95°C, is used in the polymerase chain reaction (*q.v.*),

target number   *See* **extrapolation number.**

target organ   the receptor organ upon which a hormone has its effect.

target theory   a theory developed to explain some biological effects of radiation on the basis of ionization occurring in a very small sensitive region within the cell. One or more "hits," that is, ionizing events, within the sensitive volume are postulated to be necessary to bring about the effect. *See* **extrapolation number.**

target tissue   1. the tissue against which antibodies are formed. 2. the tissue responding specifically to a given hormone.

tassel   the staminate inflorescence of corn.

taste blindness   the inability in man to taste the chemical phenylthiocarbamide (PTC). Taste-blind individuals are homozygous for an autosomal recessive gene. PTC tastes bitter to individuals carrying the dominant allele.

TATA box   Hogness box (*q.v.*); pronounced "tah-tah."

tautomeric shift   a reversible change in the location of a hydrogen atom in a molecule that alters it from one to another isomer. Thymine and guanine are normally in *keto* forms, but when in the rare *enol* forms (see diagram on page 308) they can join by three hydrogen bonds with keto forms of guanine or thymine, respectively. Likewise, adenine and cytosine are normally in amino forms, but when in the rare *imino* forms they can join by two hydrogen bonds with amino forms of cytosine or adenine, respectively.

tautomerism   the phenomenon in which two isomeric forms of molecules exist in equilibrium.

taxon   (*plural* **taxa**) the general term for a taxonomic group whatever its rank.

taxonomic category   the rank of a taxon in the hierarchy of classification. *See* **classification.**

taxonomic congruence   the degree to which different classifications of the same organisms postulate the same groupings. When the classifications compared are based on different sources of information (independent sets of data), congruence provides a measure of the degree to which the classifications remain stable as various lines of evidence are considered.

taxonomic extinction   nonsurvival of a taxon, either by extinction or by pseudoextinction (*q.v.*).

taxonomist   a specialist in taxonomy (*q.v.*).

taxonomy   the study of the classification of living things. Classically taxonomy is concerned with description, naming, and classification on the basis of morphology. More recently taxonomists have been concerned with the analyses of patterns of variation in order to discover how they evolved, with the identification of evolutionary units, and with the determination by experiment of the genetic interrelationships between such units.

Tay-Sachs disease   A lethal hereditary disease due to a deficiency of hexosamidase A. This deficiency results in storage of its major substrate (Gm2 gan-

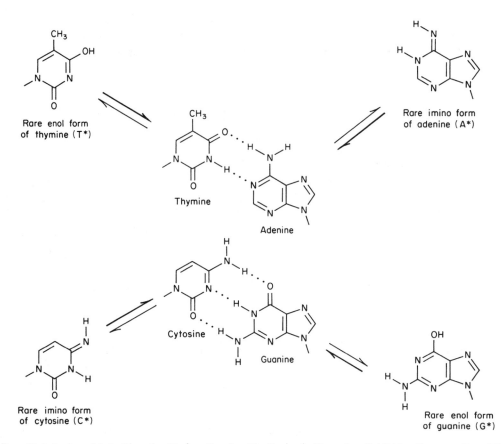

From F. J. Ayala and J. A. Kiger, Jr., *Modern Genetics,* The Benjamin/Cummings Publishing Company, Inc., 1980

Tautomeric shift.

glioside). Progressive accumulation of this compound causes developmental retardation, followed by paralysis, mental deterioration, and blindness. The normal gene, HEXA, encodes the alpha chain of hexosamidase A. About 2–4% of all Askenazi Jews are heterozygous for a defective HEXA allele. A similarly high-carrier frequency has been reported for the non-Jewish, French Canadian population living in eastern Quebec. Here the HEXA gene contains a deletion of 5 to 8 kilobases. *See* **Appendix C,** 1935, Klenk; **ganglioside, hexosaminidase, Sandhoff disease.**

TCA trichloroacetic acid (*q.v.*).

TψC loop the hairpin loop nearest the 3′ end of tRNA molecules, containing the modified base pseudouridine (ψ). This loop is thought to interact with ribosomal RNA. *See* **transfer RNA.**

T cell T lymphocyte. *See* **lymphocyte.**

T cell receptor a heteromeric protein on the sur-

face of T lymphocytes (*q.v.*) that specifically recognizes histocompatibility molecules (*q.v.*). T cell receptors are made up of two different polypeptide chains that are joined by disulfide bonds and are embedded in the plasmalemma with their carboxyl ends extending into the cytoplasm and their amino ends reaching outside the cell. The membrane portion of the T cell receptor is associated with a collection of CD3 proteins that transmit, from the outside of the cell to the inside, information as to whether or not the T cell receptor is occupied. The receptor recognizes as nonself the histocompatibility molecules on foreign cells, and it can also recognize antigenic sites on smaller molecules, provided these are presented in association with self histocompatibility molecules.

T cell receptor genes genes that encode the component polypeptides of T cell receptors (*q.v.*). There are two types of receptors: those containing an alpha and a beta chain, and those containing a delta and a gamma chain. In humans, both the alpha and

308

the gamma chains are encoded by genes on the long arm of chromosome 14. The beta chain gene is located on the long arm of chromosome 7, and the gamma chain gene resides on the short arm of chromosome 7. As in the case of the immunoglobulins, the T cell receptor polypeptide chains are encoded by gene segments that are reshuffled during the differentiation of the precursor cells. The rearrangement of segments occurs in thymocytes before the genes encoding the polypeptides are expressed. As a result, T cell receptors have more than 10^7 different amino acid sequences. *See* **Appendix C, 1984,** Davis and Mak.

T4, T8 cells classes of helper and suppressor T lymphocytes, respectively, characterized by antigenic markers that react with monoclonal antibodies designated anti-T4 and anti-T8, respectively. *See* **lymphocyte.**

T complex a region on chromosome 17 of the mouse containing genes affecting tail length and cell surface antigens. Mutations in some of these genes may arrest development at several embryonic stages.

T-DNA a group of seven genes (collectively referred to as transferred DNA) of the Ti plasmid (*q.v.*) that integrates into the nuclear DNA of the host plant during tumor induction. T-DNA is always present in crown gall cells of plants. *See* **Agrobacterium tumefaciens.**

T4 DNA ligase an enzyme encoded by *E. coli* phage T4 that not only seals nicks in double-stranded DNA, but has the unique ability to join two DNA molecules having completely base-paired (blunt) ends. This latter property is useful in forming recombinant DNA molecules.

T4 DNA polymerase an enzyme encoded by coliphage T_4 that catalyzes the synthesis of DNA in the 5' to 3' direction and also has 3' to 5' exonuclease activity. If DNA is incubated with T_4 DNA polymerase in the absence of deoxyribonucleotide triphosphates, the DNA will be partially degraded by the exonuclease. If the four dNTPs are now added, the degraded strand will be resynthesized by the polymerase. Thus, if the alpha phosphates of the added nucleotides are ^{32}P-labeled, a highly radioactive product can be obtained. The technique serves as an alternative to nick translation (*q.v.*).

teleology the explanation of a phenomenon such as evolution by the purposes or goals it serves. Teleological explanations usually invoke supernatural powers and are therefore nonscientific.

teleonomy the doctrine that the existence in an organism of a structure or function implies that it has conferred an advantage on its possessor during evolution.

telestability destabilization of a DNA double helix at a site distant from the site of binding of a protein. For example, binding of the cAMP-CAP complex to the *lac* operon of *E. coli* facilitates the distal formation of an open promoter site in which RNA polymerase can initiate transcription. *See* **catabolite activating protein.**

telocentric chromosome a chromosome with a terminal centromere.

telolecithal egg one in which the yolk spheres are accumulated in one hemisphere. *See* **centrolecithal egg, isolecithal egg, vegetal hemisphere.**

telomerase a telomere terminal transferase that adds telomeric sequences back onto the chromosome ends, one base at a time. The first telomerase was isolated from *Tetrahymena* (*q.v.*). It is a large ribonucleoprotein complex weighing about 500 kd. Telomerases bind to telomeres using G-rich strands as primers. *See* **Appendix C, 1985,** Greider and Blackburn.

telomere the natural unipolar end of linear eukaryotic chromosomes. The first telomeres to be sequenced belonged to *Tetrahymena thermophila.* Here the telomeres contained an A_2C_4 segment in one DNA strand and a T_2G_4 segment in the other, repeated in tandem 30 to 70 times. The telomeres from all species subsequently studied show the same pattern: a short DNA sequence, one strand-G-rich and one C-rich, that is tandemly repeated many times. *See* **Appendix C, 1978,** Backburn and Gall.

telophase *See* **mitosis.**

telotrophic meroistic ovary *See* **insect ovary types.**

telson the most posterior arthropod somite in which the posterior opening of the alimentary canal is located. *See* **maternal polarity mutants.**

tem triethylene melamine, an aziridine mutagen (*q.v.*).

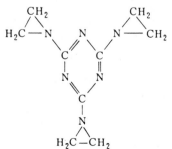

TEM transmission electron microscope. *See* **electron microscope.**

temperate phage a nonvirulent bacterial virus that infects but rarely causes lysis. It can become a prophage and thereby lysogenize the host cell.

temperature sensitive mutation a mutation that is manifest in only a limited temperature range. The product of such a gene generally functions normally, but is unstable above a certain temperature. Thus, the mutant when reared at the lower (permissive) temperature is normal, but when placed at the higher (restrictive) temperature shows the mutant phenotype. *See* **Appendix C,** 1951, Horowitz and Leupold; 1971, Suzuki *et al.*

template the macromolecular mold for the synthesis of a negative antitemplate macromolecule. The antitemplate then serves as a mold for the template. Thus the duplication of the template requires two steps. A single strand of DNA serves as a template for the complementary strand of DNA or mRNA. *See* **coding strand.**

template strand the strand of a DNA segment that is transcribed into mRNA. The anticoding strand. *See* **coding strand, RNA polymerase.**

template switching in *E. coli,* a bizarre *in vitro* reaction often accompanying strand displacement in which DNA polymerase I shifts from the original template strand to the displaced strand.

temporal isolation *See* **seasonal isolation.**

teosinte a wild Mexican grass that hybridizes with maize. *See* ***Euchlaena mexicana.***

tepa an aziridine mutagen (*q.v.*).

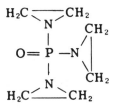

teratocarcinoma embryonal tumors originating in the yolk sac or gonads of amniotes and capable of differentiating into a variety of cell types. These tumors are used to study the regulatory mechanisms involved in embryological development. *See* **Appendix C,** 1975, Mintz and Illmensee.

teratogen any agent that raises the incidence of congenital malformations.

teratoma a tumor composed of an unorganized aggregation of different tissue types.

terminal chiasmata the end-to-end association of homologous chromosome arms resulting from terminalization.

terminal deletion *See* **deletion.**

terminalization in cytology, the progressive shift of chiasmata from their original to more distal positions as meiosis proceeds through diplonema and diakinesis. *See* **Appendix C,** 1931, Darlington.

terminal redundancy referring to the repetition of the same sequence of nucleotides at both ends of a DNA molecule.

terminal taxa the groups that occur at the ends of branches in a cladogram.

terminal transferase a deoxyribonucleotidyltransferase that is used by molecular biologists to add a homopolymer tail, e.g., polydeoxyadenylate, to each end of a vehicle DNA. The enzyme is then used to add poly T tails to a passenger DNA. The passenger and vehicle are then annealed via their complementary termini, ligated, and cloned. *See* **Appendix C,** 1972, Lobban and Kaiser.

termination codon a codon that signals the termination of a growing polypeptide chain. *See* **Appendix C,** 1965, Brenner *et al.;* **amber mutation, ochre mutation, opal codon, stop codon.**

termination factors *See* **release factors.**

termination hairpin, termination sequence *See* **terminators.**

terminators nucleotide sequences in DNA that function to stop transcription; not to be confused with terminator codons that serve as stop signals for translation. In the illustration below, the lower DNA strand is being transcribed from left to right. The RNA segment transcribed from the underlined DNA forms a hairpin-shaped loop because the two blocks of nucleotides have complementing base sequences. This tends to force the adjacent region of the DNA/RNA hybrid to open up. Since it consists of polyribo-U and polydeoxy-A regions that bind weakly, the mRNA molecule will detach at this point. *See* **attenuator, exon.**

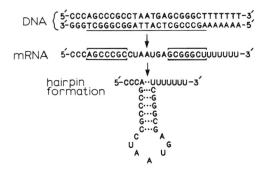

territoriality the defense by an animal or group of an area against members of the same species.

territory an area of the habitat occupied by an individual or group. If members belonging to the same species enter the territory, they are attacked as trespassers.

Tertiary the older of the two geologic periods making up the Cenozoic era. *See* **geologic time divisions.**

tertiary base pairs the specific base pairs of a tRNA molecule responsible for its three-dimensional folding. Most of these base pairs are evolutionarily conserved in all tRNA molecules.

tertiary nucleic acid structure the three-dimensional conformation of a nucleic acid strand (chain) formed by folding of the strand and formation of intrastrand complementary base pairing (e.g., transfer RNA, *q.v.*).

tertiary protein structure *See* **protein structure.**

tesserae functionally different patches of endoplasmic reticulum, each bearing a characteristic set of enzymes.

test cross a mating between an individual of unknown genotype, but showing the dominant phenotype for one or more genes, with a tester individual known to carry only the recessive alleles of the genes in question. The test cross reveals the genotype of the tested parent. For example, an individual showing the *A* and *B* phenotypes is crossed to an *aabb* tester. The F₁ contains only individuals of *AB* phenotype. This reveals the genotype of the tested parent to be *AABB*.

tester strain a multiply recessive strain that provides the genotypically known mate used in a test cross.

testicular feminization a condition in which XY individuals develop as females. Spermatogenesis does not occur in the testes, which are generally located within the abdomen. The syndrome is due to a defect in androgen receptors. The condition shows X-linked inheritance in both humans and the mouse.

testis (*plural* **testes**) the gamete-producing organ of a male animal.

testosterone a masculinizing, steroid hormone secreted by interstitial cells of the testis.

test-tube baby the production of a child by *in vitro* fertilization, followed by embryo transplantation to complete gestation in a normal uterus. This may be provided by the biological or surrogate mother.

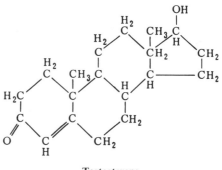

Testosterone.

tetra-allelic referring to a polyploid in which four different alleles exist at a given locus. In a tetraploid $A_1A_2A_3A_4$ would be an example.

tetracyclines a family of antibiotics obtained from various species of *Streptomyces*. Tetracyclines bind to the 30 S subunit of prokaryotic ribosomes and prevent the normal binding of aminoacyl– tRNA at the A site. *See* **translation.**

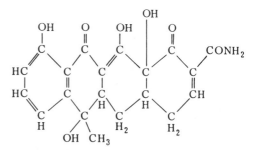

tetrad 1. four homologous chromatids (two in each chromosome of a bivalent) synapsed during first meiotic prophase and metaphase. *See* **meiosis.** 2. four haploid products of a single meiotic cycle.

tetrad analysis the analysis of crossing over by the study of all the tetrads arising from the meiotic divisions of a single primary gametocyte. To perform such an analysis one must use an organism in which the meiotic products are held together, as for example, in the case of meiospores confined in an ascus sac. Genera suitable for such analyses include *Ascobolus, Aspergillus, Bombardia, Neurospora, Podospora, Saccharomyces, Schizosaccharomyces, Sordaria,* and *Sphaerocarpus.*

tetrad segregation types For a bivalent containing the genes *A* and *B* on one homologue and *a* and *b* on the other, three patterns of chromatid segregation are possible: *AB, AB, ab, ab* (referred to as the parental ditype); *AB, Ab, aB, ab,* where two chromatids are recombinant (the tetratype); and *Ab, Ab, aB, aB,* where all chromatids are recombinant (the nonparental ditype).

tetrahydrofolate   *See* **folic acid.**

Tetrahymena   a genus containing *T. pyriformis,* the species for which the most genetic information is available and *T. thermophila,* the species where UAA and UAG were shown to encode glutamine rather than serving as stop codons. The nuclear reorganization that takes place following conjugation (*q.v.*) in these ciliates makes them a rich source of telomeres and the enzymes that work on them. This is because during the regeneration of a new macronucleus, the DNA of the micronucleus is split at specific sites into hundreds of thousands of pieces. New telomeres are synthesized at each new end, and each chromsome fragment undergoes many cycles of replication. *See* **Appendix A,** Protoctista, Ciliphora; **genetic code, telomerase, telomere.**

tetramer   a structure resulting from an association of four subunits. If the subunits are all identical, they form a homotetramer; if the subunits are not all identical, they form a heterotetramer.

tetramine   an aziridine mutagen (*q.v.*).

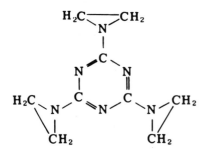

tetraparental mouse   a mouse developed by artificial fusion of embryonic cells from two genetically different blastulas.

tetraploid   having four haploid sets of chromosomes in the nucleus.

tetrasomic   having one chromosome in the complement represented four times in each nucleus.

tetratype   *See* **tetrad segregation types.**

tetravalent   *See* **quadrivalent.**

thalassemias   a group of human anemias due to imbalance in the ratio of alpha and/or beta hemoglobin subunits. Since there are four alpha genes per genome, deletions (commonly produced by unequal crossing over) can result in an individual having any number of alpha genes from zero to four. The complete absence of alpha genes produces a condition lethal at or before birth, called *hydrops fetalis.* With only one alpha gene, excess beta chains form a tetramer (β_4), resulting in hemoglo-

bin H disease. Individuals with two or three alpha genes are almost indistinguishable from normal. Incomplete beta chains can be produced by nonsense codons. Deletions in beta genes are commonly produced by unequal crossing over, as are the hybrid chains containing δ and β segments (Hb Lepore) or A_γ and β segments (Hb Kenya). Beta thalassemia (also called Cooley's anemia) is a hemoglobinopathy in which few functional beta globin chains are made. A point mutation, within an intron that alters the cutting and splicing signal, causes an extra piece of intron RNA to be present in processed mRNA; the extra piece shifts the reading frame and causes translation to stop prematurely, yielding a truncated and nonfunctional beta globin molecule. *See* **Appendix C,** 1976, Kan *et al.*; 1986, Costantini *et al.*; **hemoglobin fusion genes, hemoglobin homotetramers.**

thelytoky   a type of parthenogenesis in which diploid females are produced from unfertilized eggs and males are absent or rare. There are two types of thelytoky, meiotic (automictic) and ameiotic (apomictic). In automictic thelytoky, meiosis takes place, but the reduction in chromosome number is compensated for later in the life cycle. The most widespread method of doing this is to have a haploid polar body nucleus fuse with a haploid egg nucleus (autofertilization). In apomictic thelytoky, the maturation division in the egg is equational and therefore the egg nucleus remains diploid.

Theobroma cacao   the cacao tree, source of chocolate.

theobromine   a mutagenically active purine analogue. It is the main alkaloid stimulant in chocolate.

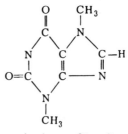

thermal denaturization profile   *See* **melting profile.**

thermal neutron   a fast neutron from uranium fission that has been slowed down by elastic collision with a moderator such as graphite to energies equivalent to those of gas molecules at room temperature (approximately 0.025 eV). The biological effect of thermal neutrons is attributable to the summation of capture and decay radiations. In biological material the reactions ^{1}H (n, γ) ^{2}H and ^{14}N (n, p) ^{14}C are the most important sources of tissue

312

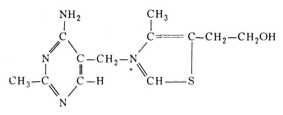

Thiamine.

ionization. The relative importance of these reactions depends on the size of the organism. Protons from nitrogen capture are the major cause of the biological effects of thermal neutrons in an organism the size of *Drosophila.*

thermoacidophiles bacteria that live in extremely acidic hot springs. Species belonging to the genus *Thermoplasma* are examples. They are placed in the archaebacteria (*q.v.*) on the basis of the nucleotide sequences of their 16S rRNAs.

thermophilic heat loving. Said of bacteria that grow at temperatures between 45°C and 65°C (found in fermenting manure and hot springs).

theta replication a bidirectional mode of replication of circular DNA molecules from a replication origin in which, midway through the replication cycle, the structure produced resembles the Greek letter θ. Also calles a *Cairns molecule. See* **Appendix C,** 1963, Cairns; **D loop, sigma replication.**

thiamine vitamin B$_1$, the anti-beriberi factor.

thiamine pyrophosphate a coenzyme of the carboxylases and aldehyde transferases.

thin layer chromatography *See* **chromatography.**

thioglycolic acid treatment a procedure used to rupture disulphide bridges linking adjacent protein chains.

Thio-tepa trade name for triethylenethiophosphoramide, a mutagenic, alkylating agent.

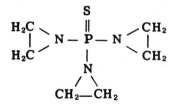

third cousin *See* **cousin.**

thirty-seven percent survival dose the radiation dose at which the number of hits equals the number of targets. The dose at which there is an average of one hit per target. *See* **target theory.**

thr threonine. *See* **amino acid.**

three-point cross a series of crosses designed to determine the order of three, nonallelic, linked genes upon a single chromosome on the basis of their crossover behavior.

three-strand double exchange *See* **inversion.**

threonine *See* **amino acid.**

threshold dose the dose of radiation below which the radiation produces no detectable effect.

threshold effect hypothesis the notion that certain traits with a polygenic basis develop only if the additive effects of contributory alleles exceed a critical value. This hypothesis is often used to explain many all-or-none phenomena with a polygenic mode of inheritance (e.g., susceptibility vs. resistance to certain diseases).

thrombin *See* **blood clotting.**

thrombocyte blood platelet. *See* **platelets.**

thrum the type of flower characterized by short styles and high anthers found among distylic species such as seen in the genus *Primula. See* **pin.**

Thy-1 antigen an antigen on the plasma membrane of thymocytes (*q.v.*) that can be used to distinguish them from other lymphocyte groups.

thylakoid *See* **chloroplast.**

thymidine the deoxyriboside of thymine. *See* **nucleoside.**

thymidine kinase an enzyme catalyzing the phosphorylation of thymidine to thymidine monophosphate.

thymidylate kinase an enzyme catalyzing the phosphorylation of thymidine monophosphate and thymidine diphosphate to thymidine diphosphate and thymidine triphosphate, respectively.

thymidylic acid *See* **nucleotide.**

thymine *See* **bases of nucleic acids.**

thymine dimer two thymine molecules joined as shown in the accompanying illustration. The reaction of greatest importance when ultraviolet radiation interacts with DNA is the formation within single polynucleotide chains of bonds between adjacent thymine molecules. Such thymine dimers block future DNA replication.

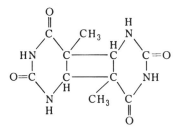

thymocyte a thymus-derived lymphocyte or T lymphocyte.

thymus an organ lying in the chest of mammals, formed embryologically from gill pouches. This organ functions to populate the body with lymphoid cells. It reaches a maximum size about the time of sexual maturity and then atrophies.

thyroglobulin *See* **thyroid hormones.**

thyroid hormones thyroxine and triiodothyronine are synthesized by the thyroid gland in response to thyrotropin, a hormone from the anterior pituitary gland. The synthesis of thyroid hormones begins with the selective accumulation of inorganic iodide by the epithelial cells of the gland. The trapped iodide is then oxidized to iodine, and tyrosine residues of the glycoprotein thyroglobulin are then iodinated, converting them to monoiodotyrosine residues (a). The iodotyrosyl residues are then coupled with the elimination of the alanine side chain

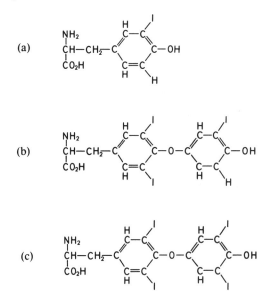

to form triiodothyronine (b) and thyroxine (c). *See* **cretinism.**

thyroid-stimulating hormone (TSH) a glycoprotein hormone stimulating secretion by the thyroid. TSH is produced by the adenohypophysis.

thyrotropin *See* **thyroid hormones.**

thyroxine *See* **thyroid hormones.**

tiller a grass side shoot produced at the base of a stem.

timber line the line in high latitudes and in high elevations in all latitudes beyond which trees do not grow.

time-lapse microcinematography a technique of photographing living cells under phase contrast with a motion picture camera taking exposures at selected intervals (1 frame per minute, for example) and then projecting the film at a more rapid speed (24 frames per second, for example). Under such circumstances time will be speeded up almost 1500 times, and one can gain a much clearer understanding of the dynamic processes the cells undergo.

Ti plasmid a *t*umor-*i*nducing (hence the acronym) plasmid found in the bacterium *Agrobacterium tumefaciens* (*q.v.*) that is responsible for crown gall disease of dicotyledonous plants; used as a vector in genetic engineering of plants. The wild-type plasmid produces tumor cells, but it can be modified so that it can carry foreign genes into cells without making the recipient cells tumorous. During tumor induction, a specific segment of the Ti plasmid, called the T-DNA (transferred DNA), integrates into the host plant nuclear DNA. *See* **Appendix C**, 1974, Zaenen *et al.;* 1981, Kemp and Hall.

Tiselius apparatus electrophoresis apparatus.

tissue a population consisting of cells of the same kind performing the same function.

tissue culture the maintenance or growth of tissue cells *in vitro* in a way that may allow further differentiation and preservation of cell architecture or function or both. *Primary cells* are those taken directly from an organism. Treating a tissue with the proteolytic enzyme trypsin dissociates it into individual primary cells that grow well when seeded onto culture plates at high densities. Cell cultures arising from multiplication of primary cells in tissue culture are called *secondary cell cultures.* Most secondary cells divide a finite number of times and then die. A few secondary cells may pass through this "crisis period" and be able to multiply indefinitely to form a continuous *cell line.* Cell lines have extra chromosomes and are usually abnormal in other respects as well. The immortality of these cells is a feature shared in common with cancer cells. *See* **Appendix C,** 1907, Harrison; 1965, Hayflick.

tissue typing identification of the major histocompatibility antigens of transplant donors and potential recipients, usually by serological tests. Donor and recipient pairs should be of identical ABO blood group, and in addition should be matched as closely as possible for H antigens in order to minimize the likelihood of allograft rejection. *See* **histocompatibility molecules.**

titer the amount of a standard reagent necessary to produce a certain result in a titration (*q.v.*).

TLC thin-layer chromatography. *See* **chromatography.**

T lymphocyte the lymphocyte responsible for cell-mediated immunological reactions, such as graft rejection, and characterized by the possession of T cell receptors (*q.v.*). T lymphocytes differentiate within the microenvironment of the thymus gland. Mature T cells can be divided into two groups (CD4 and CD8) on the basis of their ability to recognize certain classes of histocompatibility molecules (HCMs). CD4+ cells, which recognize class II HCMs, function as helper T lymphocytes (*q.v.*). CD8+ cells, which recognize class I HCMs, function as cytotoxic T lymphocytes (*q.v.*). *See* **histocompatibility molecules.** *Compare* **B lymphocyte.**

T_m the temperature at which a population of double-stranded nucleic acid molecules becomes half-dissociated into single strands; referred to as the *melting temperature* for that system.

TMV tobacco mosaic virus.

Tn symbol for transposon.

tobacco *See* *Nicotiana.*

tobacco mosaic virus a virus that produces lesions on the leaves of tobacco plants. The virus contains a central core of RNA and a coat made up of 2,130 protein subunits, each of Mr 17,500. *See* **Appendix C,** 1899, Beijerinck; 1935, Stanley; 1937, Bawden and Pirie; 1955, Fraenkel-Conrat and Williams; 1956, Gierer, Schramm and Fraenkel-Conrat.

tolerance *See* **immunological tolerance.**

toluidine blue a metachromatic basic dye used in cytochemistry.

tomato *See* *Lycopersicon esculentum.*

T24 oncogene a gene isolated from a line of human bladder carcinoma cells. This oncogene appears to be homologous to the oncogene carried by the Harvey murine sarcoma virus (*q.v.*). The

change that leads to the activation of the T24 ocogene is due to a single base substitution. *See* **Appendix C,** 1982, Reddy *et al.*

tonofilaments synonymous with keratin filaments. *See* **intermediate filaments, keratin.**

topoisomerase an enzyme that can interconvert topological isomers of DNA. These enzymes alter DNA topology by changing the linking number of circular duplex DNAs or by interconverting knotted and catenated forms. Topoisomerase has replaced a number of earlier terms (DNA relaxing enzyme, swivelase, untwisting enzyme, nick-closing enzymes). Topoisomerases are divided into two classes: Type 1 enzymes make a transient break in one strand of the duplex, whereas type 2 enzymes introduce transient double-strand breaks. During the relaxation of DNA by type 1 topoisomerases, an intact strand of the helix is passed through the break in its complementary strand.

topological isomers *See* **linking number.**

topology the study of the properties of geometrical figures that are subjected to deformations such as bending or twisting.

tormogen cell *See* **trichogen cell.**

tortoiseshell cat a cat showing patches of orange and more darkly pigmented fur in its coat. The sex-linked gene *O* is responsible for the conversion of eumelanin to phaeomelanin, which gives the fur an orange coloration. The *O* gene is epistatic to those autosomal genes, which give the coat a black or agouti color. Since the X chromosome is randomly inactivated in somatic cells during development, females heterozygous for *O* will show the tortoiseshell phenotype. The term *calico* is sometimes applied to tortoiseshell females that also have patches of white fur. Such females also contain the dominant spotting gene *S.*

totipotency the capacity of a cell to differentiate into all of the cells of the adult organism. A zygote is normally totipotent, but most cells of the embryo become progressively restricted in this capacity as development progresses. *See* **pluripotent.**

toxin a poisonous substance elaborated by a mi-

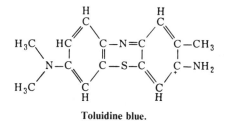

Toluidine blue.

315

croorganism, as well as some fungal, plant, and animal species. An example is alpha amanatin (*q.v.*).

toxoid a poisonous protein that has been detoxified without harm to its antigenic properties.

T phages a group of virulent viruses attacking *E. coli.* A long DNA molecule is tightly packed into each polyhedral capsid. Each phage attaches to its host by a hollow tail through which its DNA molecule is inserted into the bacterium. T4 is the best-known phage in the group. Its total genetic information is stored in a linear, double-stranded DNA molecule containing 166,000 base pairs. About 70 of its genes have been well characterized. T2 was the first phage observed under the electron microscope, and it was utilized in the famous Hershey-Chase experiment. *See* **Appendix C,** 1942, Luria and Anderson; 1952, Hershey and Chase; 1963, Epstein and Edgar; **bacteriophages, cyclically permuted sequences, rII locus.**

tracer *See* **radioactive isotope, labeled compound.**

traction fiber one of the fibers connecting the centromeres of the various chromosomes to either centriole. *See* **mitotic apparatus.**

trailer sequence a nontranslated segment at the 3′ end of mRNA following the signal that terminates translation, but exclusive of the poly-A tail. The trailer contains the binding site for the polyadenylating enzyme. Some mRNAs contain blocks of nucleotides in their trailers that bind to receptor molecules that are localized within specific regions of cells. *See* **Appendix C,** 1988, Macdonald and Stuhl; **bicoid, exon, leader sequence, polyadenylation.**

trait *See* **character.**

trans *See* **cis-trans configuration.**

trans-acting locus a genetic element, such as a regulator gene (*q.v.*) that encodes a diffusible product that can influence the activity of other genes. Trans-acting genes can be on different DNA molecules from the genes they control. *Contrast with* **cis-acting locus.**

transcribed spacer that part of a primary rRNA transcript that is discarded during the formation of functional RNAs of the ribosome (*q.v.*).

transcriptase RNA polymerase (*q.v.*).

transcription the formation of an RNA molecule upon a DNA template by complementary base pairing; mediated by RNA polymerase (*q.v.*).

transcription unit the segment of DNA between the sites of initiation and termination of transcription by RNA polymerase; more than one gene may reside in a transcription unit. A polycistronic mes-

sage (*q.v.*) may be translated as such and the translational product may be enzymatically cleaved later into two or more functional polypeptide chains. RNA is transcribed in a 5′ to 3′ direction from the coding strand of the gene. However, when describing the nucleotide sequence of a specific gene, the convention has been adopted to give it the same nucleotide sequence as the RNA transcript, except that each uridine is replaced by thymidine. Any element to the left of the initiation site is said to be "5′ to" or "upstream of" the gene. Any element to the right is "3′ to" or "downstream of" the gene. Nucleotides are numbered starting at the initiation site and receive positive values to the right and negative values to the left. Thus, in a specific gene the binding site for RNA polymerase II might include nucleotides -80 to -5, and the first intron might contain nucleotides $+154$ to $+688$. *See* **coding strand, Miller trees, polyprotein, RNA polymerase.**

transcripton a unit of genetic transcription.

transdetermination change in developmental fate of a cell or group of cells. *See* **in vivo culturing of imaginal discs.**

transduced element the chromosomal fragment transferred during transduction.

transductant a cell that has been transduced. *See* **transduction.**

transduction the transfer of bacterial genetic material from one bacterium to another using a phage as a vector. In the case of *restrictive* or *specialized transduction* only a few bacterial genes are transferred. This is because the phage has a specific site of integration on the host chromosome, and only bacterial genes close to this site are transferred. In the case of *generalized transduction* the phage can integrate at almost any position on the host chromosome, and therefore almost any host gene can be transferred with the virus to a second bacterium. Transducing phage are usually defective in one or more normal phage functions, and may not be able to replicate in a new host cell unless aided by a normal "helper" phage. *See* **Appendix C,** 1952, Zinder and Lederberg; **abortive transduction.**

transfection a term originally coined to describe the incorporation by a cell or protoplast of DNA or RNA isolated from a virus and the subsequent production of virus particles by the transfected cell. Transfection was used subsequently to refer to the incorporation of foreign DNA into cultured eukaryotic cells by exposing them to naked DNA. Such transfection experiments are directly analogous to those performed with bacteria during transformation experiments. However, the term *transfection* has been adopted rather than *transformation*

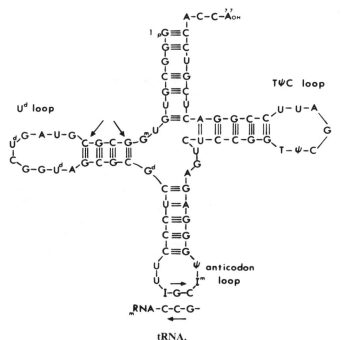

tRNA.

because transformation is used in another sense in studies involving culture animal cells (i.e., the conversion of normal cells to a state of unregulated growth by oncogenic viruses).

transfectoma a myeloma cell into which immunoglobulin genes, either wild-type or altered *in vitro,* have been transfected and expressed. Novel chimeric immunoglobulin molecules can be produced by this technique, including unique combinations of heavy and light chains, or combinations of variable regions with different constant regions (both within and between species). *Contrast with* **hybridoma.** *See* **immunoglobulin.**

transferases enzymes that catalyze the transfer of functional groups between donor and acceptor molecules. The most common molecules transferred are amino, acyl, phosphate, and glycosyl groups.

transfer factor a dialyzable extract (lymphokine) from sensitized T lymphocytes that can transfer some types of cell-mediated immunity from one individual to another.

transferred immunity *See* **adoptive transfer.**

transferrins *See* **plasma transferrins.**

transfer RNA (tRNA) an RNA molecule that transfers an amino acid to a growing polypeptide chain during translation (*q.v.*). Transfer RNA mol-

ecules are among the smallest biologically active nucleic acids known. For example, an alanine transfer RNA isolated from yeast (shown above) contains 77 nucleotides and is folded back upon itself and kept in a "clover leaf" configuration by the characteristic pairings of the bases G to C and A to U. All transfer RNAs attach to their amino acids by the 3' end which contains a terminal adenylic followed by two cytidylic acids. The 5' end always carries a terminal guanylic acid. Transfer RNAs contain several purines and pyrimidines not generally encountered in other RNA molecules. These *rare bases* (*q.v.*) are formed following transcription, since nuclei are known to contain enzymes that are capable of modifying certain bases on preformed RNA. The site recognized by tRNA synthetase is believed to be located in the neck region adjacent to the dihydrouridine loop (see arrows). The anticodon occupies positions 36–38. In the drawing, A,U,C, and G have their usual meanings. The rare bases are symbolized as follows: ψ, pseudouridylic acid; T, ribothymidylic acid; U^d, dihydrouridylic acid; G^m, methylguanylic acid; G^d, dimethylguanylic acid; I, inosinic acid; and I^m, methylinosinic acid. *See* **Appendix C,** 1958, Crick, Zamecnik; 1965, Holley *et al.;* 1970, Khorana *et al.;* 1971, Dudock *et al.;* 1973, Kim *et al.;* **isoacceptor transfer RNA.**

trans-filter induction *in vitro* inductions using organizer tissues separated from reactive cells by a Millipore filter (*q.v.*). The trans-filter induction system allows one to interrupt induction at any time by removing the inducing cells from the surface of the filter.

transformants   cells that show an inherited modification after exposure to a transforming principle or after incorporating plasmid DNA.

transformation   1. in microbial genetics, the phenomenon by which genes are transmitted from one bacterial strain to another in the form of soluble fragments of DNA. These may originate from live or dead cells. The DNA fragments dissolved in the external medium can penetrate cells only if they have receptor sites for the DNA on their surfaces. Once inside, a fragment usually replaces, by recombination, a short section of the DNA of the receptor cell that contains a zone of homology. Also called *bacterial transformation. See* **Appendix C,** 1928, Griffith; 1944, Avery *et al.;* 1964, Fox and Allen; 1970, Mandel and Higa; 1972, Cohen *et al.* **2.** the conversion of normal animal cells to a state of unregulated growth by oncogenic viruses (*q.v.*). Such viral transformation is often accompanied by alterations in cell shape, changed antigenic properties, and loss of contact inhibition (*q.v.*). Also called *cellular transformation* or *transfection. See* **Appendix C,** 1980, Capecchi.

transformation series   the various expressions of a character ordered in a hypothesized sequence from the most primitive, plesiomorphic state to the most derived, apomorphic state. This sequence may be linear (e.g., $A^0 \rightarrow A^1 \rightarrow A^2$) or it may be branched.

transgene   a foreign gene that is introduced into an organism by injecting the gene into newly fertilized eggs. Some of the animals that develop from the injected eggs will carry the foreign gene in their genomes and will transmit it to their progeny.

transgenic animals   animals into which cloned genetic material has been experimentally transferred. In the case of laboratory mice, one-celled embryos have been injected with plasmid solutions, and some of the transferred sequences were retained throughout embryonic development. Some sequences became integrated into the host genome and were transmitted through the germ line to succeeding generations. A subset of these foreign genes expressed themselves in the offspring. *See* **Appendix C,** 1980, Gordon *et al.;* 1986, Costantini *et al.*

transgressive variation   progeny phenotypes outside the range of that which occurs in the parents; usually attributed to polygene segregation.

transient diploid   a relatively short stage in the life cycles of predominantly haploid fungi or algae during which meiosis occurs.

transient polymorphism   polymorphism existing in a population during the period when an allele is being replaced by a superior one.

transition   *See* **base-pair substitution.**

translation   the formation of a protein directed by a specific messenger RNA (mRNA) molecule. Translation occurs in a ribosome (*q.v.*). A ribosome begins protein synthesis once the 5′ end of a mRNA tape is inserted into it. As the mRNA molecule moves through the ribosome, much like a tape through the head of a tape recorder, a lengthening polypeptide chain is produced. Once the leading (5′) end of the messenger tape emerges from the first ribosome, it can attach to a second ribosome, and so a second identical polypeptide can start to form. When the 3′ end of the mRNA molecule has moved through the first ribosome the newly formed protein is released, and the vacant ribosome is available for a new set of taped instructions. The assembly of amino acids into a peptide starts at the amino end (N terminus) and finishes at the carboxyl end (C terminus). There are two binding sites for transfer RNA (tRNA) in the ribosome. The P site (peptidyl-tRNA binding site) holds the tRNA molecule that is attached to the growing end of the nascent polypeptide. The A site (aminoacyl-tRNA binding site) holds the incoming tRNA molecule charged with the next amino acid. The tRNAs are held so that their anticodons form base pairs with adjacent complementary codons of the mRNA moving through the ribosome. In the diagram on page 319, an mRNA molecule is shown progressing through a ribosome. At time T_0, codon 5 of the messenger tape occupies the P site and codon 6 the A site. About half a second later the mRNA has advanced to the left by one codon. The bond between tRNA I and the 5th amino acid of the nascent polypeptide has been split, and it has been linked to amino acid 6. The atoms involved in the rearrangements are in boldface type. This reaction is catalyzed by *peptide transferase,* an enzyme that is bound tightly to the ribosome. Transfer RNA molecule I is discharged from the ribosome, tRNA II has entered the P site and is now attached to a polypeptide chain six amino acids long. A new tRNA (III) carrying an appropriate anticodon enters the A site. Note that in the diagram letters represent atoms, whereas circled letters represent molecules (i.e., nucleotides or amino acid residues). *See* **Appendix C,** 1959, McQuillen *et al.;* 1961, Dintzis; 1963, Okamoto and Takanami, Noll *et al.;* 1964, Gilbert; 1974, Shine and Dalgarno; 1976, Pelham and Jackson; **elongation factors, initiation factors, leader sequence peptide, N-formylmethionine, receptor-mediated translocation.**

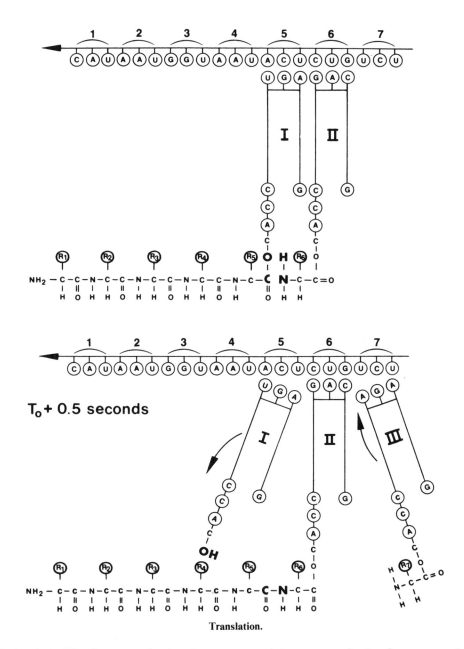

$T_0 + 0.5$ seconds

Translation.

translational amplification a mechanism for producing large amounts of a polypeptide based upon prolonged mRNA lifetime. Since there is only a single copy of the ovalbumin gene per genome, large numbers of ovalbumin molecules produced by the cells of the chicken oviduct are generated by translational amplification.

translational control the regulation of gene expression through determining the rate at which a specific RNA message is translated.

translocase a protein that forms a complex with GTP and the ribosome. Translocation of the charged tRNA from the A site to the P site is coupled with the hydrolysis of GTP to GDP and with the release of the translocase. *See* **elongation factors, erythromycin, translation.**

translocation 1. the movement of mRNA through a ribosome during translation (*q.v.*). Each translocation exposes a mRNA codon in the A site for base pairing with a tRNA anticodon.

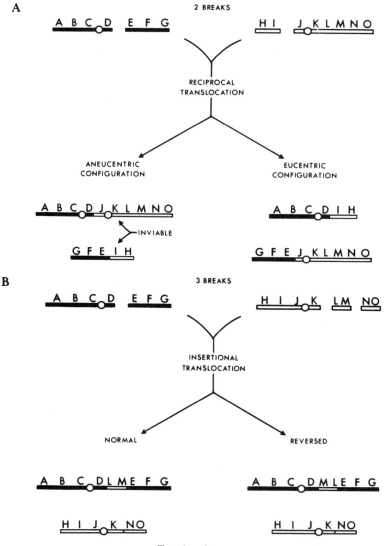

Translocation.

2. a chromosome aberration which results in a change in position of a chromosomal segment within the genome, but does not change the total number of genes present. An *intrachromosomal translocation* is a three-break aberration that results in the transposition of a chromosomal segment to another region of the same chromosome. Such an aberration is often called a *shift. Interchromosomal translocations* involve interchanges between non-homologous chromosomes. A *reciprocal translocation* is a two-break aberration that results in an exact interchange of chromosomal segments between two nonhomologous chromosomes and produces two monocentric translocated chromosomes.

A *nonreciprocal or aneucentric translocation* is a two-break aberration that results in a dicentric and an acentric translocated chromosome. A three-break interchromosomal translocation can generate a deficient chromosome and a recipient chromosome containing an intercalated segment of the other nonhomologous chromosome. This would be called an *insertional translocation. See* **Appendix C,** 1923, Bridges.

translocation Down syndrome familial Down syndrome, caused by having three copies of chromosome 21—two as separate chromosomes and one translocated to another chromosome, usually num-

320

ber 14. In families where one parent is a translocation heterozygote, the probability of having a child with Down syndrome is 0.33.

translocation heterozygote   an individual or cell in which two pairs of homologous chromosomes have reciprocally exchanged nonhomologous segments between one member of each pair. Each chromosome pair contains both homologous and nonhomologous segments, i.e., one normal (untranslocated chromosome) and one translocated chromosome. A translocation heterozygote forms a quadrivalent chromosomal association during pachynema (*see* **meiosis**), and the subsequent segregation of the four chromosomes is determined by their centromere orientations.

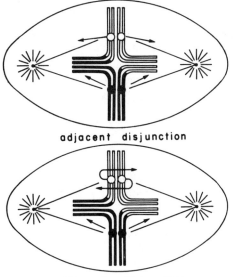

adjacent disjunction

alternate disjunction

In the case of *adjacent disjunction,* (or *adjacent segregation*), each daughter cell receives a normal and a translocated chromosome. The gametes produced from such cells are inviable because they contain certain genes in duplicate and are deficient for others. Two kinds of adjacent disjunction are recognized. In *adjacent-1 segregation,* homologous centromeres go to opposite anaphase poles. In *adjacent-2 segregation,* homologous centromeres go to the same anaphase pole. In the case of *alternate disjunction,* both translocated chromosomes go to one cell and both normal ones to the other. The gametes that result are viable because they contain all the genetic material.

translocation of proteins   *See* **receptor-mediated translocation.**

translocation mapping   gene mapping by the use of translocation chromosomes as markers. The

semisterility that is usually associated with structural translocation heterozygotes is a phenotypic marker that can be used to locate the position of the breakpoint of the translocation relative to other more conventional gene markers.

transmission electron microscope   *See* **electron microscope.**

transmission genetics   that part of genetics concerning the mechanisms involved in the transfer of genes from parents to offspring.

transmutation   the transformation of one element into another accompanying radioactive decay (*q.v.*).

transplantation   transfer of a part of an organism to another organism or to another position in the same organism. In zoology, the term is used interchangeably with graft (*q.v.*). In botany, graft is used in the above sense, and transplantation is used in the sense of planting again in a different place. *See* **rejection.**

transplantation antigen   a protein coded by genes in the major (e.g., human HLA, mouse H2) or minor histocompatibility loci and present on most vertebrate cells. Transplantation antigens are targets for T lymphocytes if the graft bears antigens different from that of the host. *See* **histocompatibility molecules.**

transposable elements   a class of DNA sequences that can move from one chromosomal site to another. This movement requires a transposase and a resolvase that recognize short nucleotide sequences that are repeated in inverted order at both ends of the element. Transposable elements were first identified in maize by Barbara McClintock (*see* **Activator-Dissociation system**) and called *controlling elements.* It is now known that McClintock's *Ac* locus is a 4.6 kb transposable element and that *Ds* is a defective DNA segment, derived from an *Ac* element by a short deletion in its transposase gene. Transposable elements were next detected in bacteria as *insertion sequences* (*q.v.*) and as mobile elements that transferred antibiotic resistance between plasmids. More recently transposable elements have been found in yeast, *Drosophila,* and *Caenorhabitis. See* **Appendix C,** 1950, McClintock; 1969, Shapiro; 1974, Hedges and Jacob; 1979, Cameron *et al.;* 1982, Bingham *et al.;* 1984, Pohlman *et al.;* **copia elements, lysogenic conversion, P elements, Ty elements.**

transposase   an enzyme that catalyzes the insertion of a transposon.

transposition   **1.** the process of insertion of a replica of a transposable element at a second site; another replica remains at the original insertion site. 2. movement of a chromosomal segment to a new location within the same chromosome, or to a different chromosome, without reciprocal exchange. *See* **translocation.**

transposons one kind of transposable element in both prokaryotes and eukaryotes that is immediately flanked by inverted repeat sequences, which in turn are immediately flanked by direct repeat sequences. Transposons usually possess genes in addition to those needed for their insertion (e.g., genes for resistance to antibiotics, sugar fermentation, etc.). *See* **retroposon.**

trans splicing joining of an exon from one gene to an exon of a different gene. This phenomenon has been observed in test-tube experiments and may occur *in vivo*; e.g., in trypanosomes, the same short RNA segment is found at the end of several mRNA molecules, but does not appear in the corresponding genes. *See* **exon shuffling.**

transvection effect the ability of a gene to influence the activity of an allele on the opposite homologue only when the two homologous chromosomes are synapsed. Chromosomal rearrangements that prevent synapsis between homologous chromosome regions tend to prevent the expression of the wild-type allele in heterozygotes. In *Drosophila melanogaster,* when trans-heterozygotes for certain mutants of the *bithorax* complex are made structurally heterozygous for chromosomal rearrangements, a cytologically observable disruption in pairing of homologous chromosomes is produced in the *bithorax* region. Under such circumstances the mutant phenotype is made more extreme. Not to be confused with transfection. *See* **Appendix C, 1945, Lewis; cis-trans configurations.**

trend an apparently directional evolutionary change in a character within a lineage; a chronocline. For example, many mammalian lineages show a tendency to increase in size over part of their evolutionary history. *See* **orthogenesis.**

triallelic referring to a polyploid in which three different alleles exist at a given locus. In a tetraploid, $A_1A_2A_3A_3$ and $A_1A_2A_2A_3$ would be examples.

Triassic the most ancient of the Mesozoic periods, during which the first dinosaurs and mammals arose. Gymnosperms and ferns were the dominant plants. Pangaea began to break apart. *See* **continental drift, geologic time divisions.**

Tribolium a genus of flour beetles containing *T. castaneum* and *T. confusum,* genetically well known species. At present about 125 mutations are known for *castaneum* and 75 for *confusum.*

tricarboxylic acid cycle *See* **citric acid cycle.**

trichloroacetic acid a compound commonly used to precipitate proteins during biochemical extractions.

$$Cl-\overset{\overset{\displaystyle Cl}{|}}{\underset{\underset{\displaystyle Cl}{|}}{C}}-COOH$$

trichocyst the minute protrusible, spindle-shaped organelles in the ectoplasm of many ciliates.

trichogen cell a large cell that secretes the long, tapering hair of the insect bristle. A smaller tormogen cell forms the circular chitinous socket around the base of the bristle.

trichogyne a receptive hypha extending from the ascogenous mycelium of a fungus such as *Neurospora.*

triethylenethiophosphoramide *See* **Thio-tepa.**

triiodothyronine *See* **thyroid hormones.**

trimester one-third of the human nine-month gestation period; cited as first, second, and third trimesters.

trimethylpsoralen a low-molecular-weight, planar molecule that undergoes photochemical reactions with pyrimidines. Trimethylpsoralen molecules intercalate into double-stranded DNA. Upon exposure to ultraviolet light, the molecules attach covalently to pyrimidines forming both monoadducts and interstrand cross links. Trimethylpsoralen has no effects on proteins.

trioecy a sexual trimorphism in plants that can exist either as male, female, or bisexual individuals.

triparental recombinant a progeny phage containing marker genes derived from each of three different phages that simultaneously infected the host cell. The observation of triparental recombinants demonstrates that repeated recombinational events occur in the infected cell between replicating nucleic acid molecules derived from all parent viruses. *See* **Visconti-Delbrück hypothesis.**

tripartite ribbons synaptonemal complexes (*q.v.*).

triplet a unit of three successive bases in DNA or RNA that codes for a specific amino acid. *See* **amino acid, genetic code, translation.**

triplet code a code in which a given amino acid is specified by a set of three nucleotides. *See* **amino acid, genetic code, reading frame, translation.**

triplex *See* **autotetraploid.**

triploid an organism having three haploid sets of chromosomes in each nucleus.

triskelion *See* **receptor-mediated endocytosis.**

trisomic an organism that is diploid but contains one extra chromosome, homologous with one of the existing pairs, so that one kind of chromosome is present in triplicate. *See* **Appendix C,** 1920, Blakeslee *et al.;* **Down syndrome, metafemale.**

trisomy 13 syndrome Patau syndrome (*q.v.*).

trisomy 18 syndrome Edward syndrome (*q.v.*).

trisomy 21 syndrome Down syndrome (*q.v.*).

tritanomaly *See* **color blindness.**

tritanopia *See* **color blindness.**

tritiated thymidine, uridine *See* **tritium.**

Triticum the genus containing the various species of wheat (*q.v.*).

tritium the radioactive isotope of hydrogen, with a half-life of 12.46 years. Tritium-labeled thymidine and uridine are often used to tag newly synthesized DNA and RNA, respectively. Tritium is the radioisotope of choice in radioautography because it emits an extremely weak beta particle when it undergoes radioactive decay. In a medium of unit density, the average tritium beta particle will penetrate only 1 μm. Therefore, in autoradiographs of tritium-labeled cells, silver grains of the photographic emulsion will be localized within 1 μm of the decaying atoms. Symbolized ^{3}H.

Triturus a genus of salamanders that are studied for their lampbrush chromosomes (*q.v.*). Species for which working chromosome maps are available include *T. alpestris, T. cristatus, T. helveticus, T. italicus, T. marmoratus,* and *T. vulgaris.* Another favorite, *T. viridescens,* has been renamed *Notophthalmus viridescens* (*q.v.*).

trivalent an association of three homologous chromosomes in meiosis.

tRNA transfer RNA (*q.v.*).

tRNA genes genes that are transcribed into tRNAs. Most tRNA genes are present in multiple copies. For example, the average repetition frequency for each tRNA gene species in yeast, *Drosophila,* humans, and *Xenopus,* has been estimated to be 5, 10, 15, and 200, respectively, per haploid genome.

tRNA isoacceptors those tRNAs that accept the same amino acid, yet differ in primary sequence, either in the anticodon, or in other regions of the molecule, or both. Such tRNA isoacceptors that dif-

fer in nucleotide sequences are generally encoded by different genes.

T4 RNA ligase an enzyme from bacteriophage T4 that catalyzes an ATP-dependent covalent joining of 5' phosphate and 3' hydroxyl termini of oligoribonucleotides.

tRNA-modifying enzymes enzymes that function to alter the primary tRNA transcript, generally by adding a single methyl group to a base of the 2'-hydroxyl group of ribose. There are at least 45 different tRNA-modifying enzymes in *E. coli,* and about 1% of its genome is devoted to encoding these enzymes. *See* **rare bases, transfer RNA.**

tRNA suppressor *See* **nonsense suppressor.**

tRNA synthetase recognition site the site on the tRNA molecule that is bound to the aminoacetyl synthetase. In the case of yeast phenylalanine tRNA, this region is located adjacent to the dihydrouridine loop and consists of the nucleotides

$$3' - CUCGA - 5'$$
$$| \; | \; | \; |$$
$$5' - GAGC - 3'$$

See **Appendix C,** 1971, Dudock *et al.;* **amino acid activation, transfer RNA.**

trophoblast the nonembryonic part of the blastocyst, that attaches to the uterine wall and later develops into the fetal portion of the placenta.

trophocyte nurse cell (*q.v.*).

tropomyosin a protein dimer made up of subunits of Mr 35,000 found associated with actin in striated muscles. Tropomyosin is involved in the regulation of muscle contraction by Ca^{++} ions. Tropomyosins exist as a number if isoforms (*q.v.*). The differences in amino acid composition between isoforms generally involve a small segment of the molecule. In *Drosophila,* tropomyosin isoforms are known that result from alternative splicing (*q.v.*).

trp tryptophan. *See* **amino acid.**

true breeding line a group of genetically identical homozygous individuals that, when intercrossed, produce only offspring that are identical to their parents. *See* **pure line.**

truncation selection a method employed by breeders in which members of a population are chosen for mating, saving only those whose phenotypic merit (on a quantitative scale) is either above or below a certain value (the truncation point).

Trypanosoma a genus of zoomastigotes that pursue a life cycle in two different hosts, a mammal and a tsetse fly. *Trypanosoma brucii* causes African sleeping sickness in humans. *See* **Appendix A,** Protoctista, Zoomastigina; **antigenic variation, Chagas disease, *Glossina*, kinetoplast.**

trypsin   a proteolytic enzyme that cleaves peptide chains on the carboxyl side of lysine and arginine residues only; initially secreted from the pancreas in the inactive form of trypsinogen, and activated to trypsin by the enzyme enterokinase in intestinal juice.

tryptophan   *See* **amino acid.**

tryptophan synthetase   the enzyme that catalyzes the union of indole and serine to form tryptophan. In *E. coli*, tryptophan synthetase is a tetrameric aggregate of two alpha chains and two beta chains. *See* **Appendix C,** 1948, Mitchell and Lein.

TSH   thyroid-stimulating hormone (*q.v.*).

***ts* mutation**   temperature sensitive mutation (*q.v.*).

T suppressor cell   *See* **suppressor T cell.**

***t* test**   Student's *t* test (*q.v.*).

tube nucleus   the vegetative nucleus that resides in a growing pollen tube.

tubulin   the principal protein component of microtubules. Tubulin is a dimer composed of α and β subunits, each of Mr 55,000. Microtubules are polymerized from $\alpha\beta$ dimers. The drug colchicine (*q.v.*) binds to $\alpha\beta$ dimers and prevents the addition of subunits to elongating microtubules. There are isoforms of both α and β tubulins, and at least some of these are products of different genes. In *Drosophila*, for example, four different genes have been identified that encode α tubulins; another four encode β tubulins.

tumor   a clump of cells (usually disfunctional) due to abnormal proliferation. Benign tumors are not life-threatening (e.g., warts); malignant tumors are potentially lethal cancers.

tumor specific transplantation antigen   an antigen found on a tumor cell but not on the normal cells or tissues of the individual in which it arose, and that can lead to immune rejection of the tumor *in situ* or if transplanted.

tumor virus   *See* **oncogenic virus.**

Turbatrix aceti   a free-living nematode whose developmental genetics is under study, primarily for comparison with *Caenorhabditis elegans* (*q.v.*). Its cell lineage is known in part.

turbid plaque   *See* **plaque.**

Turner syndrome   a group of abnormalities in humans due to monosomy for the X chromosome. Such individuals are female in phenotype, but are sterile. The ovaries are rudimentary or missing. In the mouse, the equivalent condition (XO) does not have detrimental effects except for a possible shortening of the female breeding life. *See* **Appendix C,** 1959, Ford *et al.*

turnover   the dynamic replacement of atoms in a tissue or organism without any net change in the total number of atoms. *See* **Appendix C,** 1942, Schoenheimer.

turnover number   the number of molecules of a substrate transformed per minute by a single enzyme molecule under optimal conditions.

twins   pairs of individuals produced at one birth. Monozygotic (MZ) or identical twins have identical sets of nuclear genes. MZ twins result from the separation of blastomeres and thus represent a form of asexual or clonal reproduction. Dizygotic (DZ) or fraternal twins arise when two eggs are released and fertilized. DZ twins, therefore, are no more similar genetically than are siblings that share only half of their genes, on the average. In humans, the frequency of monozygotic twinning is relatively constant (1/240 births). The frequency of dizygotic twinning varies in different races. In Caucasian populations, it is about double the frequency of MZ twinning. *See* **Appendix C,** 1869, Galton; 1874, Dareste; 1875, Galton; 1927, Bauer.

twin spots   paired patches of tissue genetically different from each other and from the background tissue, produced by mitotic crossing over in an individual of heterozygous genotype during development of the organism. In the *Drosophila* example diagrammed on page 325, twin spots occur on a female of genotype $y\ +/+\ sn$. Most of the thorax and abdomen show wild-type pigment and long, black bristles, but there are adjacent patches of yellow (drawn to appear pale) and singed tissue. These twin spots are clones of cells derived from the reciprocal products of an exchange between the singed gene and the centromere. Twin spots or sectors can also result from transposition of a controlling element to an unreplicated recipient site. If the recipient site has been replicated, twin spots result only if the target lies on the chromatid opposite that of the donor site. For example, in maize, transposition of *Ds* (*see* ***Activator-Dissociation* system**) usually is accompanied by its disappearance from the donor site (contrary to the mode of transposition in bacteria). Transposition of *Ds* almost always occurs after the donor element has been replicated. Therefore, twin sectors normally result from transposition because one chromatid has either one or

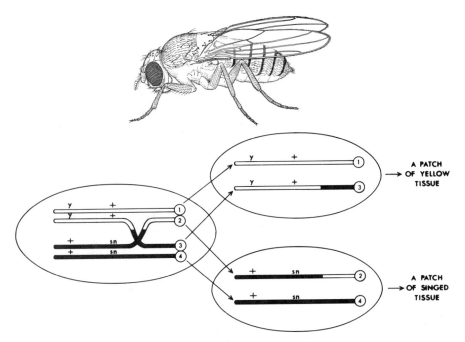

Twin spots.

no element and the sister chromatid has two elements.

twisting number the number of base pairs in duplex DNA divided by the number of base pairs per turn of the helix.

twofold rotational symmetry the situation in which the two DNA strands of a duplex have the same base sequence when read with the same polarity. For example, the Eco RI (*q.v.*) endonuclease recognizes a DNA segment six base pairs long that has two fold rotational symmetry. *See* **palindrome.**

two genes–one polypeptide chain participation of more than one gene in the coding of a given polypeptide chain, as occurs in the production of immunoglobulins. *See* **immunoglobulin genes, one gene–one polypeptide hypothesis.**

two-point cross a genetic recombination experiment involving two linked genes.

two-strand double exchange *See* **inversion.**

Ty elements transposable elements of the yeast *Saccharomyces cerevisiae.* The symbol *ty* comes from *t*ransposon-*y*east. There are 30 to 35 Ty elements per haploid yeast genome, and each consists of a central region of about 5.6 kb pairs of DNA flanked by direct repeats about 330 base pairs long. *See* **Appendix C,** 1979, Cameron *et al.;* **transposable elements.**

type specimen the individual chosen by the taxonomist to serve as the basis for naming and describing a new species.

typological thinking the consideration of the members of a population as replicas of or deviations from a hypothetical type.

tyr tyrosine. *See* **amino acid.**

tyrosinase an enzyme that converts tyrosine to dopa and oxidizes this to dopa quinone. Dopa quinone molecules undergo spontaneous polymerization to form melanin. *See* **albinism.**

tyrosine *See* **amino acid.**

tyrosinemia a hereditary disease in man arising from a deficiency of the enzyme p-hydroxyphenylpyruvic oxidase.

U

U 1. uracil or uridine. 2. uranium.

ubiquinone coenzyme Q (*q.v.*).

ubiquitin an acidic protein whose name reflects its ubiquitous presence in cells of prokaryotes and eukaryotes. The protein contains 76 amino acids, and glycine occupies its C terminus. Between 5 and 15% of the H2A histone in nucleosomes is joined to ubiquitin. The ubiquitin-H2A conjugate is abbreviated UH2A. In UH2A, the glycine at the C terminus of ubiquitin is joined to the free NH_2 group of the lysine occupying position 119 of H2A. Usually only one of the two H2A molecules in a histone octomer carries ubiquitin. During mitosis, ubiquitin is released from the chromatin. Ubiquitin is the most evolutionarily conserved of all eukaryotic proteins. The functional roles attributed to it include participation in regulation of intracellular protein degradation, gene transcription, and organization of chromatin structure, mitosis, and cell-to-cell interactions and adhesions. The process of attaching ubiquitin to other molecules is termed *ubiquitination*. Some ubiquitin genes have an unusual organization consisting of a variable number of repeated ubiquitin-coding elements that are linked head-to-tail and are not separated by introns or other spacer sequences. The protein products of these genes are thus "polyubiquitins" that must be precisely split to produce free ubiquitin. *See* **histone, nucleosome.**

UDPG uridine diphosphate glucose (*q.v.*).

UH2A *See* **ubiquitin.**

ultracentrifuge a powerful centrifuge that can attain speeds as high as 60,000 rpm and generate sedimenting forces 500,000 times that of gravity. The instrument is used to sediment macromolecules. *See* **Appendix A,** 1923, Svedberg; **centrifugation separation techniques, centrifuge, sedimentation constant.**

ultramicrotome an instrument that cuts ultrathin sections (50–100 nm thick) of plastic embedded tissues using knives made from polished diamonds or from triangles of broken plate glass. *See* **Appendix C,** 1950, Latta and Hartmann; 1953, Porter-Blum and Sjöstrand; **knife breaker.**

ultrasonic pertaining to ultrasound.

ultrasound sound waves of frequency higher than the human audible limit of about 20,000 vibrations per second.

ultraviolet absorption curve the curve showing the relation between the relative amount of ultraviolet radiation absorbed by a solution of molecules and the wave length of the incident light.

ultraviolet microscope an optical system utilizing ultraviolet radiation. Since glass filters out UV, quartz transmitting or glass reflecting lenses must be used in the UV microscope. Such a microscope has double the resolving power of the light microscope. Furthermore, if monochromatic UV of a wavelength absorbed by nucleic acids (260 mμ) is used, nucleic acid–rich structures may be photographed in unstained cells. In combination with a spectrophotometer, the UV microscope provides a method for the quantitative estimation of nucleic acids in cells.

ultraviolet radiation that part of the invisible electromagnetic spectrum just beyond the violet with wavelengths between 1,000 and 4,000 Å. Wavelengths around 260 nm are absorbed by DNA. *See* **Appendix C,** 1939, Knapp and Schreiber; **thymine dimer, UV reactivation.**

uncharged tRNA a tRNA molecule to which no amino acid is attached.

underdominance the unusual situation where a heterozygote shows an attribute, such as viability or fertility, that is lower than either homozygote. For example, the New Zealand Black (NZB) strain of mouse spontaneously develops a disease that resembles lupus erythematosis in humans (*q.v.*). New Zealand White (NZW) mice are normal in this regard. The hybrid offspring from crossing these inbred strains of mice (NZB × NZW) develop a more severe disease than that of the NZB strain.

underwinding coiling of a DNA molecule in a left-handed direction, i.e., opposite to that of the double helix; negative supercoiling.

undulipodium any cellular projection surrounding a cylindrical shaft containing a bundle of eleven microtubules, nine of which form a circle around the central pair, while the cortical microtubules are doublets. Cilia and flagellae (*q.v.*) are examples. *See* **axoneme.**

unequal crossing over the phenomenon first described in *Drosophila* at the *Bar* locus. Improper pairing of the duplicated segments is thought to occur followed by crossing over. The result is one crossover chromatid with one copy of the segment and another with three copies.

ungulate a hoofed mammal. Perissodactyla plus Artiodactyla. *See* **Appendix A.**

uniformitarianism a geological theory that "the present is the key to the past." In other words, the phenomena of volcanism, crustal movements, erosion, glaciation, etc., that can be seen today have been operating through billions of years of earth history, and they are the primary forces that have made the earth what it is today. *Contrast with* **catastrophism.**

unineme hypothesis the concept that a newly formed chromatid contains only one DNA duplex extending from one end to the other. *Contrast with* **polyneme hypothesis.** *See* **Appendix C,** 1973, Kavenoff and Zimm.

uniovular twins monozygotic twins (*q.v.*).

uniparental inheritance a phenomenon in which all offspring of a given mating seem to have received certain phenotypes from only one of the parents (usually the female) regardless of the genotype or phenotype of the other parent; such inheritance is usually the result of macromolecules or organelles stored in the cytoplasm.

unique DNA a class of DNA determined by C_0t analysis to represent sequences that are present only once in the genome. Most structural genes and their introns are unique DNAs. *See* **reassociation kinetics.**

unisexual flower a flower having only stamens or only carpels. A plant can bear either one or both kinds of unisexual flowers. *See* **flower.**

unit character a term used by early geneticists for traits that segregate according to Mendel's laws.

unit evolutionary period the time in millions of years during which a divergence of 1% occurs between the initially identical nucleotide sequence in two branches of a lineage under study. The UEP for the globin gene family is 10.4.

unit membrane the trilamellar membrane seen when the sectioned plasma membrane is viewed under the electron microscope. The membrane, which has a total thickness of about 75 Å, appears as two strata each about 20 Å thick separated by a light interzone about 35 Å wide. *See* **fluid mosaic concept.**

univalent a single chromosome seen during meiosis when bivalents are also present. A univalent has no synaptic mate. An example of a univalent would be the sex chromosome of an XO male.

universal code theory the assumption that the genetic code is used exclusively by all forms of life. This is true with a few exceptions. In yeast mitochondria, CUA codes for threonine instead of leucine; and in mammalian mitochondria, AUA codes for methionine instead of isoleucine, and codons AGA and AGG signal termination instead of coding for arginine. UGA codes for tryptophan instead of signaling termination in mitochondria from both sources. In four ciliates, *Tetrahymean thermophila, Stylonychia lemnae, Paramecium primaurelia,* and *P. tetraurelia,* UAA and UAG encode the amino acid glutamine rather than serving as termination codons. In *Mycoplasma capricolum,* as in mitochondria, UGA encodes tryptophan rather than serving as a termination codon. *See* **Appendix C,** 1961, von Ehrenstein and Lipmann; 1979, Barrell *et al.;* 1985, Horowitz and Gorowsky, Yamao; **genetic code.**

universal donor an individual with type O blood, who is able to donate red blood cells to O, A, B, or AB recipients. *See* **blood group.**

universal recipient an individual with type AB blood, who can receive red blood cells from AB, A, B, or O donors. *See* **blood group.**

univoltine *See* **voltinism.**

unordered tetrad *See* **nonlinear tetrad.**

unscheduled DNA synthesis DNA synthesis that occurs at some stage in the cell cycle other than the S period, generally to repair damaged DNA. *See* **interphase cycle, repair synthesis.**

unstable equilibrium the situation where the equilibrium value for an allele in a population fluctuates because of temporary environmental changes. For example, these may lead to a sudden selection for or against the allele, or a drastic reduction in the size of the population may result in genetic drift away from the former equilibrium value.

unstable mutation a mutation with a high frequency of reversion. The original mutation may be caused by the insertion of a controlling element (*q.v.*), and its exit produces a reversion.

untwisting enzyme *See* **topoisomerase.**

unwinding proteins proteins that bind to, destabilize, and unwind the DNA helix ahead of the replicating fork. *See* **Appendix C,** 1970, Alberts and Frey; **gene 32 protein.**

u orientation *See* **n orientation.**

up promoter mutations mutations in promoter sites that increase the rate of initiation of transcription; promoters with this property are called "high level or strong promoters."

upstream *See* **transcription unit.** *Compare with* **downstream.**

uracil *See* **bases of nucleic acids.**

uracil fragments during DNA replication in *E. coli,* polymerases I and III occasionally make mistakes and incorporate dUTP instead of TTP. Several enzymes remove these uracils from both leading and lagging strands, creating "uracil fragments."

urea *See* **ornithine cycle.**

urethane a carcinogen that induces tumorous nodules in the lungs of mammals.

$$NH_2-\overset{\displaystyle O}{\underset{\displaystyle \parallel}{C}}-O-CH_2-CH_3$$

uric acid the end product of nucleic acid catabolism in mammals; the main nitrogenous constituent of the urine of reptiles and birds.

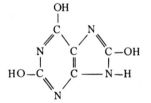

uridine *See* **nucleoside.**

uridine diphosphate galactose *See* **uridine diphosphate glucose.**

uridine diphosphate glucose (UDPG) a compound (shown below) that acts as a coenzyme and substrate to various enzymes. UDPG may be converted to uridine diphosphate galactose by the enzyme epimerase. These coenzymes play essential roles in carbohydrate metabolism.

uridylic acid *See* **nucleotide.**

Urkingdom *See* **Archaebacteria.**

Usn RNAs the U class of small nuclear RNAs (*q.v.*). These molecules range in size from 60 to 216 nucleotides and are rich in uridine. Five of Usn RNAs (U_1, U_2, U_4, U_5, and U_6) have been studied in the most detail. The first four all have a trimethylguanosine cap structure at the 5' end. Most Usn RNAs are associated with seven proteins, some of which are common to all five RNAs, while others are specific for U_1 or U_2. These snRNPs make up about a third of the mass of the spliceosomes, and they function in the excision and splicing reactions that take place in this organelle. *See* **exon, intron, lupus erythematosus, posttranscriptional modification, splice junctions.**

Ustilago a genus in the *Basidiomycota* (*see* **Appendix A**). These are yeastlike smut fungi, and two species, *U. maydis* and *U. violacea,* have been subjects of genetic research, especially in terms of recombination-defective and radiation-sensitive mutations.

uteroglobin a protein synthesized by the cells of the rabbit endometrium and present in the uterine fluids. Uteroglobin is composed of two identical subunits of 70 amino acids, and the subunits are held together by two disulphide bridges. Reduced uteroglobin binds steroids, such as progesterone.

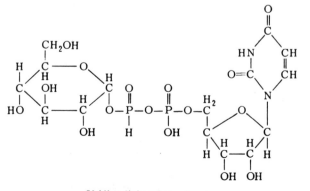

Uridine diphosphate glucose.

There is one uteroglobin gene per genome. It is 3 kb long and contains two intervening sequences and 3 exons. Some authors have proposed that uteroglobin exerts a stimulatory effect on the blastocyst (hence its other name, *blastokinin*).

UV   ultraviolet radiation (*q.v.*).

UV-induced dimers   *See* **thymine dimers.**

UV reactivation   a phenomenon in which survival of an ultraviolet-irradiated lambda phage is greater on an irradiated host than on an unirradiated host. The repair mechanism involved in UV-reactivation utilizes an error-prone replication system of the host such as that of the SOS response (*q.v.*).

V

vaccine   a suspension of dead or weakened bacteria or viruses injected into the body to immunize against the same pathogen.

vaccinia virus   the DNA virus of cowpox which serves for vaccination against smallpox. *See* **Appendix C**, 1798, Jenner.

vacuum evaporator   the vacuum chamber containing a set of electrodes through which a current can be passed to heat a metal foil placed between the electrodes. The heated metal evaporates, and atoms from it coat the specimen lying below the electrodes. The coating is cast at an angle, and as a consequence the specimen appears in relief when viewed under the electron microscope. A vacuum evaporator equipped with graphite electrodes is also employed to prepare the carbon films that sometimes are used to support ultrathin sections on specimen screens. *See* **shadow casting.**

val   valine. *See* **amino acid.**

valence   1.   in chemistry, a number representing the combining or displacing power of an atom; the number of electrons lost, gained, or shared by an atom in a compound; the number of hydrogen atoms with which an atom will combine, or which it will displace. **2.** in serology, the number of antigen-combining sites on an antibody molecule.

valine   *See* **amino acids.**

valinomycin   *See* **ionophores.**

van der Waals forces   the relatively weak, short-range forces of attraction existing between atoms or molecules, caused by the interaction of varying dipoles.

var   variety.

variable   any organismal attribute that may have different values in various situations (e.g., between members of different species, between members of the same species, within an individual at different times, etc.).

variable domain   that domain of an immunoglobulin light or heavy chain that has variable amino acid sequences within an individual.

variable number of tandem repeats locus (VNTR locus)   any gene whose alleles contain different numbers of tandemly repeated oligonucleotide sequences. Such alleles when cleaved by a specific restriction endonuclease will produce fragments that differ in length. Such restriction-length polymorphisms (*q.v.*) serve as convenient markers in linkage studies. *See* **DNA fingerprint technique.**

variable region   the N-terminal portion of an immunoglobulin chain that binds to antigen.

variance   when all values in a population are expressed as plus and minus deviations from the population mean, the variance is the mean of the squared deviations.

variant   an individual that is different from an arbitrary standard type (usually wild type) for that species. Variants are not necessarily mutants; for example, many birth defects are simply developmental accidents or environmentally induced. *See* **phenocopy.**

variate   a specific quantitative value of a variable (*q.v.*).

variation   divergence among individuals of a group, specifically a difference of an individual from others of the same species that cannot be ascribed to a difference in age, sex, or position in the life cycle. The variations of evolutionary significance are gene-controlled phenotypic differences of adaptive significance.

variegated position effect   *See* **position effects.**

variegation   1.   irregularity in the pigmentation of plant tissues due to a variety of causes (virus infection, segregation of normal and mutant plastids, bridge-breakage-fusion-bridge cycles, transposable elements, etc.). **2.** irregularity in the pigmentation of animal tissues or their products (hair, feathers, etc.) due to a variety of causes (X-chromosome inactivation, defective embryonic migration of melanocytes, localized physiological conditions, mitotic recombination, etc.).

variety   *See* **strain.**

vasopressin   a peptide hormone secreted by the hypothalamus and stored in the neurohypophysis, which constricts arterioles and promotes resorption of water by the kidney tubules. Also called the antidiuretic hormone.

vector **1.** a self-replicating DNA molecule that transfers a DNA segment between host cells; also called a *vehicle. See* **DNA vector, lambda cloning vector, plasmid cloning vector, shuttle vector. 2.** an organism (such as the malaria mosquito) that transfers a parasite from one host to another.

vegetal hemisphere the surface of the amphibian egg farthest from the nucleus, the yolk-rich hemisphere of the egg.

vegetative designating a stage or form of growth, especially in a plant, distinguished from that connected with reproduction.

vegetative cell an actively growing cell, as opposed to one forming spores.

vegetative nucleus **1.** the macronucleus of a ciliate. **2.** the tube-nucleus of a pollen grain.

vegetative petites *See* **petites.**

vegetative reproduction in plants, the formation of a new individual from a group of cells, without the production of an embryo or seed. More generally, asexual reproduction. *See* **agamospermy, apomixis.**

vegetative state the noninfective state during which a phage genome multiplies actively and controls the synthesis by the host of the materials necessary for the production of infective particles.

vehicle a plasmid or bacteriophage possessing a functional replicator site, and containing a genetic marker to facilitate its selective recognition, used to transport foreign genes into recipient cells during recombinant DNA experiments; also called a *vector.*

vermilion a sex-linked, recessive eye-color mutation in *Drosophila melanogaster.* This was the first *Drosophila* mutation to be understood biochemically. Flies lacking the plus allele of *vermilion* have an enzymatic defect that makes them unable to convert tryptophan to formylkynurenine (*q.v.*). If the latter compound is supplied in the larval diet, the adults that develop show the normal eye color. Formylkynurenine is converted through a series of steps to hydroxykynurenine (*q.v.*), and eventually to xanthommatin (*q.v.*). *See* **Appendix C,** 1935, Beadle and Ephrussi; *Drosophila* **eye pigments.**

vermilion plus substance formylkynurenine; so called because the synthesis of this compound in *Drosophila melanogaster* is controlled by the plus or wild-type allele of the gene *vermilion* (*q.v.*).

vernalization the treatment of germinating seeds with low temperatures to effect their flowering. Winter varieties of certain cereals, if vernalized, can be sown in the spring and harvested in the summer.

Veronica a large genus of hardy herbs belonging to the family Scrophulariaceae. Classic studies on the genetic control of self-sterility were performed on this species.

vertical classification a system of classification that recognizes taxa corresponding to clades and groups transitional forms with their descendants rather than with their ancestors; the opposite of *horizontal classification* (*q.v.*).

vertical evolution the process whereby an ancestral species changes through time (without splitting) to become distinctively different, and therefore recognized as a new species; phyletic evolution. *See* **anagenesis, speciation.**

vertical transmission **1.** passage of genetic information from one cell or individual organism to its progeny by conventional heredity mechanisms (mitosis, meiosis), in contrast to transfer of genetic information by infection-like processes between contemporary individuals (horizontal transmission). **2.** transmission of a parasite from parent to offspring via the egg or *in utero.*

V gene one of many (perhaps hundreds) of genes coding for the variable (N-terminus) region of an immunoglobulin chain.

viability a measure of the number of individuals surviving in one phenotypic class relative to another class, taken as standard, under specified environmental conditions.

vicariance distribution a discontinuous biogeographical distribution of organisms that previously inhabited a continuous range. The current gaps in the distribution were caused by some extrinsic factor (geologic or climatic).

Viceroy butterfly *Basilarchia archippus* (*q.v.*).

Vicia faba the broad bean; also called the fava bean, the horse bean, the Windsor bean. A plant often used in cytogenetics because its cells contain a small number (N = 6) of large chromosomes. The semiconservative nature of DNA replication was first demonstrated by analyzing autoradiographs of ^{3}H thymidine-labeled chromosomes from cells of *Vicia* root tips. *See* **Appendix C,** 1957, Taylor *et al.;* **favism.**

villus a finger-like projection extending from an epithelium. Such a villus is composed of many cells. *Contrast with* **microvillus.**

vimentin a 55,000 d cytoskeletal protein commonly found in fibroblasts. In glial cells it is co-polymerized with an acidic protein of 50,000 d molecular weight, while in muscle cells it is combined with desmin (*q.v.*).

vinblastine   a vinca alkaloid (*q.v.*).

vinca alkaloid   any of a group of mitotic poisons isolated from *Vinca rosea.*

vincaleukoblastine   a vinca alkaloid (*q.v.*).

Vinca rosea   the Madagascar periwinkle, source of the vinca alkaloids.

vincristine   a vinca alkaloid (*q.v.*).

vinculin   a fibrous protein responsible for anchorage of actin filaments to the inner side of the cell membrane. Vinculin is located in patches called *adhesion plaques* on the cell membrane that are thought to be responsible for intercellular adhesion. Cells infected with Rous sarcoma virus (*q.v.*) produce a kinase that phosphorylates the tyrosine residues of vinculin. It is hypothesized that the phosphorylation of vinculin both destabilizes actin linkages (allowing transformed cells to become rounded) and weakens intercellular adhesion (allowing metastasis).

viral-specific enzyme   any enzyme produced in the host cell after viral infection and encoded by a viral gene.

viral transformation   *See* **transformation.**

virion   a completed virus particle consisting of a nucleic acid core and a protein coat.

viroid   a disease-causing agent of plants consisting of single-stranded RNA molecules typically 270–380 nucleotides long and therefore thousands of times smaller than the most diminutive virus. Viroids are not encapsulated in a protein coat. Viroids do not pass through a DNA stage in their life cycle; they are replicated directly as RNA and are not integrated into the host genome. The pathogenic effects of viroids on their hosts result from the fact that the RNA of the virus contains segments that

are complementary to the 7SRNAs of their hosts. Thus, the virus behaves like an antisense RNA (*q.v.*) and blocks the formation and functioning of signal recognition particles (*q.v.*). Typical viroids are the apical stunt and planta macho viroids of tomatoes and the cadang cadang viroid of coconut palms. *See* **Appendix C,** 1967, Diener and Raymer.

virulence   the relative ability of an organism to produce disease.

virulent phage   a phage that causes lysis of the host bacterium. *Contrast with* **temperate phage.**

virus   an ultramicroscopic, obligate, intracellular parasite incapable of autonomous replication. Viruses can reproduce only by entering a host cell and using its translational system. Viruses probably originated as nucleic acids that escaped from cells. Viruses are generally classified according to the type of nucleic acid they contain and the morphology of the nucleocapsid (*q.v.*). Examples are given below.

virus receptors   sites on the cell membrane to which viruses attach. Such sites contain neuraminic acid (*q.v.*).

viscoelastic molecular weight determination   a method using a viscometer that allows the determination of the molecular weights of the largest molecules present in a solution. The technique is very useful in determining the molecular weights of very long DNA molecules, since a fraction of these are fragmented during the isolation procedure. *See* **Appendix C,** 1973, Kavenoff and Zimm.

Visconti-Delbrück hypothesis   according to this proposal, bacteriophages multiply upon entering a host, and the replicating units so formed mate repeatedly. Mating occurs in pairs and is at random with respect to the pairing partner. During any given mating cycle, a segment of genetic material from one parent can exchange with that from a second parent phage, yielding recombinant units. *See* **Appendix C,** 1953, Visconti and Delbrück.

visibles   referring to phenotypically observable

NUCLEIC ACID TYPE	CAPSID MORPHOLOGY	EXAMPLES
RNA viruses	helical symmetry	tobacco mosaic virus, influenza viruses, potato virus Y
	cubical symmetry	polio virus, reoviruses, retroviruses
DNA viruses	helical symmetry	smallpox
	cubical symmetry	polyoma virus, ϕX174, Shope papilloma virus
	with head and tail	T phages, λ, P22

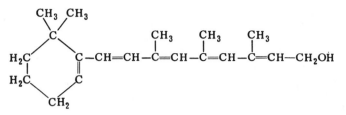

Vitamin A.

mutants, as opposed to lethals, which are scored by the absence of an expected class of individuals in a cross designed to detect induced mutants.

visual purple   *See* **rhodopsin.**

vitalism   a philosophy holding that the phenomena exhibited in living organisms are the result of special forces distinct from chemical and physical ones. *See* **mechanistic philosophy.**

vital stain   a dye used to stain living cells (Janus green, methylene blue, trypan blue, etc.).

vitamin   an organic compound (often functioning as a coenzyme) that is required in relatively minute amounts in the diet for the normal growth of a given organism.

vitamin A   a fat-soluble vitamin functioning as a precursor to retinene (*q.v.*) and retinoic acid (*q.v.*). Vitamin A is generated by the splitting in two of a molecule of beta carotene. *See* **carotenoids.**

vitamin B complex   a family of water-soluble vitamins, including thiamin (B_1), riboflavin (B_2), nicotinic acid, pantothenic acid, pyridoxin (B_6), and cobalamin (B_{12}).

vitamin C   ascorbic acid, an important regulator of the oxidation-reduction state of protoplasm.

vitamin D   calciferol; a fat-soluble vitamin required in man for the prevention of rickets. The vitamin affects the absorption and deposition of calcium and phosphate.

vitamin D–resistant rickets   a group of hereditary diseases in which patients show a reduction in the levels of calcium and phosphorous in their blood and skeletal changes characteristic of rickets, although they have adequate dietary vitamin D. An autosomal recessive form of the disease has been shown to result from mutations in a gene that encodes a vitamin D receptor (VDR). This is a zinc finger protein (*q.v.*), and some mutant VDRs have amino acid substitutions in their zinc fingers. The more common form of vitamin D–resistant rickets is X-linked, and it also results from a defect in the vitamin D receptor system.

vitamin E   alpha tocopherol, a vitamin functioning as an antioxidant.

vitamin H   biotin (*q.v.*).

vitellarium   the portion of the insect ovariole posterior to the germarium. Egg chambers complete development within the vitellarium.

vitelline membrane   a membrane that surrounds

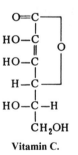

Vitamin C.

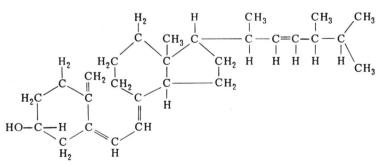

Vitamin D.

333

the ovum. In *Drosophila* the term is used specifically for the membrane that immediately surrounds the oolemma and is formed by the fusion of deposits in the intercellular space between the oocyte and the columnar follicle cells that invest it.

vitellogenesis the formation of yolk.

vitellogenic hormone *See* **allatum hormones.**

vitellogenin a protein synthesized by vitellogenic females and incorporated into the yolk spheres of the developing oocyte. In *Xenopus laevis,* vitellogenins are synthesized by the liver. In *Drosophila melanogaster,* vitellogenins are synthesized by abdominal and thoracic fat bodies and by the columnar follicle cells surrounding the oocyte. *See* **lipovitellin, phosphovitin.**

viviparous 1. producing living young rather than eggs. Embryogenesis occurs within the mother's body, as with most mammals. **2.** bearing seeds that germinate within the fruit, as in the mangrove.

Viviparus malleatus a prosobranchiate snail showing a bizarre type of spermatogenesis with the production of oligopyrene sperm. *See* **sperm polymorphism.**

v-myc *See myc.*

VNTR locus *See* **variable number of tandem repeats locus.**

voltinism a polymorphism in terms of whether or not the embryos produced by an insect enter diapause (*q.v.*). For example, in *Bombyx mori* univol-tine strains produce only diapause embryos. Bivoltine strains produce a nonhibernating brood, then diapause embryos.

von Gierke disease a hereditary glycogen storage disease in humans arising from a deficiency of the enzyme glucose-6-phosphatase. Inherited as an autosomal recessive. Prevalence 1/100,000.

von Willebrand disease the most common blood-clotting disorder of humans. It is due to a deficiency of the von Willebrand factor (vWF), which is encoded by a gene on the short arm of chromosome 12. The von Willebrand factor is an adhesive glycoprotein synthesized by endothelial cells and megakaryocytes, and it serves as a carrier in plasma for the antihemophilic factor (*q.v.*). Most von Willebrand disease patients have one normal gene for vWF and one abnormal gene. They make about half the usual amount of vWF, and their disease is mild. Such heterozygotes occur at a frequency of about 5 per 10,000 humans. Homozygotes are very rare (less than one in a million), and they bleed uncontrollably when injured. *See* **blood clotting, hemophilia.**

v-sis gene *See* **simian sarcoma virus.**

v-src gene *See* **Rous sarcoma virus.**

vulgare wheats *See* **wheat.**

Vulpes vulpes the red fox, a species bred on a large scale on ranches for its pelt. Numerous mutations influencing fur color are known.

W

Wahlund effect   the tendency for a large population that is subdivided into many small panmictic units with different gene frequencies to exhibit a deficiency of heterozygotes and an excess of homozygotes in comparison with expectations from applying the Hardy-Weinberg formula utilizing gene frequencies averaged from all subgroups. The phenomenon, discovered by S. Wahlund, is the result of inbreeding within each subgroup.

Wallace effect   the hypothesis put forth by Alfred Russel Wallace that natural selection favors the evolution of mechanisms that ensure the reproductive isolation of sexual populations that have reached the level of elementary biological species. Reproductive isolation prevents the production of sterile hybrids which compete for food reserves.

waltzer   one of the many neurological mutants in the laboratory mouse. Homozygotes are deaf and characteristically show a circling and head shaking behavior.

Waring blender   an electric kitchen appliance used to homogenize mixtures, but used in the laboratory to generate the shearing forces required to detach conjugating bacteria, to strip bacteriophages or their "ghosts" from host cell surfaces, to homogenize tissue samples, etc. *See* **interrupted mating experiment, shearing.**

warning coloration   conspicuous colors or markings on an animal that is poisonous, distasteful, or similarly defended against predators. Such coloration is presumed to facilitate learning, on the part of predators, to avoid the possessor of such markings.

Watson-Crick model   *See* **deoxyribonucleic acid.**

wax   any esters of fatty acids and long-chain monohydroxyalcohols.

weak interactions   forces between atoms, such as ionic bonds, hydrogen bonds, and van der Waals forces, which are weak relative to covalent bonds (*q.v.*).

weighted mean   the mean obtained when different classes of observations or quantities are given different weights (are multiplied by different factors) in the calculation.

Weismannism   the generally accepted concept proposed by August Weismann that acquired characters are not inherited and that only changes in the germ plasm are transmitted from generation to generation. *See* **Appendix C,** 1883, Weismann.

western blotting   *See* **Southern blotting.**

wheat   the world's most important grain crop, produced by species of the genus *Triticum.* Wheat species are generally grouped together on the basis of their chromosome numbers. The einkorn wheats are diploids (N = 7), the emmer wheats are tetraploids (N = 14), and the vulgare wheats are hexaploids (N = 21). The hexaploids, of which *T. aestivum* is the most common example, contain three genomes, A, B, and D, each of which is composed of a set of seven pairs of nonhomologous chromosomes. The A, B, and D genomes were derived from *Triticum monococcum, T. searsii,* and *T. tauschii,* respectively. See **einkorn, emmer, durum, spelt, wild einkorn,** and **wild emmer wheat,** *Aegilops.*

white leghorn   *See* **plumage pigmentation genes.**

white plymouth rock   *See* **plumage pigmentation genes.**

whole-arm fusion, whole-arm transfer   *See* **centric fusion.**

wild type   the most frequently observed phenotype, or the one arbitrarily designated as "normal."

wild-type gene   the allele commonly found in nature or arbitrarily designated as "normal."

Williston's rule   during evolution of the serially repeated parts of an organism, the number of such parts tends to diminish as the remaining parts diversify in response to division of labor.

Wilms tumor   a malignant renal tumor of young children associated with a deletion of band p13 on chromosome 11.

Wilson disease   a disease of human beings inherited as an autosomal recessive in which the biliary excretion of copper and its incorporation into ceruloplasmin is impaired. Copper accumulation leads to liver and brain damage. Prevalence 1/75,000.

winter variety a variety of cereal which must be sown in the autumn of the year preceding that in which the plants should flower. If sown in the spring, they will not flower in the same growing season.

wobble hypothesis a hypothesis developed to explain how one tRNA may recognize two codons. The anticodon in each tRNA is a base triplet. The first two bases of the mRNA codon pair according to the base pairing rules. The third base in the anticodon, however, has a certain amount of play or wobble that permits it to pair with any one of a variety of bases occupying the third position of different codons. Thus, U in the third position would recognize A or G, for example, and transfer RNA with a CUU anticodon would bind to either of two codons (GAA or GAG). *See* **Appendix C,** 1966, Crick.

Wolman disease a hereditary disease in man arising from a deficiency of the lysosomal enzyme acid lipase; due to a recessive gene on chromosome 10.

woodpecker finch *Camarhynchus pallidus,* the famous finch of the Galapagos Islands that uses a tool (a cactus spine) to dislodge insects from bark crevices. *See* **Geospizinae.**

working hypothesis a hypothesis that serves as the basis for future experimentation.

woolly monkey *Lagothrix lagothricha,* a species inhabiting the rain forests of the Amazon basin. The source of the simian sarcoma virus (*q.v.*).

Wright's equilibrium law an expression of the zygotic proportions expected in a population experiencing a certain amount of inbreeding. For a pair of alleles A and a with frequencies p and q, respectively, the zygotic proportions are expected to be $AA = p^2 + Fpq : Aa = 2pq (1 - F) : aa = q^2 + Fpq$, where F is Wright's inbreeding coefficient (*q.v.*). The Hardy-Weinberg Law (*q.v.*) is a special case of Wright's equilibrium law in which F = 0.

Wright's inbreeding coefficient (F) the probability that two allelic genes united in a zygote are both descended from a gene found in an ancestor common to both parents. Also, the proportion of loci at which an individual is homozygous.

Wright's polygene estimate in the case of quantitative inheritance (*q.v.*), the number of segregating pairs of polygenes (n) can be estimated from the variances computed for the F_1 and F_2 populations. If these are symbolized s_{F1}^2 and s_{F2}^2 respectively, and the means for the high and low P_1 strains are X_h and X_l, respectively, then

$$n = \frac{(X_h - X_l)}{8(s_{F2}^2 - s_{F1}^2)}$$

See **quantitative inheritance.**

writhing number the number of times the axis of a DNA molecule crosses itself by supercoiling.

W, Z chromosomes the sex chromosomes of an animal in which the female is the heterogametic sex (*Bombyx mori,* for example). In such cases the W chromosome is female-determining and the male is ZZ. *See* **Bkm sequences.**

X

x *See* **basic number.**

X "crossed with" (as in A ♀ × B ♂).

X₂ the offspring of an F₁ test cross.

xantha any of many chloroplast mutations in various cereal species. The *xantha* 3 mutant of barley, for example, is characterized by chloroplasts that accumulate excessive numbers of pigment granules and never develop orderly arrays of grana.

xanthommatin *See* **Drosophila** **eye pigments.**

X chromosome the sex chromosome found in double dose in the homogametic sex and in single dose in the heterogametic sex.

X-chromosome inactivation in mammalian development, the repression of one of the two X chromosomes in the somatic cells of females as a method of dosage compensation. At an early embryonic stage in the normal female, one of the two X chromosomes undergoes inactivation, apparently at random. From this point on, all descendant cells will be clonal in that they will have the same X chromosome inactivated as the cell from which they arose. Thus, the mammalian female is a mosaic composed of two types of cells—one that expresses only the paternal X chromosome, and another that expresses only the maternal X chromosome. In some cells and tissues, the inactivated X chromosome can be seen as a dense body in the nucleus (referred to as a Barr body or sex chromatin). In abnormal cases where more than two X chromosomes are present, only one X remains active and the others are inactivated. In marsupials, the paternal X is selectively inactivated during female development. Recent studies have shown that certain genes located distally on the short arm of the human X chromosome escape inactivation. *See* **Appendix C,** 1963, Russell, Davidson *et al.*

xenia referring to the situation in which the genotype of the pollen influences the developing embryo or the maternal tissue of the fruit so as to produce a phenotypically demonstratable effect upon the seed.

xenogeneic transplantation xenoplastic transplantation (*q.v.*).

xenoplastic transplantation the transplantation between individuals of different genera or widely distant species.

Xenopus a genus of African aquatic anurans containing two species, *X. laevis* and *X. borealis,* that are favorites for research in molecular genetics. Studies on the nucleolar mutants of *X. laevis* led to the conclusion that the nucleolus contains about 450 genes that function to transcribe rRNA. The 5S component of rRNA is coded for by genes that cluster near the tips of the long arms of all the chromosomes. *X. laevis* and *X. borealis* have about 24,000 and 9,000 5S rRNA genes, respectively. *See* **Appendix C,** 1966, Wallace and Birnsteil; 1967, Birnsteil; 1968, Davidson *et al.;* 1973, Ford and Southern.

xeroderma pigmentosum a group of hereditary diseases inherited as autosomal recessives in which the skin is extremely sensitive to sunlight or ultraviolet light, and death is usually due to skin cancer. Normal skin cells can repair UV damage to DNA by cut-and-patch repair (*q.v.*). Skin cells from patients with xeroderma pigmentosum cannot perform such repair and apparently lack a UV-specific endonuclease. It appears that patients with the symptoms of xeroderma pigmentosum may be homozygous for mutations occurring at any one of at least seven different chromosomal sites. *See* **Appendix C,** 1968, Cleaver.

Xg blood group a blood group defined by an antigen controlled by a gene located distally on the short arm of the human X chromosome.

X-inactivation *See* **X chromosome inactivation.**

Xiphophorus maculatus the platyfish, and *X. helleri,* the swordtail. Pigment cell genetics and the genetics of sex determination have been intensively studied in laboratory strains of these freshwater species.

X linkage the presence of a gene located on the X chromosome; usually termed "sex linkage" (*q.v.*).

XO the symbolic designation of the situation in some heterogametic organisms in which the X chromosome is present and the Y chromosome is absent.

XO monosomy *See* **Turner syndrome.**

x radiation radiations produced when high-speed electrons strike a metallic target. X-rays have wavelengths in the range between ultraviolet and gamma radiation and are ionizing radiations.

x-ray crystallography the use of the diffraction patterns produced by x-ray scattering from crystals to determine the three-dimensional structure of the atoms or molecules in the crystal. *See* **Appendix C,** 1913, Bragg and Bragg; **large angle x-ray diffraction, small angle x-ray diffraction.**

XXY trisomy *See* **Klinefelter syndrome.**

XYY trisomy a human karyotype observed in about 1 in 1000 male births. Most adult XYY males are over six feet tall. A few are sterile, and some are mentally retarded or have behavioral disorders. XYY individuals make up a greater-than-average proportion of the patients in mental-penal institutions. This may be accounted for in part by their diminished intelligence, which may make it easier for them to be apprehended.

Y

Y   the single-letter symbol for pyrimidine. *See* **R3.**

Y chromosome   the sex chromosome found only in the heterogametic sex. In *Drosophila,* the Y chromosome bears genes controlling spermiogenesis. In mammals, it contains genes that control the differentiation of the testis. *See* **Appendix C,** 1968, Hess and Meyer; 1982, Goldstein *et al.;* 1987, Page *et al.;* **dynein.**

yeast   any fungus that generally exists as single cells and reproduces by budding. The most important yeast is *Saccharomyces cerevisiae* (*q.v.*).

Y fork   the point at which a DNA molecule is being replicated; the two template strands of the parental molecule separate, forming the arms of a Y-shaped structure. The unreplicated double-stranded DNA distal to the arms forms the base of the Y. *See* **replication of DNA.**

Y linkage   genes located on the Y sex chromosome, exhibiting holandric (*q.v.*) inheritance. *See* ***Oryzias latipes.***

yolk   the complex collection of macromolecules and smaller nutrient molecules with which the oocyte is preloaded prior to fertilization. *See* **lipovitellin, phosphovitin, vitellogenin.**

Y-suppressed lethal   a sex-linked, recessive lethal that causes death of XO *Drosophila melanogaster* but allows survival of normal males.

Z

Z atomic number, the number of protons plus electrons in the nucleus of the neutral atom.

Z chromosome the sex chromosome found in both heterogametic females and homogametic males. *See* **W, Z chromosomes.**

Z DNA *See* **deoxyribonucleic acid.**

Zea mays maize or Indian corn, the plant species for which the most genetic information is available. The haploid chromosome number is 10 and cytological maps are available for pachytene chromosomes. The 10 linkage maps contain over 200 genes. *See* **Appendix C,** 1909, Shull; 1928, Stadler; 1931, 1933, 1934, 1938, McClintock; 1938, Rhoades; 1950, McClintock; 1964, Mertz *et al.;* 1984, Pohlman *et al.;* **corn, teosinte.**

zeatin *See* **cytokinins.**

Zebu the Brahman (*q.v.*) breed of cattle.

zein a group of alcohol soluble proteins that function as storage proteins in maize kernels. The proteins are encoded by a multigene family, are synthesized in the developing endosperm, and account for more than 50% of the protein in mature seeds. *See* ***Zea mays.***

zero-order kinetics the progression of an enzymatic reaction in which the formation of product proceeds at a linear rate with the time. This rate is not increased if additional substrate is added. *See* **first-order kinetics.**

zero sum assumption an aspect of the Red Queen hypothesis (*q.v.*) proposing that the beneficial effect enjoyed by a species in evolutionary advance is precisely matched by the sum of the negative effects experienced by all other species in the community. The zero sum assumption derives in part from the notion that the total resources available in the system is constant and the rate of evolution will also be constant.

zero time binding DNA strands of DNA containing intramolecular repeats that form duplexes at the start of a reassociation reaction.

Zimmermann cell fusion a technique developed by Ulrich Zimmermann in which cells are exposed to a low-level, high-frequency electric field that orients them into chains. A direct current pulse is then used to open micropores in adjoining cell membranes. These micropores allow mixing of the cytoplasms, and the cells may eventually fuse. The Zimmermann technique may also alter the permeability of the plasmalemma so that DNA fragments the size of genes can enter the cell.

zinc a biological trace element. Atomic number 30; atomic weight 65.37; valence 2^+; most abundant isotope ^{64}Zn, radioisotope ^{65}Zn, half-life 250 days, radiation emitted—positrons.

Zn zinc.

zinc finger proteins proteins possessing tandemly repeating segments that bind zinc atoms. Each segment contains two closely spaced cysteine molecules followed by two histidines. Each segment folds upon itself to form a fingerlike projection. The zinc atom is linked to the cysteines and histidines at the base of each loop as shown below, where the circles containing C's or H's represent cysteine or histidine molecules and the other amino acids of the polypeptide finger are represented by unlabeled circles. The zinc fingers serve in some way to enable the proteins to bind to DNA molecules, where they regulate transcription. *See* **Appendix C,** 1985, Miller *et al.;* 1987, Page *et al.*

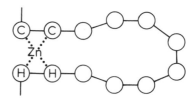

zonal electrophoresis a technique that makes possible the separation of charged macromolecules and the characterization of each molecule in terms of its electrophoretic mobility. *See* **electrophoresis.**

zona pellucida one of the envelopes surrounding the mammalian ovum that produces various substances that attract homologous sperm, prevent entry of foreign sperm, and prevent polyspermy (*q.v.*).

zoogeographic realms the divisions of the land masses of the world according to their distinctive faunas. *See* **biogeographic realms.**

zoogeography the study of the geographical distribution of animals.

ZPG zero population growth; a population status in which birth and death rates are equivalent.

ZR515 a synthetic juvenile hormone analogue that mimics the effects of JH and is more resistant to breakdown by the esterases normally found in insect hemolymph. *See* **allatum hormone.**

Z, W chromosomes *See* **W, Z chromosomes.**

zwitterion a dipolar ion. For example, amino acids in solution at neutral *p*H are in dipolar form, with the amino group protonated ($-NH_3^+$) and the carboxyl group dissociated ($-COO^-$).

Zy DNA DNA replicated during the zygotene stage of meiosis. During the premeiotic S phase, only about 99.7% of the DNA is replicated. The remainder replicates during zygotene and is intimately involved in synapsis. *See* **Appendix C,** 1971, Hotta and Stern.

zygonema *See* **meiosis.**

zygote the diploid cell resulting from the union of the haploid male and female gametes.

zygotene stage *See* **meiosis.**

zygotic induction the induction of vegetative replication in a prophage that is transferred during conjugation to a nonlysogenic F^- bacterium.

zygotic lethal in *Drosophila,* a lethal gene whose effect is apparent in the embryo, larva, or adult, but that does not render inviable any gamete that carries it.

zygotic meiosis *See* **meiosis.**

zygotic segmentation mutants mutations in *Drosophila melanogaster* that are zygotically expressed and control the spatial pattern of development of the embryo. The mutations fall into three classes that are defined by the pattern of cuticular defects they produce. The *gap genes* are active in contiguous domains along the anteroposterior axis of the embryo and regulate segmentation within each domain. The *pair rule genes* are expressed in stripes along the blastoderm with a periodicity that corresponds to every other segment. The *segment polarity genes* regulate the spatial pattern within each segment. *See* **Appendix C,** 1980, Nüsslein-Volhard and Wieschaus; 1989, Driever and Nüsslein-Volhard; *hunchback,* **maternal polarity mutants.**

zymogen the enzymatically inactive precursor of a proteolytic enzyme. Zymogens usually become activated by posttranslational modifications. For example, the zymogen pepsinogen is converted to the digestive enzyme pepsin by cleavage in a particular peptide sequence.

zymogen granules enzyme-containing particles elaborated by the cells of the pancreas.

Appendix A
Classification

Classification: the subdivision of organisms into an evolutionary hierarchy of groups. The formal hierarchy proceeding from the largest to the smallest group is Kingdom, Phylum, Class, Order, Family, Genus, and Species. To allow further subdivisions, the names Grade or Division are sometimes placed between Kingdom and Phylum, the name Branch is placed between Phylum and Class, the name Cohort between Class and Order, and the name Tribe between Family and Genus. In addition, the prefixes Super- and Sub- may be added to any group name.

Classification of living organisms

SUPERKINGDOM   Prokaryotes *(q.v.)*
  KINGDOM 1   Prokaryotae
    SUBKINGDOM   Archaebacteria *(q.v.)*
            PHYLUM   Methanocreatrices *(Methanobacterium, Methanococcus)*
            PHYLUM   Nonmethanogenic archaebacteria
              SUBPHYLUM   Halobacteria *(Halobacterium)*
              SUBPHYLUM   Thermoacidophilic bacteria *(Thermoplasma)*
              SUBPHYLUM   Sulfur-dependent thermophiles *(Sulfolbus)*

    SUBKINGDOM   Eubacteria
            PHYLUM   Aphragmabacteria *(Mycoplasma, Spiroplasma)*
            PHYLUM   Spirochaetae *(Spirochaeta)*
            PHYLUM   Thiopneutes (Sulfur-reducing bacteria, *Desulfovibrio*)
            PHYLUM   Anaerobic photosynthetic bacteria, *(Rhodospirillum)*
            PHYLUM   Cyanobacteria (blue-green "algae," *Anabaena*)
            PHYLUM   Chloroxybacteria *(Prochloron)*
            PHYLUM   Nitrogen-fixing aerobic bacteria *(Azotobacter, Rhizobium)*
            PHYLUM   Pseudomonads *(Pseudomonas)*
            PHYLUM   Omnibacteria *(Escherichia, Hemophilus, Salmonella, Serratia, Shigella)*
            PHYLUM   Chemoautotrophic bacteria *(Nitrobacter)*
            PHYLUM   Myxobacteria *(Myxococcus)*
            PHYLUM   Fermenting bacteria *(Diplococcus, Streptococcus)*
            PHYLUM   Aeroendospora (aerobic endospore-forming bacteria, *Bacillus*)
            PHYLUM   Micrococci *(Staphylococcus)*
            PHYLUM   Actinobacteria *(Actinomyces, Streptomyces)*

SUPERKINGDOM   Eukaryotes *(q.v.)*
  KINGDOM 2   Protoctista *(q.v.)*
            PHYLUM   Caryoblastea *(Pelomyxa)*
            PHYLUM   Dinoflagellata *(Gonoyaulax)*
            PHYLUM   Rhizopoda *(Amoeba)*
            PHYLUM   Chrysophyta *(Ochromonas)*
            PHYLUM   Haptophyta *(Prymnesium)*
            PHYLUM   Euglenophyta *(Euglena)*
            PHYLUM   Cryptophyta *(Cyathomonas)*

PHYLUM   Zoomastigina (animal flagellates, *Trypanosoma, Giardia*)

PHYLUM   Xanthophyta *(Vaucheria)*

PHYLUM   Eustimatophyta *(Vischeria)*

PHYLUM   Bacillariophyta (diatoms, *Diatoma*)

PHYLUM   Phaeophyta (brown algae, *Fucus, Nereocystis*)

PHYLUM   Rhodophyta (red algae, *Polysiphonia*)

PHYLUM   Gamophyta (conjugating green algae and desmids, *Spirogyra* and *Micrasterias*)

PHYLUM   Chlorophyta (green algae forming flagellated gametes, *Acetabularia, Chlamydomonas, Volvox*)

PHYLUM   Actinopoda (radiolarians and heliozoans, *Acanthocystis, Sticholonche*)

PHYLUM   Foraminera *(Fusulina, Globigerina)*

PHYLUM   Ciliophora (ciliates, *Paramecium, Tetrahymena*)

PHYLUM   Apicomplexa (sporozoans, *Plasmodium*)

PHYLUM   Cnidosporidia (microsporidians, *Nosema*)

PHYLUM   Labyrinthulomycota (slime nets, *Labryrinthula*)

PHYLUM   Acrasiomycota (cellular slime molds, *Dictyostelium, Polysphondylium*)

PHYLUM   Myxomycota (plasmodial slime molds, *Echinostelium*)

PHYLUM   Plasmodiophoromycota *(Plasmidiophora)*

PHYLUM   Hypochytridiomycota *(Hypochytrium)*

PHYLUM   Chytridiomycota *(Blastocladiella)*

PHYLUM   Oomycota *(Saprolegnia)*

KINGDOM 3   Fungi *(q.v.)*

PHYLUM   Zygomycota (conjugating fungi, *Mucor, Phycomyces, Pilobolus, Rhizopus*)

PHYLUM   Ascomycota (sac fungi, *Neurospora, Saccharomyces*)

PHYLUM   Basidiomycota (club fungi, *Agaricus, Schizophyllum, Ustilago*)

PHYLUM   Deuteromycota *(Aspergillus, Penicillium)*

PHYLUM   Mycophycophyta (lichens, *Cladonia*)

KINGDOM 4   Animalia *(q.v.)*

SUBKINGDOM   Parazoa *(q.v.)*

PHYLUM   Placozoa *(Trichoplax)*

PHYLUM   Porifera (sponges, *Euplectella*)

SUBKINGDOM   Mesozoa *(q.v.)*

PHYLUM   Mesozoa *(Dicyema)*

SUBKINGDOM   Eumetazoa *(q.v.)*

GRADE   Radiata *(q.v.)*

PHYLUM   Cnidaria (Coelenterates)

CLASS   Hydrozoa (hydroids, *Hydra*)

CLASS   Scyphozoa (true jelly fish, *Physalia*)

CLASS   Anthozoa (corals and sea anemones, *Metridium*)

PHYLUM   Ctenophora (comb jellies, *Mnemiopsis*)

GRADE   Bilateria *(q.v.)*

SUBGRADE   Protostomia *(q.v.)*

SUPERPHYLUM   Acoelomata *(q.v.)*

PHYLUM   Platyhelminthes (flatworms)

CLASS   Turbellaria (planarians)

CLASS   Trematoda (flukes, *Schistosoma*)

CLASS   Cestoda (tapeworms)

PHYLUM   Nemertina (Rhynchocoela) (ribbon worms)

PHYLUM   Gnathostomulida

SUPERPHYLUM   Pseudocoelomata *(q.v.)*

PHYLUM   Acanthocephala (spiny-headed worms)

PHYLUM   Entoprocta (entoprocts)

PHYLUM   Aschelminthes *(q.v.)*

SUBPHYLUM   Rotifera (rotifers)

SUBPHYLUM   Gastrotricha (gastrotrichs)

PHYLUM   Loricifera (loriciferans)

SUBPHYLUM   Kinorhyncha (kinorhynchs)

SUBPHYLUM   Priapulida (priapulids)
SUBPHYLUM   Nematoda (round worms, *Ascaris, Caenorhabditis*)
SUBPHYLUM   Nematomorpha (Gordiacea) (horsehair worms)
SUPERPHYLUM   Coelomata (*q.v.*)
  DIVISION   Tentaculata
   PHYLUM   Phoronida (phoronids)
   PHYLUM   Ectoprocta (bryozoa)
   PHYLUM   Brachiopoda (brachiopods)
  DIVISION   Inarticulata (*q.v.*)
   PHYLUM   Sipunculoidea (sipunculids)
   PHYLUM   Mollusca (molluscs)
      CLASS   Amphineura (chitins)
      CLASS   Scaphopoda (tooth shells)
      CLASS   Gastropoda (snails, *Cepaea, Lymnea*)
      CLASS   Pelecypoda (bivalves, *Crassostrea, Mytilus*)
      CLASS   Cephalopoda (squids, octopuses, *Nautilus*).
  DIVISION   Articulata (*q.v.*)
   PHYLUM   Echiuroidea (echiuroids)
   PHYLUM   Annelida (segmented worms)
      CLASS   Polychaeta (marine worms)
      CLASS   Oligochaeta (earthworms)
      CLASS   Hirudinea (leeches)
   PHYLUM   Tardigrada (water bears)
   PHYLUM   Onychophora (*Peripatus*)
   PHYLUM   Pentastomida (tongue worms) (parasites)
   PHYLUM   Arthropoda
    BRANCH   Chelicerata
      CLASS   Merostomata (king crabs)
      CLASS   Pycnogonida (sea spiders)
      CLASS   Arachnida (scorpions, opilionids, mites, spiders)
    BRANCH   Mandibulata
      CLASS   Crustacea
          ORDER   Branchiopoda (shrimps)
          ORDER   Ostracoda (ostracods)
          ORDER   Copepoda (copepods)
          ORDER   Cirripedia (barnacles)
          ORDER   Malacostraca (lobsters and crabs)
      CLASS   Myriapoda
          ORDER   Diplopoda (millipedes)
          ORDER   Chilopoda (centipedes)
      CLASS   Hexapoda
     SUBCLASS   Entognatha
          ORDER   Collembola (springtails)
          ORDER   Protura (proturans)
          ORDER   Diplura (campodeans)
     SUBCLASS   Insecta
      COHORT   Apterogota (primitively wingless insects)
          ORDER   Archaeognatha (machilids)
          ORDER   Zygentoma (silverfish)
      COHORT   Pterygota (winged insects)
       SUBCOHORT   Paleoptera (extended wing)
          ORDER   Ephemeroptera (Mayflies)
          ORDER   Odonata (dragonflies)
       SUBCOHORT   Neoptera (hinged wing)
        SUPERORDER   Hemimetabola (no pupal stage)
          ORDER   Embioptera (embiids)
          ORDER   Phasmida (stick insects)

ORDER   Orthoptera (grasshoppers)
ORDER   Grylloblatteria (grylloblatids)
ORDER   Dictyoptera (roaches, *Blattella*)
ORDER   Dermaptera (earwigs)
ORDER   Psocoptera (booklice)
ORDER   Phthiraptera (sucking lice)
ORDER   Hemiptera (true bugs)
ORDER   Plecoptera (stone flies)
ORDER   Thysanoptera (thrips)
SUPERORDER   Holometabola
ORDER   Coleoptera (beetles, *Tribolium*)
ORDER   Raphidioptera (snakeflies)
ORDER   Megaloptera (alder flies)
ORDER   Neuroptera (lacewings)
ORDER   Hymenoptera *(Apis, Mormoniella, Habrobracon)*
ORDER   Trichoptera (caddis flies)
ORDER   Lepidoptera (moths, *Bombyx, Ephestia*)
ORDER   Mecoptera (scorpion flies)
ORDER   Siphonaptera (fleas)
ORDER   Diptera *(Aedes, Anopheles, Chironomus, Culex, Drosophila, Glyptotendipes, Lucilia, Musca, Rhynchosciara, Sciara)*

SUBGRADE   Deuterostomia (*q.v.*)
PHYLUM   Echinodermata
CLASS   Crinoidea (sea lilies)
CLASS   Asteroidea (starfish)
CLASS   Ophiuroidea (brittle stars)
CLASS   Echinoidea (sea urchins)
CLASS   Holothuroidea (sea cucumbers)
PHYLUM   Chaetognatha (arrow worms)
PHYLUM   Pogonophora (beard worms)
PHYLUM   Chordata (notochord-bearing animals)
SUBPHYLUM   Acraniata (*q.v.*)
BRANCH   Hemichordata (acorn worms and pterobranchs)
BRANCH   Urochordata (tunicates)
BRANCH   Cephalochordata (lancelets)
SUBPHYLUM   Craniata (*q.v.*)
BRANCH   Agnatha (jawless vertebrates)
CLASS   Cyclostomata (lampreys)
BRANCH   Gnathostoma (jawed vertebrates)
CLASS   Chondrichthyes (elasmobranchs)
CLASS   Osteichthyes (bony fish)
SUBCLASS   Palaeopterygii (ancient fishes)
ORDER   Acipenseriformes (sturgeons)
ORDER   Semionotiformes (gars)
SUBCLASS   Neopterygyii (modern fishes)
ORDER   Salmoniformes (*Salmo*)
ORDER   Anguilliformes (eels)
ORDER   Cypriniformes *(Carassius, Oryzias)*
ORDER   Cyprinidontiformes *(Fundulus, Lebistes)*
ORDER   Perciformes *(Tilapia)*
ORDER   Siluriformes (catfishes)
ORDER   Elopiformes (tarpons)
ORDER   Antherinoformes (*Xiphophorus*)
ORDER   Clupeiformes (herrings)
ORDER   Gasterosteiformes (sea horses)
ORDER   Pleuronectiformes (flounders)
SUBCLASS   Crossopterygii (coelacanths, *Latimeria*)

CLASS Amphibia
    ORDER Apoda (caecilians)
    ORDER Urodela (salamanders, *Ambystoma, Triturus, Pleurodeles*)
    ORDER Anura (frogs and toads, *Rana, Xenopus*)
CLASS Reptilia (turtles, alligators, lizards, snakes)
CLASS Aves (birds)
  SUBCLASS Palaeognathae (flightless birds, ostriches, emus)
  SUBCLASS Neognathae (modern birds)
    ORDER Anseriformes (ducks, *Anas*)
    ORDER Galliformes (quail, turkeys, *Coturnix, Gallus*)
    ORDER Columbiformes (pigeons and doves, *Columba*)
    ORDER Psittaciformes (parrots)
    ORDER Passeriformes (song birds)
CLASS Mammalia
  SUBCLASS Protheria (egg-laying mammals)
  SUBCLASS Metatheria (marsupials, *Potorous*)
  SUBCLASS Eutheria (placental mammals)
    ORDER Insectivora (moles, shrews)
    ORDER Chiroptera (bats)
    ORDER Edentata (sloths)
    ORDER Rodentia (rodents, *Cavia, Chinchilla, Mesocricetus, Mus, Peromyscus, Rattus*)
    ORDER Lagomorpha (rabbits, *Oryctolagus*)
    ORDER Carnivora (carnivores, *Canis, Felis, Vulpes, Mustella*).
    ORDER Cetacea (whales)
    ORDER Proboscidea (elephants)
    ORDER Pinnipedia (seals)
    ORDER Perissodactyla (odd-toed ungulates, *Equus*)
    ORDER Artiodactyla (even-toed ungulates, *Bos, Camelus, Ovis, Sus*).
    ORDER Primates (lemurs, tarsiers, monkeys, apes, *Homo sapiens*)

KINGDOM 5 Plantae

PHYLUM Bryophyta (*q.v.*)
  SUBPHYLUM Hepaticae (liverworts, *Sphaerocarpos*)
  SUBPHYLUM Anthocerotae (hornworts)
  SUBPHYLUM Musci (mosses)
PHYLUM Tracheophyta (vascular plants)
  SUBPHYLUM Psilophyta *(Psilotum)*
  SUBPHYLUM Lycopodophyta (clubmosses and quillworts)
  SUBPHYLUM Sphenopsida (horsetails)
  SUBPHYLUM Pteropsida (ferns and seed plants)
    SUPERCLASS Filicinae (ferns)
    SUPERCLASS Gymnospermae (cone-bearing, seed plants)
      CLASS Pteridospermophyta (seed ferns)
      CLASS Cycadophyta (cycads)
      CLASS Ginkgophyta (ginkgos)
      CLASS Coniferophyta (conifers)
      CLASS Gnetophyta (gnetophytes)
    SUPERCLASS Angiospermae (flowering plants)
      CLASS Dicotyledoneae
        ORDER Magnoliales (magnolias)
        ORDER Rosales (rose, apple, plum, strawberry, etc.)
        ORDER Legumiales *(Pisum, Phaseolus)*
        ORDER Salicales (willows)
        ORDER Fagales (beech, oak, etc.)
        ORDER Geraniales (geranium, flax, citrus fruits)
        ORDER Cactales (cactuses)
        ORDER Scrophulariales *(Antirrhinum, Collinsia)*

ORDER Ranales (buttercups, water lilies, *Ranunculus*)
ORDER Myrtales (myrtle, eucalyptus, *Oenothera*)
ORDER Cruciales (cabbage, turnip, *Arabidopsis*)
ORDER Cucurbitales *(Cucurbita, Cucumis)*
ORDER Caryophyllales (carnations)
ORDER Gentianales (gentians, olives, lilacs)
ORDER Primulales (*Primula*)
ORDER Malvales (*Gossypium,* linden, elm, hemp)
ORDER Rubiales (coffee, quinine)
ORDER Saxifragales (current, gooseberry)
ORDER Umbellales (carrot, parsnip)
ORDER Sapindales (horse chestnut, maple)
ORDER Solanales *(Datura, Lycopersicon, Nicotiana)*
ORDER Celastrales (grapes, ivy)
ORDER Personales (snapdragon, *Antirrhinum*)
ORDER Boraginales (heliotrope, etc.)
ORDER Lamiales (lavender, mint, etc.)
ORDER Asterales (composites, *Haplopappus*)
CLASS Monocotyledoneae
ORDER Palmales (palms)
ORDER Graminales (grasses, *Hordeum, Oryzea, Triticum, Zea*)
ORDER Liliales (lilies, tulips, amaryllis, iris, *Colchicum*)
ORDER Commelinales *(Tradescantia)*
ORDER Arales (callas, taros)
ORDER Zingiberales (bananas)
ORDER Orchidales (orchids)
ORDER Bromeliales (pineapple)

Adapted from Lynn Margulis, 1974, *The Classification and Evolution of Prokaryotes and Eukaryotes. Handbook of Genetics* **1**:1–41, Plenum Press, New York and Lynn Margulis and Karlene V. Schwartz, 1988, *Five Kingdoms: An Illustrated Guide to the Phyla of Life on Earth,* Second Edition, W. H. Freeman, San Francisco.

Appendix B
Domesticated Species

Domesticated species, organisms that have been trained to live with or be of service to humans, include agricultural plants, livestock, household pets, laboratory animals, etc. Below is a listing of the common and scientific names of a variety of economically important domesticated organisms. Species that are given their own entries in the dictionary (i.e., barley, chicken, corn, etc.) are omitted from this list.

alfalfa *Medicago sativa*
almond *Prunus amygdalus*
alpaca *Lama pacos*
amaryllis *Amaryllis belladonna*
anise *Pimpinella anisum*
apple *Malus pumila*
apricot *Prunus armeniaca*
artichoke *Cynara scolymus* (globe)
ash *Fraxinus americana* (white)
asparagus *Asparagus officinalis*
aster *Aster novaeangliae* (New England)
avocado *Persea americana*
balsam fir *Abies balsamea*
bamboo *Bambusa vulgaris*
banana *Musa paradisaica*
bass *Micropterus salmoides* (large mouth)
bean *Vicia faba* (broad): *Ricinus communis* (castor); *Phaseolus limensis* (lima); *P. aureus* (Mung); *P. vulgaris* (string)
beech *Fagus grandifolia* (American); *F. sylvatica* (European)
beet *Beta vulgaris*
begonia *Begonia rex*
birch, paper *Betula papyrifera*
blackberry Cultivated blackberries are derived chiefly from three species of *Rubus: R. argutus, R. alleghaniensis,* and *R. frondosus.*
blueberry *Vaccinium corymbosum*
bluegrass *Poa pratensis*
Brazil nut *Bertholletia excelsa*
breadfruit *Artocarpus communis*
broccoli *Brassica oleracea italica*
broomcorn *Sorghum vulgare technicum*
Brussels sprouts *Brassica oleracea gemmifera*
buckwheat *Fagopyrum sagittatum*
cabbage *Brassica oleracea capitata*
calabash *Lagenaria siceraria*

camellia *Camellia japonica*
canary *Serinus canaria*
cantaloupe *Cucumis melo cantalupensis*
cardamom *Elettaria cardamomum*
carnation *Dianthus caryophyllus*
carp *Cyprinus carpio*
carrot *Daucus carota sativa*
cashew nut *Anacardium occidentale*
cassava *Manihot esculenta*
cauliflower *Brassica oleracea botrytis*
cedar, eastern red *Juniperus virginiana*
celery *Apium graveolens dulce*
cherry *Prunus cerasus* (sour), *P. avium* (sweet)
chestnut (European) *Castenea sativa*
chick pea *Cicer arietinum*
chili pepper *Capsicum annuum*
Chinese cabbage *Brassica rapa*
chive *Allium schoenoprasum*
chrysanthemum *Chrysanthemum morifolium*
cinnamon *Cinnamomum zeylanicum*
clove *Syzygium aromaticum*
clover *Trifolium pratense* (red), *T. repens* (white)
coconut *Cocos nucifera*
coffee *Coffea arabica*
coriander *Coriandrum sativum*
cowpea *Vigna sinensis*
crabapple *Pyrus ioensis*
cranberry *Vaccinium macrocarpon*
crocus *Crocus susianus* (cloth of gold)
daffodil *Narcissus pseudo-narcissus*
date palm *Phoenix dactilifera*
dill *Anethum graveolens*
Douglas fir *Pseudotsuga taxifolia*
duck *Anas platyrhynchos* (mallard)
ebony *Diospyros ebenum*
eggplant *Solanum melongena esculentum*
elephant *Elephas maximus* (Indian)
endive *Cichorium endivia*

ermine   *Mustela erminea*

fig   *Ficus carica*

flax   *Linum usitatissimum*

foxglove   *Digitalis purpurea*

geranium   *Pelargonium graveolens*

gerbil   *Merinoes unguiculatis* (Mongolian)

ginger   *Zingiber officinale*

gladiolus   *Gladiolus communis* (one of hundreds of species)

goat   *Capra hircus*

goose   *Cygnopsis cygnoid* (Chinese)

grape   *Vitis vinifera* (common wine)

grapefruit   *Citrus paradisi*

guava   *Psidium guajava*

hazelnut   *Corylus americana*

hemlock   *Tsuga heterophylla* (western)

hemp   *Cannabis sativa* (marijuana), *Agave sisalana* (sisal)

hickory   *Carya ovata* (shagbark)

holly   *Ibex opaca* (American), *I. aquifolium* (European)

hollyhock   *Althea rosea*

honey locust   *Gleditsia triacanthos*

hop   *Humulus lupulus*

huckleberry   *Gaylussacia baccata* (black)

hyacinth   *Hyacinthus orientalis*

iris   *Iris versicolor grandiflorum* (blue flag)

ivy   *Hedera helix* (English)

jasmine   *Jasminum officinale*

juniper   *Juniperus communis*

kapok   *Ceiba pentandra*

laburnum   *Laburnum anagyroides*

lavender   *Lavandula officinalis*

lemon   *Citrus limon*

lentil   *Lens esculenta*

lettuce   *Lactuca sativa*

licorice   *Glycyrrhiza glabra*

lime   *Citrus aurantifolia*

lingonberry   *Vaccinium vitis-idaea* (also called mountain cranberry or lowbush cranberry)

llama   *Lama glama*

lotus   *Nelumbo lutea* (yellow)

maguey   *Agava cantala* (Manila), *A. atrovirens* (pulque)

mahogany   *Swietenia mahagoni* (West Indian)

mango   *Mangifera indica*

marigold   *Tagetes erecta* (garden)

marten   *Martes americana*

millet   *Setaria italica* (foxtail), *Pennisetum glaucum* (pearl), *Panicum miliaceum* (broomcorn)

morning-glory   *Ipomoea purpurea*

mushrooms

  common edible   *Agaricus bisporus*

  Chinese   *Volvariella volvacea*

  shiitake   *Lentinus edodes*

musk ox   *Ovibos moschatus*

mustard   *Brassica hirta* (white)

narcissus   *Narcissus poeticus*

nasturtium   *Tropaeolum majus*

oak   *Quercus suber* (cork), *Q. alba* (white)

okra   *Hibiscus esculentus*

olive   *Olea europaea*

orange   *Citrus aurantium* (bitter); *C. sinensis* (sweet); the naval orange is a cultivar of this species.

oyster   *Crassostraea virginica*

Pak choy (Chinese cabbage)   *Brassica campestris*

papaya   *Carica papaya*

parakeet   *Melopsittacus undulatus*

parsnip   *Pastinaca sativa*

peach   *Prunus persica*

peanut   *Arachis hypogaea*

pear   *Pyrus communis*

pecan   *Carya illinoensis*

peony   *Paeonia officinalis*

pepper   *Capsicum frutescens* (red), *Piper nigrum* (black)

peppermint   *Mentha piperita*

perch   *Perca flavescens* (yellow)

persimmon   *Diospyros kaki* (Japanese)

philodendron   *Philodendron cordatum*

phlox   *Phlox drummondii*

pickerel   *Esox niger* (chain)

pike   *Esox lucius* (northern)

pine   *Pinus lambertiana* (sugar)

pineapple   *Ananas comosus*

pistachio nut   *Pistacia vera*

plum   *Prunus domestica*

poinsettia   *Euphorbia pulcherrima*

poppy   *Papaver somniferum* (opium)

quince   *Cydonia oblonga*

radish   *Raphanus sativus*

raspberry   *Rubus occidentalis* (black)

redbud   *Cercis canadensis*

reindeer   *Rangifer tarandus*

rhubarb   *Rheum officinale*

rose   Many species of *Rosa* are grown commercially. The most common are *R. centifolia, R. damascena,* and *R. multiflora.*

rubber   *Hevea brasiliensis*

rutabaga   *Brassica napobrassica*

rye   *Secale cereale*

sandalwood   *Santalum album*

sequoia   *Sequoia gigantia* (big tree), *S. sempervirens* (coastal redwood)

sesame   *Sesamum indicum*

sorghum   *Sorghum vulgare*

soybean   *Glycine max*

spearmint   *Mentha spicata*

spinach   *Spinacia oleracea*

spruce   *Picea pungens* (Colorado blue)

strawberry   *Fragaria virginiana* (American), *F. chiloensis* (Chilean), *F. vesca* (European)

strawflower   *Helichrysum bracteatum*

sugar cane   *Saccharum officinarum*

sunflower   *Helianthus annuus*

sweet potato   *Ipomoea batatas*

Swiss chard   *Beta vulgaris cicla*

tangerine   *Citrus reticulata*
taro   *Colocasia esculenta*
tea   *Camellia sinensis (Thea sinensis)*
teak   *Tectona grandis*
thyme   *Thymus vulgaris*
timothy   *Phleum pratense*
trout   *Salvelinus frontinalis* (brook), *S. namaycush*
   (lake), *Salmo gairdneri* (rainbow)
tulip   *Tulipa gesneriana*
tulip tree   *Liriodendron tulipifera*
turkey   *Meleagris gallopavo*
turnip   *Brassica rapa*

vanilla   *Vanilla plantifolia*
vicuña   *Vicugna vicugna*
violet   *Saintpaulia ionantha* (African)
walnut   *Juglans regia* (English)
water buffalo   *Bubalus bubalis*
watermelon   *Citrullus vulgaris*
wild rice   *Zizania aquatica*
willow   *Salix babylonica* (weeping)
yak   *Bos grunniens*
yam   *Dioscorea alata*
yucca   *Yucca brevifolia* (Joshua tree)
zinnia   *Zinnia elegans*

Appendix C
Chronology

Genetics, cytology, and evolutionary biology have received stimulation from both related and quite independent sciences. In many cases, the development of a particular physical instrument or technique has led to a golden age of discovery. Often research in various areas has advanced in nonsynchronous spurts, and consequently it is difficult to develop courses in genetics, cytology, and evolutionary biology from a strictly historical standpoint. The student, however, should have some idea of the chronological order in which certain events having a bearing on these sciences took place. The following chronology will fill this need, even though many experts will complain about the inclusion of some events and the omission of others. Furthermore, a decade from now some of the recent discoveries may be relegated to less prominent positions. The student should keep the following thought in mind when perusing this catalogue. In science, a great unifying concept generally does not spring full-blown from the mind of a single individual. Rather, when the time is ripe perhaps a dozen authorities may grope about for an explanation, and all may be on the verge of the answer. However, often one scientist may first express the unifying concept in a clear fashion, and as a matter of convenience he or she is the one listed as the progenitor of the idea. Following this chronology is an index that contains an alphabetical listing of the scientists and the years their works appear. In a bibliography that follows the index, one can find references to some of the epoch-making books listed in the chronology, as well as appreciations of great geneticists, collections of important scientific papers, and histories of certain scientific breakthroughs. By sampling this additional literature, the interested reader can put flesh upon the skeleton provided by this chronology.

1590	Z. and H. Janssen combine two double convex lenses in a tube and produce the first compound microscope.
1651	W. Harvey puts forward the concept that all living things (including man) originate from eggs.
1657	R. de Graaf discovers follicles in the human ovary, but interprets them incorrectly as eggs.
1665	R. Hooke publishes *Micrographia,* in which he gives the first description of cells.
1668	F. Redi disproves the theory of spontaneous generation of maggots.
1677	A. van Leeuwenhoek observes sperm of man and other mammals.
1694	J. R. Camerarius conducts early pollination experiments and reports the existence of sex in flowering plants.
1735	C. V. Linné (better known today in the Latinized form Linnaeus) publishes the first edition of the *Systema Naturae.* Sixteen editions of this taxonomic work are completed during his lifetime. It is the 10th edition of this book that serves as the starting point for the modern scientific naming of animals, as his *Species Plantarum* is for plants. Linné originated the "Linnean" system of binary nomenclature used to this day. His insistence on the constancy and objective classification of species posed the problem of the method of the origin of species.
1761–67	J. G. Kölreuter carries out crosses between various species of *Nicotiana* and finds that the hybrids are quantitatively intermediate between their parents in appearance. The hybrids from reciprocal crosses are indistinguishable. He concludes that each parent contributes equally to the characteristics of the offspring.
1769	L. Spallanzani demonstrates that the "spontaneous generation" of microorganisms in a nutrient medium can be prevented, provided the vessel is sealed and subjected to the temperature of boiling water for 30 minutes or more. In 1780 he performs artificial insemination experiments with amphibians and demonstrates that physical contact of the egg with spermatic fluid is necessary for fertilization and development.

1798	T. R. Malthus publishes anonymously *An Essay on the Principle of Population*. This essay suggests to Darwin in 1838 the concept of the struggle for existence and the survival of the fittest.

1798   T. R. Malthus publishes anonymously *An Essay on the Principle of Population*. This essay suggests to Darwin in 1838 the concept of the struggle for existence and the survival of the fittest.

Edward Jenner publishes *An Inquiry into the Causes and Effects of the Variolae Vaccinae, a Disease Discovered in Some of the Western Countries of England, Particularly Gloucestershire and Known by the Name of Cow Pox*. In it he gave the first account of vaccination with cowpox virus to prevent smallpox. He thus establishes the principle of active immunization and initiates the science of immunology.

1809   J. B. de Monet Lamarck puts forward the view that species can change gradually into new species through a constant strengthening and perfecting of adaptive characteristics, and that these acquired characteristics are transmitted to the offspring.

1818   W. C. Wells suggests that human populations in Africa have been selected for their relative resistance to local diseases. He is thus the first to enunciate the principle of natural selection.

1820   C. F. Nasse describes the sex-linked mode of inheritance of hemophilia in man.

1822–24   T. A. Knight, J. Goss, and A. Seton all independently perform crosses with the pea and observe dominance in the F_1 and segregation of various hereditary characters in the F_2. However, they do not study later generations or determine the numerical ratios in which the characters are transmitted.

1825   F. V. Raspail founds the science of histochemistry by using the iodine reaction for starch.

1827   K. E. von Baer gives the first accurate description of the human egg.

1830   G. B. Amici shows that the pollen tube grows down the style and into the ovule of the flower.

1831   R. Brown notes nuclei within cells.

On December 27, H.M.S. *Beagle* sets sail from Plymouth for a voyage around the world. It carries as naturalist the 22-year-old Charles Darwin. The *Beagle* reaches the Galapagos Islands, September 15, 1833, and Darwin spends five weeks surveying the plant and animal life.

1837   Darwin realizes, after he and other experts go over the collections from the Galapagos, that many species are unique to various islands. This suggests that the islands were colonized by a few species from the mainland, and from these evolved new species specialized to live on each of the many new environments provided. These conclusions stimulated Darwin to start accumulating data to support a theory of evolution through natural selection.

Hugo von Mohl provides the first description of chloroplasts (chlorophyllkörnen) as discrete bodies within the cells of green plants.

1838   The word *protein* first appears in the chemical literature in a paper by G. J. Mulder. The term, however, was invented by J. J. Berzelius.

1838–39   M. J. Schleiden and T. Schwann develop the cell theory. Schleiden notes nucleoli within nuclei.

1841   A. Kölliker shows spermatozoa to be sex cells that arise by transformation of cells in the testis.

1845   J. Dzierzon reports that drones hatch from unfertilized eggs, worker and queen bees from fertilized eggs.

1855   A. R. Wallace accumulates evidence in favor of geographical speciation during his studies of the fauna of the Malay Archipelago. He is led to doubt the dogma of the constancy of species and begins to develop a theory of evolution identical to Darwin's.

R. Virchow states the principle that new cells come into being only by division of previously existing cells.

1856   Gregor Mendel, a monk at the Augustinian monastery of St. Thomas in Brünn, Austria (now Brno, Czechoslovakia), begins breeding experiments with the garden pea, *Pisum sativum*.

1858   Darwin and Wallace give a joint presentation to the Linnean Society of London that contains a theory of evolution based upon natural selection.

1859   Darwin publishes *On the Origin of Species*.

1860   T. A. E. Klebs introduces paraffin embedding.

1861   L. Pasteur disproves the theory of spontaneous generation of microorganisms.

1865   Gregor Mendel presents the results and interpretations of his genetic studies on the garden pea at the Brünn Society for the Study of Natural Science at their monthly meetings held February 8 and March 8.

1866   G. Mendel's *Versuche über Pflanzenhybriden (Experiments on Plant Hybridization)* is published and ignored.

1869   F. Galton publishes *Hereditary Genius*. In it he describes a scientific study of human pedigrees from which he concludes that intelligence has a genetic basis.

1870   W. His invents the microtome.

1871   F. Miescher publishes a technique for the isolation of nuclei and reports the discovery of nuclein (now known to be a mixture of nucleic acids and proteins).

1872   J. T. Gulick describes variations in shell coloration among natural populations of land snails living in valleys of Oahu. He suggests that geographical isolation of small populations of such animals may be a necessary prerequisite to the formation of new species.

1873	A. Schneider gives the first account of mitosis.

1873   A. Schneider gives the first account of mitosis.

1874   C. Dareste draws attention to the distinction between monozygotic and dizygotic twins.

1875   F. Galton demonstrates the usefulness of twin studies for elucidating the relative influence of nature (heredity) and nurture (environment) upon behavioral traits.

O. Hertwig concludes from a study of the reproduction of the sea urchin that fertilization in both animals and plants consists of the physical union of the two nuclei contributed by the male and female parents.

E. Strasburger describes cell division in plants.

1876   O. Bütschli describes nuclear dimorphism in ciliates.

1877   H. Fol reports watching the spermatozoan of a starfish penetrate the egg. He was able to see the transfer of the intact nucleus of the sperm into the egg, where it became the male pronucleus.

E. Abbe begins to publish important contributions to the theory of microscopic optics.

1878   W. Kuhne coins the word *enzyme.*

1879   W. Flemming studies mitosis in the epithelium of the tail fin of salamanders. He shows that nuclear division involves a longitudinal splitting of the chromosome and a migration of the sister chromatids to the future daughter nuclei. He also coins the term *chromatin.*

1881   E. G. Balbiani discovers "cross-striped threads" within the salivary gland cells of *Chironomus* larvae. However, he does not realize that these are polytene chromosomes.

R. Koch devises the methods used to this day for isolating pure cultures of bacteria.

1882   W. Flemming discovers lampbrush chromosomes and coins the term *mitosis.*

1883   E. van Beneden studies meiosis in a species of the round worm, *Ascaris,* which (fortunately) has a diploid chromosome number of only four. He shows that the gametes contain half as many chromosomes as the somatic tissues and that the characteristic somatic number is reestablished at fertilization. He also describes fertilization in mammals.

W. Roux suggests that the filaments within the nucleus that stain with basic dyes are the bearers of the hereditary factors.

A. Weismann points out the distinction in animals between the somatic cell line and the germ cells, stressing that only changes in germ cells are transmitted to further generations.

A. F. W. Schimper proposes that chloroplasts are capable of division and that green plants owe their origin to a symbiotic relationship between chlorophyll-containing and colorless organisms.

1887   A. Weismann postulates that a periodic reduction in chromosome number must occur in all sexual organisms.

1888   T. Boveri describes the centriole.

W. Waldeyer coins the word *chromosome* for the filaments referred to by Roux (1883).

1889   F. Galton publishes *Natural Inheritance.* In it he describes the quantitative measurement of metric traits in populations. He thus founds biometry and the statistical study of variation.

1890   R. Altmann reports the presence of "bioblasts" within cells and concludes that they are "elementary organisms" that live as intracellular symbionts and carry out processes vital to their hosts. Later (1898), C. Benda names these organelles *mitochondria.*

E. von Behring shows that blood serum from previously immunized animals contains factors that are specifically lethal to the organisms used for the immunization. These factors are now called antibodies.

1896   E. B. Wilson publishes *The Cell in Development and Heredity.* This influential treatise distills the information gained concerning cytology in the half-century since Schleiden and Schwann put forth the cell theory.

1898   T. Boveri describes chromatin diminution in *Parascaris equorum.*

1899   M. W. Beijerinck demonstrates that tobacco mosaic disease is due to a self-reproducing agent that will pass through bacterial filters and can neither be seen with light microscope nor grown upon bacteriological media. He proposes that this *virus* is a self-reproducing subcellular form of life which represents a hitherto unknown class of organisms.

1900   H. de Vries, C. Correns, and E. Tschermak independently rediscover Mendel's paper. Using several plant species, de Vries and Correns had performed breeding experiments that paralleled Mendel's earlier studies and had independently arrived at similar interpretations of their results. Therefore, upon reading Mendel's publication they immediately recognized its significance. W. Bateson also stresses the importance of Mendel's contribution in an address to the Royal Society of London.

K. Pearson develops the chi-square test.

K. Landsteiner discovers the blood-agglutination phenomenon in man.

P. Ehrlich proposes that antigens and antibodies bind together because they have structural complementarity.

1901	H. de Vries adopts the term *mutation* to describe sudden, spontaneous, drastic alterations in the hereditary material of *Oenothera*.

1901   H. de Vries adopts the term *mutation* to describe sudden, spontaneous, drastic alterations in the hereditary material of *Oenothera*.

T. H. Montgomery studies spermatogenesis in various species of Hemiptera. He concludes that maternal chromosomes only pair with paternal chromosomes during meiosis.

K. Landsteiner demonstrates that humans can be divided into three blood groups: A, B, and C. The designation of the third group was later changed to O.

E. von Behring wins the Nobel Prize for his studies on antiserum therapy.

1902   C. E. McClung notes that in various insect species, equal numbers of two types of spermatozoa are formed; one type contains an "accesssory chromosome" and the other does not. He suggests that the extra chromosome is a sex determinant, and he next argues that sex must be determined at the time of fertilization, not just in insects, but perhaps in other species (including man).

T. Boveri studies the development of haploid, diploid, and aneuploid sea urchin embryos. He finds that in order to develop normally, the organism must have a full set of chromosomes, and he concludes that the individual chromosomes must carry *different* essential hereditary determinants.

W. S. Sutton advances the chromosome theory of heredity, which proposes that the independent assortment of gene pairs stems from the behavior of the synapsed chromosomes during meiosis. Since the direction of segregation of the homologs in a given bivalent is independent of those belonging to any other bivalent, the genes they contain will also be distributed independently.

F. Hofmeister and E. Fischer propose that all proteins are formed by the condensation of amino acids bound through regularly recurrent peptide linkages.

1902–09   W. Bateson introduces the terms *genetics, alleomorph, homozygote, heterozygote,* F_1, F_2, and *epistatic genes.*

1903   W. Waldeyer defines *centromeres* as chromosome regions with which the spindle fibers become associated during mitosis.

1904   A. F. Blakeslee discovers heteromixis in fungi.

1905   L. Cuénot performs crosses between mice carrying a gene that gives them yellow fur. Since they always produce yellow furred and agouti offspring in a 2:1 ratio, he concludes they are heterozygous. W. E. Castle and C. C. Little show in 1910 that yellow homozygotes die *in utero*. This dominant allele in the agouti series (A^y) is thus the first gene shown to behave as a homozygous lethal.

1906   W. Bateson and R. C. Punnett report the first case of linkage (in the sweet pea).

1907   R. G. Harrison cultures fragments of the central nervous systems of frogs in hemolymph and observes the outgrowth of nerve fibers. In so doing, he invents tissue culture.

E. F. Smith shows that a specific bacterium *Agrobacterium tumefaciens* is responsible for crown gall disease.

1908   G. H. Hardy and W. Weinberg, working independently, formulate the so-called Hardy-Weinberg law of population genetics.

1909   G. H. Shull advocates the use of self-fertilized lines in production of commercial seed corn. The hybrid corn program that resulted, created an abundance of foodstuffs worth billions of dollars.

A. E. Garrod publishes *Inborn Errors of Metabolism*, the earliest discussion of the biochemical genetics of man (or any other species).

F. A. Janssens suggests that exchanges between nonsister chromatids produce chiasmata.

C. C. Little initiates a breeding program that produces the first inbred strain of mice (the strain now called DBA).

W. Johannsen's studies of the inheritance of seed size in self-fertilized lines of beans leads him to realize the necessity of distinguishing between the appearance of an organism and its genetic constitution. He invents the terms *phenotype* and *genotype* to serve this purpose, and he also coins the word *gene*.

C. Correns and E. Bauer study the inheritance of chloroplast defects in variegated plants, such as *Mirabilis jalapa* and *Pelargonium zonale*. They find that the inability to form healthy chloroplasts is in some cases inherited in a non-Mendelian fashion.

H. Nilsson Ehle puts forward the multiple-factor hypothesis to explain the quantitative inheritance of seed-coat color in wheat.

1910   T. H. Morgan discovers white eye and consequently sex linkage in *Drosophila. Drosophila* genetics begins.

W. Weinberg develops the methods used for correcting expectations for Mendelian segregation from human pedigree data under different kinds of ascertainment applied to data from small families.

P. Rous shows that injection of a cell-free filtrate from chicken sarcomas induces new sarcomas in recipient chickens.

| 1911 | T. H. Morgan proposes that the genes for white eyes, yellow body, and miniature wings in *Drosophila* are linked together on the X chromosome.
| | W. R. B. Robertson points out that a metacentric chromosome in one orthopteran species may correspond to two acrocentrics in another and concludes that during evolution metacentrics may arise by the fusion of acrocentrics. Whole-arm fusions are called *Robertsonian translocations* in his honor.
| 1912 | A. Wegener proposes the continental drift concept.
| | F. Rambousek suggests that the "cross-striped threads" within the salivary gland cells of fly maggots are chromosomes.
| | T. H. Morgan demonstrates that crossing over does not take place in the male of *Drosophila melanogaster*. He also discovers the first sex-linked lethal.
| 1913 | Y. Tanaka reports that crossing over does not take place in the female of *Bombyx mori*. In this species, the female is the heterogametic sex.
| | W. H. Bragg and W. L. Bragg demonstrate that the analysis of x-ray diffraction patterns can be used to determine the three-dimensional atomic structure of crystals.
| | A. H. Sturtevant provides the experimental basis for the linkage concept in *Drosophila* and produces the first genetic map.
| 1914 | C. B. Bridges discovers meiotic nondisjunction in *Drosophila*.
| | C. C. Little postulates that the acceptance or rejection of transplanted tumors in mice has a genetic basis.
| 1915 | F. W. Twort isolates the first filterable bacterial virus.
| | R. B. Goldschmidt coins the term *intersex* to describe the aberrant sexual types arising from crosses between certain different races of the gypsy moth, *Lymantria dispar*.
| | J. B. S. Haldane, A. D. Sprunt, and N. M. Haldane describe the first example of linkage in vertebrates (mice).
| 1916 | H. J. Muller discovers interference in *Drosophila*.
| 1917 | F. d'Herelle coins the term *bacteriophage* and develops methods for assaying virus titre.
| | O. Winge calls attention to the important role of polyploidy in the evolution of angiosperms.
| | C. B. Bridges discovers the first chromosome deficiency in *Drosophila*.
| 1918 | H. Spemann and H. Mangold demonstrate that a living part of an embryo can exert a morphogenetic stimulus upon another part, bringing about its morphological differentiation (embryonic induction). They thus discover and name the *organizer*.
| | H. J. Muller discovers the balanced lethal phenomenon in *Drosophila*.
| 1919 | T. H. Morgan calls attention to the equality in *Drosophila melanogaster* between the number of linkage groups and the haploid number of chromosomes.
| | C. B. Bridges discovers chromosomal duplications in *Drosophila*.
| 1920 | A. F. Blakeslee, J. Belling, and M. E. Farnham describe trisomics in the Jimson weed, *Datura stramonium*.
| 1921 | F. G. Banting and C. H. Best isolate insulin and study its physiological properties.
| | R. B. Goldschmidt publishes the first genetic analysis and discussion of the evolutionary implications of industrial melanism.
| | C. B. Bridges reports the first monosomic (haplo-4) in *Drosophila*.
| 1922 | L. V. Morgan discovers attached-X chromosomes in *Drosophila*.
| | A. F. Blakeslee, J. Belling, M. E. Farnham, and A. D. Bergner discover a haploid *Datura*.
| 1923 | C. B. Bridges discovers chromosomal translocations in *Drosophila*.
| | R. Feulgen and H. Rossenbeck describe the cytochemical test that currently is most used for DNA localization.
| | T. Svedberg builds the first ultracentrifuge.
| | A. E. Boycott and C. Diver describe "delayed" Mendelian inheritance controlling the direction of the coiling of the shell in the snail *Limnea peregra*. A. H. Sturtevant suggests that the direction of coiling of the *Limnea* shell is determined by the character of the ooplasm, which is in turn controlled by the mother's genotype.
| | The XX-XY type of sex determination is demonstrated for certain dioecious plants: for *Elodea* by J. K. Santos, for *Rumex* by H. Kihara and T. Ono, and for *Humulus* by O. Winge.
| 1925 | C. B. Bridges completes his cytogenetic analysis of the aneuploid offspring of triploid *Drosophila*, and he defines the relations between the sex chromosomes and autosomes that control sexual phenotype.
| | E. M. East and A. J. Mangelsdorf propose the first satisfactory interpretation of the phenomenon of self-sterility in flowering plants.
| | A. H. Sturtevant analyzes the *Bar* phenomenon in *Drosophila* and discovers position effect.

F. Bernstein suggests that the A B O blood groups are determined by a series of allelic genes.

T. H. Goodspeed and R. E. Clausen produce an amphidiploid in *Nicotiana*.

1926   E. G. Anderson establishes that the centromere of the X chromosome of *Drosophila* is at the end opposite the locus of *yellow*.

S. S. Chetverikov initiates the genetic analysis of wild populations of *Drosophila*.

J. B. Sumner isolates the first enzyme in crystalline form and shows it (urease), to be a protein.

A. H. Sturtevant finds the first inversion in *Drosophila*.

R. E. Clausen and T. H. Goodspeed describe the first analysis of monosomics in a plant *(Nicotiana)*.

1927   K. M. Bauer reports that the rejection of skin grafts does not occur when skin is transplanted from one monozygous twin to another.

J. Belling proposes that interchanges between nonhomologous chromosomes result in ring formations at meiosis.

J. B. S. Haldane suggests that the genes known to control certain coat colors in various rodents and carnivores may be evolutionarily homologous.

J. Belling introduces the acetocarmine technique for staining chromosome squashes.

B. O. Dodge initiates genetic studies on *Neurospora*.

H. J. Muller reports the artificial induction of mutations in *Drosophila* by x-rays.

1928   L. J. Stadler reports the artificial induction of mutations in maize, and demonstrates that the dose-frequency curve is linear.

F. Griffith discovers type-transformation of pneumococci. This lays the foundation for the work of Avery, MacLeod, and McCarthy (1944).

L. F. Randolph distinguishes supernumerary chromosomes from the normal chromosomes of the plant cell. He calls the normal ones "A chromosomes" and the supernumerary ones "B chromosomes."

E. Heitz introduces the term *heterochromatin*.

1929   A. Fleming reports that a mold of genus *Penicillium* secretes a substance that prevents the growth of certain bacteria. He names this antibacterial substance *penicillin*.

C. D. Darlington is the first to suggest that chiasmata function to hold homologs together at meiotic metaphase I and so ensure that they pass to opposite poles at anaphase I.

R. C. Tryon demonstrates successful selection for rate of maze learning in the rat.

1930   R. E. Cleland and A. F. Blakeslee demonstrate that the peculiar patterns of the transmission of groups of genes in various *Oenothera* races result from a system of balanced lethal and reciprocal translocation complexes.

K. Landsteiner receives a Nobel Prize for his studies in immunology.

From 1930 to 1932, a group of books and papers were published by R. A. Fisher, J. B. S. Haldane and S. Wright that constitute the mathematical foundation of population genetics.

1931   C. Stern, and independently H. B. Creighton and B. McClintock, provide the cytological proof of crossing over.

C. D. Darlington suggests that chiasmata can move to the ends of bivalents without breakage of the chromosomes. This process of *terminalization*, as he called it, is now known to occur in some species but not in others.

B. McClintock shows in maize that if a segment of a chromosome has become inverted, individuals heterozygous for such an inversion often show reversed pairing at pachynema.

1932   M. Knoll and E. Ruska invent the prototype of the modern electron microscope.

1933   T. S. Painter initiates cytogenetic studies on the salivary gland chromosomes of *Drosophila*.

H. Hashimoto works out the chromosomal control of sex determination for *Bombyx mori*.

A. W. K. Tiselius reports the invention of an apparatus that permits the separation of charged molecules by electrophoresis.

B. McClintock demonstrates in maize that a single exchange within the inversion loop of a paracentric inversion heterozygote generates an acentric and a dicentric chromatid.

T. H. Morgan receives a Nobel Prize for his development of the theory of the gene.

1934   M. Schlesinger reports that certain bacteriophages are composed of DNA and protein.

P. L'Héritier and G. Teissier experimentally demonstrate the disappearance of a deleterious gene from populations of *Drosophila melanogaster* maintained in population cages for many generations.

A. Følling discovers phenylketonuria, the first hereditary metabolic disorder shown to be responsible for mental retardation.

H. Bauer postulates that the giant chromosomes of the salivary gland cells of fly larvae are polytene.

B. McClintock shows that the nucleolus organizer of *Zea mays* can be split by a translocation and that each piece is capable of organizing a separate nucleolus. She thus sets the stage for the later demonstration (1965) that the genes for rRNA are present in multiple copies.

1935   J. B. S. Haldane is the first to calculate the spontaneous mutation frequency of a human gene.

E. Klenk identifies the glycolipid that accumulates in the brain of patients with the Tay-Sachs disease as a ganglioside.

F. Zernicke describes the principle of the phase microscope.

G. W. Beadle and B. Ephrussi and A. Kuhn and A. Butenandt work out the biochemical genetics of eye-pigment synthesis in *Drosophila* and *Ephestia,* respectively.

W. M. Stanley succeeds in isolating and crystallizing the tobacco mosaic virus.

C. B. Bridges publishes the salivary gland chromosome maps for *Drosophila melanogaster.*

H. Spemann receives a Nobel Prize for his studies on embryonic induction.

1936   J. Schultz notes the relation of the mosaic expression of a gene in *Drosophila* to its position relative to heterochromatin.

T. Caspersson uses cytospectrophotometric methods to investigate the quantitative chemical composition of cells.

J. J. Bittner shows that mammary carcinomas in mice can be caused by a viruslike factor transmitted through the mother's milk.

A. H. Sturtevant and T. Dobzhansky publish the first account of the use of inversions in constructing a chromosomal phylogenetic tree.

C. Stern discovers somatic crossing over in *Drosophila.*

R. Scott-Moncrieff reviews the inheritance of plant pigments. The major part of this work was done by a group of English geneticists at the John Innes Horticultural Institute, and these early workers established that gene substitutions resulted in chemical changes in certain flavanoid and carotenoid pigments.

1937   T. Dobzhansky publishes *Genetics and the Origin of Species.* a milestone in evolutionary genetics.

A. F. Blakeslee and A. G. Avery report that colchicine induces polyploidy.

T. M. Sonneborn discovers mating types in *Paramecium.*

F. C. Bawden and N. W. Pirie show that the tobacco mosaic virus, although being made mostly of protein, also contains a small amount (about 5%) of RNA.

P. A. Gorer discovers the first histocompatibility antigens in the laboratory mouse.

H. Karström points out that the synthesis of certain bacterial enzymes is stimulated when the substrates these enzymes attack are added to the medium. He coins the term "adaptive enzymes" for these and differentiates them from the "constitutive enzymes" that are always formed irrespective of the composition of the medium.

P. L'Héritier and G. Teissier demonstrate frequency-dependent selection of mutants in laboratory populations of *Drosophila melanogaster.*

E. Chatton stresses the fundamental differences between the group of organisms comprising the bacteria and blue-green algae, which he named *prokaryotes,* and all other living organisms, which he called *eukaryotes.*

1938   B. McClintock describes the bridge-breakage-fusion-bridge cycle.

T. M. Sonneborn discovers the killer factor of *Paramecium.*

M. M. Rhoades describes the mutator gene *Dt* in maize.

H. Slizynska makes a cytological analysis of several overlapping *Notch* deficiencies in the salivary gland X chromosomes of *Drosophila melanogaster* and determines the band locations of the *w* and *N* genes.

1939   E. L. Ellis and M. Delbrück perform studies on coliphage growth that mark the beginning of modern phage work. They devise the "one-step growth" experiment, which demonstrates that after the phage adsorbs onto the bacterium, it replicates within the bacterium during the "latent period," and finally the progeny are released in a "burst."

P. Levine and R. E. Stetson discover maternal immunization by a fetus carrying a new blood group antigen inherited from the father. This antigen is subsequently identified with the human Rh blood group system as the cause of erythroblastosis fetalis.

A. W. K. Tiselius and E. A. Kabat demonstrate that antibodies belong to the gamma class of serum globulins.

E. Knapp and H. Schreiber demonstrate that the effectiveness of ultraviolet light in inducing mutations in *Sphaerocarpus donnelli* corresponds to the absorption spectrum of nucleic acid.

1940   W. Earle establishes the strain L permanent cell line from a C3H mouse.

H. W. Florey, E. Chain, and five colleagues successfully extract and purify penicillin. They show in experiments with mice that it is by far the most powerful chemotherapeutic agent then known against bacterial infections.

1941   G. W. Beadle and E. L. Tatum publish their classic study on the biochemical genetics of *Neurospora* and promulgate the one gene-one enzyme theory.

J. Brachet and T. Caspersson independently reach the conclusion that RNA is localized in the nucleoli and cytoplasm and that a cell's content of RNA is directly linked with its protein synthesizing capacity.

A. J. P. Martin and R. L. M. Synge develop the technique of partition chromatography and use it to determine the amino acids in protein hydrolyzates.

A. H. Coons, H. J. Creech, and R. N. Jones develop immunofluorescence techniques to demonstrate the presence of antibody-reactive sites on specific cells.

K. Mather coins the term *polygenes* and describes polygenic traits in various organisms.

1942   R. Schoenheimer publishes *The Dynamic State of Body Constituents* and describes the use of isotopically tagged compounds in metabolic studies. He introduces the concepts of "metabolic pools" and the "turnover" of the organic compounds in cells.

S. E. Luria and T. F. Anderson publish the first electron micrographs of bacterial viruses. T2 has a polyhedral body and a tail!

G. D. Snell sets out to develop highly inbred strains of mice to study the genes responsible for graft rejection.

1943   A. Claude isolates and names the microsome fraction and shows that it contains the majority of the RNA of cells.

S. E. Luria and M. Delbrück initiate the field of bacterial genetics when they demonstrate unambiguously that bacteria undergo spontaneous mutation.

1944   O. T. Avery, C. M. MacLeod, and M. McCarty describe the pneumococcus transforming principle. The fact that it is rich in DNA suggests that DNA and not protein is the hereditary chemical.

T. Dobzhansky describes the phylogeny of the gene arrangements in the third chromosome of *Drosophila pseudoobscura* and *D. persimilis.*

E. L. Tatum, D. Bonner, and G. W. Beadle use mutant strains of *Neurospora crassa* to work out the intermediate steps in the synthesis of tryptophan.

1945   R. R. Humphrey demonstrates that the female is the heterogametic sex in urodeles.

M. J. D. White publishes *Animal Cytology and Evolution,* the first monograph to review progress in the study of the evolutionary cytogenetics of animals.

S. E. Luria demonstrates that mutations occur in bacterial viruses.

E. B. Lewis describes the stable position effect phenomenon in *Drosophila.*

R. D. Owen reports that in cattle dizygotic twins are born with, and often retain throughout life, a stable mixture of each other's red blood cells. This chimerism, which results from vascular anastomoses within the chorions of the fetuses, provides the first example of immune tolerance.

A. Fleming, E. B. Chain, and H. W. Florey receive the Nobel Prize for the discovery, purification, and chemical characterization of penicillin.

1946   A. Claude introduces cell fractionation techniques based upon differential centrifugation and works out methods for characterizing the fractions biochemically.

Genetic recombination in bacteriophage is demonstrated by M. Delbrück and W. T. Bailey and by A. D. Hershey.

J. Lederberg and E. L. Tatum demonstrate genetic recombination in bacteria.

C. Auerbach and J. M. Robson report that mustard gas induces mutations in *Drosophila.* J. A. Rapoport demonstrates the mutagenic effectiveness of formaldehyde in *Drosophila.*

J. Oudin develops the gel-diffusion, antigen-antibody precipitation test that bears his name.

Nobel prizes are awarded to H. J. Muller for his contributions to radiation genetics; to J. B. Sumner for crystallizing enzymes; and to W. M. Stanley for his studies on the purification and chemical characterization of viruses.

1948   A. Boivin, R. Vendrely, and C. Vendrely show that in the different cells of an organism the quantity of DNA for each haploid set of chromosomes is constant.

H. K. Mitchell and J. Lein show that tryptophan synthetase is missing in certain mutant strains of *Neurospora.* This finding constitutes the first direct evidence for the one gene–one enzyme theory.

P. A. Gorer, S. Lyman, and G. D. Snell discover the major histocompatibility locus in the mouse. It resides on chromosome 17 and is named H2.

O. Ouchterlony develops the double-diffusion antigen-antibody precipitation test that bears his name.

H. J. Muller coins the term *dosage compensation.*

J. Lederberg and N. Zinder, and, independently, B. D. Davis develop the penicillin selection technique for isolating biochemically deficient bacterial mutants.

J. Clausen, D. D. Keck, and W. M. Hiesey describe the genetic structure of ecotypes among species of herbaceous plants along an altitudinal transect in the Sierra Nevada mountains of California.

G. D. Snell introduces the term *histocompatibility gene* and formulates the laws of transplantation acceptance and rejection.

1949   B. Ephrussi, H. Hottinger, and A. M. Chimenes discover the *petite* cytoplasmic mutation in yeast.

A. D. Hershey and R. Rotman demonstrate that genetic recombination occurs in bacteriophage.

M. M. Green and K. C. Green demonstrate that the *lozenge* locus in *Drosophila* can be divided into three subloci.

A. Kelner discovers photoreactivation of potential UV damage by visible light in *Saccharomyces.*

J. V. Neel provides genetic evidence that the sickle-cell disease is inherited as a simple Mendelian autosomal recessive.

L. Pauling, H. A. Itano, S. J. Singer, and I. C. Wells show that the H^S gene produces an abnormal hemoglobin.

M. L. Barr and E. G. Bertram demonstrate that the sex chromatin is morphologically different in the neurons of male and female cats.

1950   B. McClintock discovers the *Ac, Ds* system of transposable elements in maize.

E. Chargaff lays the foundations for nucleic acid structural studies by his analytical work. He demonstrates for DNA that the numbers of adenine and thymine groups are always equal and so are the numbers of guanine and cytosine groups. These findings later suggest to Watson and Crick that DNA consists of two polynucleotide strands joined by hydrogen bonding between A and T and between G and C.

A. Lwoff and A. Gutman study a lysogenic strain of *Bacillus megatherium* and demonstrate that each bacterium harbors a noninfectious form of a virus that gives the host the capacity to generate new phage without the intervention of exogenous bacteriophages. They propose the term "prophage" for this noninfectious phase. Together with L. Siminovitch and N. Kjeldgaard, Lwoff shows that prophage can be induced by ultraviolet light to produce infective virus.

E. M. Lederberg discovers lambda, the first viral episome of *E. coli.*

H. Latta and J. F. Hartmann introduce glass knives for ultramicrotomy.

1951   G. Gey establishes the human HeLa permanent cell line.

J. Mohr is the first to demonstrate autosomal linkage in humans (between the genes specifying the Lewis and Lutheran blood groups).

C. Stormont, R. D. Owen, and M. R. Irwin describe serological cross-reactions in the multiple allelic *B* and *C* blood group systems of cattle.

Y. Chiba demonstrates the presence of DNA in chloroplasts using the Feulgen-staining cytochemical technique.

N. H. Horowitz and U. Leupold generate large populations of temperature-sensitive mutations to determine what percentage of the genes in *E. coli* and *N. crassa* perform functions that are indispensable. The values obtained were 23% and 46%, respectively.

C. Petit reports the existence of a minority-genotype advantage in populations of *Drosophila melanogaster* and points out that this phenomenon can lead to frequency-dependent selection and stable polymorphism.

1952   G. E. Palade publishes the first high-resolution electron micrographs of mitochondria.

R. Dulbecco adapts the techniques of bacterial virology to study animal viruses. He counts plaques made by western equine encephalomyelitis virus on monolayers of cells obtained from chick embryos.

D. Mazia and K. Dan isolate the sea urchin mitotic apparatus and start work on its biochemical characterization.

N. D. Zinder and J. Lederberg describe transduction in *Salmonella.*

J. Lederberg and E. M. Lederberg invent the replica plating technique.

J. T. Patterson and W. S. Stone publish *Evolution in the Genus Drosophila,* which summarizes an encyclopedic body of information dealing with the chromosomal evolution of this most studied genus of flies.

A. H. Bradshaw reports that certain populations of grasses living near mine entrances in Great Britain can tolerate high concentrations of heavy metals (copper, lead, zinc). This is evidence for the recent natural selection of tolerant genotypes.

R. Briggs and T. J. King transplant living nuclei from blastula cells into enucleated frogs' eggs. They later demonstrate that transplanted nuclei undergo chromosome changes.

W. Beermann observes stage and tissue specificities in the puffing patterns of polytene chromosomes and suggests that these are the phenotypic reflections of differential gene activities.

F. Sanger and his colleagues work out the complete amino acid sequence for the protein hormone insulin, and show that it contains two polypeptide chains held together by disulfide bridges.

A. D. Hershey and M. Chase demonstrate that the DNA of phage enters the host, whereas most of the protein remains behind.

D. M. Brown and A. Todd demonstrate that DNA and RNA are 3′-5′ linked polynucleotides.

G. Pontecorvo and J. A. Roper describe the parasexual cycle in *Aspergillus nidulans.*

A.J.P. Martin and R.L.M. Synge receive the Nobel Prize in chemistry for their invention of chromatographic separation techniques.

1953   J. D. Watson and F.H.C. Crick propose a model for DNA comprised of two helically intertwined chains tied together by hydrogen bonds between the purines and pyrimidines.

C. C. Lindegren discovers gene conversion in *Saccharomyces.*

A. Howard and S. R. Pelc demonstrate by autoradiography that during the cell-division cycle of plants there exists a period following mitosis during which DNA synthesis does not take place (G_1), a subsequent period of DNA synthesis during which the DNA content of the interphase nucleus is doubled (S), a second growth period (G_2), and then mitosis.

W. Hayes discovers polarized behavior in bacterial recombinations. He isolates the *Hfr H* strain of *E. coli* and shows that certain genes are readily transferred from *Hfr* to F⁻ bacteria, whereas others are not.

R. E. Billingham, L. Brent, and P. B. Medawar show that immunological tolerance can be produced experimentally.

Porter-Blum and Sjöstrand ultramicrotomes become commercially available.

J. B. Finean, F. S. Sjöstrand, and E. Steinmann publish the first electron micrographs of sectioned chloroplasts.

K. R. Porter discovers and names the *endoplasmic reticulum* and identifies it as the source of cytoplasmic basophilia.

G. D. Snell finds that the major histocompatibility complex of the mouse (H-2) is composed of multiple loci.

N. Visconti and M. Delbrück put forth a hypothesis to explain genetic recombination in bacteriophages.

1954   A. J. Dalton and M. D. Felix provide the first detailed description of the ultrastructure of the Golgi complex.

A. C. Allison provides evidence that individuals heterozygous for the sickle-cell gene are protected against subtertian malaria infection. This is the first case of genetic balanced polymorphism described in a human population.

J. Dausset observes that some patients who had received multiple blood transfusions produced antibodies against antigens found on the white blood cells of other individuals but not against those of their own cells. These antibodies defined the first HLA antigens and led to the definition of the human histocompatibility system.

E. S. Barghoorn and S. A. Tyler report finding fossils of filamentous and coccoid microorganisms in sedimentary rocks over two billion years old. This discovery demonstrates that life existed in the Proterozoic era.

E. Mayr advances the peripatric speciation concept.

1955   M. B. Hoagland obtains cell-free preparations that synthesize protein.

S. Benzer works out the fine structure of the *r* II region of phage T_4 of *E. coli,* and coins the terms *cistron, recon,* and *muton.*

H. Fraenkel-Conrat and R. C. Williams reconstitute "hybrid" tobacco mosaic virus from nucleic acid and protein components arising from different sources.

O. Smithies uses starch-gel electrophoresis to identify plasma protein polymorphisms.

N. K. Jerne puts forth the natural-selection theory of antibody formation. According to this proposal, antibody molecules are already present in the host, having developed during fetal life. An invading foreign antigen selects the antibody molecule that provides the best fit and binds to it. The formation of this complex stimulates the further production of the selected antibody. These concepts are incorporated into later clonal selection theories.

M. Grunberg-Manago and S. Ochoa isolate the first enzyme involved in the synthesis of a nucleic acid–polynucleotide phosphorylase.

R. H. Pritchard studies the linear arrangement of a series of allelic adenine-requiring mutants of *Asper-*

gillus. He concludes that crossing over can occur between different alleles of the same gene, provided they are characterized by mutations at different subsites.

C. de Duve and four colleagues describe intracellular vesicles that contain hydrolytic enzymes and name them *lysosomes.*

P. Grabar and C. A. Williams devise the technique of immunoelectrophoresis to analyze complex mixtures of antigenic molecules.

1956   H. B. Kettlewell studies industrial melanism in the peppered moth. He demonstrates that moths that are conspicuous in their habitats are indeed eaten by birds more often than inconspicuous forms.

F. Jacob and E. L. Wollman are able experimentally to interrupt the mating process in *E. coli* and show that a piece of DNA is inserted from the donor bacterium into the recipient.

Groups led by S. Ochoa and A. Kornberg succeed in the *in vitro* enzymatic synthesis of polymers of ribonucleotides and deoxyribonucleotides, respectively.

J. H. Tjio and A. Levan demonstrate that the diploid chromosome number for man is 46.

C. O. Miller and his co-workers isolate and determine the chemical structure of kinetin, a substance promoting cell division in plants.

T. T. Puck, S. J. Cieciura, and P. I. Marcus succeed in growing clones of human cells *in vitro.*

G. E. Palade and P. Siekevitz isolate ribosomes.

M. J. Moses and D. Fawcett independently observe synaptonemal complexes in spermatocytes.

A. Gierer and G. Schramm and H. Fraenkel-Conrat demonstrate independently that a chemically pure nucleic acid, namely tobacco mosaic virus RNA, is infectious and genetically competent.

1957   S. A. Berson and R. S. Yalow report the first use of the radioimmunoassay procedure for the detection of insulin antibodies developed by patients in response to the administration of exogenous insulin.

J. H. Taylor, P. S. Woods, and W. L. Hughes are the first to use tritiated thymidine in high resolution autoradiography in experiments that demonstrate the semiconservative distribution of label during chromosome replication in *Vicia faba.*

V. M. Ingram reports that normal and sickle-cell hemoglobin differ by a single amino acid substitution.

A. Todd receives the Nobel Prize for his studies on the structure of nucleosides and nucleotides.

1958   F. Jacob and E. L. Wollman demonstrate that the single linkage group of *E. coli* is circular and suggest that the different linkage groups found in different *Hfr* strains result from the insertion at different points of a factor in the circular linkage group that determines the rupture of the circle.

F.H.C. Crick suggests that during protein formation the amino acid is carried to the template by an adaptor molecule containing nucleotides and that the adaptor is the part that actually fits on the RNA template. Crick thus predicts the discovery of transfer RNA.

P. C. Zamecnik and his colleagues characterize amino acid-transfer RNA complexes.

H. G. Callan and H. G. MacGregor demonstrate that the linear integrity of chromatids of amphibian lampbrush chromosomes is maintained by DNA, not protein.

M. Okamoto and, independently, R. Riley and V. Chapman discover genes that control the pairing of homoeologous chromosomes in wheat.

F. C. Steward, M. O. Mapes, and K. Mears succeed in rearing sexually mature plants from single diploid cells derived from the secondary phloem of roots of the wild carrot, *Daucus carota.* They conclude that each cell of the multicellular organism has all the ingredients necessary for the formation of the complete organism.

M. Meselson and F. W. Stahl use the density gradient equilibrium centrifugation technique to demonstrate the semiconservative distribution of density label during DNA replication in *E. coli.*

Nobel Prizes are awarded to G. W. Beadle, E. L. Tatum, and J. Lederberg for their contributions to genetics and to F. Sanger for his contributions to protein chemistry.

1959   J. Lejeune, M. Gautier, and R. Turpin show that Down syndrome is a chromosomal aberration involving trisomy of a small telocentric chromosome.

C. E. Ford, K. W. Jones, P. E. Polani, J. C. de Almeida, and J. Briggs discover that females suffering from Turner syndrome are XO.

P. A. Jacobs and J. A. Strong demonstrate that males suffering from Klinefelter syndrome are XXY.

S. J. Singer conjugates ferritin with immunoglobulin to produce a labeled antibody that is readily recognized under the electron microscope.

R. L. Sinsheimer demonstrates that bacteriophage phiX174 of *E. coli* contains a single-stranded DNA molecule.

F. M. Burnet improves Jerne's selective theory of antibody formation by suggesting that the antigen stimulates the proliferation of only those cells that are genetically programmed to synthesize the complementary antibodies.

G. M. Edelman resolves immunoglobulin G into heavy and light chains.

A. Lima-de-Faria demonstrates by autoradiography that heterochromatin replicates later than euchromatin.

M. Chèvremont, S. Chèvremont-Comhaire, and E. Baeckeland demonstrate DNA in mitochondria using a combination of autoradiographic and Feulgen-staining techniques.

K. McQuillen, R. B. Roberts, and R. J. Britten demonstrate in *E. coli* that ribosomes are the sites where protein synthesis takes place.

E. Freese proposes that mutation can occur as the result of single, base-pair changes in DNA. He coins the terms *transitions* and *transversions*.

C. Pelling finds selective labeling of puffed regions of polytene chromosomes after they are incubated in a nutrient solution containing ^{3}H uridine.

R. H. Whittaker suggests the grouping of organisms into five kingdoms: the bacteria, the eukaryotic microorganisms, animals, plants, and fungi.

S. Brenner and R. W. Horne develop the negative staining procedure for electron microscopy of subcellular particles.

S. Ochoa and A. Kornberg receive Nobel Prizes for their studies on the *in vitro* synthesis of nucleic acids.

1960   P. Nowell discovers phytohemagglutinin and demonstrates its use in stimulating mitoses in human leukocyte cultures.

P. Siekevitz and G. E. Palade describe the synthesis of secretory proteins on membrane-bound ribosomes.

P. Doty, J. Marmur, J. Eigner, and C. Schildkraut demonstrate that complementary strands of DNA molecules can be separated and recombined.

G. Barski, S. Sorieul, and F. Cornefert report the first successful *in vitro* hybridization of mammalian cells.

U. Clever and P. Karlson experimentally induce specific puffing patterns in polytene chromosomes by injecting *Chironomus* larvae with ecdysone.

J. C. Kendrew *et al.* determine the three-dimensional structure of myoglobin at 2 Å resolution.

M. F. Perutz *et al.* determine the three dimensional structure of hemoglobin at 5.5 Å resolution.

P. B. Medawar and F. M. Burnet receive a Nobel Prize for their studies on immunological tolerance.

1961   F. Jacob and J. Monod publish "Genetic regulatory mechanisms in the synthesis of proteins," a paper in which the theory of the operon is developed.

F. Jacob and J. Monod suggest that ribosomes do not contain the template responsible for the orderly assembly of amino acids. They propose that instead each DNA cistron causes synthesis of an RNA molecule of limited life span that harbors the amino acid sequence information in its nucleotide sequence. This molecule subsequently enters into temporary association with a ribosome and so confers upon it the ability to synthesize a given protein. This messenger RNA is subsequently demonstrated by S. Brenner, F. Jacob, and M. Meselson and by F. Gros, W. Gilbert, H. Hiatt, C. G. Kurland, and J. D. Watson.

M. F. Lyon and L. B. Russell independently provide evidence suggesting that in mammals one X chromosome is inactivated in some embryonic cells and their descendants, that the other is inactivated in the rest, and that mammalian females are consequently X-chromosome mosaics.

J. Josse, A. D. Kaiser, and A. Kornberg demonstrate that there is a difference in polarity between the complementary strands of the DNA helix, so that the sugars of one strand are oriented in a direction opposite to those in the other strand.

V. M. Ingram presents a theory explaining the evolution of the four known kinds of hemoglobin chains from a single primitive myoglobinlike heme protein by gene duplication and translocation.

B. D. Hall and S. Spiegelman demonstrate that hybrid molecules can be formed containing one single-stranded DNA and one RNA molecule which are complementary in base sequence. Their technique opens the way to the isolation and characterization of messenger RNAs.

S. B. Weiss and T. Nakamoto isolate RNA polymerase.

G. von Ehrenstein and F. Lipmann combine messenger RNA and ribosomes from rabbit reticulocytes with amino acid-transfer RNA complexes derived from *E. coli*. Since this cell-free system synthesized a protein similar to rabbit hemoglobin, they conclude that the genetic code is universal.

F.H.C. Crick, L. Barnett, S. Brenner, and R. J. Watts-Tobin show that the genetic language is made up of three-letter words.

W. Beermann demonstrates that a puffing locus on a *Chironomus* polytene chromosome is inherited in a Mendelian fashion.

M. W. Nirenberg and J. H. Matthaei develop a cell-free system from *E. coli* that incorporates amino acids into protein when supplied with template RNA preparations. They show that the synthetic polynucleotide, polyuridylic acid, directs the synthesis of a protein resembling polyphenylalanine.

M. Meselson and J. J. Weigle demonstrate in phage lambda that recombination involves breakage and reunion (but not replication) of the chromosome.

H. Dintzis shows that the direction of synthesis of the hemoglobin molecule is from amino to carboxyl termini.

H. Moor, K. Muhlenthaler, H. Waldner, and A. Frey-Wyssling develop the first freeze-fracture procedure that permits ultrastructual observations impossible with conventional sectioning methods.

U. Z. Littauer shows that ribosomes contain only two high-molecular-weight species of RNA, with sedimentation values of 16S and 23S in bacteria and 18S and about 28S in animals.

C. Tokunaga demonstrates that the *engrailed* gene of *Drosophila melanogaster* causes a shift from one developmental prepattern to a different but related prepattern.

J. P. Waller and J. I. Harris find that bacterial ribosomes contain a large number of different proteins.

1962   H. Ris and W. Plaut show by electron microscopy that chloroplasts contain DNA.

E. Zuckerkandl and L. Pauling calculate the approximate times of derivation of different hemoglobin chains from their common ancestors during eukaryotic evolution.

F. M. Ritossa reports that the salivary gland chromosomes of *Drosophila buskii* respond to heat shocks by puffing.

The distinction between T and B lymphocytes is shown in publications by J. F.A.P. Miller, R. A. Good *et al.,* and N. L. Warner *et al.*

R. R. Porter uses enzymes to cleave immunoglobulin molecules. He demonstrates that each molecule has two antigen-binding portions (F_{ab}) and a crystallizable segment (F_c) that does not bind antigen. He shows that the heavy and light chains are present in 1:1 ratio and suggests the four-chain model.

D. A. Rodgers and G. E. McClearn discover differences between mouse strains in alcohol preference.

U. Henning and C. Yanofsky show that amino acid replacements can arise from crossing over within triplets.

J. B. Gurdon reports that a normal fertile frog can develop from an enucleated egg injected with a nucleus from an intestinal cell. This experiment demonstrates that somatic and germinal nuclei are qualitatively equivalent.

Polyribosomes are discovered independently in three laboratories (by A. Gierer, by J. R. Warner, A. Rich, and C. E. Hall, and by T. Staehelin and H. Noll).

A. M. Campbell proposes that episomes become integrated into host chromosomes by a crossover event resembling the exchanges that were previously reported between synapsed ring- and rod-shaped chromosomes in eukaryotes.

Nobel prizes are awarded to J. D. Watson, F.H.C. Crick, and M.H.F. Wilkins for their studies on the structure of DNA and to M. F. Perutz and J. C. Kendrew for their studies on the structure of hemoglobin and myoglobin.

1963   B. B. Levine, A. Ojida, and B. Benacerraf publish the first paper on the immune-response genes of guinea pigs.

R. Rosset and R. Monier discover 5S RNA.

T. Okamoto and M. Takanami show that mRNA binds to the small ribosomal subunit.

H. Noll, T. Staehelin, and F. O. Wettstein demonstrate the tape mechanism of protein synthesis.

J. G. Gall produces evidence that the lampbrush chromatid contains a single DNA double helix.

B. J. McCarthy and E. T. Bolton use their DNA-agar technique to measure genetic relatedness between diverse species of organisms.

E. Hadorn demonstrates allotypic differentiation in cultured imaginal discs of *Drosophila*.

J. Monod and S. Brenner publish the replicon model.

R. Sager and M. R. Ishida isolate chloroplast DNA from *Chlamydomonas.*

R. H. Epstein, R. S. Edgar, and their collaborators introduce the use of conditional lethal mutations in T_4 phage for studying the action of indispensable genes.

J. Cairns demonstrates by autoradiography that the genophore of *Escherichia coli* is circular and that during its semiconservative replication Y-shaped, replicating forks proceed in opposite directions from a starting point and generate two circular offspring genophores.

E. Margoliash determines the amino acid sequences for cytochrome c derived from a wide variety of species and generates the first phylogenetic tree for a specific gene product.

L. B. Russell shows in the mouse that, when an X chromosome containing a translocated autosomal segment undergoes inactivation in somatic cells, the autosomal genes closest to the breakpoint are also inactivated. Thus the X inactivation spreads into the attached autosomal segment.

E. Mayr publishes *Animal Species and Evolution.* This volume provides a synthesis of modern ideas concerning the mechanism of speciation, and it has a profound influence on scientists working in this area.

I. R. Gibbons first isolates dynein from the arms on the microtubules of ciliary axonemes.

R. G. Davidson, H. M. Nitowski, and B. Childs demonstrate somatic mosaicism in a human female carrying on her X chromosomes the gene for the A and B forms of glucose-6-phosphate dehydrogenase.

1964　　R. B. Setlow and W. L. Carrier and, independently, R. P. Boyce and P. Howard-Flanders describe the mechanism of excision repair in bacteria.

A. S. Sarabhai, A.O.W. Stretton, S. Brenner, and A. Bolle establish the colinearity of gene and protein product in the case of the protein coating the head of virus T_4 of *Escherichia coli.* C. Yanofsky, B. C. Carlton, J. R. Guest, D. R. Helinski, and U. Henning establish the colinearity of gene and protein product in the case of tryptophan synthetase for *Escherichia coli.*

M. S. Fox and M. K. Allen show that transformation in *Diplococcus pneumoniae* involves incorporation of segments of single-stranded donor DNA into the DNA of the recipient.

E. T. Mertz, L. S. Bates, and O. E. Nelson show that the *opaque-2* mutation modifies the amino acid composition of the mature endosperm, resulting in a striking improvement in the nutritional quality of maize seed.

J. G. Gorman, V. J. Freda, and W. Pollack demonstrate that the sensitization of Rh-negative mothers can be prevented by administration of Rh antibody immediately after delivery of their first Rh-positive baby.

D.J.L. Luck and E. Reich isolate mitochondrial DNA from *Neurospora.* They demonstrate subsequently (1966) that this DNA replicates by the classical semiconservative mechanism.

G. Marbaix and A. Burny isolate a 9S RNA from mouse reticulocytes and suggest that it may be mRNA.

R. Holliday puts forth a model that defines a sequence of breakage and reunion events which must occur during crossing over between the DNA molecules of homologous chromosomes.

J. W. Littlefield develops a method for selecting somatic cell hybrids utilizing HGPRT⁻ and TK⁻ fibroblasts cultured on HAT medium.

D. D. Brown and J. B. Gurdon show that no synthesis of the 18S and 28S rRNAs occurs in *Xenopus* tadpoles homozygous for a deficiency covering the nucleolus organizer.

W. D. Hamilton puts forth the genetical theory of social behavior.

W. Gilbert finds that nascent proteins bind to the large ribosomal subunit, as do the tRNAs.

1965　　R. B. Merrifield and J. Stewart develop an automated method for synthesizing polypeptides on a solid supporting polymeric matrix. Some of the same automation principles will later be adopted for automated nucleic acid synthesis by instruments called "gene machines."

D. D. Sabatini, Y. Tashiro, and G. E. Palade show that the large subunit of ribosome attaches to the ER membrane.

L. Hayflick discovers that the *in vitro* life span of human diploid cells in tissue culture is about 50 doubling cycles.

R. W. Holley and his colleagues determine the complete sequence of alanine transfer RNA isolated from yeast.

N. Hilschmann and L. Craig report that immunoglobulin molecules are made up of carboxyl-terminal segments that are constant in their amino acid composition and amino-terminal segments that are variable. This finding poses the problem of how a gene can code for those portions of the protein that vary in their amino acid compositions.

P. Karlson, H. Hoffmeister, H. Hummel, P. Hocks, and G. Spiteller determine the complete structural configuration of ecdysone.

S. Spiegelman, I. Haruna, I. B. Holland, G. Beaudreau, and D. R. Mills succeed in demonstrating the *in vitro* synthesis of a self-propagating infectious RNA (bacteriophage Qβ of *E. coli*) using a purified enzyme (Qβ replicase).

S. Brenner, A.O.W. Stretton, and S. Kaplan deduce that UAG and UAA are the codons that signal the termination of a growing polypeptide.

F. M. Ritossa and S. Spiegelman demonstrate that multiple cistrons producing the ribosomal RNAs of *Drosophila* reside in the nucleolus organizer regions of each X and Y chromosome.

H. Harris and J. F. Watkins use the Sendai virus to fuse somatic cells derived from man and mouse and produce artificial interspecific heterokaryons.

A. J. Clark and A. D. Margulies report for *E. coli* that many mutants selected as deficient in recombination are also abnormally sensitive to ultraviolet radiation. This finding suggests that similar enzyme systems function both in repairing damaged DNA and in recombination.

F. Sanger, G. G. Brownlee, and B. G. Barrell describe a method for fingerprinting oligonucleotides from partially hydrolyzed RNA preparations.

R. Rothman demonstrates that lambda phage has a specific attachment site on the *E. coli* chromosome.

W. J. Dreyer and J. C. Bennett propose that antibody light chains are encoded by two distinct DNA sequences, one for the variable region and the other for the constant region. They suggest that there is only one constant region, but that the variable region contains hundreds of different minigenes.

F. Jacob, J. Monod, and A. Lwoff receive a Nobel Prize for their contributions to microbial genetics.

1966   B. Weiss and C. C. Richardson isolate DNA ligase.

M.M.K. Nass reports that mitochondrial DNA is a circular double-stranded molecule.

F.H.C. Crick puts forward the wobble hypothesis to explain the general pattern of degeneracy found in the genetic code.

J. Adams and M. Cappecchi show that N-formylmethionyl-tRNA functions as the initiator of the polypeptide chain forming on a ribosome.

W. Gilbert and B. Müller-Hill demonstrate that the lactose repressor of *E. coli* is a protein.

M. Ptashne shows that the phage lambda repressor is a protein and that it binds directly to the lambda DNA molecule.

H. Röller, K. H. Dahm, C. C. Sweely, and B. M. Trost determine the structural formula for the juvenile hormone of *Hyalophora cecropia*.

M. Waring and R. J. Britten demonstrate that vertebrate DNAs contain repetitious nucleotide sequences.

R. S. Edgar and W. B. Wood analyze the genetically controlled steps in the assembly of the T₄ bacteriophage.

V. A. McKusick publishes *Mendelian Inheritance in Man,* a catalogue listing some 1,500 genetic disorders of *Homo sapiens.*

E. Terzaghi, Y. Okada, G. Streisinger, J. Emrich, M. Inouye, and A. Tsugita confirm that the genetic code is translated by the sequential reading of triplets of bases starting at a defined point in phage T₄ lysozyme.

H. Wallace and M. L. Birnstiel demonstrate that an anucleolate deletion in *Xenopus laevis* removes more than 99% of the rDNA.

R. C. Lewontin and J. L. Hubby use electrophoretic methods to survey gene-controlled protein variants in natural populations of *Drosophila pseudoobscura.* They demonstrate that between 8 and 15% of all loci in the average individual genome are in the heterozygous condition. Using similar techniques, H. Harris demonstrates the existence of extensive enzyme polymorphisms in human populations.

P. Rous receives the Nobel Prize for his studies on oncogenic viruses.

1967   S. Spiegelman, D. R. Mills, and R. L. Peterson report the results of serial transfer experiments in which they select those Qβ bacteriophage molecules that replicate most rapidly *in vitro.* As these experiments in extracellular evolution progressed, the molecule became smaller as its replication rate increased. By the 74th transfer, the replicating RNA molecule was only 20% of its original length and was the smallest known self-duplicating molecule.

H. G. Khorana and his co-workers use polynucleotides with known repeating di- and trinucleotide sequences to solve the genetic code.

K. Taylor, Z. Hredecna, and W. Szybalski show that in phage lambda transcription can proceed in opposite directions in different genes on the same chromosome. Therefore, mRNA can originate from transcription units residing in the + and in the − strand of the same double helix.

B. Mintz uses allophenic mice to demonstrate that melanocytes that provide color to the fur of the mouse are derived from 34 cells that have been determined at an early stage in embryogenesis.

J. B. Gurdon transplants somatic nuclei into frog eggs at different developmental states. The synthesis of RNA and DNA of transplanted nuclei changes to the kind of synthesis characteristic of the host cell nucleus.

L. Goldstein and D. M. Prescott perform nuclear transplantations in *Amoeba.* These show there are specific proteins that move from the cytoplasm to the nucleus, and these presumably control the nucleic acid metabolism of the nuclei they enter.

C. B. Jacobson and R. H. Barter report the use of amniocentesis for intrauterine diagnosis and management of genetic defects.

C.C.F. Blake and four colleagues publish the three-dimensional structure of lysozyme at 2Å resolution. This gives the first indication as to how an enzyme molecule is shaped to accommodate its substrate.

M. Goulian, A. Kornberg, and R. L. Sinsheimer report the successful *in vitro* synthesis of biologically active DNA. The template they presented to the purified *Escherichia coli* DNA polymerase was single-stranded DNA from phiX174.

M. L. Birnstiel reports the isolation of pure rDNA from *Xenopus laevis.*

T. O. Diener and W. B. Raymer show that the potato spindle tuber disease is caused by a viroid.

M. C. Weiss and H. Green use the HAT selection procedure to localize the gene for thymidine kinase. This was the first use of somatic cell genetics to localize human genes.

V. M. Sarich and A. C. Wilson contrast the immunological properties of protein albumen between chimpanzees, gorillas, and humans. They conclude that the African apes and man shared a common ancestor 4 to 6 million years ago.

1968   R. T. Okazaki and four colleagues report that newly synthesized DNA contains many fragments. These represent short lengths of DNA that are replicated in a discontinuous manner and then spliced together.

W. Gilbert and D. Dressler put forth the "rolling circle" model of DNA replication.

J. Morgan, D. P. McKenzie, and X. Le Pinchon develop the concept of plate tectonics to explain continental drift.

Differential synthesis of genes for ribosomal DNA during amphibian oogenesis is reported by J. G. Gall and by D. D. Brown and I. B. Dawid.

M. Kimura proposes the neutral gene theory of molecular evolution.

H. O. Smith, K. W. Wilcox, and T. J. Kelley isolate and characterize the first specific restriction endonuclease (Hind II).

D. Y. Thomas and D. Wilkie demonstrate recombination of yeast mitochondrial genes.

R. P. Donahue, W. B. Bias, J. H. Renwick, and V. A. McKusick assign the Duffy blood group locus to chromosome 1 in man. This is the first gene localized in a specific autosome.

J. A. Huberman and A. D. Riggs demonstrate that mammalian chromosomes contain serially arranged, independently replicating units each about 30 μm long.

S. Wright publishes volume I of his four-volume series *Evolution and the Genetics of Populations.* The final volume will be completed ten years later.

E. H. Davidson, M. Crippa, and A. E. Mirsky show that more than 60% of the RNA labeled during oogenesis in *Xenopus laevis* is synthesized during the lampbrush stage and stored during the remaining months of oocyte maturation. This RNA is presumably a long-lived mRNA stored for use in early embryogenesis.

O. Hess and G. Meyer report extensive studies on structural modifications of the Y chromosome in various *Drosophila* species that demonstrate that the Y chromosome contains genes that control stage-specific steps in the development of sperm.

S. A. Henderson and R. G. Edwards demonstrate that the number of chiasmata per oocyte declines with increasing maternal age in the mouse and that the number of univalents increases with age. If the same occurs in human females, one would expect (as has been demonstrated) an increase in aneuploid offspring with advancing maternal age.

J. E. Cleaver shows that the repair replication of DNA is defective in patients with xeroderma pigmentosum.

R. J. Britten and D. E. Kohne demonstrate that Cot curves can be used to determine the relative abundances of repetitive and nonrepetitive DNA sequences in the genomes of different species.

R. W. Davis and N. Davidson visualize deletion mutations of bacteriophage lambda utilizing experimentally produced heteroduplex DNA molecules.

R. W. Holley, H. G. Khorana, and M. W. Nirenberg receive Nobel Prizes for discoveries concerning the interpretation of the genetic code and its function in protein synthesis.

1969   J. Abelson, L. Barnett, S. Brenner, M. Gefter, A. Landy, R. Russell, and J. D. Smith provide proof of the proposed mechanism of nonsense suppression by determining the actual nucleotide sequence of mutant tyrosine transfer ribonucleic acids.

The *in situ* hybridization techniques for the cytological localization of specific nucleotide sequences are developed by J. G. Gall and M. L. Pardue and by H. John, M. L. Birnstiel, and K. W. Jones.

B. C. Westmoreland, W. Szybalski, and H. Ris develop an electron microscopic technique for physically mapping genes in lambda phage. They photograph heteroduplex DNA molecules obtained by annealing of the − strand of one parent and the + strand of a second parent that has deletions, insertions, substitutions, or inversions.

R. Burgess, A. A. Travers, J. J. Dunn, and E. K. Bautz isolate and identify the sigma factor from RNA polymerase.

O. L. Miller and B. R. Beatty publish electron micrographs showing amphibian genes in the process of transcribing RNA molecules.

J. R. Beckwith (together with five associates) reports the isolation of pure *lac* operon DNA from *Escherichia coli*.

G. M. Edelman (together with five associates) publishes the first complete amino acid sequence for human gamma G_1 immunoglobulin.

Y. Hotta and S. Benzer and W. L. Pak and J. Grossfield independently induce and physiologically characterize neurological mutants in *Drosophila*.

C. Boon and F. Ruddle correlate the loss of particular chromosomes from a somatic hybrid cell line containing both human and mouse chromosomes with the loss of specific phenotypic characters. This approach permits assignment of specific loci to certain human chromosomes.

R. E. Lockard and J. B. Lingrel purify the 9S RNA fraction obtained from polysomes of mouse reticulocytes and show that it directs the synthesis of mouse hemoglobin β chains. They thus confirm the suggestion of Marbaix and Burny (1964).

A. Ammermann reports that in the hypotrichous ciliate *Stylonychia mytilus* macronuclear anlage undergo endomitotic DNA replication to form polytene chromosomes. Subsequently, the major portions of these are destroyed, and over 90% of the macronuclear DNA is degraded and excreted into the medium.

R. I. Huebner and G. I. Todaro put forward the oncogene theory.

H. A. Lubs describes a fragile site on the human X chromosome and shows that it is present in mentally retarded males. Subsequent studies show that this locus (Xq27) is associated with a common form of X-linked mental retardation.

J. A. Shapiro detects mutations of the galactose operon of *E. coli* caused by insertion sequences.

M. Delbrück, S. E. Luria, and A. D. Hershey receive a Nobel Prize for their contributions to viral genetics.

1970   B. M. Alberts and L. Frey isolate the protein product of gene 32 of phage T_4 and demonstrate that this protein binds cooperatively to single-stranded DNA. They suggest that 32 protein functions to initiate unwinding of the DNA molecule so that replication can begin.

H. G. Khorana (together with twelve associates) reports the total synthesis of the gene for an alanine tRNA from yeast.

M. Mandel and A. Higa develop a general method for introducing DNA into *E. coli*. They demonstrate that placing the cells in a cold calcium chloride solution renders them permeable to nucleic acid fragments.

J. Yourno, T. Kohno, and J. R. Roth succeed in fusing two bacterial enzymes into one large protein molecule that combines the functions of both. The enzyme fusion was accomplished by fusing the *his D* and *his C* genes in the histidine operon of *Salmonella,* using a pair of frame shift mutations.

D. Baltimore and H. M. Temin report the existence of an RNA-dependent DNA polymerase in two oncogenic RNA viruses (Rauscher mouse leukemia and Rous fowl sarcoma).

M. L. Pardue and J. G. Gall demonstrate that pericentric heterochromatin is rich in repetitious DNA.

D. E. Wimber and D. M. Steffensen localize the 5S RNA cistrons on the right arm of chromosome 2 of *Drosophila melanogaster.*

T. Caspersson, L. Zech, and C. Johansson use quinacrine dyes in chromosomal cytology and demonstrate specific fluorescent banding patterns in human chromosomes.

R. Sager and Z. Ramanis publish the first genetic map of non-Mendelian genes. This group of eight genes resides on a chloroplast chromosome of *Chlamydomonas.*

R. T. Johnson and P. N. Rao induce premature chromosome condensation by fusing mitotically active cells with interphase cells *in vitro.*

M. Smös and R. B. Inman use the denaturation mapping technique to study chromosome replication in bacteriophage lambda. They show that replication begins at a unique origin and that both forks are growing points that progress in opposite directions around the circular molecule.

H. O. Smith and K. W. Wilcox discover that certain restriction endonucleases can generate DNA termini, in one step, having projecting single-stranded ends.

1971   M. L. O'Riordan, J. A. Robinson, K. E. Buckton, and H. J. Evans report that all 22 pairs of human autosomes can be identified visually after staining with quinacrine hydrochloride. They demonstrate that the Philadelphia chromosome is an aberrant chromosome 22.

Y. Hotta and H. Stern characterize the DNA that is synthesized during meiotic prophase in the lily. Synthesis during zygonema represents the delayed replication of a small fraction of the DNA that failed to replicate during the previous S phase. The DNA synthesized during pachynema has the characteristics of repair replication.

369

S. H. Howell and Stern demonstrate that an endonuclease present in microspores reaches its highest concentration early in pachynema, the stage when crossing over is thought to occur.

B. Dudock, C. Di Peri, K. Scileppi, and R. Reszelbach present evidence for the phenylalanyl tRNA synthetase recognition site being adjacent to the dihydrouridine loop.

C. R. Merril, M. R. Geier, and J. C. Petricciani infect fibroblasts cultured from a patient suffering from galactosemia with transducing lambda phage carrying the galactose operon. The cells then make the missing transferase and survive longer in culture than uninfected galactosemic cells.

R. J. Konopka and S. Benzer report recovery of induced clock mutants of *Drosophila.*

D. T. Suzuki, T. Grigliatti, and R. Williamson isolate a temperature-sensitive paralytic mutant of *Drosophila.*

J. E. Manning and O. C. Richards detect circular DNA molecules in lysates of *Euglena* chloroplasts.

J. E. Darnell, L. Philipson, R. Wall, and M. Adesnik suggest that during the posttranslational processing of premessenger RNA, a polyadenylic acid segment is added and that this poly-A tail somehow stabilizes the mRNA.

C. Kung induces and isolates behavioral mutants of *Paramecium aurelia* and shows that many mutants have electrophysiological defects in their plasma membranes.

K. Dana and D. Nathans use restriction endonucleases to cleave the circular DNA of simian virus 40 into a series of fragments and then deduce their physical order.

1972   P. Lobban and A. D. Kaiser develop a general method for joining any two DNA molecules, employing terminal transferase to add complementary homopolymer tails to passenger and vehicular DNA molecules.

A. F. Zakharov and N. A. Egolina develop the BUDR labeling technique to produce harlequin chromosomes.

G. H. Pigott and N. G. Carr show that ribosomal RNAs from cyanobacteria hybridize with DNA from the chloroplasts of *Euglena gracilis.* This genetic homology provides strong support to the theory that chloroplasts are the descendants of endosymbiotic cyanobacteria.

S. J. Singer and G. L. Nicholson put forth the fluid mosaic model of the structure of cell membranes.

B. Benacerraf and H. O. McDevitt show for the mouse that *Ir* genes are linked with the H2 complex.

Y. Suzuki and D. D. Brown isolate and identify the mRNA for silk fibroin from *Bombyx mori,* and Suzuki, L. P. Gage, and Brown characterize the fibroin gene.

R. Silber, V. G. Malathi, and J. Hurwitz discover RNA ligase.

D. A. Jackson, R. H. Symons, and P. Berg report splicing the DNA of SV40 virus into the DNA of the lambda virus of *E. coli.* They are thus the first to join the DNAs of two different organisms *in vitro.*

J. Mendlewicz, J. L. Fleiss, and R. R. Fieve demonstrate that manic-depressive psychosis is transmitted by a dominant gene located on the short arm of the X chromosome.

M. L. Pardue, E. Weinberg, L. H. Kedes, and M. L. Birnstiel locate the histone genes on chromosome 2 of *Drosophila melanogaster.*

P. S. Carlson, H. H. Smith, and R. D. Dearing succeed in producing interspecific plant hybrids by parasexual means.

J. Hedgpeth, H. M. Goodman, and H. W. Boyer identify the nucleotide sequence in the DNA of coliphage lambda that is recognized by a specific endonuclease.

S. N. Cohen, A.C.Y. Chang, and L. Hsu show that *E. coli* can take up circular plasmid DNA molecules and that transformants in the bacterial population can be identified and selected utilizing antibiotic resistance genes carried by the plasmids.

U. Kuhnlein and W. Arber report the isolation of recognition-site mutations in coliphages. This confirms Arber's restriction and modification proposal to explain the host-controlled restriction of virus growth.

J. Mertz and R. W. Davis show that cleavage of DNA by R1 restriction endonuclease generates cohesive ends.

D. E. Kohne, J. A. Chisson, and B. H. Hoyer use DNA-DNA hybridization data to study the evolution of primates. They conclude that man's closest living relative is the chimpanzee.

G. M. Edelman and R. R. Porter receive a Nobel Prize for their studies on the chemical structure of antibodies.

1973   D. R. Mills, F. R. Kramer, and S. Spiegelman publish the sequence for the 218 nucleotides in a replicating RNA molecule. The molecule is MDV-1, a variant derived from the RNA of Qβ phage exposed to experimental selection and thus forced to undergo "extracellular evolution" to a shorter length. (*See* first entry for 1967.)

S. H. Kim, G. J. Quigley, F. L. Suddath, A. McPherson, D. Sneden, J. J. Kim, J. Weinzierl, and A. Rich propose a three-dimensional structure for yeast phenylalanine transfer RNA.

R. Kavenoff and B. H. Zimm use a newly developed viscoelastic method for measuring the molecular weights of DNA molecules isolated from cells from different *Drosophila* species. They conclude that a chromosome contains one long molecule of DNA and that it is not interrupted in the centromere region.

P. Debergh and C. Nitsch succeed in culturing haploid tomato plants directly from microspores.

W. G. Hunt and R. K. Selander analyze a zone of hybridization between two subspecies of the house mouse, using gel electrophoresis to trace the boundary.

P. J. Ford and E. M. Southern show for *Xenopus laevis* that different 5S RNA genes are transcribed in the oocyte than in somatic cells.

W. Fiers, W. M. Jou, G. Haegerman, and M. Ysebaert are the first to sequence a gene coding for a protein (the coat protein of the male-specific phage MS2).

B. E. Roberts and B. M. Patterson report the preparation of a wheat germ cell-free system for the *in vitro* translation of experimentally supplied mRNAs.

A. Garcia-Bellido, P. Ripoll, and G. Morata report the developmental compartmentalization of the wing disc of *Drosophila*.

S. N. Cohen, A.C.Y. Chang, H. W. Boyer, and R. B. Helling construct the first biologically functional, hybrid bacterial plasmids by *in vitro* joining of restriction fragments from different plasmids.

1974   J. Shine and L. Dalgarno show that the 3′ terminus of *E. coli* 16S rRNA contains a stretch of nucleotides that is complementary to ribosome binding sites of various coliphage mRNAs. They suggest that this region of the 16S rRNA may play a base-pairing role in the termination and initiation of protein synthesis on mRNA.

I. Zaenen, N. van Larebeke, H. Teuchy, M. van Montagu, and J. Schell discover the tumor-inducing plasmid of the crown gall bacterium.

K. M. Murray and N. E. Murray manipulate the recognition sites for restriction endonucleases in lambda phage so that its chromosome can be used as a receptor site for restriction fragments from foreign DNAs. Lambda thus becomes a cloning vehicle.

A. Tissieres, H. K. Mitchell, and U. M. Tracy find that heat shocks result in the synthesis of six new proteins in *Drosophila*. These are also synthesized by tissues that do not have polytene chromosomes.

B. Dujon, P. P. Slonimski, and L. Weill propose a model for recombination and segregation of mitochondrial genomes in *Saccharomyces cerevisiae*. According to it, mtDNA molecules are present in the zygote cell in multiple copies. These pair at random, and during any mating cycle a segment from one parent can exchange with that from a second parent mtDNA yielding recombinant units.

R. D. Kornberg proposes that chromatin is built up of repeated structural units of 200 base pairs of DNA and two each of the histones H2A, H2B, H3, and H4. These structures, which are later called nucleosomes, are isolated by M. Noll. A. L. Olins and D. E. Olins publish the first electron micrographs of chromatin spreads from nuclei that show nucleosomes.

B. Ames develops a rapid screening test for detecting mutagenic and possibly carcinogenic compounds.

S. Brenner describes methods for inducing, isolating, and mapping mutations in the nematode *Caenorhabditis elegans*.

R. W. Hedges and A. E. Jacob discover mobile elements that transfer penicillin resistance between plasmids.

G. L. Stebbins publishes *Flowering Plants, Evolution Above the Species Level*.

C. A. Hutchison, J. E. Newbold, S. S. Potter, and M. H. Edgell demonstrate the maternal inheritance of mitochondrial DNA in horse-donkey hybrids.

A. Claude, C. de Duve, and G. Palade receive Nobel Prizes for their contributions to cell biology.

1975   G. Köhler and C. Milstein perform experiments with mouse cells that show that somatic cell hybridization can be used to generate a continuous "hybridoma" cell line producing a monoclonal antibody.

Molecular biologists from around the world meet at Asilomar, California, to write an historic set of rules to guide research in recombinant DNA experiments.

The NIH Recombinant DNA Committee issues guidelines aimed at eliminating or minimizing the potential risks of recombinant DNA research.

L. L. Goldstein and M. S. Brown demonstrate that normal fibroblasts have binding sites for low-density lipoproteins, whereas fibroblasts from humans homozygous for the hypercholesterolemia gene lack these receptors.

M. Grunstein and D. S. Hogness develop the colony hybridization method for the isolation of cloned DNAs containing specific DNA segments or genes.

D. Pribnow determines the nucleotide sequences of two independent bacteriophage T7 promoters, and compares these and other known promoter sequences to form a model for promoter structure and function.

E. M. Southern describes a method for transferring DNA fragments from agarose gels to nitrocellulose filters. The filters are subsequently hybridized to radioactive RNA and the hybrids detected by autoradiography.

W. D. Benton and R. W. Davis describe a rapid and direct method for screening plaques of recombinant λgt phages involving transfer of phage DNA to a nitrocellulose filter and detection of specific DNA sequences by hybridization to complementary labeled nucleic acids.

F. Sanger and A. R. Coulson develop the "plus and minus" method for determining the nucleotide sequences in DNA by primed synthesis with DNA polymerase.

G. Morata and P. A. Lawrence show in *Drosophila* that the *engrailed* mutation allows cells of the posterior wing compartment to mix with those of the anterior compartment. Therefore, the normal allele of this gene functions to define the boundary conditions between the sister compartments of the developing wing.

B. Mintz and K. Illmensee inject XY diploid cells from a malignant mouse teratocarcinoma into mouse blastocysts that then are transferred to foster mothers. Cells derived from the carcinoma appear in both somatic and germ cells of some F_1 males. When these are mated, some F_2 mice contain marker genes from the carcinoma. The experiments demonstrate that the nuclei of teratocarcinoma cells remain developmentally totipotent, even after hundreds of transplant generations during which they functioned in malignant cancers.

S. L. McKenzie, S. Henikoff, and M. Meselson isolate mRNAs for heat shock proteins and show that they hybridize to specific puff sites on the *Drosophila* polytene chromosomes.

L. H. Wang, P. H. Duesberg, K. Beemon, and P. K. Vogt locate within the RNA genome of the Rous sarcoma virus the segment responsible for its oncogenic activity.

G. Blobel and B. Dobberstein put forth the signal hypothesis.

R. Dulbecco, H. Temin, and D. Baltimore receive Nobel Prizes for their studies on oncogenic viruses.

1976   H.R.B. Pelham and R. J. Jackson describe a simple and efficient mRNA-dependent *in vitro* translation system using rabbit reticulocyte lysates.

R. V. Dippell shows in *Paramecium* that kinetosomes contain RNA (not DNA) and that RNA (not DNA) synthesis accompanies kinetosome reproduction.

N. Hozumi and S. Tonegawa demonstrate that the DNA segments coding for the variable and constant regions of an immunoglobulin chain are distant from one another in the chromosomes isolated from mouse embryos, but the segments are adjacent in chromosomes isolated from mouse plasmacytomas. They conclude that somatic recombination during the differentiation of B lymphocytes moves the constant and variable gene segments closer together.

W. Y. Kan, M. S. Golbus and A. M. Dozy are the first to use recombinant DNA technology in a clinical setting. They develop a prenatal test for alpha thalassemia utilizing molecular hybridization techniques.

M. F. Gellert and his colleagues discover DNA gyrase to be the enzyme that converts a relaxed, closed, circular DNA molecule into a negatively supercoiled form.

W. Y. Chooi shows that ferritin-labeled antibodies raised against proteins (isolated from rat ribosomes) bind to the terminal knobs of fibers extending from Miller trees (isolated from the ovarian nurse cells of *Drosophila*). This observation proves that Miller trees are rRNA transcription units and shows that at least some ribosomal proteins attach to a precursor rRNA molecule before its transcription is completed.

B. G. Burrell, G. M. Air, and C. A. Hutchison report that phage phiX174 contains overlapping genes.

Formal guidelines regulating research involving recombinant DNA are issued by the National Institutes of Health in the United States.

The first genetic-engineering company is formed and named Genentech.

A. Efstratiadis, F. C. Kafatos, A. Maxam, and T. Maniatis are the first to enzymatically generate eukaryotic gene segments *in vitro*. They synthesize double-stranded DNA molecules that contain the sequences transcribed into the mRNAs for the alpha and beta chains of rabbit hemoglobin.

J. T. Finch and A. Klug propose that the 300Å threads seen in electron micrographs of fragmented chromatin are formed by the folding of DNA-nucleosome filaments into solenoids.

L. H. Miller, S. J. Mason, D. F. Clyde, and M. H. McGinnis conclude that the Duffy blood group antigens (Fy^a and Fy^b) serve as receptors for the merozoites of *Plasmodium vivax* and that individuals of blood group Fy^-/Fy^- are resistant to *P. vivax* infections because their red cells lack these receptors.

1977   A. Knoll and E. S. Barghoorn find microfossils which they interpret as undergoing cell division in rocks 3,400 million years old. This discovery pushes back the age of life on earth to the lower Archean eon.

C. Jacq, J. R. Miller, and G. G. Brownlee describe the presence of "pseudogenes" within the 5S DNA cluster of *Xenopus laevis* oocytes.

J. C. Alwine, D. J. Kemp, and G. R. Stark prepare diazobenzyloxymethyl (DBM) paper and describe methods for transferring electrophoretically separated bands of RNA from an agarose gel to the DBM paper. Specific RNA bands are then detected by hybridization with radioactive DNA probes, followed by autoradiography. Since this method is the reverse of that described by Southern (1975) in that RNA rather than DNA is transferred to a solid support, it has come to be known as "northern blotting."

F. Sanger and eight colleagues report the complete nucleotide sequence for the DNA genome of bacteriophage phiX174.

M. Leffak, R. Grainger, and H. Weintraub show that "old" histone octamers remain intact during DNA replication and that "new" octamers consist entirely of proteins synthesized during the time of replication.

W. Gilbert induces bacteria to synthesize useful nonbacterial proteins (insulin and interferon).

A.M. Maxam and W. Gilbert publish the "chemical method" of DNA sequencing.

R. W. Old, H. G. Callan, and K. W. Gross use labeled cloned histone genes from sea urchins to localize, by *in situ* hybridization, histone mRNAs being transcribed on the lampbrush chromosomes of salamander oocytes.

D. S. Hogness, D. M. Glover, and R. L. White report the occurrence of intervening segments in some of the 28S rRNA genes of *Drosophila melanogaster.* Intervening noncoding segments are then described for genes that encode proteins, namely, the rabbit beta-globin gene (A. Jeffreys and R. A. Flavell) and the chicken ovalbumin gene (R. Breathnach, J. L. Mandel, and P. Chambon).

J. Weber, W. Jelinek, and J. E. Darnell report that alternative splicing of nonconsecutive DNA segments in the adenovirus-2 genome can produce multiple mRNAs.

J. F. Pardon and five colleagues use neutron contrast matching techniques to demonstrate that in nucleosomes the DNA segment that attaches to the histone octamer is on the outside of the particle.

J. Sulston and H. R. Horvitz work out the postembryonic cell lineages for *Caenorhabditis elegans.*

J. Collins and B. Holm develop cosmids for cloning large DNA fragments.

R. S. Yalow receives a Nobel Prize for developing the radioimmunoassay procedure.

1978   R. M. Schwartz and M. O. Dayhoff compare sequence data for a variety of proteins and nucleic acids from an evolutionarily diverse assemblage of prokaryotes, eukaryotes, mitochondria, and chloroplasts. Their computer-generated evolutionary trees identify the times during evolution when protoeukaryotic organisms entered into symbiosis with mitochondria and chloroplasts (about 2 and 1 billion years ago, respectively).

W. Gilbert coins the terms *introns* and *exons.*

T. Maniatis, R. C. Hardison, E. Lacy, J. Lauer, C. O'Connell, D. Quon, G. K. Sim, and A. Efstratiadis develop a procedure for gene isolation, which involves construction of cloned libraries of eukaryotic DNA and screening of these libraries for individual sequences by hybridization to specific nucleic acid probes.

D. J. Finnegan, G. M. Rubin, M. W. Young, and D. S. Hogness make detailed analyses of dispersed repetitive DNAs in *Drosophila.* Their discoveries represent a new beginning in the understanding of mutability, transposition, transformation, hybrid dysgenesis, and retroviruses in eukaryotes.

E. B. Lewis concludes that the component genes in the *bithorax* complex have related functions in *Drosophila* segmentation and that they evolved from a smaller number of ancestral genes by their duplication and subsequent specialization.

V. B. Reddy and eight colleagues publish the complete nucleotide sequence for simian virus 40 and correlate the sequence with the known genes and mRNAs of the virus.

Y. W. Kan and A. M. Dozy demonstrate the value of using restriction-fragment-length polymorphisms as linked markers for the prenatal diagnosis of sickle-cell anemia.

C. A. Hutchison and five colleagues demonstrate that it is possible to introduce specific mutations at specific sites in a DNA molecule.

E. H. Blackburn and J. G. Gall demonstrate that telomeres from *Tetrahymena pyriformis* consist of short DNA sequences (one strand containing AACCCC, the other TTGGGG) repeated tandemly 30 to 70 times.

R. T. Schimke, R. J. Kaufman, F. W. Alt, and R. F. Kellems show that cultured mouse cells exposed to methotrexate develop resistance by amplifying the genes that encode the enzymes that serve as the target for the drug.

W. Arber, H. O. Smith, and D. Nathans receive Nobel Prizes for the development of techniques utilizing restriction endonucleases to study the organization of genetic systems.

1979   J. G. Sutcliffe determines the complete 4,362 nucleotide pair sequence of the plasmid cloning vector pBR322.

J. C. Avise, R. A. Lansman, and R. O. Shade successfully use restriction endonucleases to measure mitochondrial DNA sequence relatedness in natural populations.

The National Institutes of Health relax guidelines on recombinant DNA to allow viral DNA to be studied.

B. G. Barrell, A. T. Bankier, and J. Drouin report that the genetic code of human mitochondria has some unique, non-universal features.

E. F. Fritsch, R. M. Lawn, and T. Maniatis determine the chromosomal arrangement and structure of human globin genes utilizing recombinant DNA technology.

J. R. Cameron, E. Y. Loh, and R. W. Davis discover transposable elements in yeast.

D. V. Goeddel and nine colleagues construct a gene that encodes human growth hormone (HGH) using recombinant DNA technology. The synthesized gene is expressed in *E. coli* under the control of the *lac* promoter and a polypeptide having the properties of HGH is synthesized.

M. R. Lerner and J. A. Steitz report the discovery of small nuclear ribonucleoproteins (snurps).

1980   L. Olsson and H. S. Kaplan produce the first human hybridomas that manufacture a pure antibody in laboratory culture.

The United States Supreme Court rules that genetically modified microorganisms can be patented. General Electric company, on behalf of A. Chakrabarty, obtains a patent for a genetically engineered microorganism capable of consuming oil slicks.

J. W. Gordon, G. A. Scangos, D. J. Plotkin, J. A. Barbosa, and F. H. Ruddle produce the first transgenic mice by direct injection of cloned DNA into the pronucleus of a fertilized egg.

M. R. Capecchi describes a technique for efficient transformation of cultured mammalian cells by direct microinjection of DNA into cells with glass micropipettes.

H. Gronemeyer and O. Pongs demonstrate that, in *Drosophila melanogaster* salivary glands, β ecdysone binds directly to sites on polytene chromosomes where ecdysone-inducible puffs occur.

C. Nüsslein-Volhard and E. Wieschaus describe the isolation and characterization of zygotic segmentation mutations of *Drosophila melanogaster.*

A. R. Templeton provides a new theoretical framework for speciation by the founder principle.

Nobel Prizes in Physiology and Medicine go to G. D. Snell, J. Dausset, and B. Benacerraf for their contributions to immunogenetics.

P. Berg, W. Gilbert, and F. Sanger receive Nobel Prizes in Chemistry for their contributions to the experimental manipulation of DNA.

1981   R. C. Parker, H. E. Varmus, and J. M. Bishop demonstrate that the tumorigenic properties of the Rous sarcoma virus are due to a protein encoded by the *v-src* gene. Cells from various vertebrates contain a homologous gene, *c-src.* The two genes differ in that *v-src* has an uninterrupted coding sequence, whereas *c-src* contains seven exons separated by six introns.

L. Margulis publishes *Symbiosis in Cell Evolution.* Here she summarizes the evidence for the theory that organelles such as mitochondria, chloroplasts, and kinetosomes evolved from prokaryotes that lived as endosymbionts in the ancestors of modern-day eukaryotes.

R. Lande proposes a new model of speciation based on sexual selection on polygenic traits. This model results in a revival of interest in sexual selection.

J. D. Kemp and T. H. Hall transfer the gene of a major seed storage protein (phaseolin) from beans to the sunflower via a plasmid of the crown gall bacterium *Agrobacterium tumefaciens,* creating a "sunbean."

T. R. Cech, A. J. Zaug, and P. J. Grabowski report the discovery of a self-splicing rRNA in *Tetrahymena thermophila.* This is the first demonstration that a macromolecule other than a protein can act as a biological catalyst.

W. F. Anderson, D. H. Ohlendorf, Y. Takeda, and B. F. Matthews determine the three-dimensional structure of the *cro* repressor at 2.8Å resolution.

G. Hombrecher, N. J. Brewin, and A.W.B. Johnson demonstrate that the ability of *Rhizobium* bacteria to nodulate legumes and fix atmospheric nitrogen is due to plasmid-linked genes.

P. R. Langer, A. A. Waldrop, and D. C. Ward develop a procedure for synthesizing biotinylated DNA probes that hybridize normally with complementary DNA, providing an anchor for streptavidin-linked, color-generating systems.

The complete nucleotide sequence and genetic organization of the human mitochondrial genome are worked out by a research group composed of S. Anderson, B. G. Barrell, F. Sanger, and eleven other scientists.

M. E. Harper and G. F. Saunders demonstrate that single copy genes can be mapped on human mitotic chromosomes utilizing an improved *in situ* hybridization technique.

J. Banerji, S. Rusconi, and S. Schaffner show that the transcription of the beta-globin gene is enhanced hundreds of times when this gene is linked with certain SV40 nucleotide sequences that they name "enhancer sequences."

M. Chalfie and J. Sulston identify among the touch-insensitive mutants of *Caenorhabditis elegans* five genes that affect a specific set of six sensory neurons.

K. E. Steinbeck, L. McIntosh, L. Bogorad, and C. J. Arntzen demonstrate that the resistance of a weed, *Amaranthus hybridus*, to triazine herbicides is controlled by a chloroplast gene that encodes a polypeptide to which the herbicide binds. Resistant strains of the weed produce a modified gene product that fails to bind triazine.

1982   Eli Lilly International Corporation is the first to market a drug made by recombinant DNA techniques. The drug is human insulin, sold under the trade name "Humulin."

E. P. Reddy, R. K. Reynolds, E. Santos, and M. Barbacid report that the genetic change that leads to the activation of an oncogene carried by a line of human bladder carcinoma cells is due to a single base substitution in this gene. The result is the incorporation of valine instead of lysine in the 12th amino acid of the protein encoded by the oncogene.

P. M. Bingham, M. G. Kidwell, and G. M. Rubin show that P strains of *Drosophila* contain 30 to 50 copies per genome of a transposable P element. This is the cause of hybrid dysgenesis. Then A. C. Spradling and Rubin demonstrate that cloned P elements, when microinjected into *Drosophila* embryos, become integrated into germ-line chromosomes and that P elements can be used as vectors to carry DNA fragments of interest into the *Drosophila* germ line.

L.S.B. Goldstein, R. W. Hardy and D. L. Lindsley demonstrate that the Y chromosome of *Drosophila* contains structural genes that encode proteins that are structural components of the sperm axoneme.

A. Klug receives the Nobel Prize for his contributions to the analysis of crystalline structures of biological importance, especially virus particles, tRNA, and nucleosomes.

1983   E. A. Miele, D. R. Mills, and F. R. Kramer succeed in constructing the first recombinant RNA molecule, involving the insertion of a synthetic foreign deca-adenylic acid into a variant of the RNA genome of a small bacterial virus called Qβ via the action of the Qβ replicase.

H. J. Jacobs (and six colleagues) report the presence of promiscuous DNA in the sea urchin.

I. S. Greenwald, P. W. Sternberg, and H. R. Horvitz demonstrate that the *lin*-12 mutant of *Caenorhabditis* functions as a developmental control gene.

The first papers on the molecular genetics of the segment-identity genes of *Drosophila* are published; M. P. Scott *et al.* for the *Antennapedia* cluster and W. Bender *et al.* for the *bithorax* cluster.

The gene encoding the circumsporozoite protein of *Plasmodium knowlesi* is cloned by G. N. Godson, J. Ellis, P. Svec, D. H. Schlessinger, and V. Nussenzweig. They show that the protein contains a repetitive epitope that serves as a decoy to the host immune system.

C. Guerrier-Takada, K. Gardiner, T. Marsh, N. Pace, and S. Altman report that the catalytic activity of ribonuclease P depends solely upon its RNA component.

M. Kimura and T. Ohta estimate 1.8×10^9 years as the time of divergence of eukaryotes and prokaryotes through comparative studies of the nucleotide sequences of 5S rRNAs from humans, yeasts, and bacteria.

M. Rassoulzadegan and six colleagues isolate a recombinant DNA clone from the polyoma virus that immortalizes cultured fibroblast cells from rat embryos. They also show that only the amino-terminal portion of the protein encoded by the viral gene carries the immortalizing function.

R. F. Doolittle and six colleagues demonstrate that the simian sarcoma virus oncogene, *v-sis,* is derived from the gene encoding a platelet-derived growth factor.

B. McClintock receives the Nobel Prize for her discovery of transposable genetic elements.

1984   D. C. Schwartz and C. R. Cantor show that pulsed field gradient electrophoresis can be used to separate DNA fragments as large as 2,000 kbp. This method overcomes the limitation of agarose gel electrophoresis, which can only separate molecules of much smaller sizes (50 kbp or less).

J. Gitschier and eight colleagues report the cloning of the gene encoding the antihemophilic factor in humans.

W. McGinnis, C. P. Hart, W. J. Gehring, and F. H. Ruddle demonstrate that the homeobox sequence, first identified within the homeotic genes of *Drosophila,* also exists in the mouse. The high degree of base-sequence similarity suggests that this DNA segment has an essential function in animal development.

R. F. Pohlman, N. V. Federoff, and J. Messing determine the nucleotide sequence of the maize transposable element *Activator.*

M. Davis and T. Mak identify and clone the genes for the T cell receptor.

N. K. Jerne, G. Köhler, and C. Milstein receive the Nobel Prize in medicine for their contributions to immunology.

R. B. Merrifield is awarded a Nobel Prize in Chemistry for his work in automated peptide synthesis.

1985   J. R. Miller, A. D. McLachlan, and A. Klug report the isolation and characterization of a zinc finger protein from *Xenopus* oocytes. This protein binds to the 5S RNA gene and controls its transcription.

The notion of the universality of the genetic code is amended because codons that serve as termination signals according to the "universal" genetic code are found to encode amino acids in certain ciliates and bacteria. For example, in *Stylonychia lemnae* UAA and UGA encode glutamine (S. Horowtiz and M. A. Gorowsky) and in *Myoplasma capricolum* UGA encodes tryptophan (F. Yamao).

C. M. Newman, J. E. Cohen and C. Kipnis demonstrate mathematically that the punctuated shifting equilibrium patterns of species formation seen in the fossil record are to be expected on traditional grounds and do not require special mechanisms to explain them.

C. W. Greider and E. H. Blackburn isolate a telomerase from *Tetrahymena pyriformis.*

O. Smithies and four colleagues report the successful insertion of DNA sequences into human tissue culture cells by homologous recombination at the beta globin locus.

A. J. Jeffries, V. Wilson, and S. L. Thien develop the DNA fingerprint technique and point out its potential use in forensic science.

R. K. Saiki, K. B. Mullis, and five colleagues report the use of the polymerase chain reaction to allow enzymatic amplification *in vitro* of specific β-globin gene fragments.

H. L. Carson demonstrates that sexual selection is a basis for the morphological and behavioral evolution of Hawaiian *Drosophila* species.

M. S. Brown and J. L. Goldstein receive the Nobel Prize for identifying the low-density lipoprotein receptor pathway and for demonstrating that familial hypercholesterolemia is a genetic defect in this pathway.

1986   M.-C. Shih, G. Lazar, and H. M. Goodman show that the nuclear genes that encode chloroplast glyceraldehyde-3-phosphate dehydrogenase of higher plants are direct descendants of the genes from the symbionts that gave rise to the chloroplast. Later during evolution, these genes were transferred from the chloroplast to the nuclear genome.

A. Tomlinson and D. F. Ready report the discovery of *sevenless,* a mutation in *Drosophila* that controls the developmental fate of a specific cell in the ommatidium.

A. G. Amit, A. Mariuzza, S.E.V. Phillips, and R. J. Poljak determine the three-dimensional structure of an antigen-antibody complex at a resolution of 2.8Å.

F. Costantini, K. Chada, and J. Magram demonstrate that cloned normal beta-globin genes can be experimentally substituted for defective thalassemia genes in the mouse. They inject cloned normal genes into the fertilized thalassemic eggs. The mice that develop possess red blood cells that can synthesize normal beta-globin chains. These transgenic mice transmit this ability to their offspring.

J. Nathans, D. Thomas, and D. S. Hogness isolate and characterize the human visual pigment genes.

M. Noll and four colleagues demonstrate that the *paired* gene contains multiple conserved domains and suggest that it "networks" with other genes containing homologous domains to program the early development of *Drosophila.*

E. Ruska receives the Nobel Prize for designing the first electron microscope.

1987   M. R. Kuehn and four colleagues introduce a human gene into the mouse to allow its study in a convenient laboratory rodent. They employ a mutant allele of the gene encoding HPRT and use a retrovirus as a vector to insert it into cultured mouse embryonic germ cells. These are then implanted into mouse embryos to form chimeras. Strains of mice carrying the human gene are obtained from these chimeras.

C. Nüsslein-Volhard, H. G. Frohnhöfer, and R. Lehmann show that a small group of maternal effect genes exist in *Drosophila* that determine the polarized pattern of development of the embryo.

E. P. Hoffman, R. H. Brown, and L. M. Kunkel isolate dystrophin, the protein encoded by the muscular-dystrophy locus.

D. C. Wiley and five colleagues determine the three-dimensional structure of HLA-A2, a human class I histocompatibility molecule.

D. C. Page and eight colleagues clone a segment of the human Y chromosome that contains a gene which encodes a factor influencing testis differentiation. Within the Y chromosome fragment is a 1.2kb ORF that appears to encode a zinc finger protein.

R. L. Cann, M. Stoneking and A. C. Wilson compare the extent of sequence divergence in the mtDNA of individuals belonging to geographically distinct human populations. They erect a genealogical tree that suggests that all mtDNAs can be traced back to a common African maternal ancestor.

S. Tonegawa wins the Nobel Prize for his elucidation of the genetic mechanism that generates antibody diversity.

1988   W. Driever and C. Nüsslein-Volhard demonstrate that the *bicoid* gene encodes a protein that is distributed in an exponential concentration gradient along the anteroposterior axis of the embryo.

P. M. Macdonald and G. Struhl show that a 625-nucleotide segment in the trailer of a message encoded by the maternal polarity gene *bicoid* is responsible for the anterior localization of this mRNA in the *Drosophila* oocyte.

W. H. Landschulz, P. F. Johnson, and S. L. McKnight discover the leucine zipper and propose that it functions as a DNA binding site.

V. Sorsa publishes a two-volume monograph that reviews the encyclopedic literature concerning polytene chromosomes and presents electron microscopy maps of *Drosophila* salivary-gland chromosomes.

The first U.S. patent is issued for a genetically altered animal. Harvard University receives the patent for oncomice, developed by P. Leder and T. Stewart.

S. L. Mansour, K. R. Thomas, and M. R. Capecchi describe a general strategy for gene targeting in the laboratory mouse.

1989   W. Driever and C. Nüsslein-Volard show that in *Drosophila* the protein encoded by the *bicoid* gene acts by switching on the *hunchback* segmentation gene.

B. Zink and R. Paro show by immunostaining that a protein encoded by the *Polycomb (Pc)* gene binds to a limited number of discrete sites along the *Drosophila* polytene chromosomes. The sites include the *Antennapedia* complex and the *bithorax* complex, which contain genes known to be repressed by *Pc*.

L.-C. Tsui and 24 colleagues identify the cystic fibrosis gene, predict the amino acid sequence of the protein it encodes, and determine the nature of its most common mutant allele.

J. M. Bishop and H. E. Varmus receive the Nobel Prize in Medicine for their studies on the oncogenes of retroviruses.

T. R. Cech and S. Altman receive the Nobel Prize in Chemistry for their demonstration that certain RNAs have enzymatic functions.

Index of Scientists Listed in the Chronology

Arber, W., 1972, 1978

Arntzen, C. J., 1981

Auerbach, C., 1946

Avery, A. G., 1937

Avery, O. T., 1944

Avise, J. C., 1979

Baeckeland, E., 1959

Bailey, W. T., 1946

Balbiani, E. G., 1881

Baltimore, D., 1970, 1975

Banerji, J., 1981

Bankier, A. T., 1979

Banting, F. G., 1921

Barbacid, M., 1982

Barbosa, J. A., 1980

Barghoorn, E. S., 1954, 1977

Barnett, L., 1961, 1969

Barr, M. L., 1949

Barrell, B. G., 1965, 1976, 1979, 1981

Barski, G., 1960

Barter, R. H., 1967

Bates, L. S., 1964

Bateson, W., 1900, 1902–1909, 1906

Bauer, E., 1909

Bauer, H., 1934

Bauer, K. M., 1927

Bautz, E. K., 1969

Bawden, F. C., 1937

Beadle, G. W., 1935, 1941, 1944, 1958

Beatty, B. R., 1969

Beaudreau, G., 1965

Beckwith, J. R., 1969

Beemon, K., 1975

Beermann, W., 1952, 1961

Beijerinck, M. W., 1899

Belling, J., 1920, 1922, 1927

Benacerraf, B., 1963, 1972, 1980

*Benda, C., 1890 (1898)

Bender, W., 1983

Bennett, J. C., 1965

Benton, W. D., 1975

Benzer, S., 1955, 1969, 1971

Berg, P., 1972, 1980

Bergner, A. D., 1922

Bernstein, F., 1925

Berson, S. A., 1957

Bertram, E. G., 1949

Berzelius, J. J., 1838

Best, C. H., 1921

Bias, W. B., 1968

Billingham, R. E., 1953

Bingham, P. M., 1982

Birnstiel, M. L., 1966, 1967, 1969, 1972

Bishop, J. M., 1981, 1989

Bittner, J. J., 1936

Blackburn, E. H., 1978, 1985

Blake, C.C.F., 1967

Blakeslee, A. F., 1904, 1920, 1922, 1930, 1937

Blobel, G., 1975

Bogorad, L., 1981

Boivin, A., 1948

Bolle, A., 1964

Bolton, E. T., 1963

Bonner, D., 1944

Boon, C., 1969

Boveri, T., 1888, 1898, 1902

Boyce, R. P., 1964

Boycott, A. E., 1923

Boyer, H. W., 1972, 1973

Brachet, J., 1941

Bradshaw, A. H., 1952

Bragg, W. H., 1913

Bragg, W. L., 1913

Breathnach, R., 1977

Brenner, S., 1959, 1961, 1963, 1964, 1965, 1969, 1974

Brent, L., 1953

Brewin, N. J., 1981

Bridges, C. B., 1914, 1917, 1921, 1923, 1925, 1935

Briggs, J. H., 1959

Briggs, R., 1952

Britten, R. J., 1959, 1966, 1968

Brown, D. D., 1964, 1968, 1972

Brown, D. M., 1952

Brown, M. S., 1975, 1985

Brown, R., 1831

Brown, R. H., 1987

Brownlee, G. G., 1965, 1977

Buckton, K. E., 1971

Burgess, R., 1969

Burnet, F. M., 1959, 1960

*Burny, A., 1964 (1969)

Burrell, B. G., 1976

Butenandt, A., 1935

Bütschli, O., 1876

Cairns, J., 1963

Callan, H. G., 1958, 1977

Camerarius, J. R., 1694

Cameron, J. R., 1979

Campbell, A. M., 1962

Cann, R. L., 1987

Cantor, C. R., 1984

Capecchi, M. R., 1980, 1988

*In situations where a research contribution made at a later time is referred to in an earlier entry, both years are given. Thus, Benda, 1890 (1898) indicates that his 1898 publication is referred to in a 1890 entry (on Altmann's bioblasts).

Cappecchi, M., 1966
Carlson, P. S., 1972
Carlton, B. C., 1964
Carr, N. G., 1972
Carrier, W. L., 1964
Carson, H. L., 1985
Caspersson, T., 1936, 1941, 1970
*Castle, W. E., 1905 (1910)
Cech, T. R., 1981, 1989
Chada, K., 1986
Chakrabarty, A., 1980
Chalfie, M., 1981
Chambon, P., 1977
Chain, E., 1940, 1945
Chang, A.C.Y., 1972, 1973
Chapman, V., 1958
Chargaff, E., 1950
Chase, M., 1952
Chatton, E., 1937
Chetverikov, S. S., 1926
Chèvremont, M., 1959
Chèvremont-Comhaire, S., 1959
Chiba, Y., 1951
Childs, B., 1963
Chimenes, A. M., 1949
Chison, J. A., 1972
Chooi, W. Y., 1976
Cieciura, S. J., 1956
Clark, A. J., 1965
Claude, A., 1943, 1946, 1974
Clausen, R. E., 1925, 1926
Cleaver, J. E., 1968
Cleland, R. E., 1930
Clever, U., 1960
Clyde, D. F., 1976
Cohen, J. E., 1985
Cohen, S. N., 1972, 1973
Collins, J., 1977
Coons, A. H., 1941
Cornefert, F., 1960
Correns, C., 1900, 1909
Costantini, F., 1986
Coulson, A. R., 1975
Craig, L., 1965
Creech, H. J., 1941
Creighton, H. B., 1931
Crick, F.H.C., 1953, 1958, 1961, 1962, 1966
Crippa, M., 1968
Cuénot, L., 1905

Dahm, K. H., 1966
Dalgarno, L., 1974
Dalton, A. J., 1954
Dan, K., 1952
Dana, K., 1971

Dareste, C., 1874
Darlington, C. D., 1929, 1931
Darnell, J. E., 1971, 1977
Darwin, C., 1831, 1858, 1859
Dausset, J., 1954, 1980
Davidson, E. H., 1968
Davidson, N., 1968
Davidson, R. G., 1963
Davis, B. D., 1948
Davis, M., 1984
Davis, R. W., 1968, 1972, 1975, 1979
Dawid, I. B., 1968
Dayhoff, M. O., 1978
de Almeida, J. C., 1959
Dearing, R. D., 1972
Debergh, P., 1973
de Duve, C., 1955, 1974
de Graff, R., 1657
Delbrück, M., 1939, 1943, 1946, 1953, 1969
de Vries, H., 1900, 1901
d'Herelle, F., 1917
Diener, T. O., 1967
Dintzis, H., 1961
Di Peri, C., 1971
Dippell, R. V., 1976
Diver, C., 1923
Dobberstein, B., 1975
Dobzhansky, T., 1936, 1937, 1944
Dodge, B. O., 1927
Donahue, R. P., 1968
Doolittle, R., 1983
Doty, P., 1960
Dozy, A. M., 1976, 1978
Dressler, D., 1968
Dreyer, W. J., 1965
Driever, W., 1988, 1989
Drouin, J., 1979
Dudock, B., 1971
Duesberg, P. H., 1975
Dujon, B., 1974
Dulbecco, R., 1952, 1975
Dunn, J. J., 1969
Dzierzon, J., 1845

Earle, W., 1940
East, E. M., 1925
Edelman, G. M., 1959, 1969, 1972
Edgar, R. S., 1963, 1966
Edgell, M. H., 1974
Edwards, R. G., 1968
Efstratiadis, A., 1976, 1978
Egolina, N. A., 1972
Ehrlich, P., 1900
Eigner, J., 1960
Ellis, E. L., 1939

Ellis, J., 1983
Emrich, J., 1966
Ephrussi, B., 1935, 1949
Epstein, R. H., 1963
Evans, H. J., 1971

Farnham, M. E., 1920, 1922
Fawcett, D., 1956
Federoff, N. V., 1984
Felix, M. D., 1954
Feulgen, R., 1923
Fiers, W., 1973
Fieve, R. R., 1972
Finch, J. T., 1976
Finean, J. B., 1953
Finnegan, D. J., 1978
Fischer, E., 1902
Fisher, R. A., 1930
Flavell, R. A., 1977
Fleiss, J. L., 1972
Fleming, A., 1929, 1945
Flemming, W., 1879, 1882
Florey, H. W., 1940, 1945
Fol, H., 1877
Følling, A., 1934
Ford, C. E., 1959
Ford, P. J., 1973
Fox, M. S., 1964
Fraenkel-Conrat, H., 1955, 1956
Freda, V. J., 1964
Freese, E., 1959
Frey, L., 1970
Frey-Wyssling, A., 1961
Fritsch, E. F., 1979
Frohnhöfer, H. G., 1987

Gage, L. P., 1972
Gall, J. G., 1963, 1968, 1969, 1970, 1978
Galton, F., 1869, 1875, 1889
Garcia-Bellido, A., 1973
Gardiner, K., 1983
Garrod, A. E., 1909
Gautier, M., 1959
Gefter, M., 1969
Gehring, W. J., 1984
Geier, M. R., 1971
Gellert, M. F., 1976
Gey, G., 1951
Gibbons, I. R., 1963
Gierer, A., 1956, 1962
Gilbert, W., 1961, 1964, 1966, 1968, 1977, 1978, 1980
Gitschier, J., 1984
Glover, D. M., 1977
Godson, G. N., 1983
Goeddel, D. V., 1979

Golbus, M. S., 1976
Goldschmidt, R. B., 1915, 1921
Goldstein, J. L., 1975, 1985
Goldstein, L., 1967
Goldstein, L.S.B., 1982
Good, R. A., 1962
Goodman, H. M., 1972, 1986
Goodspeed, T. H., 1925, 1926
Gordon, J. W., 1980
Gorer, P. A., 1937, 1948
Gorman, J. G., 1964
Gorowsky, M. A., 1985
Goss, J., 1822–24
Goulian, M., 1967
Grabar, P., 1955
Grabowski, P. J., 1981
Grainger, R., 1977
Green, H., 1967
Green, K. C., 1949
Green, M. M., 1949
Greenwald, I. S., 1983
Greider, C. W., 1985
Griffith, F., 1928
Grigliatti, T., 1971
Gronemeyer, H., 1980
Gros, F., 1961
Gross, K. W., 1977
Grossfield, J., 1969
Grunberg-Manago, M., 1955
Grunstein, M., 1975
Guerrier-Takada, C., 1983
Guest, J. R., 1964
Gulick, J. J., 1872
Gurdon, J. B., 1962, 1964, 1967
Gutman, A., 1950

Hadorn, E., 1963
Haegerman, G., 1973
Haldane, J. B. S., 1915, 1927, 1930, 1935
Haldane, N. M., 1915
Hall, B. D., 1961
Hall, C. E., 1962
Hall, T. H., 1981
Hamilton, W. D., 1964
Hardison, R. C., 1978
Hardy, G. H., 1908
Hardy, R. W., 1982
Harper, M. E., 1981
Harris, H., 1965, 1966
Harris, J. I., 1961
Harrison, R. G., 1907
Hart, C. P., 1984
Hartmann, J. F., 1950
Haruna, I., 1965
Harvey, W., 1651
Hashimoto, H., 1933

Hayes, W., 1953
Hayflick, L., 1965
Hedges, R. W., 1974
Hedgpeth, J., 1972
Heitz, E., 1928
Helinski, D. R., 1964
Helling, R. B., 1973
Henderson, S. A., 1968
Henikoff, S., 1975
Henning, U., 1962, 1964
Hershey, A. D., 1946, 1949, 1952, 1969
Hertwig, O., 1875
Hess, O., 1968
Hiatt, H., 1961
Hiaga, A., 1970
Hilschmann, N., 1965
His, W., 1870
Hoagland, M. B., 1955
Hocks, P., 1965
Hoffman, E. P., 1987
Hoffmeister, H., 1965
Hofmeister, F., 1902
Hogness, D. S., 1975, 1977, 1978, 1986
Holland, I. B., 1965
Holley, R. W., 1965, 1968
Holliday, R., 1964
Holm, B., 1977
Hombrecher, G., 1981
Hooke, R., 1665
Horne, R. W., 1959
Horowitz, N. H., 1951
Horowitz, S., 1985
Horvitz, H. R., 1977, 1983
Hotta, Y., 1969, 1971
Hottinger, H., 1949
Howard, A., 1953
Howard-Flanders, P., 1964
Howell, S. H., 1971
Hoyer, B. H., 1972
Hozumi, N., 1976
Hradecna, Z., 1967
Hsu, L., 1972
Hubby, J. L., 1966
Huberman, J. A., 1968
Huebner, R. I., 1969
Hughes, W. L., 1957
Hummel, H., 1965
Humphrey, R. R., 1945
Hunt, W. G., 1973
Hurwitz, J., 1972
Hutchison, C. A., 1974, 1976, 1978

Illmensee, K., 1975
Ingram, V. M., 1957, 1961
Inman, R. B., 1970
Inouye, M., 1966

Irwin, M. R., 1951
Ishida, M. R., 1963
Itano, H. A., 1949

Jackson, D. A., 1972
Jackson, R. J., 1976
Jacob, A. E., 1974
Jacob, F., 1956, 1958, 1961(3), 1965
Jacobs, H. J., 1983
Jacobs, P. A., 1959
Jacobson, C. B., 1967
Jacq, C., 1977
Janssen, H., 1590
Janssen, Z., 1590
Janssens, F. A., 1909
Jeffreys, A., 1977
Jeffries, A. J., 1985
Jelinek, W., 1977
Jenner, E., 1798
Jerne, N. K., 1955, 1984
Johannsen, W., 1909
Johansson, C., 1970
John, H., 1969
Johnson, A.W.B., 1981
Johnson, P. F., 1988
Johnson, R. T., 1970
Jones, K. W., 1959, 1969
Jones, R. N., 1941
Josse, J., 1961
Jou, W. M., 1973

Kabat, E. A., 1939
Kafatos, F. C., 1976
Kaiser, A. D., 1961, 1972
Kan, Y. W., 1976, 1978
Kaplan, H. S., 1980
Kaplan, S., 1965
Karlson, P., 1960, 1965
Karström, H., 1937
Kaufman, R. J., 1978
Kavenoff, R., 1973
Kedes, L. H., 1972, 1977
Kellems, R. F., 1978
Kelley, T. J., 1968
Kelner, A., 1949
Kemp, D. J., 1977
Kemp, J. D., 1981
Kendrew, J. C., 1960, 1962
Kettlewell, H. B., 1956
Khorana, H. G., 1967, 1968, 1970
Kidwell, M. G., 1982
Kihara, H., 1923
Kim, J. J., 1973
Kim, S. H., 1973
Kimura, M., 1968, 1983

King, T. J., 1952
Kipnis, C., 1985
Kjeldgaard, N., 1950
Klebs, T. A. E., 1860
Klenk, E., 1935
Klug, A., 1976, 1982, 1985
Knight, T. A., 1822–24
Knapp, E., 1939
Knoll, A., 1977
Knoll, M., 1932
Koch, R., 1881
Köhler, G., 1975, 1984
Kohne, D. E., 1968, 1972
Kohno, T., 1970
Kölliker, A., 1841
Kölreuter, J. G., 1761–67
Konopka, R. J., 1971
Kornberg, A., 1956, 1959, 1961, 1967
Kornberg, R. D., 1974
Kramer, F. R., 1973, 1983
Kuehn, M. R., 1987
Kuhn, A., 1935
Kuhne, W., 1878
Kung, C., 1971
Kunkel, L. M., 1987
Kuhnlein, U., 1972
Kurland, C. G., 1961

Lacy, E., 1978
Lamarck, J. B. de Monet, 1809
Lande, R., 1981
Landschulz, W. H., 1988
Landsteiner, K., 1900, 1901, 1930
Landy, A., 1969
Langer, P. R., 1981
Lansman, R. A., 1979
Latta, H., 1950
Lauer, J., 1978
Lawn, R. M., 1979
Lawrence, P. A., 1975
Lazar, G., 1986
Leder, P., 1988
Lederberg, E. M., 1950, 1952
Lederberg, J., 1946, 1948, 1952(2), 1958
Leffak, M., 1977
Lehmann, R., 1987
Lein, J., 1948
Lejeune, J., 1959
LePinchon, X., 1968
Lerner, M. R., 1979
Leupold, U., 1951
Levan, A., 1956
Levine, B. B., 1963
Levine, P., 1939
Lewis, E. B., 1945, 1978
Lewontin, R. C., 1966

L'Héritier, P., 1934
Lima-de-Faria, A., 1959
Lindegren, C. C., 1953
Lindsley, D. L., 1982
Lingrel, J. B., 1969
Linné, C. V. (Linnaeus), 1735
Lipmann, F., 1961
Littauer, U. Z., 1961
*Little, C. C., 1905 (1910), 1909, 1914
Littlefield, J. W., 1964
Lobban, P., 1972
Lockard, R. E., 1969
Loh, E. Y., 1979
Lubs, H. A., 1969
Luck, D.J.L., 1964
Luria, S. E., 1942, 1943, 1945, 1969
Lwoff, A., 1950, 1965
Lyman, S., 1948
Lyon, M. F., 1961

Macdonald, P. M., 1988
MacGregor, H. G., 1958
MacLeod, C. M., 1944
Magram, J., 1986
Mak, T., 1984
Malathi, V. G., 1972
Malthus, T. R., 1798
Mandel, J. L., 1977
Mandel, M., 1970
Mangelsdorf, A. J., 1925
Mangold, H., 1918
Maniatis, T., 1976, 1978, 1979
Manning, J. E., 1971
Mansour, S. L., 1988
Mapes, M. O., 1958
*Marbaix, G., 1964 (1969)
Marcus, P. I., 1956
Margoliash, E., 1963
Margulies, A. D., 1965
Margulis, L., 1981
Mariuzza, R. A., 1986
Marmur, J., 1960
Marsh, T., 1983
Martin, A.J.P., 1941, 1952
Mason, S. J., 1976
Matthaei, J. H., 1961
Matthews, B. F., 1981
Maxam, A. M., 1976, 1977
Mayr, E., 1954, 1963
Mazia, D., 1952
McCarthy, B. J., 1963
McCarty, M., 1944
McClearn, G. E., 1962
McClintock, B., 1931, 1933, 1934, 1938, 1950, 1983
McClung, C. E., 1902
McDevitt, H. O., 1972

McGinnis, M. H., 1976
McGinnis, W., 1984
McIntosh, L., 1981
McKenzie, D. P., 1968
McKenzie, S. L., 1975
McKnight, S. L., 1988
McKusick, V. A., 1966, 1968
McLachlan, A. D., 1985
McPherson, A., 1973
McQuillen, K., 1959
Mears, K., 1958
Medawar, P. B., 1953, 1960
Mendel, G., 1856, 1865, 1866
Mendlewicz, J., 1972
Merrifield, R. B., 1965, 1984
Merril, C. R., 1971
Mertz, E. T., 1964
Mertz, J., 1972
Meselson, M., 1958, 1961 (2), 1975
Messing, J., 1984
Meyer, G., 1968
Miele, E. A., 1983
Miescher, F., 1871
Miller, C. O., 1956
Miller, J. F. A. P., 1962
Miller, J. R., 1977, 1985
Miller, L. H., 1976
Miller, O. L., 1969
Mills, D. R., 1965, 1967, 1973, 1983
Milstein, C., 1975, 1984
Mintz, B., 1967, 1975
Mirsky, A. E., 1968
Mitchell, H. K., 1948, 1974
Mohr, J., 1951
Monier, R., 1963
Monod, J., 1961(2), 1963, 1965
Montgomery, T. H., 1901
Moor, H., 1961
Morata, G., 1973, 1975
Morgan, J., 1968
Morgan, L. V., 1922
Morgan, T. H., 1910, 1911, 1912, 1919, 1933
Moses, M. J., 1956
Muhlenthaler, K., 1961
Mulder, G. J., 1838
Muller, H. J., 1916, 1918, 1927, 1946, 1948
Müller-Hill, B., 1966
Mullis, K. B., 1985
Murray, K. M., 1974
Murray, N. E., 1974

Nakamoto, T., 1961
Nass, M.M.K., 1966
Nasse, C. F., 1820
Nathans, D., 1971, 1978
Nathans, J., 1986
Neel, J. V., 1949

Nelson, O. E., 1964
Newbold, J. E., 1974
Newman, C. M., 1985
Nicholson, G. L., 1972
Nilsson Ehle, H., 1909
Nirenberg, M. W., 1961, 1968
Nitsch, C., 1973
Nitowsky, H. M., 1963
Noll, H., 1962, 1963
Noll, M., 1974, 1986
Nowell, P., 1960
Nussenzweig, V., 1983
Nüsslein-Volhard, C., 1980, 1987, 1988, 1989

O'Connell, C., 1978
Ochoa, S., 1955, 1956, 1959
Ohlendorf, D. H., 1981
Ohta, T., 1983
Ojida, A., 1963
Okada, Y., 1966
Okamoto, M., 1958
Okamoto, T., 1963
Okazaki, R. T., 1968
Old, R. W., 1977
Olins, A. L., 1974
Olins, D. E., 1974
Olsson, L., 1980
Ono, T., 1923
O'Riordan, M. L., 1971
Ouchterlony, O., 1948
Oudin, J., 1946
Owen, R. D., 1945, 1951

Pace, N., 1983
Page, D. C., 1987
Painter, T. S., 1933
Pak, W. L., 1969
Palade, G. E., 1952, 1956, 1960, 1965, 1974
Pardon, J. F., 1977
Pardue, M. L., 1969, 1970, 1972
Parker, R. C., 1981
Paro, R., 1989
Pasteur, L., 1861
Patterson, B. M., 1973
Patterson, J. T., 1952
Pauling, L., 1949, 1962
Pearson, K., 1900
Pelc, S. R., 1953
Pelham, H.R.B., 1976
Pelling, C., 1959
Perutz, M. F., 1960, 1962
Peterson, R. L., 1967
Petit, C., 1951
Petricciani, J. C., 1971
Philipson, L., 1971

Phillips, S.E.V., 1986
Pigott, G. H., 1972
Pirie, N. W., 1937
Plaut, W., 1962
Plotkin, D. J., 1980
Pohlman, R. F., 1984
Polani, P. E., 1959
Poljak, R. J., 1986
Pollack, W., 1964
Pongs, O., 1980
Pontecorvo, G., 1952
Porter, K. R., 1953
Porter, R. R., 1962, 1972
Potter, S. S., 1974
Prescott, D. M., 1967
Pribnow, D., 1975
Pritchard, R. H., 1955
Ptashne, M., 1966
Puck, T. T., 1956
Punnett, R. C., 1906

Quigley, G. J., 1973
Quon, D., 1978

Ramanis, Z., 1970
Rambousek, F., 1912
Randolph, L. F., 1928
Rao, P. N., 1970
Rapoport, J. A., 1946
Raspail, F. V., 1825
Rassoulzadegan, M., 1983
Raymer, W. B., 1967
Ready, D. F., 1986
Reddy, E. P., 1982
Reddy, V. B., 1978
Redi, F., 1668
Reich, E., 1964
Renwick, J. H., 1968
Reszelbach, R., 1971
Reynolds, R. K., 1982
Rhoades, M. M., 1938
Rich, A., 1962, 1973
Richards, O. C., 1971
Richardson, C. C., 1966
Riggs, A. D., 1968
Riley, R., 1958
Ripoll, P., 1973
Ris, H., 1962, 1969
Ritossa, F. M., 1962, 1965
Roberts, B. E., 1973
Roberts, R. B., 1959
Robertson, W.R.B., 1911
Robinson, J. A., 1971
Robson, J. M., 1946
Rodgers, D. A., 1962

Röller, H., 1966
Roper, J. A., 1952
Rossenbeck, H., 1923
Rosset, R., 1963
Roth, J. R., 1970
Rothman, R., 1949, 1965
Rous, P., 1910, 1966
Roux, W., 1883
Rubin, G. M., 1978, 1982
Ruddle, F., 1969, 1980, 1984
Rusconi, S., 1981
Ruska, E., 1932, 1986
Russell, L. B., 1961, 1963
Russell, R., 1969

Sabatini, D. D., 1965
Sager, R., 1963, 1970
Saiki, R. K., 1985
Sanger, F., 1952, 1958, 1965, 1975, 1977, 1980, 1981
Santos, E., 1982
Santos, J. K., 1923
Sarabhai, A. S., 1964
Sarich, V. M., 1967
Saunders, G. F., 1981
Scangos, G. A., 1980
Schaffner, S., 1981
Schell, J., 1974
Schildkraut, C., 1960
Schimke, R. T., 1978
Schimper, A.F.W., 1883
Schleiden, M. J., 1838–39
Schlesinger, M., 1934
Schlessinger, D. H., 1983
Schneider, A., 1873
Schoenheimer, R., 1942
Schramm, G., 1956
Schreiber, H., 1939
Schultz, J., 1936
Schwann, T., 1838–39
Schwartz, D. C., 1984
Schwartz, R. M., 1978
Scileppi, K., 1971
Scott, M. P., 1983
Scott-Moncrieff, R., 1936
Selander, R. K., 1973
Setlow, R. B., 1964
Seton, A., 1822–24
Shade, R. O., 1979
Shapiro, J. A., 1969
Shih, M.-C., 1986
Shine, J., 1974
Shull, G. H., 1909
Siekevitz, P., 1956, 1960
Silber, R., 1972
Sim, G. K., 1978
Siminovitch, L., 1950

Singer, S. J., 1949, 1959, 1972
Sinsheimer, R. L., 1959, 1967
Sjöstrand, F. S., 1953
Slizynska, H., 1938
Slonimski, P. P., 1974
Smith, E. F., 1907
Smith, H. H., 1972
Smith, H. O., 1968, 1970, 1978
Smith, J. D., 1969
Smithies, O., 1955, 1985
Smös, M., 1970
Sneden, D., 1973
Snell, G. D., 1942, 1948, 1953, 1980
Sonneborn, T. M., 1937, 1938
Sorieul, S., 1960
Sorsa, V., 1988
Southern, E. M., 1973, 1975
Spallanzani, L., 1769
Spemann, H., 1918, 1935
Spiegelman, S., 1961, 1965, 1967, 1973
Spiteller, G., 1965
Spradling, A. C., 1982
Sprunt, A. D., 1915
Stadler, L. J., 1928
Staehelin, T., 1962, 1963
Stahl, F. W., 1958
Stanley, W. M., 1935, 1946
Stark, G. R., 1977
Stebbins, G. L., 1974
Steffensen, D. M., 1970
Steinbeck, K. E., 1981
Steinmann, E., 1953
Steitz, J. A., 1979
Stern, C., 1931, 1936
Stern, H., 1971
Sternberg, P. W., 1983
Stetson, R. E., 1939
Steward, F. C., 1958
Stewart, J., 1965
Stewart, T., 1988
Stone, W. S., 1952
Stoneking, M., 1987
Stormont, C., 1951
Strasburger, E., 1875
Streisinger, G., 1966
Stretton, A.O.W., 1964, 1965
Strong, J. A., 1959
Struhl, G., 1988
Sturtevant, A. H., 1913, 1923, 1925, 1926, 1936
Suddath, F. L., 1973
Sulston, J., 1977, 1981
Sumner, J. B., 1926, 1946
Sutcliffe, J. G., 1979
Sutton, W. S., 1902
Suzuki, D. T., 1971
Suzuki, Y., 1972
Svec, P., 1983

Svedberg, T., 1923
Sweely, C. C., 1966
Symons, R. H., 1972
Synge, R.L.M., 1941, 1952
Szybalski, W., 1967, 1969

Takanami, M., 1963
Takeda, Y., 1981
Tanaka, Y., 1913
Tashiro, Y., 1965
Tatum, E. L., 1941, 1944, 1946, 1958
Taylor, J. H., 1957
Taylor, K., 1967
Teissier, G., 1934
Temin, H. M., 1970, 1975
Templeton, A. R., 1980
Terzaghi, E., 1966
Teuchy, H., 1974
Thien, S. L., 1985
Thomas, D., 1986
Thomas, D. Y., 1968
Thomas, K. R., 1988
Tiselius, A.W.K., 1933, 1939
Tissieres, A., 1974
Tjio, J. H., 1956
Todaro, G. I., 1969
Todd, A., 1952, 1957
Tokunaga, C., 1961
Tomlinson, A., 1986
Tonegawa, S., 1976, 1987
Tracy, U. M., 1974
Travers, A. A., 1969
Trost, B. M., 1966
Tryon, R. C., 1929
Tschermak, E., 1900
Tsugita, A., 1966
Tsui, L.-C., 1989
Turpin, R., 1959
Twort, F. W., 1915
Tyler, S. A., 1954

van Beneden, E., 1883
van Larebeke, N., 1974
van Leeuwenhoek, A., 1677
van Montagu, M., 1974
Varmus, H. E., 1981, 1989
Vendrely, C., 1948
Vendrely, R., 1948
Virchow, R., 1855
Visconti, N., 1953
Vogt, P. K., 1975
von Baer, K. E., 1827
von Behring, E., 1890, 1901
von Ehrenstein, G., 1961
von Mohl, H., 1837

Bibliography

Adelberg, E. A., editor 1960 *Papers on Bacterial Genetics*. Little, Brown, Boston.

Angier, N. 1988 *Natural Obsessions: The Search for the Oncogene*. Houghton Mifflin Co., Boston.

Babcock, E. B. 1950 *The Development of Fundamental Concepts in the Science of Genetics*. American Genetics Assoc., Washington, D. C.

de Beer, G. 1964 *Charles Darwin*. Doubleday, New York.

Bowler, P. J. 1984 *Evolution: The History of an Idea*. Harvard University Press, Cambridge, Mass.

Boyer, S. H., editor 1963 *Papers in Human Genetics*. Prentice-Hall, Englewood Cliffs, N.J.

Brock, T., editor 1961 *Milestones in Microbiology*. Prentice-Hall, Englewood Cliffs, N.J.

Brosseau, G. E., Jr. 1967 *Evolution*. W. C. Brown, Dubuque, Iowa. (A collection of classic papers on evolution).

Cairns, J., G. Stent, and J. D. Watson 1966 *Phage and the Origins of Molecular Biology*. Cold Spring Harbor Laboratory of Quantitative Biology, New York.

Carlson, E. A. 1966 *The Gene: A Critical History*. W. B. Saunders, Philadelphia.

Carter, G. S. 1957 *A Hundred Years of Evolution*. Macmillan, New York.

Corwin, H. O. and J. B. Jenkins, editors 1976 *Conceptual Foundations of Genetics: Selected Readings*. Houghton Mifflin Company, Boston.

Crick, F., 1988 *What Mad Pursuit: A Personal View of Scientific Discovery*. Weidenfeld and Nicholson, New York.

Dampier, W. C. 1943 *A History of Science*. Cambridge University Press, Cambridge.

Darwin, C. 1839 *Journal of Researches into the Natural History and Geology of the Countries Visited by HMS Beagle, Under the Command of Captain Fitzroy, R.N. from 1832 to 1836*. Henry Colburn Publishers, London.

Darwin, C. 1951 *The Origin of Species by Means of Natural Selection or the Preservation of Favoured Races in the Struggle for Life*. Philosophical Library, New York. (Reprint of the first edition published November 24, 1859).

Darwin, C. and A. R. Wallace 1958 *Evolution by Natural Selection*. Cambridge University Press, Cambridge, England. (This collection of essays includes the papers read by Darwin and Wallace before the Linnaean Society of London on July 1, 1858).

Dawes, B. 1952 *A Hundred Years of Biology*. Duckworth, London.

Dobzhansky, T. 1951 *Genetics and the Origin of Species*. Columbia University Press, New York.

Dobzhansky, T. 1970 *Genetics and the Evolutionary Process*. Columbia University Press, New York.

Dunn, L. C., editor 1951 *Genetics in the 20th Century*. Macmillan, New York.

Dunn, L. C. 1965 *A Short History of Genetics*. McGraw-Hill, New York.

Fischer, E. P. and C. Lipson 1988 *Thinking About Science: Max Delbrück and the Origins of Molecular Biology*. W. W. Norton & Company, New York.

Fisher, R. A. 1930 *The Genetical Theory of Natural Selection*. Oxford University Press, Oxford.

Fruton, J. S. 1972 *Molecules and Life. Historical Essays on the Interplay of Chemistry and Biology*. Wiley-Interscience, New York.

Gabriel, M. L. and S. Fogel, editors 1955 *Great Experiments in Biology*. Prentice-Hall, Englewood Cliffs, N.J.

Gall, J. G., K. R. Porter, and P. Siekevitz, editors 1981 *Discovery in Cell Biology*. Rockefeller University Press, New York.

Galton, F. 1869 *Hereditary Genius*, Macmillan & Co., London.

Gardner, E. J. 1960 *History of Life Science*. Burgess Publ. Co., Minneapolis.

Garrod, A. E. 1909 *Inborn Errors of Metabolism*. Reprinted by H. Harris (see below).

Glass, B., O. Temkin, and W. L. Straus, Jr. 1959 *Forerunners of Darwin, 1745–1859*. Johns Hopkins Press, Baltimore.

Goldschmidt, R. B. 1956 *Portraits from Memory*. University of Washington Press, Seattle.

Goldstein, L. 1966 *Cell Biology*. W. C. Brown, Dubuque, Iowa. (A collection of classic papers on cytology).

Grant, V. 1956 The Development of a Theory of Heredity. *Am. Sci.* **44**:158–179.

Haldane, J.B.S. 1932 *The Causes of Evolution*. Harper and Row, New York.

Hamburger, V., 1988 *The Heritage of Experimental Embryology: Hans Spemann and the Organizer.* Oxford University Press, New York.

Harris, H. 1963 *Garrod's Inborn Errors of Metabolism.* Oxford University Press, London.

Hsu, T. C. 1979 *Human and Mammalian Cytogenetics: An Historical Perspective.* Springer Verlag, New York.

Hughes, A. 1959 *A History of Cytology.* Abelard-Schuman, New York.

Irvine, W. 1955 *Apes, Angels, and Victorians.* McGraw-Hill, New York.

Judson, H. F. 1979 *The Eighth Day of Creation: Makers of the Revolution in Biology.* Simon & Schuster, New York.

Keller, E. F. 1983 *A Feeling for the Organism: The Life and Work of Barbara McClintock.* W. H. Freeman, San Francisco.

Kimura, M. 1983 *The Neutral Theory of Molecular Evolution.* Cambridge University Press, Cambridge, England.

Kohn, D. editor 1986 *The Darwinian Heritage.* Princeton University Press, Princeton, N.J.

Levine, L. editor 1971 *Papers on Genetics: A Book of Readings.* C. V. Mosby, St. Louis.

Lewis, E. B., editor 1961 *Selected Papers of A. H. Sturtevant on Genetics and Evolution.* W. H. Freeman, San Francisco.

Lwoff, A. and A. Ullman 1979 *Origins of Molecular Biology: A Tribute to Jacques Monod.* Academic Press, New York.

Manger, L. N. 1979 *A History of the Life Sciences.* Marcel Dekker, New York.

Margulis, L. 1981 *Symbiosis in Cell Evolution.* W. H. Freeman and Co., San Francisco.

Mayr, E. 1963 *Animal Species and Evolution.* Harvard University Press, Cambridge, Mass.

McCarty, M. 1985 *The Transforming Principle: Discovering That Genes Are Made of DNA,* W. W. Norton, New York.

McKusick, V. A. 1988 *Mendelian Inheritance in Man.* 8th edition, Johns Hopkins University Press, Baltimore, MD.

Moore, J. A. editor 1972 *Readings in Heredity and Development.* Oxford University Press, New York.

Moore J. A. editor, 1987 *Genes, Cells and Organisms: Great Books in Experimental Biology,* vol. 17. The collected papers of Barbara McClintock. The Discovery and Characterization of Transposable Elements. Garland Publishing Inc., New York.

Moore, R. 1961 *The Coil of Life. The Story of the Great Discoveries in Life Sciences.* Knopf, New York.

Morgan, T. H. 1926 *The Theory of the Gene.* Yale University, New Haven, Conn.

Muller, H. J. 1962 *Studies in Genetics.* Indiana University Press, Bloomington.

National Academy of Sciences of the USA; *Biographical Memoirs.* Columbia University Press, New York. Published at periodic intervals (Vol. 57 appeared in 1987).

Nobel Prize; *Lectures in Physiology/Medicine, 1942–1962, 1922–1941.* (2 vols published in 1964 and 1965, respectively), American Elsevier, New York.

Nordenskjöld, E. 1928 *The History of Biology, A Survey.* Knopf, New York.

Olby, R. 1974 *The Path to the Double Helix.* University of Washington Press, Seattle.

Oppenheimer, J. 1963 Theodor Boveri: The Cell Biologists' Embryologist. *Quart. Rev. Biol.* **38**:245–249.

Orel, V. 1984 *Mendel.* Oxford University Press, Oxford.

Patterson, J. T. and W. S. Stone 1952 *Evolution in the Genus* Drosophila. Macmillan, New York.

Peters, J. A., editor 1961 *Classical Papers in Genetics.* Prentice-Hall, Englewood Cliffs, N.J.

Provine, W. B. 1971 *The Origins of Theoretical Population Genetics.* University of Chicago Press, Chicago.

Portugal, F. H. and J. S. Cohen 1977 *A Century of DNA: A History of the Discovery of the Structure and Function of the Genetic Substance.* The MIT Press, Cambridge, Mass.

Provine, W. B. 1986 *Sewall Wright and Evolutionary Biology.* University of Chicago Press, Chicago.

Punnett, R. C. 1950 The Early Days of Genetics. *Heredity* **4**:1–10.

Royal Society; *Biographical Memoirs of Fellows of the Royal Society.* Burlington House, Picadilly, London, W 1. Published at periodic intervals (Vol. 34 appeared in 1988).

Singer, C. 1959 *A History of Biology.* Abelard-Schuman, New York.

Schoenheimer, R. 1942 *The Dynamic State of Body Constituents.* Harvard University Press, Cambridge, Mass.

Sorsa, V., 1988 *Chromosome Maps of Drosophila.* Vols. 1 and 2. CRC Press, Boca Raton, Florida.

Srinivasan, P. R., J. S. Fruton, and J. T. Edsall, editors 1979 *The Origin of Modern Biochemistry.* Ann. N.Y. Acad. Sci. **325**:1–375.

Stebbins, G. L. 1974 *Flowering Plants, Evolution Above the Species Level.* Harvard University Press, Cambridge, Massachusetts.

Stent, G. S., editor 1960 *Papers on Bacterial Viruses.* Little, Brown, Boston.

Stent, G. S. editor 1981 *The Double Helix: A Personal Account of the Discovery of the Structure of DNA by James D. Watson. Text, Commentary, Reviews and Original Papers.* W. W. Norton, New York.

Stern, C. and E. R. Sherwood 1966 *The Origin of Genetics, A Mendel Source Book.* W. H. Freeman, San Francisco. (contains a translation of Mendel's paper).

Stubbe, H. 1972 *History of Genetics from Prehistoric Times to the Rediscovery of Mendel's Laws.* Translated from the revised second German edition of 1965 by T. R. W. Walters. The MIT Press, Cambridge, Mass.

388

Sturtevant, A. H. 1965 *A History of Genetics.* Harper and Row, New York.

Taylor, J. H. editor 1965 *Selected Papers on Molecular Genetics.* Academic Press, New York.

Terzaghi, E. A., A. S. Wilkins, and D. Penny, editors 1984 *Molecular Evolution: An Annotated Reader.* Jones and Bartlett Publishers, Boston.

Thomas, L. 1975 *The Lives of a Cell: Notes of a Biology Watcher.* Bantam Books, New York.

Voeller, B. R. editor 1968 *The Chromosome Theory of Inheritance: Classic Papers in Development and Heredity.* Appleton-Century-Crofts, New York.

Watson, J. D. and J. Tooze 1981 *The DNA Story: A Documentary History of Gene Cloning.* W. H. Freeman, San Francisco.

White, M.J.D. 1973 *Animal Cytology and Evolution.* 3rd edition. Cambridge University Press, Cambridge, England.

Williams, T. I. editor 1982 *A Biographical Dictionary of Scientists,* 3rd edition. John Wiley and Sons, New York.

Willier, B. J. and J. M. Oppenheimer, editors 1964 *Foundations of Experimental Embryology.* Prentice-Hall, Englewood Cliffs, N.J.

Wilson, E. B. 1925 *The Cell in Development and Heredity.* 3rd edition. Macmillan, New York.

Wright, S. 1931 Evolution in Mendelian Populations. *Genetics* **16:**97–159.

Wright, S. 1968 *Evolution and the Genetics of Populations.* University of Chicago Press, Chicago, Illinois. Volumes 2, 3, and 4 were publsihed in 1969, 1977, and 1978, respectively.

Zubay, G. L. and J. Marmur, editors 1973 *Papers in Biochemical Genetics,* 2nd edition. Holt, Rinehart and Winston, New York.

Appendix D
Periodicals Cited in the Literature of Genetics, Cytology, and Molecular Biology

A listing of titles and addresses of publishers or distributors. The addresses of those firms publishing or distributing more than one journal are placed at the end of this list.

Acta Cytologica
Science Printers & Publishers, Inc.
2 Jaclynn Court
St. Louis, Missouri 63132
Acta Geneticae Medicae et Gemellologiae
Alan R. Liss, Inc.
Acta Genetica et Statistica Medica
(superseded by *Human Heredity*)
Acta Medica Auxologica
Centro Axologico Italiano DePiancavallo
Corso Magenta, 42
Milano 20123, Italy
Acta Paediatrica Scandinavica
Almqvist and Wiksell Periodical Co.
P.O. Box 638
S101 28 Stockholm, Sweden
Acta Virologica
Academic Press
Advances in Agronomy
Academic Press
Advances in Applied Microbiology
Academic Press
Advances in Biological and Medical Physics
Academic Press
Advances in Biophysics
University of Tokyo Press
7-3-1, Hongo
Bunkyo-ku
Tokyo, Japan
Advances in Cell and Molecular Biology
Academic Press
Advances in Cell Biology
Appleton-Century-Crofts
Advances in Clinical Chemistry
Academic Press
Advances in Enzymology
John Wiley and Sons

Advances in Experimental Medicine and Biology
Plenum Publishing Co.
Advances in Genetics
Academic Press
Advances in Human Genetics
Plenum Publishing Co.
Advances in Immunology
Academic Press
Advances in Metabolic Disorders
Academic Press
Advances in Microbial Physiology
Academic Press
Advances in Morphogenesis
Academic Press
Advances in Pediatrics
Year Book Medical Publishers
Advances in Protein Chemistry
Academic Press
Advances in Radiation Biology
Academic Press
Advances in Teratology
Academic Press
Advances in Viral Oncology
Raven Press
Advances in Virus Research
Academic Press
Agri Hortique Genetica
Plant Breeding Institute
Weibullsholm
Landskrona, Sweden
Agronomy Abstracts
American Society of Agronomy
Agronomy Journal
(formerly *Journal of the American Society of Agronomy*)
American Society of Agronomy
AIDS Research and Human Retroviruses
Mary Ann Liebert Inc. Publishers

American Journal of Diseases of Children
  American Medical Association
American Journal of Human Genetics
  University of Chicago Press
American Journal of Medical Genetics
  Alan R. Liss, Inc.
American Journal of Mental Deficiency
  Boyd Printing Company
  49 Sheridan Avenue
  Albany, New York 12216
American Naturalist
  University of Chicago Press
Animal Blood Groups and Biochemical Genetics
  Superseded by Animal Genetics
Animal Breeding Abstracts
  Commonwealth Agricultural Bureaux
Animal Cell Biotechnology
  Academic Press
Animal Genetics
  (formerly Animal Blood Groups and
  Biochemical Genetics)
  Blackwell Scientific Publishers, Ltd.
Animal Production
  Longman Publishers
Annales de Génétique
  Expansion Scientifique Francaise
  15 Rue Saint-Benoit
  75278 Paris Cedex 06, France
Annales de Génétique et da Sélection Animal
  Institut National de la Recherche Agronomique
Annales d'Embryologie et de Morphogenese
  Centre National de la Recherche Scientifique
  15 Quai Anatole France
  Paris (7e), France
Annales de l'Institut Pasteur
  Elsevier, Paris
Annals of Eugenics
  (superseded by the Annals of Human Genetics)
Annals of Human Biology
  Taylor and Francis, Ltd.
  4 John Street
  London WC IN 2ET, England
Annals of Human Genetics
  Cambridge University Press
Annals Paediatrici
  (superseded by Pediatric Research)
Annual Review of Biochemistry
  Annual Reviews, Inc.
Annual Review of Biophysics and Biophysical Chemistry
  Annual Reviews, Inc.
Annual Review of Cell Biology
  Annual Reviews, Inc.
Annual Review of Entomology
  Annual Reviews, Inc.
Annual Review of Genetics
  Annual Reviews, Inc.

Annual Review of Immunology
  Annual Reviews, Inc.
Annual Review of Medicine
  Annual Reviews, Inc.
Annual Review of Microbiology
  Annual Reviews, Inc.
Arabidopsis Information Service
  Dr. G. Röbbelen
  Institute of Agronomy
  University of Göttingen
  Göttingen 3H, Germany
Archiv für Genetik
  Orell Fussli Arts Graphiques S. A.
  Imprimeries "Au Froschauer"
  Dietzingerstrasse 3
  8022 Zürich, Switzerland
Archiv der Julius Klaus Stiftung für Vererbungsforschung,
  Sozial-anthropologie und Rassenhygiene
  (superseded by Archiv für Genetik)
Archives d'Anatomie Microscopique et de Morphologie
  Experimentale
  Masson et Cie
Archiv für Mikroskopische Anatomie und
  Entwicklungsmechanik
  Springer-Verlag
Archives of Biochemistry and Biophysics
  Academic Press
Archives of Disease in Childhood
  British Medical Association
Archives of Insect Biochemistry and Physiology
  Alan R. Liss, Inc.
Archives of Microbiology
  Springer-Verlag
Archives of Pediatrics
  (superseded by Clinical Pediatrics)
Archives of Virology
  Springer-Verlag
Aspergillus Newsletter
  Dr. J. A. Roper
  Department of Genetics
  University of Sheffield
  Sheffield S102 Tn, England
Atti Associazioni Genetica Italiana
  Dipartimento di Biologia
  Via Trieste, 75
  35100, Padua, Italy
Australian Journal of Agricultural Research
  Commonwealth Scientific and Industrial
  Research Organization
Australian Journal of Biological Sciences
  Commonwealth Scientific and Industrial
  Research Organization
Bacteriological Proceedings
  American Society for Microbiology
Bacteriological Reviews
  American Society for Microbiology

Barley Newsletter
>Dr. D. R. Metcalfe
>Canada Agriculture Research Station
>25 Dafue Road
>Winnipeg 19
>Manitoba, Canada

Behavior Genetics
>Plenum Publishing Co.

Bibliographia Genetica
>Martinus Nijhoff

Biken Journal
>Osaka University
>Research Institute for Microbial Diseases
>3 Dojima Nishimachi
>Kita-ku
>Osaka, Japan

Biochemical and Biophysical Research Communications
>Academic Press

Biochemical Genetics
>Plenum Publishing Co.

Biochemical Medicine and Metabolic Biology
>Academic Press

Biochemistry and Cell Biology
>National Research Council of Canada

Biochimica et Biophysica Acta
>Elsevier/North Holland

Biocytologia
>Masson et Cie.

BioEssays
>Cambridge University Press

Biological Bulletin
>Woods Hole Oceanographic Institution
>Woods Hole, Mass. 02543

Biological Reviews of the Cambridge Philosophical Society
>Cambridge University Press

Biology of the Cell
>Elsevier, Paris

Biomedical Ethics Reviews
>Human Press Inc.
>Cresent Manor
>P.O. Box 2148
>Clifton, New Jersey 07015

Biometrics
>Biometrics
>Suite 401
>1429 Duke Street
>Alexandria, Virginia 22314

Biophysical Journal
>Rockefeller University Press

Biophysik
>Springer-Verlag

Biopolymers
>John Wiley and Sons, Inc.

BioSystems
>Elsevier/North Holland

Biotechniques
>Eaton Publishing Company
>197 West Central Street
>Natick, Mass. 01760

Biotechnology and Bioenginnering
>John Wiley and Sons

Biotechnology and Genetic Engineering News
>Intercept Ltd.
>P.O. Box 402
>Wimborne, Dorsett BH 229TZ
>England

Blood
>W. B. Saunders Company
>The Curtis Center
>Independence Square West
>Philadelphia, Penn. 19106

Blut
>Springer-Verlag

Botanical Review
>New York Botanical Garden

Brain, Behavior and Evolution
>S. Karger AG

British Journal of Haematology
>Blackwell Scientific Publishers

British Medical Bulletin
>Longman Group Ltd.

British Poultry Science
>Longman Group Ltd.

Brookhaven Symposia in Biology
>Brookhaven National Laboratory
>Biology Department
>Upton, New York 11973

Canadian Journal of Genetics and Cytology
>Superseded by *Genome*

Canadian Journal of Microbiology
>National Research Council of Canada

Cancer Genetics and Cytogenetics
>Elsevier/North Holland

Carlsburg Research Communications
>Springer-Verlag

Carnivore Genetics Newsletter
>Carnivore Genetics Research Center
>Post Office Box 5
>Newtonville, Mass. 02160

Caryologia
>Journal of Cytology, Cytosystematics, and Cytogenetics
>Via Lamarmora 4
>Firenze, Italy

Cell
>Cell Press
>50 Church Street
>Cambridge, Mass. 02138

Cell and Tissue Kinetics
>Blackwell Scientific Publishers

Cell and Tissue Research

(formerly *Zeitschrift für Zellforschung und mikroskopische Anatomie*)
Springer-Verlag

Cell Biology Communications (Amsterdam)
(superseded by *Currents in Modern Biology*)

Cell Biology International Reports
Academic Press

Cell Differentiation
Elsevier/North-Holland

Cellular and Molecular Biology
Pergamon Press

Cellular Immunology
Academic Press

la Cellule
S. A. Vander
Munstraat 10, B-3000
Louvain, Belgium

Chinese Journal of Genetics
Allerton Press, Inc.
150 5th Avenue
New York, New York 10011

Chromosomes Today
Plenum Publishing Co.

Chromosoma
Springer-Verlag

Clinical Genetics
Munksgaard Forlag

Clinical Immunology and Immunopathology
Academic Press

Clinical Pediatrics
(formerly *Archives of Pediatrics*)
J. B. Lippincott

Cold Spring Harbor Symposia on Quantitative Biology
Cold Spring Harbor Laboratory

Comparative Biochemistry and Physiology
Pergamon Press

CRC Critical Reviews in Biochemistry

CRC Critical Reviews in Immunology

CRC Critical Reviews in Microbiology
Chemical Rubber Company Press
2000 NW Corporate Blvd.
Boca Raton, Florida 33431

Crop Science
Crop Science Society of America
677 South Segoe Road
Madison, Wisconsin 53711

Current Genetics
Springer-Verlag

Current Problems in Pediatrics
Year Book Medical Publishers

Currents in Modern Biology
(formerly *Cell Biology Communications, Amsterdam*)
North-Holland Publishing Co.

Current Topics in Cellular Regulation
Academic Press

Current Topics in Developmental Biology
Academic Press

Current Topics in Microbiology and Immunology
Springer-Verlag

Current Topics in Radiation Research
Elsevier/North-Holland

Cytobiologie
(superseded by the *European Journal of Cell Biology*)

Cytobios
Faculty Press
88 Regent St.
Cambridge, England

Cytogenetics
(superseded by *Cytogenetics and Cell Genetics*)

Cytogenetics and Cell Genetics
S. Karger AG

Cytologia (Tokyo)
Botanical Society of Japan
c/o The Toyo Bunko
Honkomagome 2-chome
28-21, Bunkyo-ku
Tokyo 113, Japan

Cytology and Genetics (USSR)
See *Tsitologiya i Genetika*

Cytopathology
Blackwell Scientific Publications

Development
(formerly *Journal of Embryology and Experimental Morphology*)
Company of Biologists, Ltd.

Developmental and Comparative Immunology
Pergamon Press

Developmental and Cell Biology
Cambridge University Press

Developmental Biology
Academic Press

Developmental Genetics
Alan R. Liss, Inc.

Development, Growth and Differentiation
(formerly *Embryologia*)
Maruzen Co. Ltd.
PO Box 605
Tokyo Central
Tokyo, Japan

Differentiation
Springer-Verlag

Dissertation Abstracts
University Microfilms
Xerox Company
300 North Zeeb Road
Ann Arbor, Michigan 48106

DNA - A Journal of Molecular and Cellular Biology
Mary Ann Liebert Inc. Publishers

DNA and Protein Engineering Techniques
Alan R. Liss, Inc.

DNA Repair Reports
  Elsevier/North Holland
Drosophila Information Service
  Dr. P. W. Hedrick
  Division of Biological Sciences
  University of Kansas
  Lawrence, Kansas 66045
Dysmorphology and Clinical Genetics
  Alan R. Liss, Inc.
Egyptian Journal of Genetics and Cytology
  Egyptian Society of Genetics
  Department of Genetics
  Faculty of Agriculture
  Alexandria University
  Alexandria, Egypt
Electron Microscopy Reviews
  Pergamon Press
EMBO Journal
  IRL Press at Oxford University Press
Embryologia
  (superseded by *Development, Growth,)
  Differentiation*
Endeavor
  Pergamon Press
Environmental and Molecular Mutagenesis
  Alan R. Liss, Inc.
Environmental Mutagen Society Newsletter
  Dr. F. J. deSerres
  P.O. Box Y
  Oak Ridge National Laboratory
  Oak Ridge, Tenn. 37830
Enzyme
  (formerly *Enzymologia, Biologica et Clinica*)
  S. Karger AG
Enzymologia
  (superseded by *Molecular and Cellular
  Biochemistry*)
Enzymologia Biologica et Clinica
  (superseded by *Enzyme*)
Eugenics Quarterly
  (superseded by *Social Biology*)
Eugenics Review
  (superseded by the *Journal of Biosocial Science*)
Euphytica
  Netherlands Journal of Plant Breeding
  Lawickse Allee 166
  Wageningen, Holland
European Journal of Biochemistry
  Springer-Verlag
European Journal of Cell Biology
  (formerly *Cytobiologie*)
  Wissenschaftliche Verlagsgesellschaft
European Journal of Immunology
  Academic Press
Evolution
  Society for the Study of Evolution

Entomology Department
  University of Kansas
  Lawrence, Kansas 66045
Evolutionary Biology
  Appleton-Century-Crofts
Experimental and Molecular Pathology
  Academic Press
Experimental Cell Biology
  S. Karger AG
Experimental Cell Research
  Academic Press
FASEB Journal
  (formerly *Federation Proceedings*)
  Federation of American Societies for
  Experimental Biology
  9650 Rockville Pike
  Bethesda, Maryland 20814
*Federation of European Biochemical Societies,
  Symposia*
  Elsevier/North-Holland
Federation Proceedings
  superseded by *FASEB Journal*
Fungal Genetics Newsletter
  (formerly *Neurospora Newsletter*)
  Fungal Genetics Stock Center
  Department of Microbiology
  University of Kansas Medical School
  Rainbow Boulevard at 39th Street
  Kansas City, Kansas 66103
Gamete Research
  Alan R. Liss, Inc.
Gene
  Elsevier/North-Holland
Gene Analysis Techniques
  Elsevier/North Holland
Genen en Phaenen
  Journal of the Dutch Genetical Society
  Institute of Genetics
  State University of Utrecht
  Opaalweg 20
  Utrecht, Holland
Genes and Development
  Cold Spring Harbor Laboratory
Genes, Chromosomes and Cancer
  Alan R. Liss, Inc.
Genetica
  Kluwer Academic Publishers
Genetica Agraria
  Isitituto Sperimantale per la Cerealicoltura
  Via Cassia 176
  00191, Rome, Italy
Genética Ibérica
  Sectión de Distribución de Publicaciones
  C.S.I.C.
  Vitrubio, 8
  Madrid 6, Spain

Genetical Research
Cambridge University Press
Genetica Polonica
Institute Plant Genetics
Polish Academy of Science
ul, Wojsha Polskiego 71c
Poznan, Poland
Genetic Engineering
Academic Press
Genetic Engineering: Principles and Methods
Plenum Press
Genetic Epidemiology
Alan R. Liss, Inc.
Genetics
Genetics
428 East Preston Street
Baltimore, Maryland 21202
Genetic Stocks Inventory
National Seed Storage Laboratory
Fort Collins, Colorado 70521
Genetik
Gustav Fischer Verlag
Genetika
See *Soviet Genetics*
Genetika i Selekcija
Academy of Agricultural Sciences of Bulgaria
Bul. Dragon Cankov 6
Sofia, Bulgaria
Genome
(formerly *Canadian Journal of Genetics and Cytology*)
National Research Council of Canada
Genomics
Academic Press
Harvey Lectures
Academic Press
Héreditas
J. D. Törnqvists Bokhandel AB
Landskrona, Sweden
Heredity
Longman Group Ltd.
Histochemical Journal
Chapman and Hall
11 New Fetter Lane
London EC4P 4EE, England
Histochemistry
Springer-Verlag
Hoja Genetica
Sociedad Rioplatense de Genetica
Colonia, Uruguay
Human Biology
Wayne State University Press
5959 Woodward Ave.
Detroit, Michigan 48202
Human Genetics
(formerly *Humangenetik*)
Springer-Verlag

Humangenetik
(formerly *Zeitschift für menschliche Vererbungs-und Konstitutionslehre,* renamed *Human Genetics*)
Human Heredity
(formerly *Acta Genetica et Statistica Medica*)
S. Karger AG
Hybridoma
Mary Ann Liebert, Inc. Publishers
Immunobiology
Gustav Fischer Verlag
Immunochemistry
Pergamon Press
Immunogenetics
Springer-Verlag
Immunological Communications
Marcel Dekker, Inc.
270 Madison Avenue
New York, New York 10016
Immunological Reviews
Munksgaard Förlag
Immunology
Blackwell Scientific Publications
Immunology Letters
Elsevier/North-Holland
Immunology Today
Elsevier/North-Holland
Indian Journal of Agricultural Sciences
Indian Council of Agricultural Research
Business Manager
Krishi Bhavan
New Delhi, India
Indian Journal of Biochemistry and Biophysics
Council of Scientific and Industrial Research
Public Information Directorate
Hillside Road
New Delhi 12, India
Indian Journal of Genetics and Plant Breeding
Indian Agricultural Research Institute
New Delhi 12, India
Indian Journal of Poultry Science
Indian Poultry Science Association
Krishi Bhavan
New Delhi, India
Infection and Immunity
American Society for Microbiology
Information Newsletter of Somatic Cell Genetics
Dr. R. A. Roosa
Wistar Institute
36th Street at Spruce
Philadelphia, Penn. 19101
International Journal of Insect Morphology and Embryology
Pergamon Press
International Journal of Invertebrate Reproduction and Development
Elsevier/North-Holland

International Journal of Peptide and Protein Research
  Munksgaard Förlag
International Journal of Radiation Biology
  Taylor and Francis Ltd.
  4 John Street
  London, WC1N 2ET
  England
International Review of Cytology
  Academic Press
International Society for Cell Biology, Symposia
  Academic Press
Intervirology
  S. Karger AG
In Vitro
  Williams and Wilkins
Jackson Laboratory, Annual Report
  The Jackson Laboratory
  Bar Harbor, Maine 04609
Japanese Journal of Breeding
  Japanese Society of Breeding
  c/o Faculty of Agriculture
  University of Tokyo
  Bunkyo-ku
  Tokyo 113, Japan
Japanese Journal of Developmental Biology
  Yamashiro Publishing Co. Ltd.
  Ogawa-Nishiiru
  Teranouchidori, Kamikyo-ku
  Kyoto, Japan
Japanese Journal of Genetics
  Japan Publications Trading Co.
Japanese Journal of Human Genetics
  Institute of Medical Genetics
  Tokyo Medical and Dental University
  Yushima 1-5, Bunkyo-ku
  Tokyo, Japan
Japanese Journal of Microbiology
  Japan Bacteriologists and Virologists
  Department of Microbiology
  Keio University School of Medicine
  Shinamachi, Shinjuku
  Tokyo, Japan
Journal de Génétique Humaine
  Institute of Medical Genetics
  8, Chemin Thury
  1205 Genéve, Switzerland
Journal de Microscopie
  Société Française de Microscopie
  Electronique
  Ecole Normale Supérieure
  Laboratorie de Botanique
  4, Rue Lhomond
  Paris 5e, France
Journal of Agricultural Research
  U.S. Department of Agriculture
  14th Street & Independence Avenue S.W.
  Washington, D.C. 20250

Journal of Agricultural Science
  Cambridge University Press
Journal of Animal Breeding and Genetics
  Paul Parey Scientific Publishers
  35-37 West 38th Street
  New York, New York 10018
Journal of Applied Bacteriology
  Academic Press
Journal of Bacteriology
  American Society for Microbiology
Journal of Biochemistry
  Japanese Biochemical Society
  Department of Biochemistry
  Faculty of Medicine
  University of Tokyo
  Tokyo, Japan
Journal of Biological Chemistry
  Williams and Wilkins
Journal of Biophysical and Biochemical Cytology
  (superseded by *Journal of Cell Biology*)
Journal of Biosocial Science
  (formerly *Eugenics Review*)
  Blackwell Scientific Publishers
Journal of Cell Biology
  (formerly *Journal of Biophysical and Biochemical Cytology*)
  Rockefeller University Press
Journal of Cell Science
  (formerly *Quarterly Journal of Microscopical Science*)
  Company of Biologists, Ltd.
Journal of Cellular Physiology
  (formerly *Journal of Cellular and Comparative Physiology*)
  Alan R. Liss, Inc.
Journal of Chemical Technology and Biotechnology
  Elsevier/North-Holland
Journal of Chronic Diseases
  Pergamon Press
Journal of Cytology and Genetics
  Banaras Hindu University
  Varanasi, India
Journal of Dairy Research
  Cambridge University Press
Journal of Dairy Science
  American Dairy Science Association
  113 North Neil Street
  Champaign, Illinois 61820
Journal of Embryology and Experimental Morphology
  (superseded by Development)
Journal of Experimental Biology
  Company of Biologists, Ltd.
Journal of Evolutionary Biology
  Berkhäuser Verlag AG
  P.O. Box 133
  CH-4010 Basel, Switzerland

Journal of Experimental Zoology
    Alan R. Liss, Inc.
Journal of General and Applied Microbiology
    Journal Press
Journal of General Microbiology
    Cambridge University Press
Journal of General Virology
    Cambridge University Press
Journal of Genetic Psychology
    Journal Press
Journal of Genetics
    Indian Academy of Sciences
    P.B. 8005
    Bangalore 560080, India
Journal of Heredity
    Oxford University Press
Journal of Histochemistry and Cytochemistry
    Elsevier/North-Holland
Journal of Horitcultural Science
    Headley Brothers Ltd.
    The Invicta Press
    Ashford, Kent, England
Journal of Human Evolution
    Academic Press
Journal of Immunogenetics
    Blackwell Scientific Publications
Journal of Immunological Methods
    Elsevier/North-Holland
Journal of Immunology
    Williams and Wilkins
Journal of Inherited Metabolic Disease
    Kulwer Academic Publishers
Journal of Insect Physiology
    Pergamon Press
Journal of Medical Genetics
    Publishing Department
    British Medical Association House
    Tavistock Square
    London WC1H9JR, England
Journal of Medical Primatology
    Alan R. Liss, Inc.
Journal of Medical Virology
    Alan R. Liss, Inc.
Journal of Mental Deficiency Research
    Blackwell Scientific Publications
Journal of Molecular and Applied Genetics
    Raven Press
Journal of Molecular Biology
    Academic Press
Journal of Molecular and Cellular Immunology
    Springer-Verlag
Journal of Molecular Evolution
    Springer-Verlag
Journal of Morphology
    Alan R. Liss, Inc.
Journal of Neurogenetics
    Harwood Academic Publishers

Journal of Obstetrics and Gynaecology of the British Commonwealth
    Royal College of Obstetricians and
    Gynecologists
    27 Sussex Place
    Regents Park
    London N.W. 1, 4RG
    England
Journal of Pediatrics
    C. V. Mosby Co.
Journal of Submicroscopic Cytology
    Editrice Compositori Bologna
    Viale XII Giugno 3
    40124 Bologna, Italy
Journal of the American Society of Agronomy
    (superseded by the *Agronomy Journal*)
Journal of the Australian Institute of Agricultural Science
    Australian Institute of Agricultural Science
    191 Royal Parade
    Parkville, Victoria 3052
    Australia
Journal of the European Society for Animal Blood Group Research
    listed under *Animal Blood Groups and Biochemical Genetics*
Journal of the National Cancer Institute
    National Cancer Institute
    Bethesda, Maryland 20892
Journal of Theoretical Biology
    Academic Press
Journal of the Sericultural Society of Japan
    Sericultural Experiment Station
    3-55-30 Wada
    Suginami-ku
    Tokyo, Japan 166
Journal of Ultrastructure and Molecular Structure Research
    Academic Press
Journal of Virology
    American Society for Microbiology
Kihara Institute for Biological Research, Reports
    Kihara Institute for Biological Research
    Mutsukawa, Minami-ku
    Yokahama, Japan
Laboratory Primate Newsletter
    Dr. A. M. Schrier
    Psychology Department
    Brown University
    Providence, Rhode Island 02912
Lancet
    Lancet Ltd.
    46 Bedford Square
    London WCI8 3SL, England
Life Sciences
    Pergamon Press
Maize Genetics Corporation Newsletter
    Dr. M. M. Rhoades

Botany Department
Indiana University
Bloomington, Indiana 47401
Mammalian Chromosome Newsletter
D. T. C. Hsu
Section of Cell Biology
M.D. Anderson Hospital
Houston, Texas 77025
Mankind Quarterly
American Philosophical Society
Library
105 South 5th St.
Philadelphia, Penn. 19106
Medical Genetics
Pergamon Press
Metabolism
Grune and Stratton
Methods in Cell Biology
Academic Press
Methods in Cell Physiology
Academic Press
Methods In Enzymology
Academic Press
Methods in Immunology and Immunochemistry
Academic Press
Methods in Medical Research
Year Book Medical Publishers
Methods in Virology
Academic Press
Microbial Genetics Bulletin
Dr. H. Adler
P.O. Box Y
Oak Ridge National Laboratory
Oak Ridge, Tenn. 37830
Microbiological Reviews
American Society for Microbiology
Microbiological Sciences
Blackwell Scientific Publications
Mikrobiologiya
Mezhdunarodnaya Kniga
Modern Cell Biology
Alan R. Liss, Inc.
Molecular and Cellular Biochemistry
(formerly *Enzymologia*)
Kluwer Academic Publishers
Molecular and Cellular Biology
American Society for Microbiology
Molecular and General Genetics
(formerly *Zeitschrift für
Verebungslehre*)
Springer-Verlag
Molecular Biology and Evolution
University of Chicago Press
Molecular Biology and Medicine
Academic Press
Molecular Biology Reports
Kluwer Academic Publishers

Molecular Biology SSSR
Mezhdunarodnaya Kniga
Molecular Carcinogenesis
Alan R. Liss, Inc.
Molecular Immunology
Pergamon Press
Molecular Microbiology
Blackwell Scientific Publications
Molecular Microbiology and Medicine
Academic Press
Molecular Pharmacology
Academic Press
Molecular Reproduction and Development
Alan R. Liss, Inc.
Molekulyarnaya Biologiya
See *Molecular Biology SSSR.*
Monatshefte für Chemie
Springer-Verlag
Monatsschrift für Kinderheilkunde
Springer-Verlag
Monographs in Human Genetics
S. Karger AG
Mosquito News
American Mosquito Control Association
Box 278
Selma, California 93662
Mouse Newsletter
Dr. A. G. Searle, editor
M. R. C. Radiobiology Unit
Harwell, Didcot
Berkshire, England
Mutagenesis
IRL Press at Oxford University Press
Mutation Research
Elsevier/North-Holland
Mycologia
New York Botanical Garden
National Cancer Institute Monographs
(superseded by *NCI Monographs*)
National Institute of Genetics (Mishima), Annual Report
National Institute of Genetics
Yata 1, 111 Mishima
Sizuoka-ken 411, Japan
Nature
Macmillan
Naturwissenschaften
Springer-Verlag
NCI Monographs
(formerly *National Cancer Institute
Monographs*)
Superintendent of Documents
U.S. Government Printing Office
Washington, D.C. 20402
Neurospora Newsletter
(superseded by *Fungal Genetics Newsletter*)
Nucleic Acids Research
IRL Press at Oxford University Press

The Nucleus
    Cytogenetics Laboratory
    Department of Botany
    University of Calcutta
    35, Ballygunge Circular Rd.
    Calcutta 19, India
Oak Ridge National Laboratory Symposia
    (supplements to the *Journal of Cell Physiology*)
    Wistar Institute Press
Oat Newsletter
    Dr. J. A. Browning
    Department of Botany
    Iowa State University
    Ames, Iowa 50010
Obstetrics and Gynecology
    Elsevier/North-Holland
Oncogene
    Macmillan Journals Ltd.
Oncogene Research
    Harwood Academic Publishers
Origins of Life
    Kluwer Academic Publishers
Oxford Surveys in Evolutionary Biology
    Oxford University Press
Oxford Surveys on Eukaryotic Genes
    Oxford University Press
Pasteur Institute
    (see *Annales de l'Institut Pasteur*)
Pathologia et Microbiologia
    S. Karger AG
Pediatric Research
    (formerly *Annales Paediatrici*)
    Williams and Wilkins
Pediatrics
    American Academy of Pediatrics
    P.O. Box 927
    Elk Grove Village, Illinois 60007
Perspectives in Biology and Medicine
    University of Chicago Press
Philosophical Transactions of the Royal Society of London, Series B, Biological Sciences
    The Royal Society
    6 Carlton House Terrace
    London SW1Y 5AG, England
Photochemistry and Photobiology
    Pergamon Press
Phytochemistry
    Pergamon Press
Pisum Newsletter
    Weibullsholm Plant Breeding Institute
    Landskrona, Sweden
Planta
    Springer-Verlag
Plant Breeding Abstracts
    Commonwealth Agricultural Bureau
Plant Molecular Biology
    Kluwer Academic Publishers

Plasmid
    Academic Press
Poultry Science
    Poultry Science Association
    Texas A & M University
    College Station, Texas 77843
Proceedings of the National Academy of Sciences of the United States of America
    National Academy of Sciences
    2101 Constitution Avenue
    Washington, D.C. 20418
Proceedings of the Royal Society of Edinburgh, Section B (Biological Sciences)
    Royal Society of Edinburgh
    22 George St.
    University of Edinburgh
    Edinburgh EH 2-2 PQ, Scotland
Proceedings of the Royal Society of London, Series B, Biological Sciences
    Royal Society of London
    6 Carlton House Terrace
    London SW1Y SAG, England
Progress in Biophysics and Molecular Biology
    Pergamon Press
Progress in Histochemistry and Cytochemistry
    Gustav Fischer Verlag
Progress in Medical Genetics
    Grune and Stratton
Progress in Medical Virology
    S. Karger AG
Progress in Nucleic Acid Research and Molecular Biology
    Academic Press
Progress in Theoretical Biology
    Academic Press
Protein, Nucleic Acid, and Enzyme
    Kyoritsu Shippan Co., Ltd. Publishers
    4-6-19 Kohinata
    Bunkyo-ku
    Tokyo, Japan
Proteins; Structure, Function and Genetics
    Alan R. Liss, Inc.
Protoplasma
    Springer-Verlag
Quarterly Journal of Microscopical Science
    (superseded by the *Journal of Cell Science*)
Quarterly Review of Biology
    Stony Brook Foundation
    State University of New York
    Stony Brook, New York 11790
Quarterly Reviews of Biophysics
    Cambridge University Press
Radiation Botany
    Pergamon Press
Radiation Research
    Academic Press
Reports of the Tomato Genetics Cooperative

c/o Professor C. M. Rich
Department of Vegetable Crops
University of California
Davis, California 95616

Resumptio Genetica
Martinus Nijhoff

Revista Brasileira de Genetica
Sociedade Brasileira de Genetica
Departamento de Genetica
Faculdade de Medicina de Ribeirao Preto
14100 Ribeirao Preto, S.P.
Brazil

Revue Suisse de Zoologie
Revue Suisse de Zoologie
Muséum d'Histoire Naturelle
Genève, Switzerland

Science
American Association for the
Advancement of Science
1515 Massachusetts Ave. N.W.
Washington, D.C. 20005

Scientific Agriculture
Agricultural Institute of Canada
151 Slater Street
Ottawa, Ontario, Canada

Scientific American
415 Madison Ave.
New York, New York 10017

Seiken Ziho
Kihara Seibutsugaku Kenkyusho
Mazume, Muko-machi
Otokunigun, Kyoto-fu
Japan

Sequence: The Journal of DNA Mapping and Sequencing
Harwood Academic Publishers

Sexual Plant Reproduction
Springer-Verlag

Silvae Genetica
(formerly *Zeitschrift für Forstgenetik und
Forstpflanzen-Züchtung*)
J. P. Sauerlander's Verlag
Fruhenofstrasse 21
6 Frankfurt am Main,
West Germany

Social Biology
(formerly *Eugenics Quarterly*)
University of Chicago Press

Society for Developmental Biology, Symposia
Princeton University Press
Princeton, New Jersey 08540

Society for Experimental Biology, Symposia
Cambridge University Press

Society for General Microbiology, Symposia
Cambridge University Press

Somatic Cell and Molecular Genetics
Plenum Publishing Company

Soviet Genetics

(an English translation of *Genetika*)
Plenum Publishing Co.

Stadler Genetics Symposia
University of Missouri Press
Columbia, Missouri 65201

Stain Technology
Williams and Wilkins

Studies in Drosophila Genetics
See *University of Texas Publications.*

Sub-Cellular Biochemistry
Plenum Press

Symposia Genetica
Istituto di Zoologia
Piazza Botta
Pavia, Italy

Teratology, Carcinogenesis and Mutagenesis
Alan R. Liss, Inc.

Theoretical and Applied Genetics
(formerly *der Züchter*)
Springer-Verlag

Theoretical Population Biology
Academic Press

Tissue and Cell
Longmans Group, Ltd.

Tissue Antigens
Munksgaard Förlag

Transactions of the British Mycological Society
Cambridge University Press

Transactions of the New York Academy of Sciences
New York Academy of Sciences

Transplantation
Williams and Wilkins

Transplantation Proceedings
Grune & Stratton, Inc.

Transplantation Reviews
Williams and Wilkins

Trends in Biochemical Sciences
Elsevier/North-Holland

Trends in Biotechnology
Elsevier/North-Holland

Trends in Ecology and Evolution
Elsevier/North-Holland

Trends in Genetics
Elsevier/North-Holland

Tribolium *Information Bulletin*
Dr. A. Sokoloff
Natural Sciences Division
California State College
San Bernadino, Cal. 92407

Trudy Instituta Genetiki i Selektsii
Biblioteka Akademia Nauk SSR
Birgevaja Lenaja 1
Lenningrad, B-164, U.S.S.R.

Tsitologiya
Mezhdunarodnaya Kniga

Tsitologiya i Genetika
Akademia Nauk URSU, Kiev.

Institut Botaniki
Ul. Repina 4
Kiev, U.S.S.R.
UCLA Symposia on Molecular and Cellular Biology
Alan R. Liss, Inc.
Ultrastructure in Biological Systems
Academic Press
University of Texas, M. D. Anderson Hospital and
Tumor Institute—Symposia on Fundamental Cancer
Research
Texas Medical Center
Houston, Texas 77025
University of Texas Publications (Studies in the Genetics
of Drosophila, Studies in Genetics)
University Station
Austin, Texas 78712
Vector Genetics Information Service
Vector Biology and Control Unit
World Health Organization
Avenue Appia
Geneva, Switzerland
Virology
Academic Press
Virus Research
Elsevier/North-Holland
Wheat Newsletter
Dr. E. G. Heyne
Agronomy Dept.
Kansas State University
Manhattan, Kansas 66502
Wilhelm Roux's Archives of Developmental Biology
Springer-Verlag
World's Poultry Science Journal

National Poultry Tests Ltd.
Eaton, Godalming
Surrey, England
Yearbook of Obstetrics and Gynecology
Year Book Medical Publishers
Yearbook of Pediatrics
Year Book Medical Publishers
Yeast
John Wiley and Sons
Zeitschrift für Forstgenetik und Forstpflanzen-Züchtung
(superseded by *Silvae Genetica*)
Zeitschrift für Immunitätsforschung
Gustav Fischer Verlag
Zeitschrift für induktive Abstammungs-und
Verebungslehre
(superseded by *Zeitschrift für Verebungslehre*)
Zeitschrift für menschliche Vererbungs-und
Konstitutionslehre
(superseded by *Humangenetik*)
Zeitschrift für Naturforschung
Verlag der Zeitschrift für Naturforschung
Uhlandstrasse 11, P.O. Box 2645
D7400 Tübingen, West Germany
Zeitschrift für Vererbungslehre
(formerly *Z.f. induktive Abstammungs-und*
Vererbungslehre)
superseded by *Molecular and General Genetics*
Zeitschrift für Zellforschung und mikroskopische
Anatomie
(superseded by *Cell and Tissue Research*)
der Züchter
(superseded by *Theoretical and Applied*
Genetics)

Multijournal Publishers

Academic Press, 1250 Sixth Avenue, San Diego, California 92101
Alan R. Liss, Inc., 41 East 11th Street, New York, New York 10003
American Medical Association, 535 North Dearborn Street, Chicago, Illinois 60610
American Society of Agronomy, 677 South Segoe Road, Madison, Wisconsin 53711
American Society for Microbiology, 1913 I Street N. W., Washington, D.C. 20006
Annual Reviews, Inc., 4139 E. Camino Way, Palo Alto, California 94303
Appleton-Century-Crofts, 440 Park Ave. South, New York, New York 10016
Blackwell Scientific Publications Ltd., P.O. Box 88, Oxford, England
British Medical Association, Tavistock Square, London WC1H 9JR, England
Cold Spring Harbor Laboratory, P.O. Box 100, Cold Spring Harbor, New York 11724
Cambridge University Press, 32 East 57th St., New York, New York 10022
Commonwealth Agricultural Bureaux, Farnham Royal, Bucks, England
Commonwealth Scientific and Industrial Research Organization, 314 East Albert St., East Melbourne, Victoria 3002, Australia
Company of Biologists Ltd., Department of Zoology, University of Cambridge, Downing Street, Cambridge CB23EJ, England.
C. V. Mosby Co., 11830 Westline Industrial Dr., St. Louis, Missouri 63141
Elsevier, 29 Rue Buffon, 75005 Paris, France
Elsevier/North-Holland, 52 Vanderbilt Ave., New York, New York 10017
Grune & Stratton Inc., 111 5th Ave., New York, New York 10003
Gustav Fischer Verlag, P.O. Box 7- 20143, Stuttgart, Germany
Harper & Row Publishers, 2350 Virginia Ave., Hagerstown, Maryland 21740
Harwood Academic Publishers, P.O. Box 786, Cooper Station, New York, New York 10276
Institut National de la Recherche Agronomique, Service des Publication, Route de Saint Cyr, 78-Versailles, France
IRL Press at Oxford University Press, 200 Madison Avenue, New York, New York 10016
Japan Publications Trading Co. Ltd., 175 5th Avenue, New York, New York 10010
J. B. Lippincott Co., East Washington Square, Philadelphia, Pennsylvania 19105
John Wiley and Sons, 605 3rd Avenue, New York, New York 10158
Journal Press, 2 Commercial Street, Provincetown, Massachusetts 02657
Kluwer Academic Publishers, P.O. Box 322, 3300 AH Dordrecht, Holland
Longman Group Ltd., 43-45 Annandale Street, Edinburgh, Scotland
Macmillan Journals Ltd., 4 Little Essex Street, London, WC2R 3LF, England
Martinus Nijhoff, P.O. Box 269, The Hague, Holland
Mary Ann Liebert, Inc. Publishers, 1651 Third Avenue, New York, New York 10128
Masson et Cie., 120 Boulevard Saint Germain, F75280, Paris 6e, France
Mezhdunarodnaya Kniga, Moscow G-200, USSR
Munksgaard Förlag, 35 Norre Sogade, DK 1016, Copenhagen K, Denmark
National Research Council of Canada, Research Journals Publishing Dept., Ottawa 2, Ontario, Canada K1AOR6
New York Botanical Garden, Bronx, New York 10458
Oliver and Boyd Ltd., Tweeddale Court, 14 High St., Edinburgh EH1 1YL, Scotland
Oxford University Press, 200 Madison Avenue, New York, New York 10016
Pergamon Press, Maxwell House, Fairview Park, Elmsford, New York 10523
Plenum Publishing Co., 233 Spring St., New York, New York 10013
Raven Press, 1140 Avenue of the Americas, New York, New York 10036
Rockefeller University Press, 1230 York Avenue, New York, New York 10021

S. Karger AG, Allschwilerstrasse 10, CH 4009 Basel, Switzerland
Springer-Verlag, 175 Fifth Avenue, New York, New York 10010
University of Chicago Press, 5720 South Woodlawn Ave, Chicago, Illinois 60637
Williams and Wilkins Co., East Preston St., Baltimore, Maryland 21202
Wissenschaftliche Verlagesgesellschaft GMBH, P.O. Box 40, D-7000 Stuttgart 1, Germany
Wistar Institute Press, 3631 Spruce St., Philadelphia, Pennsylvania 19104
Year Book Medical Publishers, Inc., 35 East Wacker Drive, Chicago, Illinois 60601

Foreign Words Commonly Found in Scientific Titles

Abbildung -*G*- figure
Abhandlung -*G*- dissertation, transaction, treatise, paper
Abstammungslehre -*G*- theory of descent, origin of species
Abteil, Abteilung -*G*- division
Acta -*L*- chronicle
allgemein -*G*- general
angewandt -*G*- applied
Annalen -*G*- annals
Anzeiger -*G*- informer
Arbeiten -*G*- work
Atti -*I*- proceedings
Band -*G*- volume
Beiheft -*G*- supplement
Bericht -*G*- report
Bokhandel -*Sw*- bookstore
Boktryckeri -*Sw*- press
Bunko -*J*- library
Bunkyo -*J*- education
Buchbesprechung -*G*- book review
Comptes Rendus -*F*- proceedings
Daigaku -*J*- university
Doklady -*R*- proceedings
Entwicklungsmechanik -*G*- embryology
Ergänzungshefte -*G*- supplement
Ergebnis -*G*- conclusion
Folia -*L*- leaflet, pamphlet, journal
Förlag -*Sw*- publisher
Forschung -*G*- research
Fortbildung -*G*- construction
Forstgenetik -*G*- forestry genetics
Fortschritt -*G*- advance, progress
gesamt -*G*- general
Gesellschaft -*G*- association, society
hebdomadaire -*F*- weekly
Hefte -*G*- number (of a periodical)
Helvetica -*L*- Swiss
Hoja -*Sp*- paper, pamphlet, record, journal
Iberica -*L*- referring to Spain and Portugal
Idengaku -*J*- genetics
Inhalt -*G*- contents

Jahrbuch -*G*- yearbook, annual
Kenkyusho -*J*- research institute
Kniga -*R*- book
Kunde -*G*- science
Lebensmittel -*G*- nutrition
Lehrbuch -*G*- textbook
Mezhdunarodnaya -*R*- international
Monatsblätter -*G*- monthly journal
Nachrichten -*G*- news
Naturwissenschaft -*G*- natural science
Nauk -*R*- science
Österreich -*G*- Austria
Planches -*F*- plates
real -*Sp*- royal
Recueil -*F*- collection
Rendiconti -*I*- account
Resumptio -*D*- review
Revista -*Sp*- review
Rundschau -*G*- overview or survey
Sammlung -*G*- collection
Säugetier -*G*- mammal
Schriften -*G*- publication
Schweizerische -*G*- Swiss
Scripta -*I*- writing
Séance -*F*- session, meeting
Seibutsugaku -*J*- biology
Seiken -*J*- biological institute
Shokubutsugaku -*J*- botany
Silvae -*L*- forest
sperimentale -*I*- experimental
Shuppan -*J*- publication
Teil -*G*- part
Tierärtliche Medizin -*G*- veterinary medicine
Tijdschrift -*D*- magazine, periodical
Tome -*F*- volume
Toyo -*J*- East, Orient
Travaux -*F*- work
Untersuchungen -*G*- research
Vererbungslehre -*G*- genetics
vergleichen -*G*- comparative
Verhandlung -*G*- proceeding, transaction

D = Dutch; *F* = French; *G* = German; *I* = Italian; *J* = Japanese; *L* = Latin; *R* = Russian; *Sp* = Spanish; *Sw* = Swedish.

Verlag *-G-* publishing house
Verslag *-D-* report, account
Vorbericht *-G-* preliminary report
Wissenschaft *-G-* science
Wochenschrift *-G-* weekly publication
Zasshi *-J-* magazine

Zeitschrift *-G-* periodical, journal, magazine
Zeitung *-G-* newspaper
Zellforschung *-G-* cytology
Zentralblatt *-G-* overview or survey
Ziho *-J-* journal
Züchtung *-G-* breeding, culturing, rearing